UNDERWATER EXPLOSIONS

By ROBERT H. COLE

PRINCETON, NEW JERSEY · 1948

PRINCETON UNIVERSITY PRESS

London: Geoffrey Cumberlege, Oxford University Press

PRINTED IN THE UNITED STATES OF AMERICA
BY THE COLONIAL PRESS INC., CLINTON, MASS.

Preface

The content and purpose of this book are largely the result of research on underwater explosions carried out by many groups in the years 1941–46. Much of the present knowledge and understanding of this field was acquired because of the demands of these war years and the few available discussions of the subject have become inadequate or obsolete. This book is an attempt to supply a reasonably comprehensive account which will be of use both to workers in the field of underwater explosions and to others interested in the basic physical processes involved.

Because of the fact that hydrodynamics is a relatively unfamiliar branch of physical science, some attention has been given to development of necessary hydrodynamical relations from first principles. The discussions of theoretical and experimental methods have been developed with the hope both of making clear the value and limitations of the results obtained and of making available material of possible interest in other fields.

The theoretical predictions and experimental data presented have been selected primarily on a basis of fundamental interest rather than military importance; this criterion was felt desirable in most cases and was made necessary in some for reasons of security. The presentation on this basis has fortunately not been unduly restricted by the requirements of military security, although such requirements have in some cases prevented inclusion of otherwise interesting information. The writer is indebted to T. L. Brownyard of the Bureau of Ordnance, U. S. Navy, for his assistance and advice on clearance, and to Stephen Brunauer for his continued helpful interest in the work.

The writer's experience in the field of underwater explosions is the result of his association with the work of the Underwater Explosives Research Laboratory, a wartime organization established under contracts of the Woods Hole Oceanographic Institution with the Office of Scientific Research and Development, and later with the Bureau of Ordnance, U. S. Navy. Much of the illustrative material was drawn from the work of this laboratory, both because of this association and because in some cases other equally pertinent data were for one reason or another not available. The original impetus for the writing of this book came from Paul C. Cross; the early planning of its scope and content benefited greatly from the advice and criticism of Dr. Cross and of E. Bright Wilson, Jr. The actual writing of the book was begun at the end of 1945, and most of the work was completed during the summer of 1946 at Woods Hole.

The writer is greatly indebted to Columbus O'D. Iselin, director of the Oceanographic Institution, and Paul M. Fye, then director of the Underwater Explosives Research Laboratory, for their generous and unfailing cooperation in making the facilities at Woods Hole available. The help of many others should also be acknowledged, both in helpful discussion and advice and in the preparation of the manuscript. The list of all these is too long to be given here, as it should for one reason or another include virtually the entire personnel of the Underwater Explosives Research Laboratory during its existence. The writer can only hope that they will find the accounts given here of some of their work a partial recompense.

A final debt is to my wife Elisabeth for her unfailing help and encouragement, particularly in the more tedious parts of the enterprise.

Robert H. Cole

Brown University

Contents

UNDERWATER EXPLOSIONS

1. *The Sequence of Events in an Underwater Explosion*

The purpose of this introductory chapter is to sketch briefly the primary phenomena in an underwater explosion, in order to indicate their order of occurrence and suggest the physical laws and properties governing them. The discussion will appear superficial to those well acquainted with the subject, but it is hoped that the description will serve as an adequate outline and introduction[1] for readers less familiar with the field.

1.1. The Initial Conditions

An explosion is a chemical reaction in a substance which converts the original material into a gas at very high temperature and pressure, the process occurring with extreme rapidity and evolving a great deal of heat. The temperature in the product gases is of the order 3,000° C. and the pressure 50,000 atm. Any explosive material, whether solid, liquid, or gas, is thus an inherently unstable compound which, once started, undergoes chemical changes which convert it into a more stable product. A reaction of this kind can be initiated if sufficient energy is provided at some point in the explosive. This is done usually by means of a heated wire or by frictional heat from impact by a firing pin, either of which, in most cases, acts directly upon a small amount of especially sensitive material. The reaction of this material then in turn initiates the reaction in the main body of the explosive.

Once initiated, the intense heat and pressure developed are sufficient to set up the explosive reaction in adjacent material, and the reaction is propagated through the material. The way in which the disturbance proceeds depends upon the physical and chemical properties of the material, and upon external physical factors such as the container or surrounding medium, but two general types of behavior can be distinguished. The more important of these from the point of view of destructiveness is the process of detonation, in which the chemical transformation occurs so rapidly that it can keep up with the physical disturbance resulting from the reaction. A reaction occurring in this way develops a very narrow boundary between material in its initial

[1] An excellent introductory account of underwater explosions has been given by Kennard (54), although much of the data on which the discussion is based has been superseded by more recent results. Numbers in parentheses refer to the bibliography on page 427.

condition and the products at high temperatures and pressure. This clearly defined rapidly advancing discontinuity is known as a "detonation wave," and travels with a velocity of several thousand meters per second. On the other hand, the chemical reaction may take place more slowly and be unable to keep up with the advancing physical disturbance of pressure and particle motion which it causes. The final reaction state is then reached more gradually and there is not a well defined boundary. This more gradual process is called "burning," although the rate at which it occurs may still be high.

The two types of disturbance, detonation and burning, correspond closely to the two major classifications of military explosives: high explosives such as TNT, which detonate with large and rapid evolution of energy and are used for destructive purposes in bombs, depth charges, torpedo warheads; and "propellants" such as gunpowder, which burn with a gradual building up to the final state, and are used, as the name implies, to drive a shell, rocket, airplane, etc.

From the point of view of phenomena which occur as a result of an explosive set off underwater, the explosion process is of interest chiefly because one must know the physical conditions at the boundary of the explosive and surrounding water to calculate what will happen in the water. A determination of these conditions from measurable properties of the explosive material is therefore necessary, and the ways in which this is done are discussed in Chapter 3. The underlying physical relations necessary to a discussion of detonation in the explosive and propagation of disturbances in the water are derived in Chapter 2.

1.2. Dynamical Properties of Water

As a result of the explosion process, the initial mass of explosive becomes a very hot mass of gas at tremendous pressures, and it is evident that these conditions cannot persist without affecting the surrounding medium. If this is water, we must consider what changes occur in a body of water as a result of specified forces or displacements, a problem which is part of the field of physics known as hydrodynamics. If we can restrict ourselves to the concept of water as a homogeneous fluid incapable of supporting shearing stresses we have a medium in which the volume can readjust itself to displacements of its boundaries by flow. In addition, changes in pressure on a definite mass result in compression (change in volume) of the mass.

The fact that water is compressible leads to the conclusions that a pressure applied at a localized region in the liquid will be transmitted as a wave disturbance to other points in the liquid with a velocity which, though large, is finite, and that the wave involves local motion of the water and changes in pressure. If the pressure is small enough, the rate of propagation is practically independent of the magnitude of the pressure, and in sea water at 18° C. is about 4,900 ft./sec. This state of affairs is realized in underwater sound transmission, but temperature and density changes also affect the velocity. If the motion is one-dimensional so that plane waves are generated, the wave travels without significant change of magnitude or shape. If the waves are radiated from a spherical source, the amplitude decreases with distance from the source and the water motion is modified by the pressure differences resulting from this spherical divergence, a phenomenon known as the surge or afterflow.

In the regions of water surrounding an explosion the pressures are so large that the wave velocity cannot be assumed independent of pressure. This has the physical result that the form of the wave depends on the magnitude of the pressure and displacement of the water as it progresses. These complications for waves of finite amplitude are expressed in much more difficult mathematical statements than those which suffice to explain the propagation of small amplitude waves.

If, on the other hand, the disturbances affecting the water are slowly changing, and the water can accommodate itself to them before the disturbance has changed appreciably, it may suffice to neglect entirely the wave propagation by which the accommodation takes place. We may then consider the water as an incompressible medium in which the disturbance spreads instantaneously to all points in the liquid. Motion which can be accounted for in this way is usually described as incompressive flow.

1.3. The Shock Wave

The first cause of disturbance to the water in an explosion is the arrival of the pressure wave in the reacting explosive at the water boundary. Immediately upon its arrival, this pressure, which is of the order of $2 \cdot 10^6$ lb./in.^{2} for TNT, begins to be relieved by an intense pressure wave and outward motion of the water. The extremely dense mass of gas left when detonation or burning is complete then begins to expand, its pressure diminishes and the pressure in the water also falls off rapidly. In the case of a detonating high explosive, such as TNT, the pressure rise is for all practical purposes discontinuous, and is then followed by a roughly exponential decay, the duration being measured in times of a few milliseconds at most. The pressure level about a 300 pound spherical charge shortly after complete detonation is sketched in Fig. 1.1(a).

Once initiated, the disturbance is propagated radially outward as a wave of compression in the water, the steep fronted wave being described as the "shock wave." As compared to waves of infinitesimal amplitude, this shock wave has the following characteristics:

(i). The velocity of propagation near the charge is several times the limiting value of about 5,000 ft./sec., this value being approached quite rapidly as the wave advances outward and the pressure falls to "acoustic" values.

(ii). The pressure level in the spherical wave falls off more rapidly

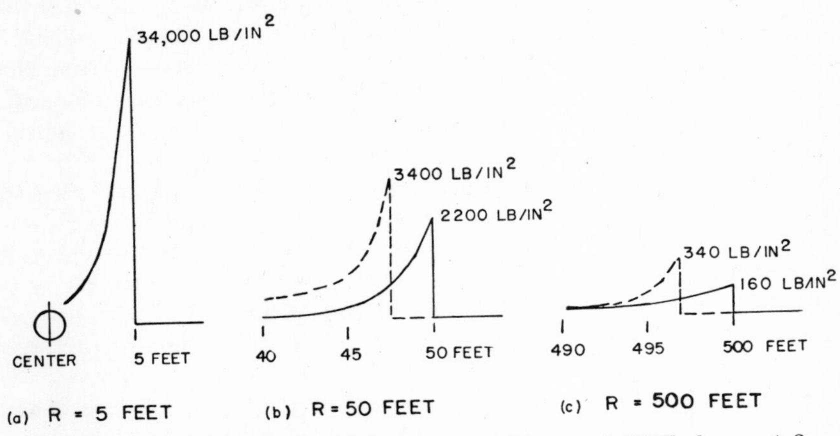

Fig. 1.1 The pressure distribution around a 300 pound TNT charge at 3 times after completion of detonation.

with distance than the inverse first power law predicted for small amplitudes, but eventually approaches this behavior in the limit of large distances.

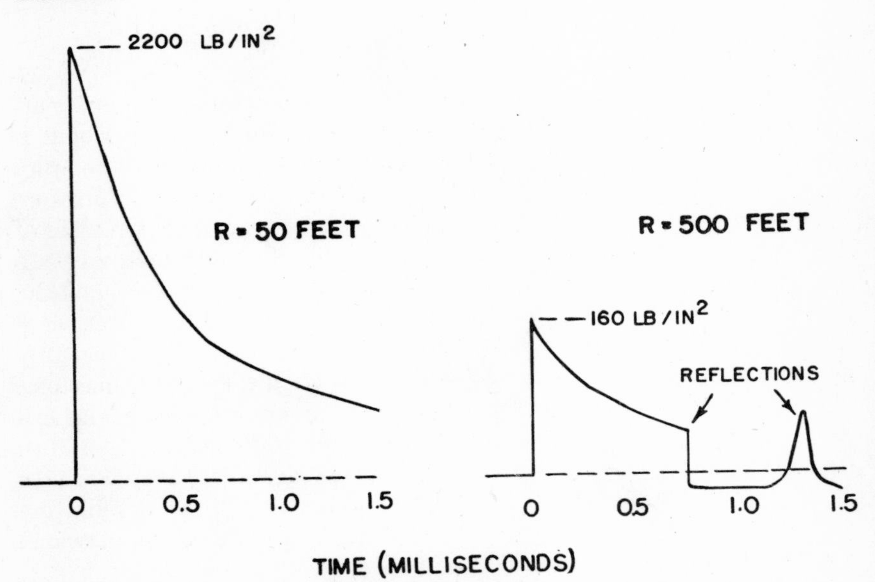

Fig. 1.2 Shock wave pressure-time curves at 2 distances from 300 pound TNT charges.

(iii). The profile of the wave broadens gradually as the wave spreads out. This spreading effect is most marked in the region of high pressures near the charge.

These properties of the shock wave are illustrated in Fig. 1.1(b, c), drawn for two later stages in the explosion of a 300 pound charge. For comparison, the pressure waves which would exist if the earlier state shown in Fig. 1.1(a) were propagated as an acoustic wave are indicated by the dashed curve. These sketches of course represent the conditions at three instants of time. The pressure-time curve at a given distance from the explosion will have the same general form, and the pressures observed at distances of 50 and 500 feet are sketched in Fig. 1.2(a, b).

The illustrations given refer to a particular size of charge, and it is natural to ask what conclusion can be drawn about another size of charge. In other words, what scaling laws may be applied? The answer for the shock wave is very simple and is provided by the "principle of similarity," which states that if the linear size of the charge be changed by a factor k, the pressure conditions will be unchanged if new distance and time scales k times as large as the original ones are used. As an example, the pressures for the 300 pound charge of Figs. 1.1 and 1.2 will be obtained also for a charge of half the linear dimensions (one-eighth the weight), provided that we make the observations at distances from the smaller charge one-half as great and divide the time scale by a factor of two.

The theoretical justification of the principle, given in Chapter 4, is not difficult, and it has been amply verified by experimental observation. The validity of the principle depends, among other things, on the assumption that no external forces act upon the system. Gravity is such an external force, and of course it is always present. It is unimportant compared with the internal forces involved in generation and propagation of the shock wave, but its effect cannot be neglected in the later behavior of the gaseous explosion products. The principle of similarity as stated above, therefore, does not apply to the phenomena following the shock wave.

1.4. Motion of the Gas Sphere

The initial high pressure in the gas sphere is considerably decreased after the principal part of the shock wave has been emitted, but it is still much higher than the equilibrium hydrostatic pressure. The water in the immediate region of the sphere or "bubble," as it is usually called, has a large outward velocity and the diameter of the bubble increases rapidly. This outward velocity is in excess of that to be expected from the magnitude of the pressure existing at the time, owing to the after-flow characteristics of spherical waves mentioned in section 1.2, an effect which has also to be considered in careful analysis of the shock wave. The expansion continues for a relatively long time, the internal gas pressure decreases gradually, but the motion persists because of the inertia of the outward flowing water. The gas pressure at later times falls below the equilibrium value determined by atmospheric plus hydrostatic pressure, the pressure defect brings the outward flow to a stop, and the boundary of the bubble begins to contract at an increasing rate. The inward motion continues until the compressibility of the gas, which is insignificant in the phase of appreciable expansion, acts as a powerful check to reverse the motion abruptly. The inertia of the water to-

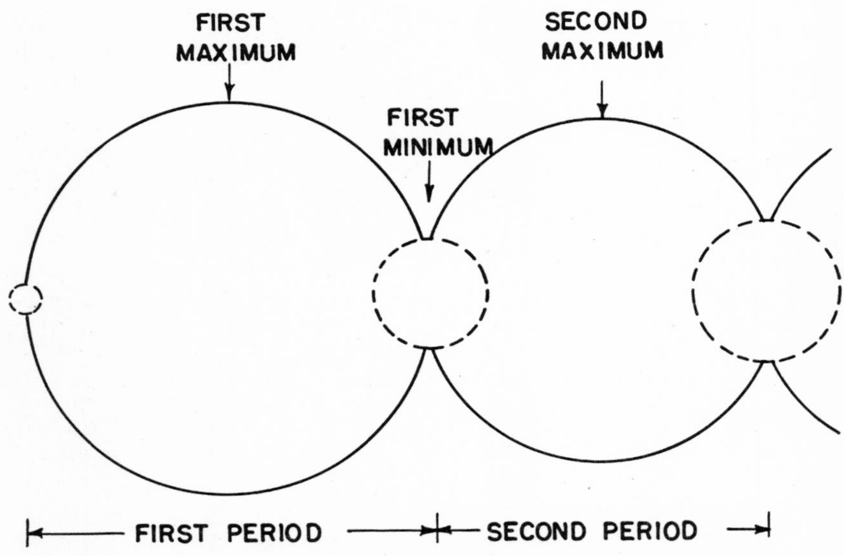

Fig. 1.3 Pulsations of the gas products from an underwater explosion.

gether with the elastic properties of the gas and water thus provide the necessary conditions for an oscillating system, and the bubble does in fact undergo repeated cycles of expansion and contraction. Ordinarily the original state of the bubble is approximately spherical and the radial nature of the later flow results in an asymmetrical oscillation about the mean diameter, the bubble spending most of its time in an expanded condition. These phases in bubble oscillation are shown schematically in Fig. 1.3, which shows the bubble size as a function of time. The period of oscillation, in the absence of disturbing effects due to boundaries, turns out to be quite simply related to the internal energy of the gas and the hydrostatic pressure (and hence depth below the surface of the water), being proportional to the cube root of energy and inverse five-sixths power of pressure.

Oscillations of the gas sphere can persist for a number of cycles, ten or more such oscillations having been detected in favorable cases. The number observable is limited by its loss of energy by radiation or turbulence, as described in the next section, and by the disturbing effects of gravity and any intervening boundary surfaces. It is perfectly evident that the gaseous products must, because of their buoyancy when in equilibrium with the surrounding pressure, eventually rise to the sur-

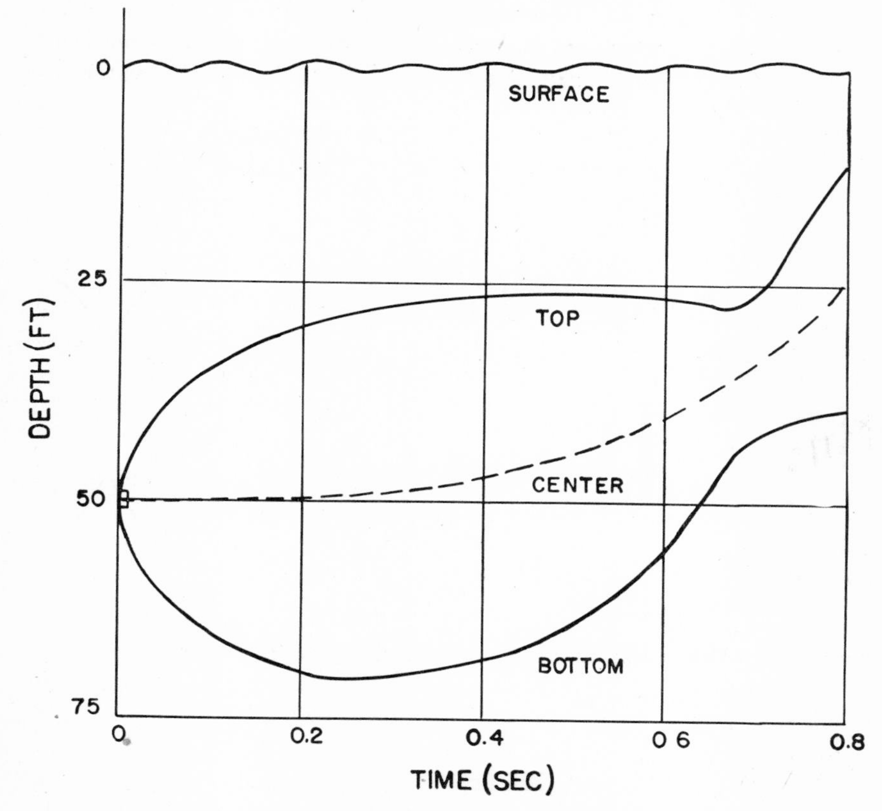

Fig. 1.4 Displacements of the gas sphere from a 300 pound TNT charge fired 50 feet below the surface.

face. It is less evident that the gas sphere will in the course of its oscillation experience a net repulsive force away from a free surface and will be attracted toward a rigid boundary. The motion of the gas sphere is thus affected by its buoyancy and by the proximity of the surface of the water and the sea bed or other boundary surfaces. These complications make detailed calculations of bubble behavior difficult in spite of the fact that during most of its motion the changes in density of the surrounding water are negligible and the motion may be described as incompressive flow.

The motions of the gas sphere from 300 pounds of TNT detonated 50 feet below the surface are sketched in Fig. 1.4. The data are obtained from somewhat inexact analysis (see Chapter 8), and are intended only to illustrate the general characteristics of the motion.

1.5. Secondary Pressure Pulses

The motion of the gas sphere has associated with it emission of energy in the form of pressure waves advancing radially outward from the bubble. If it is assumed that motion of the water around the bubble is incompressible, it can be shown that the pressure in the water should depend on the square of the rate of bubble expansion or contraction. From Fig. 1.3, it is evident that this rate is greatest when the bubble is near the point of smallest volume. One should therefore expect, both from incompressive theory and the fact that only in this region is the water under appreciable compression, that the pressures will be significant only in a small interval about the time of maximum contraction. This is the state of affairs observed; pressure pulses are emitted which build up to a maximum value at times corresponding to the minimum volumes and fall off again as the bubble expands.

The form of the bubble pulses from a given charge depends considerably on the depth of water and proximity of boundary surfaces, as would be expected from the effect of these factors on the bubble motion. The peak pressure in the first bubble pulse is no more than ten to twenty per cent of that of the shock wave, but the duration is much greater, and the areas under the two pressure-time curves are comparable. A considerable amount of the energy initially present is lost at the time of each pulse, both in the pulse and in turbulence resulting from rapid radial and vertical motion of the gas sphere (see Fig. 1.4). As a result successive pulses are progressively weaker and usually only the first pulse is of practical significance. The relation between shock wave and bubble pulse pressures and durations is shown in the sketch of Fig. 1.5 of a continuous pressure-time record at a point 60 feet from the same size charge. Fig. 1.6 shows in more detail the bubble pulses from 300 pound TNT charges detonated at various depths in 100 feet total depth of water. It will be observed that the profile of the curve becomes more irregular for initial charge positions close to the surface or bottom. It should also be noted that pressure waves reflected from the surface and bottom give rise to interference, and the later portions of observed pressure-time curves may be considerably different from the pressure-time curve which would be observed in an infinite medium. The same interference phenomenon occurs in shock waves but is less effective and more easily recognized because of the shorter duration and discontinuous front (see Fig. 1.2).

Because of the fact that gravity is an omnipresent factor in bubble motion which does not scale in the same way as the effects of internal forces and boundaries, the principle of similarity for shock waves described in section 1.3 does not hold for bubble motion in the general case. One can derive scaling factors which express the theoretical equations in an approximate form and account for the major features of

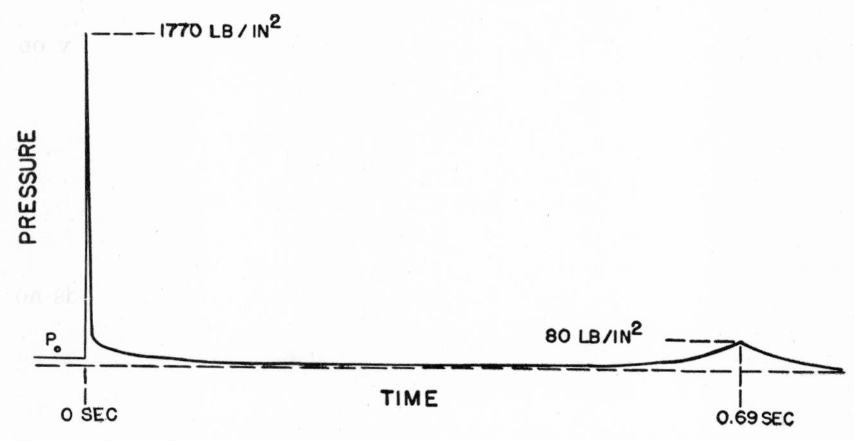

Fig. 1.5 Pressure 60 feet from a 300 pound TNT charge fired 50 feet below the surface.

bubble motion. The scaling laws and feasible numerical calculations of bubble effects are, however, only approximate, and experimental results are often not in very good agreement with tractable theoretical calculations.

1.6. Surface and Other Effects

The preceding sections have outlined the more important properties of underwater explosions as they would appear to sufficiently rugged observers below the surface equipped with a sense for pressure measure-

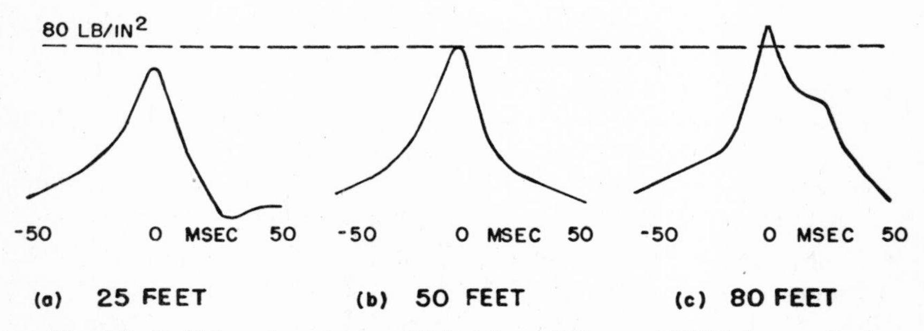

Fig. 1.6 Bubble pulse pressures 60 feet from 300 pound TNT charges fired at 3 depths in 100 feet of water.

ment. The features visible above the surface depend considerably on the initial depth of the charge, being quite spectacular for shallow explosions, but becoming virtually undetectable for great depths. The sequence of events in the shallow case may be quite complicated, but three main phases can be distinguished. These result from the arrival of the primary shock wave at the surface, the approach of the gas sphere to the surface, and the final breakthrough of the gases to the atmosphere. The motion of the water in a positive pressure wave finds no hindrance to its continuance when the shock wave reaches the surface, as the atmosphere cannot supply appreciable resistance by compression. As a result, a reflected wave of negative pressure is formed at the surface with a value such that the sum of the direct and reflected pressures is practically zero. The water in the surface layer is thrown up with a velocity proportional to the pressure existing in the direct wave, and a rounded dome of whitish water forms directly above the charge in the next fraction of a second. Regions of the surface further from the charge are disturbed much less, but a rapidly advancing ring of apparently darkened water, the "slick," can often be seen spreading out from the charge for hundreds of feet (for, say, a 300 pound charge), the extent of the region indicating the points to which the shock wave has advanced. If the bubble is sufficiently close to the surface in an early phase of its motion, it may also disturb the surface before its final breakthrough. The breakthrough occurs at a later time, determined by the depth, and shoots up plumes of spray which may rise hundreds of feet. All these effects become less pronounced for greater initial depths of water. The slick may be seen for a 300 pound charge 400 or 500 feet deep, but the dome is insignificant and no trace of plumes is evident even at much smaller depths.

As already mentioned, the pressure field in the water is usually complicated by reflections from the surface and bottom, the reflected wave from a free surface being negative and from a rigid surface positive. The resultant pressures observed are superpositions of the direct and reflected waves. A complication arises for reflected waves of negative pressure because the maximum absolute tension sea water can withstand is certainly quite small and is very probably less than an atmosphere, if not zero. As a result, while a reflected negative wave can reduce the resultant absolute pressure at any point to zero, appreciable negative pressures do not exist under any ordinary conditions. The reflected wave from the bottom is usually quite irregular and much smaller than would be expected for perfect reflection. In the extreme case of a charge fired on the bottom, the increase in pressure is of the order ten to fifteen per cent, roughly half what would be ideally realized. (Although it might seem that the pressure should be doubled in the ideal case, the increase is much less because the energy transport, which *is* doubled, depends on both pressure and duration of the wave.)

The character of pressure waves at great distances from the charge may be affected by other factors, such as attenuation by viscosity effects and refraction by velocity gradients in the water. These effects become appreciable only for such ranges that the wave is of acoustic intensity, and their detailed description belongs more properly to a treatise on underwater sound transmission.

2. *Hydrodynamical Relations*

The purpose of this chapter is to develop from first principles the equations describing the motion of a fluid, whether liquid or gas, which are the basis of the more detailed theories of the phenomena in underwater explosions, to discuss the implications of these equations, and to indicate the manner of development appropriate to problems later treated more explicitly.

2.1. The Differential Equations for Ideal Fluids

As a first step in discussing the propagation of waves in fluids, it is necessary to put the basic laws of mechanics into a suitable mathematical form. It is assumed in what follows that the fluid is ideal in the sense that viscous stresses and effects of heat conduction may be neglected. The discussion is further restricted to regions of space and instants of time for which there are no discontinuities of pressure, velocity of the fluid, or internal energy.

A. *Conservation of mass.* The simplest restriction on the motion of the fluid is the conservation of mass. If we consider a small fixed region of space in the interior of the fluid, it must be true that any change in mass of fluid contained in the volume is equal to the net quantity of fluid which flows through the boundary surface. If the region is a small cube of volume $dxdydz$, the change in mass in a time dt resulting from change in the density ρ at the point x,y,z is

$$\frac{\partial \rho}{\partial t}\, dt\, dxdydz$$

If such a change occurs it must be as a result of motion in the fluid. Let the velocity of a point moving with the fluid be described by its three components u, v, w, in the x, y, z directions, these components being functions of the space coordinates and time. The net transport of fluid into the fixed volume in time dt resulting from motion in the x direction is the difference in amounts flowing through the two faces of area $dydz$ and is given by

$$[(\rho u)_x - (\rho u)_{x+dx}]\, dt\, dydz = -\frac{\partial}{\partial x}(\rho u)\, dt\, dxdydz$$

higher order terms in the expansion of $(\rho u)_{x+dx}$ disappearing in the limit of small displacements. Similar terms are obtained for the other two components of motion and the conservation of mass requires that their sum equal the increase of mass in the volume, hence

$$\frac{\partial \rho}{\partial t} + \frac{\partial}{\partial x}(\rho u) + \frac{\partial}{\partial y}(\rho v) + \frac{\partial}{\partial z}(\rho w) = 0 \tag{2.1}$$

which is the equation of continuity. This result is more concisely expressed in vector notation as

$$\frac{\partial \rho}{\partial t} + \text{div}\,(\rho \mathbf{v}) = 0 \tag{2.2}$$

where the velocity vector $\mathbf{v}$ has the components u, v, w in Cartesian coordinates.

The equation of continuity could have been obtained equally well by considering a small volume moving with the fluid and containing the same definite mass of fluid at all times. Changes in density of the fluid as it moves therefore require compensating changes in dimensions of the element. The straightforward result obtained from this approach, originally due to Euler, comes out to be

$$\frac{d\rho}{dt} + \rho\,\text{div}\,\mathbf{v} = 0$$

if the symbol d/dt is understood to mean differentiation at a point moving with the fluid rather than a fixed point, i.e.,

$$\frac{d}{dt} = \frac{\partial}{\partial t} + \frac{\partial x}{\partial t}\cdot\frac{\partial}{\partial x} + \frac{\partial y}{\partial t}\cdot\frac{\partial}{\partial y} + \frac{\partial z}{\partial t}\cdot\frac{\partial}{\partial z} = \frac{\partial}{\partial t} + u\frac{\partial}{\partial x} + v\frac{\partial}{\partial y} + w\frac{\partial}{\partial z}$$

With this meaning, it is evident that the results of the two approaches are identical, as they must be.

B. *Conservation of momentum.* In order to express Newton's second law, or conservation of momentum, it is convenient to consider the forces acting on element of volume $dxdydz$ moving with the fluid rather than an element fixed in space. Considering the x-component of motion, the acceleration of the element is given by the total time derivative du/dt of the particle velocity, the total derivative being correct as it expresses the total change in velocity resulting from the changes at a fixed point in space and from displacement of the element in space. The product of this acceleration and the mass $\rho\, dxdydz$ of the moving element must, by Newton's second law, equal the net force acting on the element in the x-direction. This force is supposed due only to differences in the pressure P at the two faces of area $dydz$ and is given by

$$[P_x - P_{x+dx}]\, dydz = -\frac{\partial P}{\partial x}\, dxdydz$$

Equating the force and inertia terms, we have

$$\rho \frac{du}{dt} = -\frac{\partial P}{\partial x}$$

together with similar terms for the other two components. If the total acceleration is resolved into its two parts we have

$$\rho \frac{\partial u}{\partial t} + \rho u \frac{\partial u}{\partial x} + \rho v \frac{\partial u}{\partial y} + \rho w \frac{\partial u}{\partial z} = -\frac{\partial P}{\partial x} \tag{2.3}$$

$$\rho \frac{\partial v}{\partial t} + \rho u \frac{\partial v}{\partial x} + \rho v \frac{\partial v}{\partial y} + \rho w \frac{\partial v}{\partial z} = -\frac{\partial P}{\partial y}$$

$$\rho \frac{\partial w}{\partial t} + \rho u \frac{\partial w}{\partial x} + \rho v \frac{\partial w}{\partial y} + \rho w \frac{\partial w}{\partial z} = -\frac{\partial P}{\partial z}$$

These equations are equivalent to the single vector equation

$$\rho \frac{d\mathbf{v}}{dt} = \rho \frac{\partial \mathbf{v}}{\partial t} + \rho\, (\mathbf{v} \cdot \text{grad})\mathbf{v} = -\text{grad}\, P \tag{2.4}$$

C. *Conservation of energy.* As in the derivation of the equation of motion, it is convenient to consider an element of volume moving with the fluid and enclosing a fixed mass of fluid. The total energy per unit mass of the fluid consists of kinetic energy and internal energy E, which is the sum of thermal and any chemical energy. The change in time dt for the element of volume $dxdydz$ is

$$\rho \frac{d}{dt}\, [E + \tfrac{1}{2}(u^2 + v^2 + w^2)]\, dt\, dxdydz$$

where the total time derivative is again used to account for the displacement of the element during the interval. This change in energy must equal the work done on the faces of the element. The work done in time dt on an area $dydz$ in motion along x is the product of force and displacement, or $Pu \cdot dtdydz$, and the net amount of work done on the two faces of the volume element is

$$[(Pu)_x - (Pu)_{x+dx}]\, dt\, dydz = -\frac{\partial}{\partial x}\, (Pu)\, dt\, dxdydz$$

The work done on the other faces is obtained in the same way, and equating the total to the increase in energy gives

$$\rho \frac{d}{dt} [E + \tfrac{1}{2}(u^2 + v^2 + w^2)] = -\left[\frac{\partial}{\partial x}(Pu) + \frac{\partial}{\partial y}(Pv) + \frac{\partial}{\partial z}(Pw)\right]$$

which in vector notation is

$$\rho \frac{d}{dt} [E + \tfrac{1}{2}(\mathbf{v} \cdot \mathbf{v})] = -\operatorname{div}(P\mathbf{v})$$

The energy equation may be transformed into a more useful expression by combining it with the equations of continuity and motion. Solving for variation in internal energy gives

$$\rho \frac{dE}{dt} = -P \operatorname{div} \mathbf{v} - \mathbf{v} \cdot \operatorname{grad} P - \rho \mathbf{v} \cdot \frac{d\mathbf{v}}{dt} \tag{2.5}$$

But from Eq. (2.2)

$$-\operatorname{div} \mathbf{v} = \frac{1}{\rho}\frac{\partial \rho}{\partial t} + \frac{1}{\rho}(\mathbf{v} \cdot \operatorname{grad})\rho = \frac{1}{\rho}\frac{d\rho}{dt}$$

and from Eq. (2.4)

$$\operatorname{grad} P = -\rho \frac{d\mathbf{v}}{dt}$$

which gives on substitution

$$\rho \frac{dE}{dt} = \frac{P}{\rho}\frac{d\rho}{dt} \tag{2.6}$$

D. *Pressure-density relation.* In our derivation of the fundamental hydrodynamic equations, the effect of dissipation processes has been neglected. If the properties of a specified small element of the fluid are described by these equations, a further condition on the state of this element is implied which has not been explicitly stated. If no dissipative processes take place in a given period of the motion, no element moving with the fluid can exchange heat with any other element or its surroundings during this time. The changes in the physical state of the element must therefore take place at constant entropy, a situation which can be expressed by the relation

$$\frac{dS}{dt} = 0$$

where the total derivative refers to a point fixed in and moving with the liquid, not to a point in a fixed coordinate system. This is equivalent to the statement that changes of density due to applied pressure take place adiabatically. Therefore, for any point in the fluid at any time for which dissipative processes can be neglected, the pressure is a single valued function of density alone, and the law of variation is the adiabatic law found from the equations of state appropriate to the fluid. Two different elements of fluid, however, may have undergone dissipative processes at earlier times which involve significant and different changes of entropy. At later times not involving such processes, each element will have a single valued relation between pressure and density, the exact form being determined by the change in entropy in the earlier irreversible process, but these two adiabatic laws will be different.

For example, a steep fronted shock wave, in which very large pressures and pressure gradients exist, may result in considerable dissipation as it passes through an element, but the entropy changes in successive elements need not be the same, owing to loss in intensity of the disturbance. The passage of such a wave can therefore leave each successive element in a different condition.

A more explicit formulation of the pressure-density relation for a given fluid element can easily be written from the energy equation. The internal energy E can be expressed as a function of pressure and density as thermodynamic variables and its differential can then be written as

$$dE = \frac{\partial E}{\partial P}\,dP + \frac{\partial E}{\partial \rho}\,d\rho = \frac{P}{\rho^2}\,d\rho$$

from Eq. (2.6). Solving for $dP/d\rho$, we have

$$\frac{dP}{d\rho} = \left(\frac{P}{\rho^2} - \frac{\partial E}{\partial \rho}\right) \bigg/ \left(\frac{\partial E}{\partial \rho}\right)$$

If initial values of P and p are given, and the functional dependence of E on P and ρ is known from equation of state and thermochemical data, the equation can be solved explicitly for $P(\rho)$.

2.2. Waves of Small Amplitude

If the disturbances created in a liquid by external sources are sufficiently small, the fundamental equations of section 2.1 can be considerably simplified. It will be assumed that the density ρ changes insignificantly compared to its initial value ρ_0 and can therefore be treated as constant in terms of the form $(\partial/\partial x)\,(\rho u)$; similarly the particle velocity is always small and terms of the form $u\partial u/\partial x$ can be neglected. The equations of motion and equation of continuity then become

$$\frac{\partial \mathbf{v}}{\partial t} = -\frac{1}{\rho_o}\operatorname{grad} P \tag{2.7}$$

$$\frac{\partial \rho}{\partial t} = -\rho_o \operatorname{div} \mathbf{v}$$

From section 2.1, P is a definite function of density, and we may write

$$\frac{\partial P}{\partial t} = \left(\frac{dP}{d\rho}\right)_{s_o} \frac{\partial \rho}{\partial t}$$

the subscript s_o indicating changes along the adiabatic having the entropy of the undisturbed fluid. The total differential $dP/d\rho$, understood to be evaluated for an adiabatic change, will be denoted by c_o^2, and Eqs. (2.7) may be written

$$\frac{\partial \mathbf{v}}{\partial t} = -\frac{1}{\rho_o}\operatorname{grad} P \tag{2.8}$$

$$\frac{1}{c_o^2}\frac{\partial P}{\partial t} = -\rho_o \operatorname{div} \mathbf{v}$$

A. *Plane waves.* It is assumed that the motion takes place only along x, Eqs. (2.8) become

$$\frac{\partial u}{\partial t} = -\frac{1}{\rho_o}\frac{\partial P}{\partial x}, \quad \frac{1}{c_o^2}\frac{\partial P}{\partial t} = -\rho_o \frac{\partial u}{\partial x} \tag{2.9}$$

Solving for P by differentiation and elimination, we obtain

$$\frac{\partial^2 P}{\partial x^2} = \frac{1}{c_o^2}\frac{\partial^2 P}{\partial t^2}$$

together with a similar equation for u.

This one-dimensional form of the wave equation is satisfied by any function of the form $f(t \pm x/c_o)$, the double sign choice accounting for waves advancing in either positive or negative directions.

Physically, these solutions mean that any disturbance originated at some value of x travels unchanged in form with a velocity $c_o = \sqrt{dP/d\rho}$. For sea water at 20° C., this velocity is about 4,967 ft./sec. The particle velocity u corresponding to the pressure $P = f(t - x/c_o)$ may be found from the first of Eqs. (2.9):

$$\frac{\partial u}{\partial t} = -\frac{1}{\rho_o}\frac{\partial P}{\partial x} = \frac{1}{\rho_o c_o} f'\left(t - \frac{x}{c_o}\right)$$

the prime indicating differentiation with respect to the argument. Integrating, we have

$$u - u_o = \frac{1}{\rho_o c_o} \int_{t_o}^{t} f'\left(t - \frac{x}{c_o}\right) dt$$

$$= \frac{P - P_o}{\rho_o c_o}$$

If the constants of integration are chosen to make $u = 0$ when $P = P_o$, the pressure in the undisturbed fluid, we have

$$u = \frac{P - P_o}{\rho_o c_o}$$

The relation for the wave $f(t + x/c_o)$ travelling toward negative x is found in the same way to be $u = -(P - P_o)/\rho_o c_o$. As a result the point value of either P or u suffices to determine the other if the wave motion is one-dimensional.

The relative magnitudes of pressure and particle velocity are greatly different in liquids and gases initially at atmospheric pressure. For example, an excess pressure of 0.15 lb./in.^{2} (one hundredth atmospheric pressure) corresponds to a particle velocity of 0.056 ft./sec. in sea water at 20° C., but for air at 20° C. the particle velocity for the same pressure is 3,700 times as great.

B. *Spherical waves.* The simplest form of spherical wave is one in which the disturbance is a function of radial distance from a source and not of the angular position. If the radial component of particle velocity is u_r, other components being zero, and $P = P(r)$, Eqs. (2.8) become

$$\frac{\partial u_r}{\partial t} = -\frac{1}{\rho_o}\frac{\partial P}{\partial r} \tag{2.10}$$

$$\frac{1}{c_o^2}\frac{\partial P}{\partial t} = -\rho_o \frac{1}{r^2}\frac{\partial}{\partial r}\left(r^2 \frac{\partial u_r}{\partial r}\right)$$

If the second equation is differentiated with respect to t, and the particle velocity eliminated by the first equation, we obtain

$$\frac{1}{r^2}\frac{\partial}{\partial r}\left(r^2 \frac{\partial P}{\partial r}\right) = \frac{1}{c_o^2}\frac{\partial^2 P}{{}^2\partial t}$$

It is easily verified that any function of the form $P(r, t) - P_o = (1/r)f(t - r/c_o)$ is a solution, the negative sign corresponding to an outgoing wave about the center. The form of an infinitesimal spherical wave thus does not change as it spreads out with the speed c_o, but its amplitude diminishes as r increases owing to the factor $(1/r)$, which results from the greater area over which the disturbance is spread as it advances outward.

C. *The afterflow.* The relation of u_r to $P(r)$ can be obtained by integration of the first of Eqs. (2.10). Inserting the value of P, we have

$$\frac{\partial u_r}{\partial t} = \frac{1}{\rho_o c_o}\frac{1}{r} f'\left(t - \frac{r}{c_o}\right) + \frac{1}{\rho_o r^2} f\left(t - \frac{r}{c_o}\right) \tag{2.11}$$

Integrating from a time t_o to variable time t at constant distance r gives

$$u_r(t) - u_r(t_o) = \frac{P - P_o}{\rho_o c_o} + \frac{1}{\rho_o r}\int_{t_o}^{t} [P(r, t') - P_o]\, dt' \tag{2.12}$$

If the time $t_o = 0$ is chosen to precede any disturbance, it is seen that the velocity in the fluid at a later time is a function, not only of the pressure at that time but of all the previous changes in pressure after a disturbance first reaches the point. These changes are such that, in a radial disturbance, the water will be left with an outward velocity, or afterflow, following passage of a positive pressure wave. This afterflow remains, even though the pressure has returned to its equilibrium value, and will be brought to zero only after the pressure falls below the equilibrium value.

A complete evaluation of the pressures and flow velocities behind the front of an advancing pressure wave clearly can be made only by considering the properties of the spherical source, or agency by which the pressure wave is generated. The conditions at the front, no matter how far it progresses, are determined in the acoustic approximation by the initial motion of the source. The conditions behind the front, however, depend on the later behavior of the source, and any physically realizable source must in turn be affected by the motion of the fluid surrounding it.

To clarify the nature of the afterflow term, it is appropriate for underwater explosions to consider the source as a spherical boundary in the fluid containing gas initially at high pressure. The initial pressure in the pressure wave is determined by the initial gas pressure. This initial compression leaves behind it outward flowing water in an increasingly large sphere. If the compression is to be maintained in this volume, increasingly large displacements of water near the source are necessary, despite the weakening of the initial pressure at the front by spherical divergence. The pressure in the gas sphere, however, decreases as the volume determined by the spherical boundary increases, and the strength of this source must decrease. Outward accelerations of the water near the boundary will thus decrease, but as long as there is a pressure excess over hydrostatic, outward flow continues.

The change in character of the motion can be made more explicit by considering the relation between pressure at any point in the fluid, and at the gas sphere of radius a. We have that

$$P(r, t) - P_o = \frac{1}{r} f\left(t - \frac{r}{c_o}\right)$$

and on the gas sphere

$$P(a, t) - P_o = \frac{1}{a} f\left(t - \frac{a}{c_o}\right)$$

which we can also write

$$P\left(a, t - \frac{r - a}{c_o}\right) = \frac{1}{a} f\left(t - \frac{r}{c_o}\right)$$

Comparing these equations

$$P(r, t) - P_o = \frac{a}{r}\left[P\left(a, t - \frac{r - a}{c_o}\right) - P_o\right] \tag{2.13}$$

The pressure $P(r, t)$ at any point in the fluid is thus determined by the pressure $P(a)$ on the boundary of radius a at a time $(r - a)/c_o$ earlier. This difference is just the time required for a pressure at a to be transmitted to r and the time $(t - (r - a)/c_o)$ is commonly referred to as a retarded time.

In the later stages of the motion, for which the excess pressure $P(a, t) - P_o$ is small and changes slowly with time, we can write Eq. (2.13) as

$$P(r, t) - P_o \cong \frac{a}{r}\left[P(a, t) - P_o\right]$$

and the velocity u_r is, from Eq. (2.12), given by

$$u_r - u_o = \frac{a}{\rho_o r^2} \int [P(a, t) - P_o]\, dt$$

$$= \frac{a^2}{r^2} [u_a(t) - u_a(o)]$$

This result, however, simply expresses the conservation of mass in noncompressive radial flow, for which $4\pi r^2 dr/dt = 4\pi a^2 da/dt$ and hence

$u_r = dr/dt = (a^2/r^2)u_a$. Thus, in the limit of small pressure differences, the afterflow velocity is simply the velocity of noncompressive flow outward from the expanding gas sphere boundary.

In the later stages of the motion, the pressure in the gas sphere and surrounding fluid falls below the hydrostatic value P_o, the outward flow is brought to rest, and then inward flow begins. The kinetic energy of this motion is thus returned to compression of the gas sphere, rather than being radiated to infinity as a wave of compression. At points behind the shock front, for which the pressures are large and rapidly changing, the particle velocity depends on both the past history of the pressure and its value at the time, and a clear cut distinction between motion resulting directly from compression and noncompressive flow cannot be made in this region. The natural attempt to discuss the two types of motion as distinct must therefore run into difficulties, and the fact that the two approximations are not mutually exclusive under these conditions must be remembered.

2.3. Waves of Finite Amplitude

In the derivation of solutions for waves of small amplitude, a number of simplifying assumptions were made which led to well-known forms of the wave equation appropriate to the type of motion assumed possible. Although the approximations are amply justified for the small variations of density and pressure developed by sound sources in air or underwater, we have no reason to suppose that they could be for the conditions existing in the gaseous products of an explosion or in the water in the near vicinity. Before considering solutions of the exact equations, the differences which must result can profitably be considered by more qualitative arguments.

It has been assumed that the quantity $c = \sqrt{dP/d\rho}$ entering the equations could be treated as a constant c_o independent of the pressure or state of motion in the fluid, and that the velocities of the fluid were always negligible. These assumptions led to the result that any part of a wave disturbance is propagated with a velocity c relative to a fixed system of coordinates. If, however, this small amplitude velocity of propagation c depends on the density and, as we should expect qualitatively, this velocity at any point is properly measured with respect to coordinates moving with the liquid at the point, it is easy to see that matters become more complicated. Consider first the variations of c with density to be expected in real liquids and gases. The relation $P = P(\rho)$ is, of course, simply a curve of adiabatic compression and is, for all normal fluids, concave upward, with the result that the slope $dP/d\rho$ and $c = \sqrt{dP/d\rho}$ increase with increasing compression. The simplest example is the perfect gas for which $P = k\rho^\gamma$ and therefore $c = \sqrt{k\gamma\rho^{\gamma-1}}$. The ratio of specific heats γ is always greater than unity and hence c increases with increasing density. The adiabatic curve for water is another example of the same law, although the density changes are fractionally smaller. The velocity of infinitesimal waves cannot, therefore, be treated as constant, but is rather a quantity which increases in regions of greater density. In order to see what effect this has on wave propagation, suppose that, as a result of displacements of a piston in a tube, a plane wave of pressure is advancing from left to right in the tube, and at some instant in time has the form shown in Fig. 2.1(a). Compression started in the positive direction at point a will appear to travel with a speed c_a relative to the fluid at the point. If the particle velocity in the fluid is u_a, the speed with respect

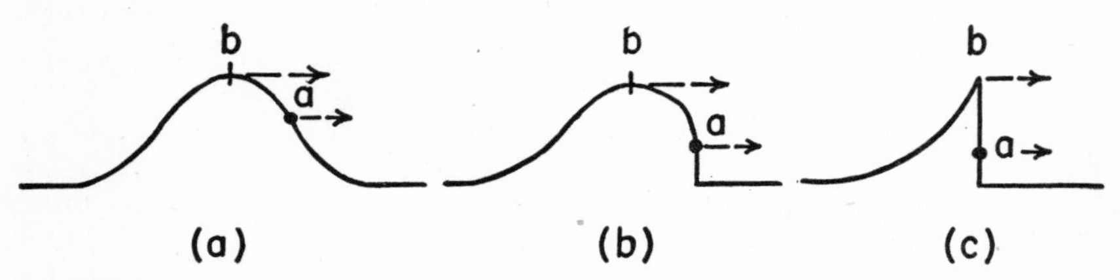

Fig. 2.1 Formation of a shock front in a plane wave of finite amplitude.

to the walls will be $c_a + u_a$. Similarly, a compression at point b will travel with a speed $c_b + u_b$ relative to the fixed wall. If the pressure set up in the fluid by the main wave is greater at b than at a, the speed of sound c and the particle velocity u will both be greater at b, and the disturbance at b will advance faster than that at a. At a later time, therefore, we have to expect that regions of higher pressure in the wave will approach those of lower pressure ahead of it, as shown in Fig. 2.1(b), the effect increasing as the pressure differences increase. The ultimate result of this overtaking effect will be to make the front of the wave very steep as shown in Fig. 2.1(c). As the condition of infinite steepness is approached, however, the pressure and temperature of closely adjacent layers will be very different; in other words, the gradients will be large. Under these circumstances, large amounts of energy can be dissipated as heat, effects which have been neglected in the fundamental equations of section 2.1. We should, therefore, not expect results based on these equations to apply to steep fronts, and will have to consider this situation in another manner, as is done in section 2.5.

Steep fronted waves of this type are known as shock waves, and while it is not physically reasonable for the ultimate slope to be infinitely steep (as this would imply infinite accelerations), shock waves found in practice have times of rise less than the resolving time of experimental measurements, less than 0.5 microseconds in some cases. It is important to note that while nothing has been said about the velocity of such shock fronts, it is always greater than c_o, the velocity of small amplitude sound waves.

It is easy to see, by reversing the argument given for waves of compression, that a wave of rarefaction, in which later portions of the wave are regions of lower pressure and the particle velocity is away from the direction of advance, will broaden out as it advances and shock waves of rarefaction cannot develop.

Although the arguments just given apply to plane waves, we should expect the same sort of effect in spherical waves except that the amplitude will be weakened by the spreading out of the wave, and the effect will become less important as the distance from the source increases. It would be erroneous, however, to conclude that effects of finite amplitude at a shock front are important only within a few feet of an explosion.

A. *Plane waves of finite amplitude.* The qualitative argument just given for plane waves can be made more explicit by means of a relatively simple argument due to Riemann.[1] The equation of continuity (2.2) and equation of motion (2.4) for motion in one dimension are

$$\frac{\partial \rho}{\partial t} + \rho \frac{\partial u}{\partial x} + u \frac{\partial \rho}{\partial x} = 0 \tag{2.14}$$

$$\rho \frac{\partial u}{\partial t} + \rho u \frac{\partial u}{\partial x} + \frac{\partial P}{\partial x} = 0$$

Riemann's treatment is based on introducing a new variable σ defined by the integral

$$\sigma = \int_{\rho_o}^{\rho} c(\rho) \frac{d\rho}{\rho} \tag{2.15}$$

where ρ_o is the initial density in the absence of a disturbance and the sound velocity $c = \sqrt{dP/d\rho}$ is a function of density ρ. If the variables σ and c are used to replace P and ρ in Eqs. (2.14), we obtain the relations[2]

[1] This development follows essentially the treatment by Lamb (65). A complete discussion for plane waves in air has been given by Rayleigh (90).

[2] The relations follow from Eqs. (2.14) by noting that

$$\frac{\partial \sigma}{\partial x} = \frac{c}{\rho}\cdot\frac{\partial \rho}{\partial x} = \frac{1}{\rho c}\cdot\frac{\partial P}{\partial x}, \quad \frac{\partial \sigma}{\partial t} = \frac{c}{\rho}\cdot\frac{\partial \rho}{\partial t}.$$

$$\frac{\partial \sigma}{\partial t} + u \frac{\partial \sigma}{\partial x} + c \frac{\partial u}{\partial x} = 0$$

$$\frac{\partial u}{\partial t} + u \frac{\partial u}{\partial x} + c \frac{\partial \sigma}{\partial x} = 0$$

A more symmetrical and useful pair of relations is obtained by addition and subtraction, which gives

$$\frac{\partial}{\partial t}(\sigma + u) + (c + u)\frac{\partial}{\partial x}(\sigma + u) = 0 \tag{2.16}$$

$$\frac{\partial}{\partial t}(\sigma - u) - (c - u)\frac{\partial}{\partial x}(\sigma - u) = 0$$

These equations mean that the quantity $(\sigma + u)$ will be unchanged in time at a point moving with velocity $(c + u)$, and the quantity $(\sigma - u)$ remains unchanged for a point of velocity $-(c - u)$. If a pressure wave is considered in which the direction of propagation and particle velocity are in the direction of increasing x, i.e., a wave of compression, a change in type of this wave will develop. For, as time increases, the propagation of $(\sigma + u)$ with positive velocity $(c + u)$ and of $(\sigma - u)$ with negative velocity $-(c - u)$ will lead to the development of an increasingly distinct region of increasing x, for which $(\sigma - u)$ approaches zero in virtue of the propagation of this quantity backward in space. The value of $\sigma + u$ will thus approach a value $2u$ in the advancing front of the disturbance. In this limit both σ and u will be propagated with a velocity $c + u$ which is greater than the local sound velocity c, as σ is, from Eq. (2.15), positive for a final density ρ greater than ρ_o. The quantity σ is a function of density alone in the absence of dissipative processes. The Riemann form of the dynamical equations thus leads to the conclusion that particle velocity and density (hence pressure also) are ultimately propagated with the velocity $c + u$. This conclusion was tacitly assumed in the preliminary discussion, and the further conclusion that plane waves of compression develop increasingly steep fronts must also follow.

It is possible to apply the Riemann approach outlined to various special forms of disturbance, as has been done by a number of authors.[3] Because the plane wave case is of limited utility in actual problems of interest, these applications will not be considered further here. One simple conclusion is of interest, however, namely that even in the plane wave case we cannot expect a shock wave to advance indefinitely unchanged in intensity if the source of the disturbance does only a finite

[3] See, for example, the report by Kistiakowsky and Wilson (64).

amount of work on the medium. This conclusion follows immediately on observing that the development of a shock front is limited ultimately by dissipative processes which abstract energy from the disturbance. Hence, unless the source continues to supply energy by doing work on the fluid, the finite amount of energy available is reduced by irreversible processes at the shock front converting it to heat, and the energy of the wave must decrease.

B. *Spherical waves of finite amplitude.* The equations of continuity and motion for a problem involving radial motion only are

$$\frac{\partial \rho}{\partial t} + \frac{2\rho}{r} u_r + \rho \frac{\partial u_r}{\partial r} + u_r \frac{\partial \rho}{\partial r} = 0$$

$$\rho \frac{\partial u_r}{\partial t} + \rho u_r \frac{\partial u_r}{\partial r} + \frac{\partial P}{\partial r} = 0$$

Introducing the function σ defined by Eq. (2.15) and dropping the subscript r, we obtain the Riemann form of these equations:

$$\frac{\partial}{\partial t}(\sigma + u) + (c + u)\frac{\partial}{\partial r}(\sigma + u) = -\frac{2cu}{r} \tag{2.17}$$

$$\frac{\partial}{\partial t}(\sigma - u) - (c - u)\frac{\partial}{\partial r}(\sigma - u) = -\frac{2cu}{r}$$

Eqs. (2.17) differ from Eqs. (2.16) for the plane wave case by the presence of the spherical divergence term $-2cu/r$ arising from the equation of continuity. As a result, the propagation of the functions $(\sigma + u)$ and $(\sigma - u)$ with velocities $(c + u)$ and $(c - u)$ in the plane wave case does not hold true for spherical waves. It is therefore not true that $\sigma - u = 0$ after a spherical wave of compression has developed, even at its front. The equations can, however, be solved by stepwise numerical integration from prescribed conditions at a given time. If we consider a sufficiently small increment of time dt and let $N = (\sigma + u)/2$, $Q = (\sigma - u)/2$ we have from Eqs. (2.17)

$$dN = \frac{\partial N}{\partial t} dt + \frac{\partial N}{\partial r} dr$$

$$= -\frac{cu}{r} dt + \Big[dr - (c + u)\, dt\Big]\frac{\partial N}{\partial r}$$

$$dQ = \frac{\partial Q}{\partial t} dt + \frac{\partial Q}{\partial r} dr$$

$$= -\frac{cu}{r} dt + [dr + (c - u)\, dt]\frac{\partial Q}{\partial r}$$

Taking the changes dr for a given dt to be $(c + u)dt$ and $-(c - u)dt$ for the two functions, the equations become

$$dN = -\frac{cu}{r}\cdot dt, \quad dr = (c + u)\,dt \tag{2.18}$$

$$dQ = -\frac{cu}{r}\cdot dt, \quad dr = -(c - u)\,dt$$

If at a time t, values of c, u, and σ are known as a function of r, increments in N, Q may be calculated for a sufficiently small interval dt and corresponding values of dr. Carrying out this process gives new values of N and Q at distances $r + dr$ and time $t + dt$. From these new values of N and Q as a function of r, at time $t + dt$, c and u can be determined if σ is known from the equation of state for the fluid and the process can be repeated.

Penney (83) and later Penney and Dasgupta (85) have carried out calculations of this kind for spherical TNT charges. The details of the process, such as the largest permissible values of dt for sufficient accuracy, and insurance that the iteration accounts for all points in the fluid from which a disturbance can be propagated to a given (r, t), evidently depends on the initial conditions. In the case of high explosives, these are evidently the distribution of pressure and particle velocity in the products when the charge is completely detonated. Penney and Dasgupta base their calculations for TNT on results of H. Jones and G. I. Taylor described in Chapter 3; their method of computation from these results is given in more detail in section 4.4. A further consideration is the fact that the Riemann equations apply only where dissipation effects can be neglected. The conditions realized for a discontinuity of compression advancing in the fluid must therefore be determined by other factors, as described in section 2.5. Although Penney's method of integrating the Riemann equations is simple and straightforward in principle, the numerical calculations are sufficiently tedious and complex that they have been carried out only to six charge radii for one explosive, TNT. It is therefore highly desirable to have a more flexible analytic theory not involving excessive approximation which can readily be applied to a number of explosives and a wide range of distances. A theoretical development of this kind is the subject of the next section.

2.4. Kirkwood-Bethe Propagation Theory

A more analytical approach to shock wave propagation in water, than that of Penney, has been developed by Kirkwood and Bethe and extended by Kirkwood and co-workers to detailed calculations of shock wave pressures for a number of explosives.[4] In this attack, the hydrodynamical equations are considered in a somewhat different way than the one already outlined.

Kirkwood and Bethe consider the solution of the fundamental equations by introduction of a new variable, the enthalpy H defined by the relation

$$H = E + \frac{P}{\rho} \tag{2.19}$$

where E is the internal energy per gram of fluid. If H_o is the enthalpy of the undisturbed fluid ahead of the shock wave, the change in enthalpy $\omega = H - H_o$, and we have

$$d\omega = dE + \frac{dP}{\rho} + Pd\left(\frac{1}{\rho}\right)$$

From the second law of thermodynamics

$$dE = TdS - Pd\frac{1}{\rho}$$

and it follows that

$$d\omega = TdS + \frac{1}{\rho}dP \tag{2.20}$$

Behind the shock front it is assumed that $dS = 0$, and numerical calculations show that the change in enthalpy at the shock front from dissipation are only a few per cent of the total (see section 2.6). The enthalpy calculated on the basis that $dS = 0$ is thus a good approximation, and it proves possible to obtain approximate solutions of the equations of motion, expressed in terms of the enthalpy, in a form to which boundary conditions at the gas sphere can practically be applied. The details of this solution will be treated more fully in Chapter 3. It is desirable, however, to outline the initial steps of the analysis at this point, as they follow immediately from the fundamental equations, and the necessary conditions for the propagation theory at the gas sphere and the shock front must also be put in suitable form for their application. With the assumption $dS = 0$, we can, from Eq. (2.20), write

[4] A series of reports on various stages of these calculations have appeared in NDRC Division 8 Interim Reports (113). The final results are collected and discussed in two OSRD reports by J. G. Kirkwood, S. R. Brinkley, Jr., and J. M. Richardson (59, V and VII). Other related reports by Kirkwood et al are also listed in Reference (59).

$$\text{grad } P = \rho \text{ grad } \omega$$

$$\frac{d\rho}{dt} = \frac{1}{\left(\frac{dP}{d\rho}\right)_s} \cdot \frac{dP}{dt} = \frac{1}{c^2}\frac{dP}{dt} = \frac{\rho}{c^2}\frac{d\omega}{dt}$$

where $c^2 = (dP)/(d\rho)_s$. With these substitutions, the equation of continuity (2.2) and equation of motion (2.4) become,

$$\text{div } \mathbf{v} = \frac{1}{c^2}\frac{d\omega}{dt} \tag{2.21}$$

$$\frac{d\mathbf{v}}{dt} = \frac{\partial \mathbf{v}}{\partial t} + (\mathbf{v}\cdot\text{grad})\,\mathbf{v} = -\text{grad } \omega$$

It is convenient here to introduce a function of ω and v, which may be called the kinetic enthalpy Ω, defined by the relation

$$\Omega = \omega + \tfrac{1}{2}\mathbf{v}\cdot\mathbf{v} = \omega + \tfrac{1}{2}v^2 \tag{2.22}$$

the name arising from the fact that the term $(1/2)\mathbf{v}^2$ is the kinetic energy per gram of fluid. It is easily shown that

$$\text{grad } \omega = \text{grad } \Omega + (\mathbf{v}\cdot\text{grad})\,\mathbf{v} + \mathbf{v}\,(\text{curl } \mathbf{v})$$

In further development, radial flow for which curl $\mathbf{v} = 0$ is assumed, and the equations are profitably examined in terms of a velocity potential ϕ defined by $\mathbf{v} = -\text{grad } \phi$. The second of Eqs. (2.21) becomes $\Omega = \partial\phi/\partial t$ and the first may then be written

$$\nabla^2\phi = \frac{1}{c^2}\frac{\partial^2\phi}{\partial t^2} + \frac{1}{c^2}\left[(\mathbf{v}\cdot\text{grad})\,\mathbf{v}^2 - \frac{d\mathbf{v}^2}{dt}\right] \tag{2.23}$$

If a solution of the form $\phi = \Phi/r$ is assumed and the Laplace operator ∇^2 expressed in spherical coordinates, this equation becomes

$$\frac{\partial^2\Phi}{\partial r^2} - \frac{1}{c^2}\cdot\frac{\partial^2\Phi}{\partial t^2} = \frac{1}{c^2}\left[\frac{u}{2}\frac{\partial u^2}{\partial r} - \frac{d}{dt}u^2\right]$$

and the kinetic enthalpy is given by $r\Omega = \partial\Phi/\partial t = G(r, t)$.

The analysis of Kirkwood and Bethe is based on determining $G(r, t)$ from the boundary conditions and equation of state for the fluid. In order to visualize this approach, it is instructive to consider the propagation of the function $G(r, t)$ in two limiting cases. For incompressible motions $c \rightarrow \infty$, and Eq. (2.23) becomes

$$\frac{\partial^2 \Phi}{\partial r^2} = 0$$

This has the solution $\Phi = \Phi(t)$ by integration, applying the condition that $G \to 0$ as $r \to \infty$. In this case, therefore, the function $r\Omega = \partial\Phi/\partial t$ is propagated outward with infinite velocity. If the disturbance is sufficiently weak, terms involving u^2 are negligible and the wave equation

$$\frac{\partial^2 \Phi}{\partial r^2} - \frac{1}{c_o^2}\frac{\partial^2 \Phi}{\partial t^2} = 0, \quad c_o^2 = \left(\frac{dP}{d\rho}\right)_{s_o}$$

has a solution of the form $\Phi = \Phi(t - r/c_o)$ for an outgoing wave, where c_o is evaluated for the undisturbed medium. In this case then, $G = r\Omega$ is propagated with a velocity c_o.

The limiting ways in which G is propagated make natural the assumption that, in the case of finite amplitude, G is propagated with a variable velocity $\bar{c}$, as expressed by the relation

$$\bar{c} = \left(\frac{\partial r}{\partial t}\right)_{r\Omega = \text{const.}} \tag{2.24}$$

The function $r(G, t)$ may be thought of as a series of curves in the r, t plane for various assigned values of G. It is reasonable, from the limiting cases considered, to assume that the slope $\bar{c}$ of these curves for an outgoing wave is finite and positive, and that these curves for the region behind the shock front do not intersect. With these assumptions, the function $G(r, t)$ may be expressed as a function of a single variable $\tau(r, t)$

$$G(r, t) = G_a(\tau) \tag{2.25}$$

the functional form of $G_a(\tau)$ being unrestricted except for the requirement that $\tau(r, t)$ be a single valued function of $G(r, t)$. With this condition, Eq. (2.24) for the propagation of G may be written

$$\left(\frac{\partial r}{\partial t}\right)_\tau = \bar{c}(r, \tau)$$

If the function G is to be a useful one, we must be able to relate its value at any point to conditions on the boundary of the gas sphere. By our assumptions, the function $G(r, t)$ has, at some earlier time t', the value $G[a(t'), t']$ at the gas sphere boundary, for which $r = a(t')$. The time required for G to be propagated to the point (r, t) is then

$$t - t' = \int_{a(t')}^{r} \frac{dr}{\bar{c}(r, \tau)}$$

the integration being along a path of constant τ. If we set the variable τ equal to t', we have

$$\tau = t - \int_{a(\tau)}^{r} \frac{dr}{\bar{c}(r, \tau)} \tag{2.26}$$

The quantity τ is thus assigned the dimension of time and, as consideration of the function $G_a(\tau)$ shows, plays the role of a retarded time. For the values $\tau = t$, $r = a(t)$, Eq. (2.25) becomes

$$G_a(t) = G[a(t), t]$$

and the function $G_a(t)$ is thus simply the value of G on the gas sphere at time t. The solution $G(r, t)$ is then expressed in terms of $G_a(\tau)$ evaluated at the retarded time τ. The kinetic enthalpy $\Omega(r, t)$ is, from the definition of G, similarly expressed by the relation

$$\Omega(r, t) = \frac{G(r, t)}{r} = \frac{G_a(\tau)}{r} = \frac{a(\tau)}{r} \Omega_a(\tau)$$

where $\Omega_a(\tau) = \Omega[a(\tau), \tau]$ is evaluated on the gas sphere boundary at time τ.

As Kirkwood and Bethe develop it, the solution of the propagation problem is therefore reduced to a determination of the boundary conditions at the gas sphere and evaluation of the retarded time τ. The explicit calculation of $\Omega_a(\tau)$ is described in Chapter 3, but it is instructive to consider qualitatively the value $G(R, t_o)$ at the front of the shock wave, which has been propagated to the point R at time t_o. The time t_o is given by

$$t_o = \int_{a_o}^{R} \frac{dr}{U(r, \tau(r))}$$

where U is the velocity of the shock front and a_o is the initial radius of the gas sphere. The time τ_o at which $G_a(\tau_o)$ determines the value $G(R, t_o)$ is, from Eq. (2.26), given by

$$\tau_o = \int_{a_o}^{R} \frac{dr}{U[r, \tau(r)]} - \int_{a(\tau_o)}^{R} \frac{dr}{\bar{c}[r, \tau_o]} \tag{2.27}$$

In the acoustic case we have shown that $\bar{c} = c_o$ and the front of the wave also travels with velocity $U = c_o$. The time τ_o is therefore zero, and conditions at the front are therefore determined by the initial condition of the gas sphere

$$G(R, t_o) = G_a(0) = a_o\Omega_a(0)$$

For waves of finite amplitude, however, neither $\bar{c}$ nor U is adequately represented by c_o. As will be shown, a propagation velocity $c + \sigma$, which is approximately equal to the local sound velocity $c + u$, is a good approximation to $\bar{c}$, as we might guess from the Riemann formulations of the hydrodynamic equations.

Detailed calculations of the next section, based on the equation of state for water boundary conditions at the shock front, show that the value $c + \sigma$ behind the shock front is greater than U, a result which is reasonable from the discussion of the overtaking effect. The retarded time τ_o is therefore positive and increases as the shock front advances. The conditions at the front are then determined by the value $G_a(\tau_o)$ at progressively later times. This value of $G_a(\tau_o)$ must, however, be expected to decrease with time as the wave is emitted, it being a measure of the energy in the fluid at the boundary. The earlier parts of the wave corresponding to times less than τ are thus progressively lost as the wave travels outward. This characteristic can be thought of as a destruction of these parts of the wave as they advance into the shock front. The effects of dissipation at the shock front are thus implicitly included in the framework of the theory, resulting from the way in which $G(r, t)$ is propagated and the boundary conditions at the front which limit its advance.

2.5. Conditions at a Shock Front in a Fluid

The discussion in section 2.3 led to the conclusion that waves of compression in a fluid develop increasingly steep fronts as they progress, until the disturbance becomes so abrupt that dissipative processes must be examined if any further conclusions are to be reached as to the exact form of such shock fronts. Experimentally, however, it is known that fronts of this kind are so steep as to be virtually discontinuous and their exact shape is ordinarily of no practical concern; indeed, experimental measurements would, by present evidence, be a matter of formidable difficulty. Reserving, until later, any discussion of the dissipation processes which must occur, we consider what can be learned on the assumption, justified by experience and such discussion, that the thickness of the shock front is negligible, and the front can be treated as a discontinuity in comparison to the changes occurring behind it. The equations applying to such a discontinuity were originally developed by

Rankine and by Hugoniot,[5] and are easily obtained by considering regions immediately ahead of and behind the discontinuity. If an observer moves with the velocity U of such a front (see Fig. 2.2) into a region of particle velocity u_o and density ρ_o, the apparent velocity of the fluid toward him is $U - u_o$ and in a time dt, a mass of fluid $\rho_o(U - u_o)dt$ will enter the unit area of the front. The apparent velocity of fluid leaving the front is $-(U - u)$, where u is the particle velocity relative

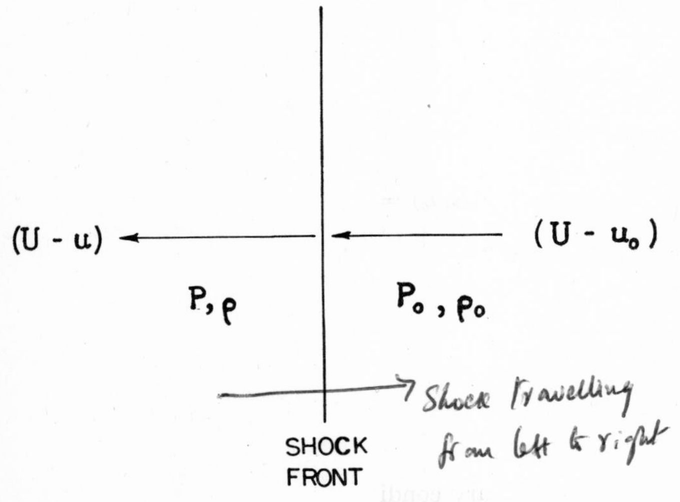

Fig. 2.2 Conditions at a shoc front moving with velocity U.

to fixed coordinates, and a mass of fluid $\rho(U - u)dt$ will leave the front in time dt, ρ being the density. If the front is discontinuous, we can shrink the time dt to infinitesimal values for which the mass flow into the front must approach that away from it. In the limit the two become equal so that

$$\rho_o(U - u_o) = \rho(U - u)$$

an expression of conservation of mass.

The mass flow into the front in time dt has momentum $\rho_o(U - u_o)u_o dt$, and the mass flow out has momentum $\rho_o(U - u_o)u dt$. The difference, which is the change in momentum, must equal the impulse of the net force per unit area in the limit $dt \rightarrow 0$. If the pressures ahead of and behind the front are P_o and P, we obtain

[5] The argument which follows was first developed by Rankine (89) in determining the conditions necessary at a discontinuous front. Rankine considered the equations of mass and momentum. Hugoniot in later work gave the equation for increase in internal energy, assuming the existence of a discontinuity. See Lamb (65), p. 484ff.

$$\rho_o(U - u_o)(u - u_o) = P - P_o$$

an expression of conservation of momentum.

A similar argument for energy requires that the net work done by the pressures P and P_o equal the increase in kinetic plus potential energy when the time increment dt becomes infinitesimal. The work done by P is $Pudt$ for unit area of the front, by P_o is $P_o u_o dt$. The kinetic energies per unit mass are $\frac{1}{2}u^2$ and $\frac{1}{2}u_o^2$, and if E and E_o denote the internal energies per unit mass, we have

$$Pu - P_o u_o = \rho_o(U - u_o)\,[E - E_o + \tfrac{1}{2}(u^2 - u_o^2)]$$

the requirement for energy conservation.

In further applications of these equations we shall usually be interested in the case of an undisturbed fluid ahead of the front for which $u_o = 0$. With this simplification and some rearrangement, we obtain more convenient relations

$$\begin{aligned} \rho(U - u) &= \rho_o U \\ P - P_o &= \rho_o U u \\ E - E_o &= \tfrac{1}{2}(P + P_o)\left(\frac{1}{\rho_o} - \frac{1}{\rho}\right) \end{aligned} \tag{2.28}$$

These relations, being obtained for the limit of negligible thickness of the shock front, should be equally valid for plane or spherical shock fronts.

If the equation of state for the fluid is known, it is possible to determine the increase in internal energy $E - E_o$ as a function of the pressures P and P_o, and densities ρ and ρ_o. For given values of P_o and ρ_o, in undisturbed fluid ahead of the shock front, the third of Eqs. (2.28) then defines a relation between P and ρ immediately behind the front. This pressure-density relationship is usually known as the Rankine-Hugoniot equation, and is a curve of somewhat the same form as adiabatic or isothermal P–V curves, although it is evidently not the same function.

The first and second of Eqs. (2.28) can be solved for the shock velocity U and particle velocity u in terms of the pressures and densities behind the front, giving

$$\begin{aligned} U &= \sqrt{\left(\frac{\rho}{\rho_o}\right)\frac{P - P_o}{\rho - \rho_o}} \\ u &= \frac{\rho - \rho_o}{\rho}\cdot U \end{aligned} \tag{2.29}$$

The first of these equations is frequently useful, as together with a knowledge of the equation of state it permits calculation of shock front velocity U in terms of the pressure P behind the shock front and the initial conditions P_o, ρ_o. Similarly, the second equation permits calculations of the particle velocity u. Considered more generally, Eqs. (2.28) provide three relations among the four variables, P, ρ, U, u, by means of which any three can be expressed in terms of the fourth, given the initial values P_o, ρ_o.

In the preliminary discussion, it was shown that shock waves of rarefaction, in which the pressure is less than that in the fluid ahead of the disturbance, cannot be expected to maintain themselves in any normal liquid or gas. This conclusion also follows from the Rankine-Hugoniot relations. It is evident that a rapid compression of any fluid must, as a result of dissipative processes, leave it at a higher temperature than would be the case for adiabatic compression to the same final volume. For any normal fluid, this irreversible process also means a higher final pressure than the adiabatic value, and it is easily shown that such a state is consistent with the energy requirements for a shock front of compression. This excess of temperature over the adiabatic value is, of course, the result of degradation of energy to heat which cannot be returned to the fluid as available energy. A reversal of the process as an expansion, which would be required to satisfy the conditions for a shock front of rarefaction, cannot therefore take place and such fronts are impossible in normal fluids.

2.6. Properties of Water at a Shock Front

We have seen from section 2.5 that the requirements for conservation of mass, momentum, and energy at a shock front lead to three equations for the pressure, density, and particle velocity behind the front in terms of the corresponding quantities ahead of it and the velocity of the front. In the case of shock waves travelling into undisturbed water, these conditions therefore provide three relations among the four variables describing the front at any point. The necessary fourth relation to solve for the properties of the water as a function of any one variable, is evidently provided by a suitable equation of state and specific heat data. One objective of such a solution is to provide relations suitable for explicit calculation of shock wave pressures from explosions on the basis of the conditions at the boundary of the gas sphere and the propagation theory of Kirkwood and Bethe. For this purpose, solutions of reasonable accuracy over a broad range of pressures, from 0 to 500,000 lb./in.^{2} or higher, are needed. A second desired type of solution is one of greater accuracy over a more limited, lower pressure range, which will be suitable for example, in comparison of experimental shock wave velocity and pressure measurements. Although the general procedure in obtaining both solutions is the same, the appropriate numerical data and approximations for analytical purposes are different.

A. *Methods of solution.* In order to express the Rankine-Hugoniot conditions in a convenient form for evaluation from experimental data, some manipulation based on purely thermodynamic considerations is necessary. Changes in internal energy E of unit mass of water can be expressed in terms of pressure and temperature variations as

$$dE = \left(\frac{\partial E}{\partial T}\right)_P dT + \left(\frac{\partial E}{\partial P}\right)_T dP$$

E is, however, a point function, its value being independent of the process by which the fluid was changed from its initial condition to the final one. The change of internal energy ΔE for water can therefore be written as a sum of changes along any suitable paths, and we have

$$\Delta E = \int_{T_o}^{T} \left(\frac{\partial E}{\partial T}\right)_{P_o} dT + \int_{P_o}^{P} \left(\frac{\partial E}{\partial P}\right)_T dP$$

the first process being carried out at the initial pressure P_o, the second being an isothermal compression at the final temperature T. From the first and second laws, the partial derivatives of E can be expressed in terms of experimentally obtained quantities as

$$\left(\frac{\partial E}{\partial T}\right)_P = \left(\frac{\partial Q}{\partial T}\right)_P - P\left(\frac{\partial V}{\partial T}\right)_P = C_P - P\left(\frac{\partial V}{\partial T}\right)_P$$

$$\left(\frac{\partial E}{\partial P}\right)_T = T\left(\frac{\partial S}{\partial P}\right)_T - P\left(\frac{\partial V}{\partial P}\right)_T = -T\left(\frac{\partial V}{\partial T}\right)_P - P\left(\frac{\partial V}{\partial P}\right)_T$$

where C_P is the specific heat at constant pressure, $(\partial V/\partial T)_P$ is the thermal expansion coefficient, and $-(\partial V/\partial P)_T$ is the compressibility (V is the specific volume, i.e., the volume of unit mass of water). Substituting, we have for ΔE

$$\Delta E = \int^{T} C_{P_o}\, dT - P_o(V - V_o) - T\int_{P_o}^{P} \left(\frac{\partial V}{\partial T}\right)_P dP$$
$$- (PV - P_oV_o) + \int_{P_o}^{P} V(T)\, dP$$

where one integral has been evaluated and a second integrated by parts. The increase ΔE across a shock front is, from Eq. (2.28), given by

$$\Delta E = \tfrac{1}{2}(P + P_o)(V - V_o)_o$$

Therefore, equating the two values of ΔE results in an implicit relation between the initial and final conditions of pressure and temperature in terms of known properties of water.

Kirkwood and co-workers[6] have devised effective methods of successive approximation to determine the temperature increase satisfying the equation for an assumed pressure increase. Knowing the temperature of the final state, the density can be computed from the equation of state. These data then determine the enthalpy ω and Riemann function σ used in Kirkwood's formulation of shock wave propagation, which are defined by the relations

$$\omega = \int_{P_o}^{P} \frac{dP}{\rho}, \quad \sigma = \int_{\rho_o}^{\rho} \frac{c}{\rho}\, d\rho \text{ where } c^2 = \left(\frac{dP}{d\rho}\right)_S$$

Although, strictly speaking, these quantities should be evaluated isentropically for a transition to the final state, the error incurred by using the Rankine-Hugoniot P–V relation is small. This can be shown by calculation of the change in entropy involved at the shock front, the results being given in the next section.

Other variables of interest, such as the particle velocity and shock front velocity, can also be computed from the Rankine-Hugoniot relations, and thermodynamic functions such as the changes in entropy or internal energy can be calculated in a straightforward manner by appropriate integrations in terms of known functions.

B. *Solutions for high pressures.* Kirkwood et al. (59, II and III) have made calculations for pressures up to 50 kilobars in pure water and pressures up to 90 kilobars in sea water (taken to be a 0.7 molal solution of NaCl) at 0, 20°, 40° C. The differences between the two conditions are not large and only the results for salt water will be described here. Extensive work has been done on the P–V–T relation for water at pressures up to 350,000 lb./in.^{2} by Bridgman and Gibson. Gibson has made measurements to determine appropriate constants for a Tait form of equation of state and the effect of salinity. The Tait equation may be written

$$V(T, P) = V(T, O)\left\{1 - \frac{1}{n}\log\left(1 + \frac{P}{B}\right)\right\}$$

or

$$-\left(\frac{\partial V}{\partial P}\right)_T = \frac{1}{n[P + B(T)]}$$

[6] See Reference (59), Reports II and III.

where n is a constant and B a function of temperature only. Use of this equation to determine $(\partial V/\partial T)_P$ and $(\partial V/\partial P)_T$, together with values of $V(T, O)$ and $C_P(T)$, then enable the calculations already outlined to be made. This procedure was used for the original calculations on pure water. For the calculations on salt water a somewhat different scheme of calculation was set up, based on a modified form of the Tait equation. It was found convenient to make calculations of shock front conditions by integrations along paths of constant pressure and entropy, and for this purpose the equation of state was taken to be of the form

$$-\frac{1}{V}\left(\frac{\partial V}{\partial P}\right)_S = \frac{1}{n[P + B(S)]} \tag{2.30}$$

or in integrated form

$$P = B(S)\left[\left(\frac{V(T, O)}{V(T, P)}\right)^n - 1\right]$$

where $B(S)$ is a function of entropy alone and n is approximately a constant. It was found that the necessary value of n to fit P–V–T data between 20° and 60° C. up to pressures of 25 kilobars deviated by less than ± 4 per cent from an average value 7.15, which was adopted for the calculations. The slowly changing constant $B(S)$ was taken to be the same as the value $B(T)$ determined for 0.7 molal salt water from data of Gibson. At 20° C., B has the value 3.047 kilobars.

This form of equation of state should give a better extrapolation of the P–V–T relations to high pressures than the isothermal Tait equation, which permits a state of zero volume for finite pressure and hence must overestimate the compressibility and related sound wave velocities at high pressures. With the adiabatic form of equation and values of $V(T, O)$ and C_p from the literature (cited in Table 2.1), the enthalpy ω, particle velocity u, sound and shock velocities c and U, and Riemann function σ can be computed by straightforward methods. Results for sea water initially at 20° C. are given in Table 2.1 for pressures up to 80 kilobars (1 kilobar = 14,513 lb./in.^{2}), and plotted in Fig. 2.3.

Although they are not directly needed, the change in entropy and temperature rise at the shock front are of interest and approximate values are given in Table 2.1 for several pressures. The entropy change is worth consideration as Kirkwood and Bethe in their propagation theory neglect its contribution to the enthalpy of the shock wave. This contribution is given by

$$\Delta\omega(S) = \int_{S_o}^{S} T(P_o, S)\, dS$$

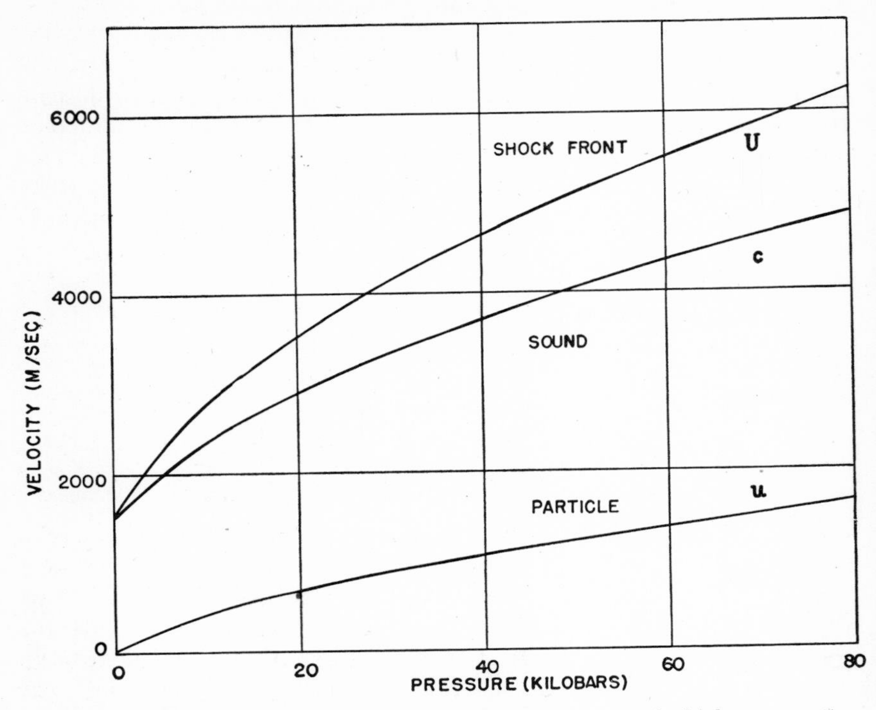

Fig. 2.3 Velocities at a shock front as a function of pressure for high pressures.

P (kilobars)	u (m./sec.)	σ (m./sec.)	U (m./sec.)	c (m./sec.)	ΔT (° C.)	ΔS (cal./gm. deg.)
0	0	0	(1465)	(1465)	0	0
5	251	249	1975	2230	15.7	.0045
10	426	416	2335	2755	35.5	.019
15	567	549	2630	3175	53	.038
20	689	663	2880	3535	69	.060
25	798	765	3110	3855	85	.082
30	898	859	3320	4140	100	.102
35	990	947	3510	4405	114	(.120)
40	1075	1030	3690	4650	127	(.138)
50	1235	1185	4020	5100	152	(.174)
60	1380	1330	4325	5505		
70	1510	1465	4610	5880		
80	1625	1600	4885	6240		

Data employed:
$V(T)$: Gibson and Loeffler, *JACS 63*, 443 (1941).
C_p (joules/gm. deg.) = $3.9644 \times 6.24 \times 10^{-4}\ T$(° C.), interpolated from ICT.

Table 2.1. Properties of strong shock fronts in salt water initially at 20° C. (calculations of Kirkwood and Richardson, except for ΔT, ΔS which are from calculations of Kirkwood and Montroll for pure water).

and for estimates of magnitude may be approximated by

$$\Delta\omega(S) \cong T_{av}\Delta S$$

where T_{av} is the mean temperature. For a final pressure of 3.10^5 lb./in.^{2} we obtain

$$\Delta\omega(S) = 330 \cdot 0.06 \cdot 4.19 = 83 \text{ joules/gm.}$$

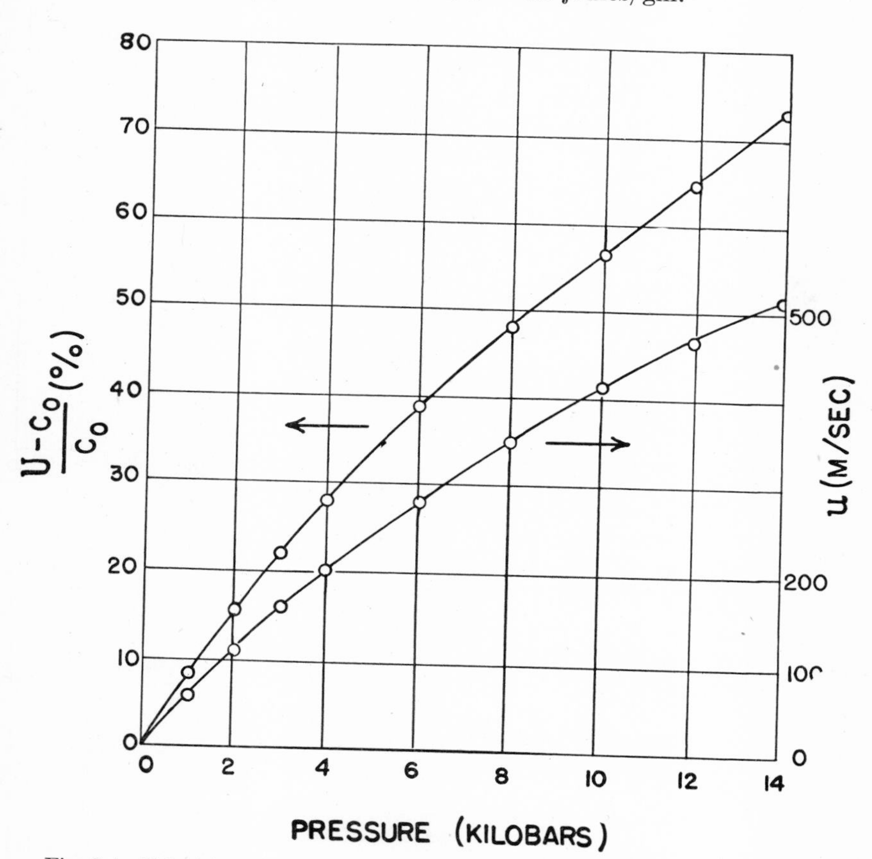

Fig. 2.4 Velocities at a shock front as a function of pressure for low pressures.

which is less than 5 per cent of the change of 1,750 joules/gm. at constant entropy.

An interesting consequence of the result is that the P–V curve for the shock wave passes, at about 360,000 lb./in.^{2} pressure, through a region found by Bridgman to be a solid state of equilibrium which he labelled "Ice VII." This order of pressure is realized at a distance of about two charge radii for TNT, but the actual formation of ice by the shock wave is not to be expected in the times that the water remains in this condition, as transitions of this kind take place very slowly. The water is rather in a metastable condition not amenable to conventional measurement.

In use of the shock front data of Table 2.1, it is to be remembered that the equation of state used is only approximate and involves considerable extrapolation for pressures above 360,000 lb./in.^{2} (25 kilobars)

Pressure (kilobars)	0.25	0.50	0.75	1.00	1.25	1.50
$(U - c_0)/c_0$ (%)	2.07	4.03	5.93	7.81	9.65	11.44
u (m./sec.)	15.1	31.2	46.2	60.5	74.4	87.7
ΔH (cal./gm.)	5.8	11.5	17.2	23.0	28.5	34.1

(a). Pressures up to 1.5 kilobars, initial temperature 15° C., acoustic sound velocity c_0 = 1500.5 m./sec.

Pressure (kilobars)	1.0	2.0	3.0	4.0	6.0	8.0	10.0	12.0	14.0
$(U - c_0)/c_0$ (%)	8.18	15.39	21.9	27.8	38.6	48.1	56.7	64.7	72.3
u (m./sec.)	59.2	110	158	201	278	346	409	467	511
ΔH (cal./gm.)	23.0	45.3	67	89	131	173	214	255	293

(b). Pressures up to 14 kilobars, initial temperature 25° C., acoustic sound velocity c_0 = 1528.5 m./sec.

Data employed:

$V(T)$: Knudsen's Tables, as given in Oceanographical Tables, compiled by Zubov, Commissariat of Agriculture USSR, Moscow.

c_0: Values given by Mathews in Report HD-282, British Admiralty Hydrographic Office, which also gives compressibility data of Ekman for pressures up to 1.5 kilobars. For pressures up to 11 kilobars, the Tait equation was fitted to data of Adams, *JACS 53*, 3769 (1931).

Table 2.2 Properties of weak shock fronts in salt water (3.79 weight per cent NaCl) from calculations of Arons and Halverson (4).

for want of experimental data. The values at higher pressures are therefore of necessity increasingly uncertain, and the values at pressures below 15 kilobars should be obtained from Table 2.2 and Fig. 2.4, which are based on more accurate data on water at lower pressures. For example, the velocity of sound at atmospheric pressure and 20° C. given by Kirkwood is 4,797 ft./sec., and the best value is about 4,967 ft./sec.

Penney (83) has also made calculations of shock front velocity, sound velocity and particle velocity as a function of pressure behind the shock front. Penney also used an adiabatic form of the Tait equation which be fitted to Bridgman's data for pure water with the result which may be written

$$-\frac{1}{V}\left(\frac{\partial V}{\partial P}\right)_S = \frac{1}{7.47(P + 2.94)}$$

where P is in kilobars. The slightly different constants, as compared to Kirkwood's values, lead to results which differ by one or two per cent only for pressures up to eighty kilobars.

It is interesting to note that any adiabatic equation of the form (2.30) leads to a simple numerical relation between the Riemann function σ and the sound velocity c behind the front, the relation being

$$\sigma = \frac{2}{n - 1}(c - c_o) \tag{2.31}$$

where c_o is the sound velocity in the undisturbed medium ahead of the shock front. A relation of this kind considerably simplifies numerical integration of the Riemann equations or analytic formulations, and both Penney and Kirkwood make use of such relations.

C. *Approximate relations for adiabatic changes.* In application of the Kirkwood-Bethe theory for propagation of spherical shock waves it is necessary to utilize relations for the various variables behind the shock front. The adiabatic modification of the Tait equation provides a relation convenient for this purpose, and it is also found useful to employ the Riemann function σ as an independent variable. Although the final state behind a shock front should properly be calculated from the Rankine-Hugoniot relations, the results so obtained differ little from those obtained assuming an adiabatic change. This is plausible from the small values of entropy change at the shock front (see Table 2.1), and will be shown explicitly from the adiabatic results which we proceed to derive.

If the pressure P_o in the undisturbed fluid is taken to be zero (a pressure of at most a few atmospheres being negligible for the present interest), the adiabatic Tait equation gives a relation for the parameter B in terms of the velocity of sound at zero pressure c_o,

$$B = \frac{\rho_o c_o^2}{n}, \quad \text{where} \quad c_o^2 = \left(\frac{dP}{d\rho}\right)_{S_o}$$

Assuming that n and B are constants, the adiabatic equation gives the following relations

$$P = \frac{\rho_o c_o^2}{n}\left[\left(\frac{\rho}{\rho_o}\right)^n - 1\right]$$

$$c^2 = \left(\frac{dP}{d\rho}\right)_S = c_o^2\left(\frac{\rho}{\rho_o}\right)^{n-1}$$

$$\sigma = \int_{\rho_o}^{\rho} c\,\frac{d\rho}{\rho} = \frac{2c_o}{n-1}\left[\left(\frac{\rho}{\rho_o}\right)^{\frac{n-1}{2}} - 1\right]$$

$$\omega = \int_{\rho_o}^{\rho} \frac{dP}{\rho} = \frac{c_o^2}{n-1}\left[\left(\frac{\rho}{\rho_o}\right)^{n-1} - 1\right]$$

If σ is used as an independent variable, we find by substitution that

$$c = c_o\left[1 + \frac{n-1}{2c_o}\,\sigma\right]$$

$$\omega = c_o\sigma\left[1 + \frac{n-1}{4c_o}\,\sigma\right]$$

From Table 2.1 for properties of water at the shock front it is evident that σ is approximately equal to u. Substituting $u = \sigma$,[7] we obtain for the kinetic enthalpy Ω (from its definition) and the shock front velocity (from the Rankine-Hugoniot conditions) the relations

$$\Omega = \omega + \frac{u^2}{2} = c_o\sigma\left[1 + \frac{n+1}{4c_o}\,\sigma\right]$$

$$U = \Omega/\sigma = c_o\left[1 + \frac{n+1}{4c_o}\,\sigma\right]$$

The propagation velocity $c + \sigma$ as a function of σ also follows immediately, and with the notation $\beta = (n + 1)/4c_o$ we obtain the relations employed by Kirkwood and Bethe in their propagation theory

(2.32) $$c + \sigma = c_o\,[1 + 2\beta\sigma]$$

$$U = c_o\,[1 + \beta\sigma], \quad \text{where} \quad \beta = \frac{n+1}{4c_o}, \quad c_o^2 = \left(\frac{dP}{d\rho}\right)_{S_o}$$

$$\Omega = c_o\sigma[1 + \beta\sigma]$$

The values of $c + \sigma$, U, and Ω obtained by these relations with $n = 7.0$ and values of σ from Table 2.1 are compared for several pres-

[7] It might appear that replacing u by σ is an additional approximation, but the errors in the propagation theory of Kirkwood and Bethe are actually reduced, as discussed in the Appendix.

sures with numerical values from the equation of state and Rankine-Hugoniot conditions in Table 2.3. It is seen that the approximate equations give reasonably good agreement with the more accurate calculations, and therefore form a good basis for development of propagation theory. The velocity $c + \sigma$ which Kirkwood and Bethe use as their fundamental propagation velocity behind the shock front is seen always to exceed the shock front velocity U. This is what the argument of section 2.3 led us to expect, and makes plausible the result that the outward motion of a shock wave involves a progressive overtaking of the front by the disturbance behind the front and dissipation at the front which limits its rate of advance.[8]

2.7. The Thickness of Shock Fronts

The Rankine-Hugoniot equations are universally employed to represent the conditions at a shock front. Since their derivation depends on the assumption that a shock front can be regarded as a discontinuity, it is important to examine whether a theoretical estimate of the shock front thickness is consistent with this assumption. A number of writers have made such estimates assuming that the mechanism of dissipation in the shock front could be represented by the macroscopic shear viscosity of the fluid. The common result of all such calculations is that they predict thicknesses of the order 10^{-5} to 10^{-6} cm., which are of course completely negligible. It is evidently meaningless to ascribe any physical significance to lengths of molecular magnitude obtained by macroscopic concepts beyond the fact that they are so exceedingly small. The failure of experimental measurements to yield more than an upper limit set by the limitations of the measuring equipment constitutes additional adequate evidence that the physical thickness of the shock front is not important in the analysis of shock wave propagation.

Pressure (kilobars)	U(m./sec.)		$c + u$ (m./sec.)		Ω (10^6 m.^{2}/sec.^{2})	
	Formula	Tabular	Formula	Tabular	Formula	Tabular
5	1962	1975	2460	2481	0.488	0.491
10	2296	2335	3127	3181	0.954	0.970
25	2995	3110	4525	4653	2.29	2.37
50	3835	4020	6205	6335	4.54	5.15

Table 2.3. Comparison of approximate expressions for shock front parameters with numerically computed values.

Of the various theoretical discussions which have been given, those of Rayleigh (90) and of Taylor and MacColl (30) should be mentioned.

[8] A more general discussion of shock wave velocities for an arbitrary equation of state has been given by Bethe (8), who shows that very generally $c_o < U < c$.

Herring[9] has also given a particularly simple derivation which illustrates the type of argument that can be used to determine the order of shock front thickness given by macroscopie viscosity considerations. If the coefficient of shear viscosity is denoted by μ, the classical theory of viscous fluids shows that the energy dissipation per unit volume and time in a plane wave is given by

$$2\mu\left(\frac{\partial u}{\partial x}\right)^2$$

For a plane wave, however, $\partial u/\partial x = -(1/\rho)\, d\rho/dt$ by the equation of continuity, and writing $\Delta t = (1/c)\,\Delta x$ where Δx is the thickness of the front, the dissipation per unit area of the front is roughly

$$2\mu\left(\frac{1}{\rho}\frac{d\rho}{dt}\right)^2 \cdot \Delta x \cong 2\mu c^2 \frac{(\Delta\rho)^2}{\rho_o{}^2\Delta x}$$

where $\Delta\rho$ represents the increase in density across the shock front. In order to form an estimate of Δx it is necesssary to obtain an alternative estimate of the energy dissipated in the front. This can be done if we assume that the Rankine-Hugoniot conditions are at least approximately true. With this assumption the dissipated energy can be estimated from the difference in area between the *R–H* curve on the *P–V* diagram and the adiabatic curve, as Herring does. Essentially the same result can be obtained by using the dissipated enthalpy computed by Kirkwood and Montroll (59, II). If the ratio of dissipated enthalpy to the total is α, we have

$$\text{Energy dissipated/gram} = \alpha\Delta H = \alpha u\left(U - \frac{u}{2}\right)$$

from the fact that $\Delta H = \Delta(E + P/\rho) = u(U - u/2)$. The mass of water traversed in unit time by unit area of the shock front is with sufficient accuracy $\rho_o c$, and equating the two expressions gives

$$\Delta x = \frac{2\mu c}{\rho_o \alpha u\left(U - \frac{u}{2}\right)}\left(\frac{\Delta\rho}{\rho_o}\right)^2$$

For a pressure of 20 kilobars Kirkwood's calculations give $\Delta\rho = 0.32$ gm./cm.^{3}, $\alpha = 0.05$, and employing a value $\mu = 0.01$ cgs. units gives $\Delta x \cong 0.6 \times 10^{-6}$ cm.

[9] Herring's calculation is given in a chapter written by him on "Explosions as a Source of Sound," for inclusion in a summary technical report for Division 6 of NDRC, and to be entitled "Physics of Sound in the Sea."

More detailed investigations of the thickness of shock fronts have been made by a number of writers. In a classic paper, Becker (7) obtained exact solutions of the hydrodynamical equations for a plane shock wave in an ideal gas, and came to the conclusion that the thickness of the shock front became comparable with the mean free path for only moderate pressure ratios on the two sides of the front. Under these circumstances the hydrodynamical equations are no longer valid and kinetic theory must be used, which reduces to hydrodynamics if fluctuations over distances of a free path can be neglected. Becker further concluded that even classical kinetic theory was inapplicable to very intense shock waves in gases. Thomas (113) has shown that this conclusion is in error because of faulty assumptions about the coefficients of viscosity and thermal conductions for gases at high temperatures and pressure, and when the proper values are used he finds that kinetic theory should be applicable even to extremely intense shocks. All these results are for gases and, of course, do not apply directly to liquids. However, it seems clear that the thickness of shock fronts is so small as to be virtually undetectable by direct measurement, and the gradients of pressure and particle velocity are so large that hydrodynamical methods are no longer applicable to their detailed description.

2.8. Conditions at a Boundary between Two Different Media

In our review of the hydrodynamic conditions in fluids, we have so far considered only the development and propagation of pressure waves in an infinite medium. Even this development is incomplete unless account is taken of the way in which conditions in the gas sphere are related to those in the water at their boundary, and in practical circumstances the medium of propagation is of finite extent. The natural boundaries which always exist are of course the surface of the sea and the bottom, and the most important use of underwater explosions has long been for attack against ships or other structures. The general problem of what happens when the disturbances set up by an underwater explosion come in contact with a different medium at a boundary is, therefore, one of considerable importance, but only a few of its ramifications can be dealt with here.

The approximation is again assumed that the motion of water or gaseous products is described sufficiently well without viscosity or other dissipation mechanisms, except at vanishingly narrow fronts or discontinuities. If we consider a plane boundary between two media having different mechanical properties (density, compressibility), it is evident that the pressures must be equal at adjacent points in the two media on either side of the boundary. Further, any motion of the fluids at right angles to the boundary must be such that the components of particle velocity perpendicular to the boundary are the same at adjacent points. With the neglect of viscosity, there is no restriction on the tangential components of particle velocity and the two fluids are free to slide parallel to the boundary. This lack of restraint is not strictly true in any fluid, and there must be a layer at the boundary where viscous friction is important if there are differences of tangential velocity near the boundary. For the cases we shall have to consider, however, there is no reason to believe that this layer is thick enough to modify the motion appreciably, and in the special case of motion normal to the boundary this question does not arise.

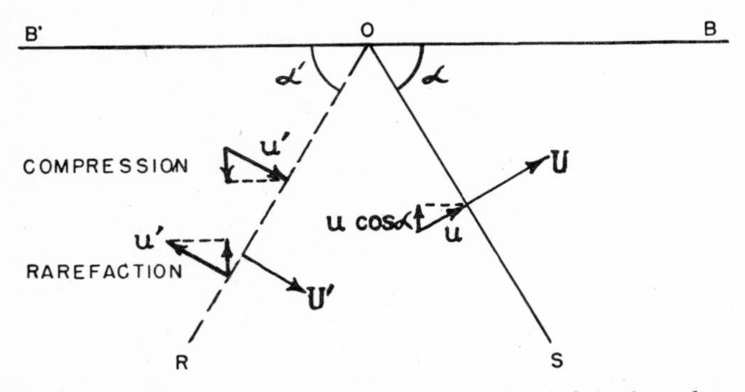

Fig. 2.5 Reflection of a plane shock front at an infinite boundary.

Before becoming involved with mathematical consequences of the boundary conditions on pressure and normal particle velocity, it is of interest to examine the physical situation when a shock wave arrives at a boundary. We consider that at some time a plane shock front of compression SO is incident on the boundary BB' at a point O as shown in Fig. 2.5, the wave front making an angle α with the boundary as shown. To the right of SO in the undisturbed fluid the particle velocity is zero, but the fluid behind the front has acquired a velocity u normal to the front in the direction of advance, the magnitudes of u and shock front velocity U in the fluid below OB' being related to the compression by the Rankine-Hugoniot conditions and equation of state. We have next to consider the state of affairs along the line OB' (actually a plane normal to the plane of the figure). Suppose, first of all, that the medium above the boundary is infinitely rigid, by which is meant that no motion of the boundary can take place. As the shock wave travels from left to right it leaves the fluid at the boundary with a component of velocity $u \cos \alpha$ normal to the boundary. This is, however, inconsistent with the assumed rigid boundary and the resistance of the boundary must therefore modify the motion of the water in such a way as to annul the normal flow set up by the incident wave. A stationary condition can be set up only if we postulate a secondary shock front originating at O of such strength and direction that its passage through the fluid behind OS introduces a normal velocity equal and opposite to that produced by the incident wave all along the line OB'. A conceivable front of this kind is indicated by the dashed line OR in Fig. 2.5 and is called regular reflection. Whether or not the necessary compensation can be achieved by such a front requires a more detailed examination but it is clear that a front of the type indicated must be one of compression.

The opposite extreme occurs if the medium above OB' offers no resistance to compression. This state of affairs is very nearly realized if the fluid below OB' is water and above OB' air, the ratio of compressibilities being of the order $1/7 \cdot 10^5$. In this case it is not possible to develop a significant compression in the medium along the boundary

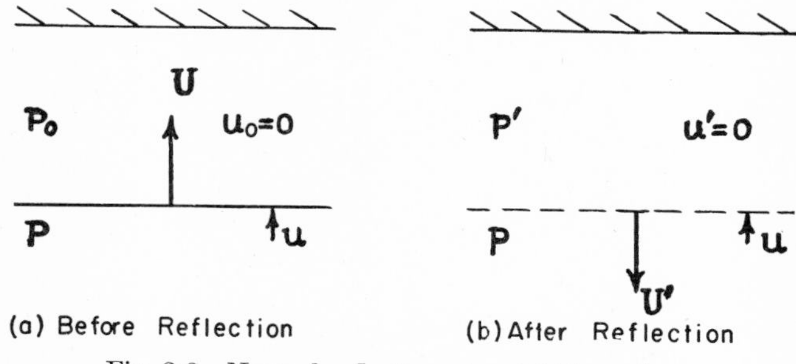

Fig. 2.6 Normal reflection at a rigid boundary.

OB', and pressure waves must be originated at O which leave the fluid along OB' in its original state of compression. There is of course no opposition to motion of the boundary and hence no restriction on the normal components of velocity. The necessary conditions are then again conceivably realized by a pressure front originating at O, but differing from the rigid boundary case, in that a rarefaction wave is required to reduce the pressure to its original value. In this case the particle velocity developed by the rarefaction has a component toward the boundary as indicated in Fig. 2.5. The fluid behind OR is therefore left with a normal component of velocity toward the boundary which is the sum of the components introduced by the two pressure fronts and the boundary will be displaced upward.

In the special case where the boundary is normal to the incident wave, incident and reflected wave fronts do not exist simultaneously. If the boundary is perfectly rigid, the situation when the wave has just reached it is one of fluid moving toward the boundary. This motion can only be destroyed at the boundary by a discontinuity travelling back through the moving fluid with intensity and velocity of such magnitude as to destroy the flow velocity, and satisfy the Rankine-Hugoniot conditions. The conditions before and after reflection are indicated schematically in Fig. 2.6.

If the medium above the boundary is a compressible fluid, the boundary conditions of continuity and pressure can be satisfied only if a wave is transmitted into the fluid. The pressure and particle velocity left behind the front at the boundary must equal the pressure and particle velocity left behind the reflected wave, which is itself travelling in the compressed upward moving fluid left by the original wave. The situation is shown in Fig. 2.7. It is to be understood that the condi-

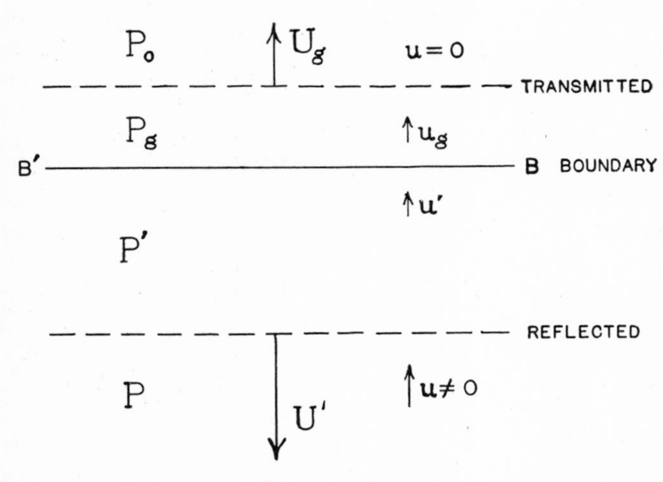

Fig. 2.7 Reflected and transmitted waves for normal incidence.

tions shown are strictly true only at the boundary, which in general moves after the original impact ($u' \neq 0$) and that the pressure P' and u' at the boundary will decay with time even for a plane shock wave.

The relative compressions and flow velocities in the different regions of Fig. 2.7 depend not only on the relative densities and compressibilities but also, for finite amplitude waves, on the intensity of the original wave. In the case of normal incidence on a boundary, we can take as the boundary conditions

$$P' = P_g'$$
$$u' = u_g'$$

where P', u' refer to the fluid below OB', P'_g, u'_g to the fluid above OB'. The equalities are true at any time after incidence and hence the total time derivatives, which measure the change at the boundary, must also be equal:

$$\frac{dP'}{dt} = \frac{dP_g'}{dt}$$

$$\frac{du'}{dt} = \frac{du_g'}{dt}$$

These are the appropriate conditions, for example, in following the boundary of the gas sphere after detonation of a charge.

2.9. Reflection of Acoustic Waves

A. *Rigid boundary.* Before considering the more general problem of waves of finite amplitude, it is worthwhile to examine the results in the limiting case of small compressions and flow velocities. Returning to the case of a rigid boundary indicated in Fig. 2.5, we observe that, for a sufficiently weak disturbance, the fluid behind the incident front OS differs infinitesimally from that ahead of it. As the front progresses parallel to the boundary, points on it at fixed distances all have the same velocity $c_o/\sin\alpha$ parallel to OB', where c_o is the normal velocity of the infinitesimal disturbance OS. The postulated front OR travels with the same velocity c_o and the point velocity is $c_o/\sin\alpha'$ where α' is the angle of the front with the boundary. In the approximations of an infinitesimal disturbance these velocities become asymptotically equal: $c_o/\sin\alpha = c_o/\sin\alpha'$, and we have the familiar law of reflection that $\alpha = \alpha'$. The condition imposed by the boundary is that there be no net component of particle velocity normal to it, which is expressed by

$$u\cos\alpha - u'\cos\alpha' = 0 \tag{2.33}$$

and since $\alpha = \alpha'$, necessarily $u = u'$. Hence the particle velocities of the two waves are equal and it follows that the two fronts are of equal strength.

The pressure at the boundary to the left of O is the sum of the two pressures and is hence $2P$, regardless of the direction of the incident shock. This independence of direction, however, leads to a familiar acoustic paradox. If taken literally, this result predicts a pressure $2P$ for $\alpha = 90°$. But $\alpha = 90°$ represents a wave parallel to the boundary for which there is no change in contact with the boundary as it advances and no way for the pressure to be doubled. The derivation, of course, breaks down for $\alpha = 90°$, as $\cos\alpha = \cos\alpha' = 0$ in Eq. (2.33), but we are nevertheless left with a discontinuity between the value $2P$ for α only slightly less than $90°$ and the value P behind a front travelling parallel to the boundary. This circumstance holds strictly only for waves of infinitesimal strength and the whole analysis and conclusions from it may be expected to break down for waves of finite amplitude. This question is examined in section 2.10, which describes the theory due to von Neumann for the more general case. These results are both curious and interesting, and have consequences of some importance in the understanding of underwater shock waves.

B. *Arbitrary boundary.* In the acoustic case of an arbitrary medium above the boundary we must add a transmitted wave in the medium, as shown in Fig. 2.8. It is again to be expected that the angles of inci-

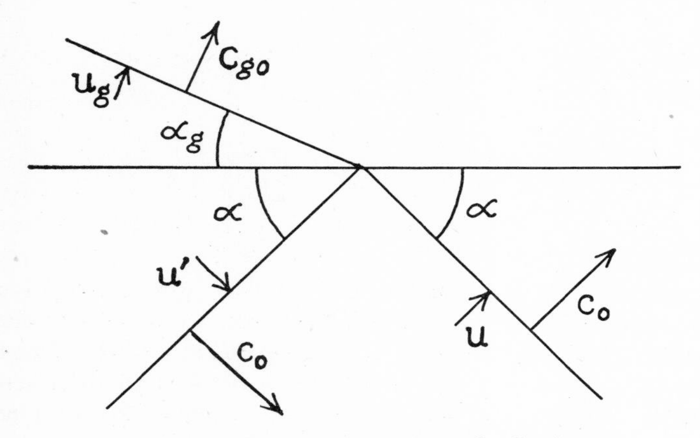

Fig. 2.8 Acoustic transmission and reflection.

dence and reflection are equal, and for a stationary condition the tangential velocity of the front of the transmitted wave must equal that for the incident wave, hence

$$c_o/\sin \alpha = c_{go}/\sin \alpha_g$$

which gives Snell's law for the direction of the transmitted wave

$$\frac{\sin \alpha}{\sin \alpha_g} = \frac{c_o}{c_{go}}$$

Continuity of pressure and particle velocity across the boundary requires that

$$u \cos \alpha - u' \cos \alpha' = u_g \cos \alpha_g$$

For sufficiently small disturbances the excess pressures and particle velocities in the waves are related by the equations (see section 2.2)

$$P = \rho_o c_o u, \quad P' = \rho_o c_o u', \quad P_g = \rho_{go} c_{go} u_g$$

in which it is assumed that the density and propagation velocity are negligibly different from the values ρ_o, c_o in the undisturbed fluid.

Solving the three sets of equations for the excess pressure and particle velocity ratios in the waves gives

$$(2.34) \qquad \frac{P'}{P} = \frac{\rho_{go}c_{go}\cos\alpha - \rho_o c_o \cos\alpha_g}{\rho_{go}c_{go}\cos\alpha + \rho_o c_o \cos\alpha_g} = \frac{u'}{u}$$

$$\frac{P_g}{P} = \frac{2\rho_{go}c_{go}\cos\alpha}{\rho_{go}c_{go}\cos\alpha + \rho_o c_o \cos\alpha_g} = \frac{\rho_{go}c_{go}}{\rho_o c_o}\cdot\frac{u_g}{u}$$

It is readily verified that for a rigid boundary ($\rho_{go}c_{go} \gg \rho_o c_o$) these relations reduce to

$$P' = P, \quad u' = u$$

$$P_g = 2P, \quad u_g = 0$$

as already found.

A question of obvious importance is what level of compression in water is compatible with the assumption that the waves are of infinitesimal amplitude so that Eqs. (2.34) can be applied, rather than more complex calculations taking into account variations of density and compressibility. The answer can only be demonstrated by such calculations but a safe rule based on them is that for pressures of less than 10,000 lb./in.^{2} and angles of incidence not greater than 30°, the departures from a regular scheme of reflection as depicted in Fig. 2.8 and calculations using Eqs. (2.34) are insignificant. In the majority of cases, reflections from the surface or bottom of the sea can be treated in the acoustic approximation. The reflection of a pressure wave from a free or nearly free surface presents, however, some problems of its own, which we consider briefly before returning to effects in waves of finite amplitude.

C. *Free surface.* In the discussion of boundary conditions it has been tacitly assumed that the pressure waves involved were of constant strength at the initial discontinuity, and that the pressure at all points behind the front had a constant value, i.e., the pressure was a step function. With these assumptions a wave of compression striking a free surface is reflected as a rarefaction of equal strength which leaves behind it fluid at the initial pressure. In actual cases, either or both of these assumptions is likely to be a poor approximation. Consider first the common case of a spherical shock wave of relatively small amplitude. In the acoustic approximation, the amplitude falls off inversely as the distance from the source and the profile behind the front is roughly an exponential decay. If this wave strikes a free surface obliquely, we have to expect from Eqs. (2.34) that at the boundary a reflected wave is set up of amplitude equal to the incident wave, but of opposite sign in order that the net pressure be zero at the boundary, and that the velocity of the surface is twice the normal component of velocity in the incident wave, which may be expressed as

$$P' = -P, \quad u_g = 2u \cos \alpha$$

where α is the angle of incidence. Initially, therefore, the surface of water above an explosion will be thrown upward and the pressure cancelled. As the wave progresses, however, the pressure in the incident wave at a point near the surface decreases and the head of the rarefaction wave moves down and away from the boundary and the source, as sketched in Fig. 2.9. It is evident that the negative change in pressure at this front will encounter regions in which there is a smaller value

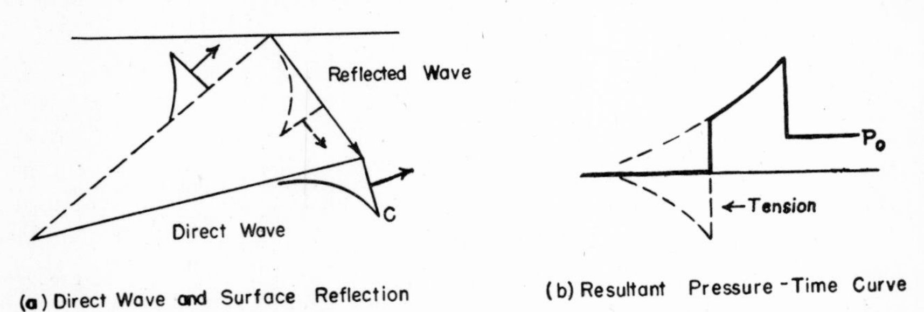

Fig. 2.9 Effect of surface reflection on observed pressure below a free surface.

of excess pressure. The net pressure behind the rarefaction will therefore become negative, and if this regime persists the water will be required to develop a state of negative absolute pressure. (It is to be remembered that the water at the surface is initially under a hydrostatic pressure of one atmosphere and this pressure increases at the rate of one atmosphere for each thirty-three feet of depth.) The pressure-time curve at a point C in Fig. 2.9(a) would, therefore, look somewhat as sketched in Fig. 2.9(b), the time of arrival of the reflected wave being readily calculable from the geometry. It is well known that water will in favorable circumstances withstand considerable tensions, values as high as 600 lb./in.^{2} having been reported, but under ordinary conditions much lower values are to be expected. Hence for sufficient negative absolute pressure we must expect that the water mass is pulled apart from itself and "holes," so to speak, are formed which prevent an increase in tension, a region of cavitation being formed instead. The pressure after this has occurred is indicated qualitatively by the solid line of Fig. 2.9(b), the dashed line being the unrealized state of tension. This "surface cutoff" effect plays a considerable role in the surface phenomena after an explosion, and must also be considered in analysis of the effects of explosions on yielding structures; these phenomena are described in more detail in Chapter 10.

2.10. Reflection of Finite Amplitude Waves

A. *Normal reflection from a rigid boundary.* In the case of acoustic pressure levels the reflection from an infinite rigid surface results in a pressure at the surface which is double the incident pressure. If the changes in density are not infinitesimal this conclusion will not hold, and an explicit calculation based on the conservation conditions developed in section 2.5 must be made. The simplest case is the one of normal incidence and the calculations for this case will be considered first. If the incident wave travels with velocity U into a fluid at rest with density ρ_o and pressure P_o and leaves behind it fluid with density ρ, pressure P and particle velocity u, we have

$$\rho(U - u) = \rho_o U, \quad P - P_o = \rho_o U u \tag{2.35}$$

When this wave strikes the boundary, the physical situation must develop in such a way that the fluid at the boundary is at rest. This boundary condition is satisfied if a reflected wave of compression travels back from the boundary with velocity U' and a strength just sufficient to bring the fluid entering the front with velocity $U' - u$ to rest. If the pressure behind this front is P' and the density ρ' we have for conservation of mass and energy

$$\rho(U' - u) = \rho' U', \quad P' - P = \rho' U' u \tag{2.36}$$

Solving Eqs. (2.35) and (2.36) for the ratio of excess pressures behind the incident and reflected waves in terms of the densities gives

$$\frac{P' - P_o}{P - P_o} = 1 + \frac{\rho'}{\rho_o} \frac{\dfrac{\rho}{\rho_o} - 1}{\dfrac{\rho'}{\rho_o} - \dfrac{\rho}{\rho_o}} \tag{2.37}$$

In order to determine the ratio a second pressure-density relation is required. Strictly, this should be computed using the Hugoniot energy condition and equation of state data, as the passage of each shock front involves a finite dissipation and leaves the fluid behind the front on a new adiabatic. For water, however, the change in entropy is small and an adequate approximation is obtained if the change is neglected. The same adiabatic equation of state can then be used for all three conditions of the fluid. From section 2.5, a suitable representation of the pressure-density relation for adiabatic compression is

$$P + B = k\rho^{\gamma}$$

where B is approximately 2.94 kilobars (= 42,670 lb./in.^{2}) and $\gamma = 7.25$ from Penney's calculations. The ratio of reflected to incident pressures is then

$$\frac{P' - P_o}{P - P_o} = \frac{\left(\frac{\rho'}{\rho_o}\right)^{\gamma} - 1}{\left(\frac{\rho}{\rho_o}\right)^{\gamma} - 1} \qquad (2.38)$$

The pair of equations, (2.37) and (2.38), permit simultaneous solution for the final pressure and density in terms of the values behind the incident wave. It can readily be seen that the ratio $(P' - P_o)/(P - P_o)$ is always greater than two, and approaches this value for weak shock waves as it should. Penney and Dasgupta (85) have computed the

Incident pressure (kilobars)	0.5	1.0	2.0	5.0	10	25	50
$\frac{\text{Final pressure}}{\text{Incident pressure}}$	2.088	2.170	2.30	2.60	2.92	3.44	3.93

Table 2.4. Pressure increase by normal reflection at a rigid boundary (from calculations of Penney and Dasgupta).

ratio P'/P_o for various values of P_o and some of their results are given in Table 2.4. It is evident that even for relatively enormous pressures the departures from acoustic "pressure doubling" are not very large. It should also be pointed out that the particle or flow velocities behind shock fronts in water are always subsonic, i.e., less than sound velocity, as the values given in section 2.5 show. These results are of course for water; in more compressible media, such as air, supersonic flow and non-acoustic differences occur for practical cases of importance.

The case considered here is clearly very special and, if the wave has a finite duration, applies only to the pressures at the boundary immediately after impact. The pressure at later times and the impulse, or time integral of pressure, for waves of finite duration have been computed by Finkelstein[10] for linear and exponential decay. He finds that the pressure at a rigid boundary decays more rapidly than in the incident wave and that the impulse is less than predicted by acoustic

[10] R. Finkelstein, U. S. Navy Bureau of Ordnance Explosives Research Report #6.

theory. The differences in durations and impulse are, however, less than 5 per cent for pressures up to 4,000 lb./in.^{2} and so can be disregarded in many cases. It is also to be remembered that a one-dimensional case has been considered and results of this kind hold only if the boundary is of infinite extent. For a surface of finite extent, the difference between incident pressure at the edge of the surface and doubled pressure at interior points must be relieved by a diffraction wave. In other words, a solution equivalent to geometrical optics does not satisfy the physical conditions, and the hydrodynamic equivalent of optical diffraction must be considered in cases where the linear dimensions of the surface are comparable with the length of the pressure wave. This situation, which must obviously be considered in analyzing the forces exerted by an explosion on a target, will be examined in Chapter 10.

B. *Oblique incidence at a rigid boundary.* The discussion in part (a) showed that no very considerable modifications of the simple acoustic law of reflections resulted for finite amplitude waves at normal incidence, but a further question, of what happens to the acoustic paradox of a discontinuity in pressure at grazing incidence, remains to be considered. One might offhand expect that for finite waves the pressure at the boundary would fall continuously to the incident pressure as the angle of incidence approached grazing values. This, however, is completely untrue and the apparently paradoxical phenomena actually observed have been the subject of considerable investigation, especially in air where the actual conditions are particularly striking. von Neumann (116) has considered the general problem for a rigid boundary in considerable detail, and his theoretical considerations have been extended to specific types of fluids by Polachek and Seeger (87). Any very complete summary of these investigations would be beyond the scope of this work, and all that is attempted here is an outline of the physical factors which lead to the complex phenomena involved and a brief discussion of the results.

Before considering in detail the incidence at a rigid boundary, it is pertinent to answer an argument which might plausibly lead one to conclude that the whole question is academic. This argument runs as follows: the non-acoustic behavior of water is significant only for very high pressures, at these pressures even materials such as steel will yield and hence a rigid boundary under these conditions is an unrealistic assumption. The argument is not without merit, and certainly the assumption may be an oversimplification of many actual circumstances. The premise that only extremely high pressures need be considered is, however, not demonstrated, and in fact we shall see that it is erroneous for nearly glancing incidence. A second point is the fact that reflection of a shock wave from a boundary is precisely the same problem as the intersection of two plane shock waves of equal intensity, the boundary condition for this problem being that the particle velocity at the plane of symmetry through the line of intersection has no component normal to the plane. The necessary result is therefore the solution for the rigid boundary plus the reflection of this solution in the boundary. The question of rigidity disappears, and the validity of the solution involves only the assumption that shearing effects can be neglected. Intersecting shock waves are at least an approximation to the circumstances involved in pressure waves from multiple charges or different parts of a single nonspherical charge, and the discussion following is thus significant in such situations.

We saw, at the beginning of this section, that an oblique shock wave striking a rigid surface leaves the fluid behind it with a component of

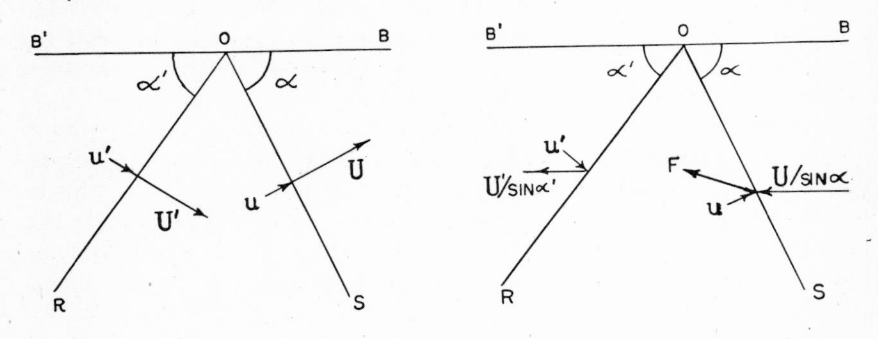

Fig. 2.10 Oblique reflection of a shock front at a rigid boundary.

flow toward the surface and that the necessity of reducing this component to zero requires that a reflected wave be developed. The simple scheme of mirror reflection was found to hold for a very weak disturbance, but must now be re-examined. In the situation of Fig. 2.10(a), the flow velocity u behind OS can no longer be considered infinitesimal. The physical situation actually realized must be one which reduces the component of u normal to the boundary to zero. Whether or not a single reflected shock OR can do this is the first consideration. If Fig. 2.10 is to apply at all times it must be true that the velocity of OR parallel to itself is just equal to that of the incident wave OS, otherwise OR will get ahead of itself, so to speak, and leave a situation at and near O incompatible with the boundary condition. Precisely this must happen, however, as the angle of incidence α approaches 90°. For in this case, the particle velocity u will be nearly normal to the front OR and the fluid ahead of OR has an increasingly large component of velocity normal to OR for any assumed angle α'. The pressure front OR can then be thought of as propagating into a fluid under compression which has velocity away from OR. Its velocity U' in fixed coordinates will therefore become increasingly supersonic for both reasons as α approaches 90°.

Regardless of the direction of OR, it encounters for sufficiently large values of α a fluid in which its displacement parallel to BB' will exceed that of OS. The validity of the conclusion is perhaps seen most easily for a frame of reference in which the incident wave OS is at rest, as shown in Fig. 2.10(b). The fluid entering OS has initially a flow velocity parallel to BB' modified by OS into a flow F which has a component normal to BB'. The component of F normal to OS is increasingly reduced as u, the flow left normal to OS in a fixed frame of reference, increases. As OS becomes increasingly oblique ($\alpha \to 90°$), this normal component represents more nearly the whole of F and the flow velocity F itself will, for sufficiently high compression, be less than sonic, i.e., less than the velocity with which even an acoustic wave is propagated. Any shock wave OR will therefore advance into this fluid, as its resultant velocity normal to its front in the moving frame, obtained by compounding the component of F normal to OR and the shock front velocity, is always directed away from BB'. The point of contact of OR must therefore move to the right in Fig. 2.10(b) if OS is fixed. In a fixed frame of reference OR advances on OS, and the point O must move downward from the wall.

The possibility of a breakdown of regular reflections having been shown, it remains to determine the critical condition for which it fails and to examine what happens on either side of this condition. If regular reflection is to hold it must be true that the resultant velocity left by the two shock waves has no normal component. It is convenient to choose a frame of reference fixed with respect to the two fronts. The condition for no normal component of velocity is then

$$u \cos \alpha + u' \cos \alpha' = 0 \tag{2.39}$$

where u' is the velocity acquired in passing through the reflected front. It is convenient to formulate the relation between the tangential components by the condition that the tangential velocity u_x, of a point in the fluid between the two shocks, must be the same whether calculated from the conditions ahead or behind the front. Any such point arrives at the initial shock with velocity $U/\sin \alpha$ if U is the velocity of this shock relative to the fluid ahead of it. It acquires a velocity u normal to the front and the tangential velocity u_x behind the front is therefore

$$u_x = \frac{U}{\sin \alpha} - u \sin \alpha = \frac{(U - u) + u \cos^2\alpha}{\sin \alpha}$$

If the reflected shock travels with a velocity U' relative to the fluid behind it, the tangential velocity ahead of this shock is $U'/\sin \alpha'$ less the tangential component of u', and

$$u_x = \frac{U'}{\sin \alpha'} - u' \sin \alpha' = \frac{(U' - u') + u' \cos^2\alpha'}{\sin \alpha'}$$

Equating these values gives

$$(2.40) \qquad (u_x) = \frac{(U - u) + u \cos^2\alpha}{\sin \alpha} = \frac{(U' - u') + u' \cos^2\alpha'}{\sin \alpha'}$$

The pair of equations (2.39) and (2.40) therefore determine the values of $(U' - u')$ and α' for given values of $(U - u)$ and α (since $(U' - u')$ and u' are not independent for any given fluid), and can be solved by numerical methods. There are, in general, two values of α' for a given intensity and direction of incidence, and corresponding values of U' and u', these values being identified with two values of pressure P'. As the value of α increases, however, these two solutions merge into one, and real solutions for α' do not exist for more oblique incidence and larger α. This failure indicates that the postulated reflection scheme cannot be realized physically and some other system of shock fronts must replace the simple one drawn in Fig. 2.10. Before considering what "irregular" reflections can satisfy the physical condition it is of interest to determine at what angle of incidence α_{extr} the regular reflection breaks down, and to consider the relation between incident and reflected shock fronts in the regular region as the extreme value is approached.

The first question involved in the regular scheme of reflection is that of which of the two values of α' for a given P and α is physically realized. General considerations as well as detailed calculations show that the larger value of α' always corresponds to a greater pressure discontinuity, and this solution thus represents a shock wave stronger than the one for smaller α', which moreover is ahead of the weaker shock. Intuitively it is to be expected that the wave realized will be the inferior one of smaller α' in which energy is degraded to a greater extent, and the existing experimental data all support (or at least do not conflict with) this conclusion. The evidence therefore excludes the larger of the two values of α'. Values of α' and P'/P_o as functions of α and P/P_o, where P_o is atmospheric pressure, have been computed for water by Polachek and Seeger (87) from the equivalent of Eqs. (2.39), (2.40). In these calculations an adiabatic law of the form of Eq. (2.30) was used with $\gamma = 7.15$. Some of their values for α' as a function of α are plotted in Fig. 2.11. The dashed parts of the curves correspond to the larger, unrealized values of α'. It will be seen that α' is always greater than α, the increase being greater for stronger shock fronts. The maximum value of α for regular reflection decreases with increasing P/P_o, the limiting value being indicated by the line marked extreme. A corresponding set of curves for the reflected pressures is plotted in Fig.

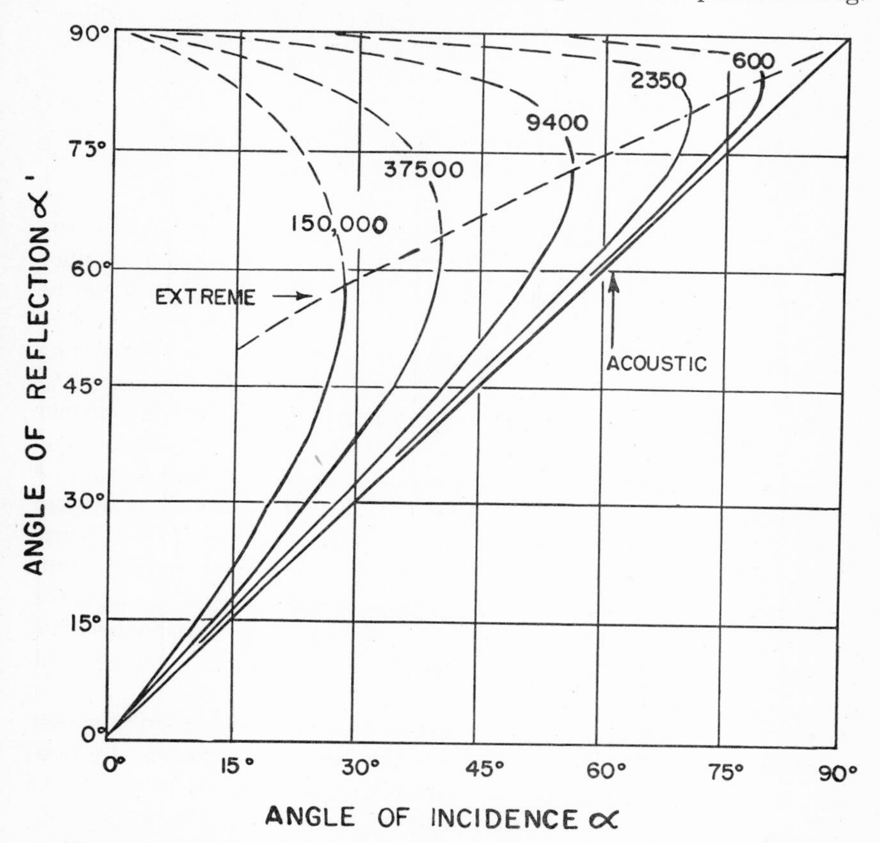

Fig. 2.11 Angle of reflection α' as a function of angle of incidence α for shock fronts of different strengths.

2.12. Rather than plotting P'/P_o it is found convenient to use the auxiliary variable ξ' defined by

$$\xi' = \frac{P' + B}{P_o + B}$$

It is seen from the plot that the strength of the reflected wave, as measured by ξ', increases significantly for sufficiently large values of α and P/P_o. For example, for $P = 37{,}500$ lb./in.^{2} ($P/P_o = 10^{3.4}$) and an angle of incidence of 40°, the pressure P' behind the reflected wave is 112,500 lb./in.[2] ($\xi' = 2.0$) or three times the incident pressure. The larger values of P' indicated by the dashed curves, which correspond to the larger value of α', are not realized; and the locus of pressures corresponding to the maximum value α_{extr} is represented by the dashed curve marked α_{extr}. It is evident that the maximum angle for regular reflection is determined by the condition $\partial\xi'/\partial\alpha = \infty$, as the curves of ξ'

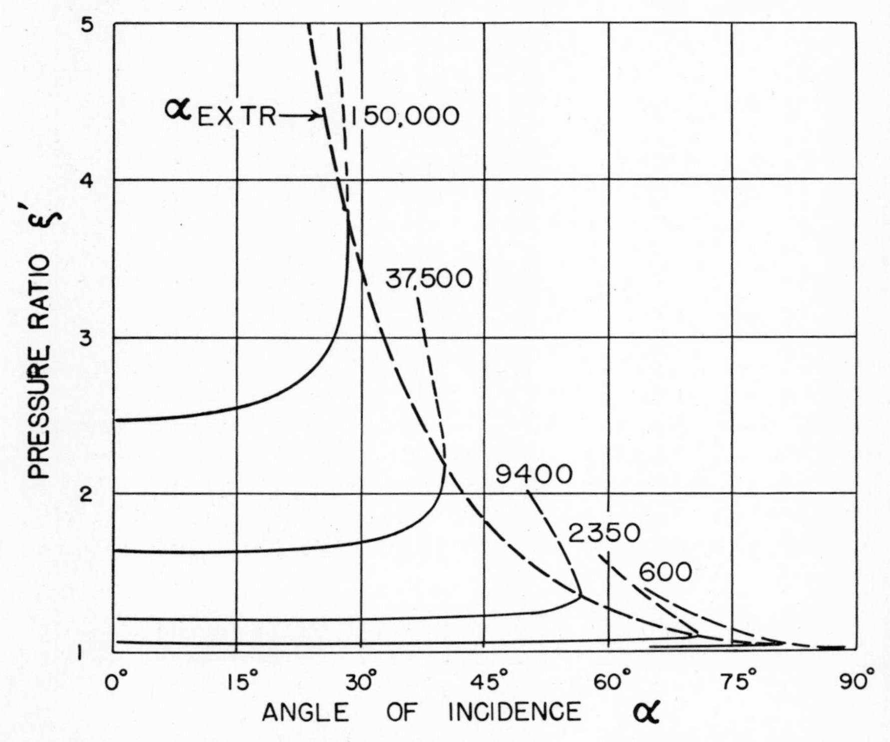

Fig. 2.12 Pressure behind a reflected shock front, expressed by the variable $\xi' = (P' + B)/(P_o + B)$, as a function of incident pressure.

vs. α are vertical for $\alpha = \alpha_{extr}$. This condition is more formally established from Eqs. (2.39), (2.40) as representing the boundary of solutions. With this condition it is readily shown from these equations that

$$\tan^2\alpha_{extr} = \frac{1}{\xi'^{1/\gamma}} \cdot \frac{\gamma\xi'(\xi'^{1/\gamma} - 1) + (\xi' - 1)}{\gamma\xi'(\xi'^{1/\gamma} - 1) - (\xi' - 1)} \tag{2.41}$$

It is of interest to compare the pressure behind the reflected wave for α_{extr} with the value for normal incidence. Several such values are given in Table 2.5, together with the values of α_{extr}, α'_{extr}. It is seen that the pressure is greater for oblique incidence in all cases, and for pressures of the order 10,000 lb./in.^{2} the increase is very appreciable. A comparison of these results on regular reflections with available experimental evidence on water is made in Chapter 7. The theory is in qualitative agreement with experiment. Other experiments under better defined conditions in air leave little doubt that the theory is essentially correct for regular reflection insofar as it is applicable. (The theory considers only the case of plane shock waves of infinite

Incident pressure (lb./in.^{2})	Reflected pressure		α_{extr}	α'_{extr}
	Normal	Extreme		
2,340	2.03	2.99	71°	80.5°
9,400	2.11	3.05	57°	73°
37,500	2.35	3.61	40.5°	64°
147,000	2.92	4.63	28.5°	57.5°

Table 2.5. Pressure increase by normal and oblique reflection at a rigid boundary (from calculations of Polachek and Seeger).

duration and so can only describe the initial pressure differences near the stationary point for waves of finite duration.)

C. *Mach reflection.* There remains to be considered the question of what happens for sufficiently oblique incidence that regular reflection cannot satisfy the physical relations. The theoretical difficulties of this problem are very great and the available experimental data, particularly in water, are not adequate as yet to clarify its various aspects.

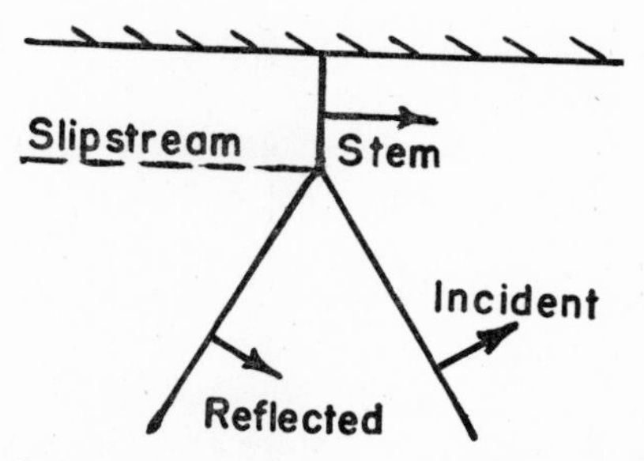

Fig. 2.13 Mach reflection at a rigid boundary.

The experimental evidence does, however, show that a reflected shock is developed, which meets the incident wave at a point in the fluid, and the two waves join to form a third shock wave extending to the wall. This form of reflection, sketched in Fig. 2.13, is known as irregular or Mach reflection and the third shock is frequently described as the Mach stem (after E. Mach who made early investigations on similar phenomena). There is one other feature of the irregular or Mach reflection which requires comment. The fluid behind the shock waves and sufficiently far from the wall (or plane of symmetry) arrives there by passing through the incident and reflected shock waves in succession while that near the wall passes through only the Mach stem. These two parts of the fluid must have velocity in the same direction and the same pressure but their other properties, in particular their density and magnitude of velocity, will not be the same, as they have different past histories. As a result a discontinuity of velocity, called a "slipstream," is formed, which is also a density discontinuity but not a shock wave.

Detailed solutions of this "three-shock" problem have been carried out by von Neumann (116) by simple assumptions as to the nature of the discontinuities at the points of intersection. The results are not, however, in agreement with experimental results for air and there is as yet no very clear understanding of what complications exist. von Neumann has suggested that the difficulties may be largely mathematical ones, and has expressed the opinion that the basic hydrodynamical equations are probably not at fault. The existing experimental data for water are much more meager and do not offer any appreciable help in the theoretical questions.

In closing this section, it may be well to point out that, although fluids in general can be expected to show finite amplitude effects of the type discussed, the details of both regular and Mach reflection depend very greatly on the equation of state for the fluid. In water and other liquids, the existence of an "internal pressure" and a large exponent γ lead to small deviations from simple acoustic behavior until pressures of the order of 40,000 lb./in.^{2} are attained, whereas in air marked deviations occur for excess pressures of 15 lb./in.^{2} or less.

D. *Reflection at a free surface.* One naturally inquires as to whether the complex results obtained for reflection from a rigid surface have their counterpart in reflection at a free surface. The necessary boundary condition is that reflected disturbances must leave the fluid at the free surface with its original pressure. This condition is appropriate for a fluid surface beneath the atmosphere, as the pressure of the air can change only insignificantly for the possible displacements of the surface. These displacements are sufficiently small in shock waves so that changes in gravitational potential are likewise small in comparison with the energy of compression. If it is also assumed that density changes by irreversible processes can be neglected, the density also remains unchanged at the boundary. With these assumptions, the function of reflected waves must be to restore the initial conditions except for velocity acquired by the fluid in the process.

In the limiting case of normal incidence, the boundary conditions can be satisfied by a reflected wave of rarefaction travelling back into the fluid which leaves the fluid behind at the equilibrium pressure and density. The rarefaction wave is thus the negative of the initial shock wave but is propagated into a fluid having a velocity u toward its front and the boundary. The rarefaction must change this velocity by an amount $-u$ relative to the direction of its advance away from the boundary, and hence leave the fluid with a velocity $2u$ toward the boundary. Relative to the undisturbed fluid, the front thus has a velocity $U - 2u$ away from the boundary, U being the velocity of the incident plane wave front, and the boundary is displaced away from the fluid with velocity $2u$.

The velocity doubling for acoustic waves is thus found for waves of finite amplitude at normal incidence because of the apparent reversibility of the Rankine-Hugoniot conditions which involve only relative velocities. The only differences are that the velocity u is related to the incident pressure by the Rankine-Hugoniot conditions, and the front of the rarefaction wave travels with a decreased velocity. The particle velocity in terms of pressure is given from Eq. (2.28) by

$$u = \frac{P - P_o}{\rho_o U} \tag{2.42}$$

and this differs only from the acoustic case by use of the shock front velocity U instead of the sound velocity c_o. The velocity $U - 2u$ of the rarefaction front is for water very nearly equal to the acoustic velocity c_o.

The derivation of Eq. (2.42) assumes the reversibility of the shock front process. This is of course not defensible, as energy is degraded into heat by the passage of a compression wave. The fluid behind the front is therefore hotter and cannot be returned to its initial pressure without a new value of density. A second associated difficulty is the fact that, even if a discontinuous rarefaction front is developed at the boundary, the advance of this front must involve a progressive decrease in the pressure gradient, as discussed in section 2.3. In view of these difficulties, it is perhaps surprising that steep rarefaction fronts are in fact observed, and that the velocity doubling for normal incidence has been confirmed by experiment at pressures up to 20,000 lb./in.^{2} (see section 10.2). It should equally be no surprise if the simple rarefaction scheme is found not to apply at higher pressures or oblique incidence.

The phenomena at oblique incidence have not been examined in any work reported at the present time. It is evident, however, that some readjustment must occur to take into account the acoustic paradox for grazing incidence on a boundary, at which the incident wave produces only tangential velocity changes and yet the fluid must be reduced to its initial pressure. This sort of complication must occur even in the idealization of a plane wave of infinite duration and negligible viscosity effects. In practical cases of a source very near a free surface the incident wave is spherical and of finite duration, and the nature of an explosive source may furthermore be significantly modified by venting of the explosion products. Very little experimental evidence is available on these questions and the whole problem of oblique incidence at a free surface is one on which much further work is needed.

3. The Detonation Process in Explosives

3.1 Explosive Materials

So far very little has been said about the explosive (the active cause of explosions), except for assuming that unstable compounds, chemical or otherwise, existed in which very rapid, violent reaction could be started and, once started, maintain itself. The chemistry, mechanical properties, testing, and production of such compounds is a very large subject which can hardly be adequately summarized in the space of this volume, let alone described in any detail.[1] The complete omission of such material would, however, be equally indefensible and the present section attempts to give a qualitative picture of the information on explosives which is pertinent to an understanding of their effects underwater.

High explosives as a class nearly always contain oxygen which is readily freed from its original molecule and made available for recombination with other atoms into more stable molecules. (That this is not always true is evidenced by lead azide, $Pb(N_3)_2$, which contains no oxygen at all but is a very sensitive and powerful material.) An explosive reaction may thus be thought of as a breakdown of the original molecules into product molecules (such as CO, CO_2, H_2O, NO, CH_4, H_2 as gases, and C, Pb, Al_2O_3 as solids) together with the evolution of large amounts of heat of the order of a kilocalorie per gram of explosive. Reactions of this kind can take place either as a process involving initially a single reaction molecule or between dissimilar molecules present in, say, a powdered mixture. The rate of reaction and violence of the process will evidently be favored by the intimate proximity in a single molecule of oxygen and other atoms with which it can combine. Thus for high explosives, organic nitrated compounds are used with a chemical formula of the form $(C_qH_rO_sN_t)$, these molecules being described as oxygen-rich or oxygen-deficient depending on whether or not enough oxygen atoms are present to react with all the carbon and hydrogen. Frequently metals such as Al, Pb in the form of powders are incorporated in the material to react with the oxygen and increase the heat involved, or conversely oxygen rich materials such as ammonium nitrate, perchlorates, may be added to provide more oxygen. These additions might therefore be expected to delay completion of the reac-

[1] An elementary account of the properties of explosives has been given by Kistiakowsky (63), which includes discussions of detonation and burning, shock waves, initiation, sensitivity, stability, and other properties. A number of books discuss the chemistry of explosives, of which may be mentioned the works by Marshall (71), Davis (25), and Meyer (73).

tion as a result of the much less intimate relation between the added reactants and the primary molecules.

If a high explosive material is to be practical it must obviously be stable under all ordinary conditions of handling and storage, and at the same time be capable of initiation by some external agency. Once a sufficient source of energy is applied, the explosive reaction can be started. The subsequent course of events depends on a large number of factors, both physical and chemical. If the reaction can and is permitted to build up, the rate and violence of the reaction will ultimately be limited by the laws of propagation of wave motion and the conditions of a steep fronted detonation wave may be realized. It is, however, possible to develop a lower and controllable rate by suitably decreasing the concentration of reacting molecules or by using mixtures such as gunpowder requiring reaction between particles. The practical application of such slow reactions is, of course, in propellants for shells, rockets, etc. where the shattering effect of a high explosive reaction (sometimes termed "brisance") would be more destructive than useful, but where large quantities of energy must be made available in a moderately short time (this property being described as the "power" of the explosive).

The slower processes in an explosion are usually described as burning, a term made natural by the importance of oxygen in the reaction. The maximum rate at which an explosion can react is the ideal condition of detonation, in which the velocity and final state are determined only by the explosive material and the initial density and temperature (the initial pressure ordinarily being negligible). This ideal condition may not be realized and low-order detonations are possible as a result of any of a number of factors: lack of confinement by the surroundings permitting expansion of the products before the reaction is complete and a falling off of pressure behind the detonation front; insufficient supply or duration of energy initiating the reaction; effects of the physical state of the explosive such as density and homogeneity of the charge.

From the point of view of underwater explosives, the most important factors to be considered are perhaps first of all that a reproducible, high order detonation is achieved, and second, that no effects occur peculiar to the particular physical size or condition of the explosive which are not characteristic of the explosive material under ideal conditions. As an example of the latter, the failure of some small charges to give results consistent with larger charges may be cited, effects of this kind being attributable to a disproportionate energy supplied for initiation, failure of the surface layer of the charge to react completely before appreciable relief of pressure by the surrounding medium, and other causes.

The process of initiation is also of importance. The more stable explosives from the point of view of shock or impact are more difficult to initiate and for satisfactory "triggering" of these relatively safest explosives it is necessary to use a small "booster" charge of more sensitive material. The booster is set off by a percussion or electrical detonator composition which can be initiated by a small quantity of heat from mechanical impact or by an electrical current.

Explosives materials, exclusive of propellants, have been roughly divided into two general classes: primary or initiating explosives for which detonation can be achieved in small quantities with a nominal supply of energy from, say, a detonator cap, and high explosives for which detonation is achieved only in larger quantities and is initiated by a booster charge. The distinction does not imply that a primary explosive is "lower" than a high explosive (the reverse may be true), but rather is made on the basis of difficulty in establishing detonation. The distinction is of course a somewhat arbitrary one, as taken literally it would make propellants, which do not detonate at all, the limiting cases of high explosives. It does, however, differentiate between the explosives used in quantity for destructive purposes and those used in small amounts to start the process.

The list of explosives which have been used or tested is a very long one and only a few will be considered here. The available data on these few are likewise numerous and all that will be attempted is to give the formulas and an indication of their primary interest for underwater explosion phenomena.

A. *TNT* ($C_7H_5O_6N_3$), *trinitrotoluene.* TNT has for many years been the standard high explosive. Its usefulness is the result of extreme lack of sensitivity (and consequent safety), considerable energy, and low melting point (80° C.) which makes it ideal as the liquid component in preparing mixtures. The density of crystalline TNT is about 1.65gm./cm.^{3}, but densities of 1.50–1.55 are ordinarily realized in cast charges prepared by melting granular TNT and pouring. Although cast TNT has been used as the standard filling for more high explosive munitions than anything else, a number of more effective mixed explosives have been developed, in which TNT is a primary component.

B. *Tetryl* ($C_7H_5O_8N_5$)2,4,6*trinitrophenylmethylnitramine.* A more sensitive explosive than TNT, tetryl is extensively used as a primary charge or booster for initiating TNT and other less sensitive explosives. For this purpose, the granular powder product is formed into a solid pellet by pressures of 2,000–4,000 kg./cm.^{2}

C. *RDX or cyclonite* ($C_3H_6N_6O_6$), *cyclomethylene trinitramine.* RDX is extensively used in mixtures with TNT and other components to produce more powerful high explosives. The mixture Composition B, containing sixty per cent RDX and forty per cent TNT, is a form in which RDX is often employed either as an explosive or in preparation of other mixtures.

D. *PETN* ($C_5H_8O_{12}N_4$), *pentaerythritol tetranitrate.* PETN is a more sensitive explosive than tetryl but is less convenient for general use as a booster charge. An important application of PETN is in primacord, a detonating high velocity fuse much used in demolition work for firing a number of charges nearly simultaneously; it consists of a waterproof, flexible cord with a small amount of PETN (approx. 3 g./ft.) as a core. The velocity of detonation in primacord is about 6,000 m./sec., the exact value depending on the production lot. This high and quite reproducible velocity has been extensively employed in experimental work to provide time intervals accurate to within a few microseconds. Mixtures of PETN and TNT are called pentolites, the most common being a 50/50 mixture. Cast pentolite charges can be initiated by a detonator cap plus a few grams at most of loose (granular) tetryl, and have been found highly reproducible and convenient charges for small scale explosion tests.

E. *Other explosives.* A number of military explosives have been developed as the result of efforts to improve upon the performance of TNT. The necessary properties of any military explosive, such as safety, stability, and ease of production, and the desirable feature of greater power, are more easily specified than achieved. It is thus understandable that data on the composition and performance of successful developments have not been made freely available. The addition to TNT of such components as aluminum, to increase the heat of reaction, and ammonium nitrate or other oxygen rich materials, to improve oxygen balance, are two possible modifications by which the performance of TNT may be improved. A further discussion of these and other explosives is ruled out either by security requirements or by the fact that little is known about their properties as underwater explosives.

F. *Initiating materials.* Many varieties and combinations of especially sensitive explosives are used as the primary material in setting off explosives. The exact composition and weight of such materials in detonation caps are trade secrets, but the various possible materials include lead azide, mercury fulminate, PETN, and tetryl. The performance of a detonation cap is rated in terms of exceeding various arbitrary tests and it is not to be expected that similarly rated detonators from different manufacturers will produce the same results in underwater measurements. Any classification of detonator caps as to function is beyond the scope of this discussion, but seismographic caps, which detonate in a much smaller and more reproducible time interval after applying a firing current, should be mentioned.

3.2. Detonation Waves

We have already mentioned that even a plane shock wave will ultimately die out as a result of dissipation processes at the shock front unless energy is continually supplied at some point behind the shock front. If, however, the passage of the wave involves a release of chemical energy in the medium it becomes possible to realize a self-sustaining wave, which after its initiation builds up to a stable limiting rate of propagation characteristic of the medium. Such a wave can develop in an explosive, which is thermodynamically an unstable substance capable of reacting to form a more stable product with release of energy, and is called a detonation wave. This self-sustaining wave differs in two important respects from a shock wave. In the first place, although the Rankine-Hugoniot conditions still relate conditions immediately behind the shock front, the chemical energy appropriate to the pressure and density of the transformed matter behind the front must be included in calculations of these conditions. The second difference is that the propagation of the wave is not controlled by the conditions behind the front, such as the motion of a boundary surface without which a shock wave could not exist, but is rather determined by the internal conditions in the material just behind the front. The three equations expressing conservation of mass, momentum, and energy at the shock front do not suffice to determine the four unknown quantities behind the detonation front: pressure P, density ρ, particle velocity u, and velocity of propagation D. A fourth condition on these variables is therefore necessary to define which of the otherwise possible waves will actually be established. Chapman (19) arbitrarily assumed that the detonation velocity in any particular case was the minimum velocity consistent with the Rankine-Hugoniot conditions for which a self-sustained wave could exist, and later Jouget (53) postulated the condition

$$D = c + u \tag{3.1}$$

where u is the particle velocity and c the velocity of small amplitude waves in the reaction products behind the shock front. These assumptions are, as we shall show, equivalent and Eq. (3.1) is the mathematical expression of what is known as the Chapman-Jouget condition.

There is little doubt that the Chapman-Jouget condition correctly predicts detonation velocities for the cases of interest to us and it is generally accepted, but as yet there is apparently no valid theoretical proof of its necessity in all practical circumstances. von Neumann (117) has made a detailed analysis of the conditions at a detonation front which throw considerable light on the matter. Before describing his results it is worth-while to consider simpler arguments which have been advanced to make the condition plausible and a natural assumption. A rarefaction wave behind the front of a detonation wave travels with the velocity $c + u$ and if the velocity D of the front were less than this any such wave could catch up with the front and weaken it. Such a rarefaction will result if, for example, the detonation is initiated by a piston which later comes to rest, and in general a rarefaction will result from any cause which abstracts energy from the reaction products.

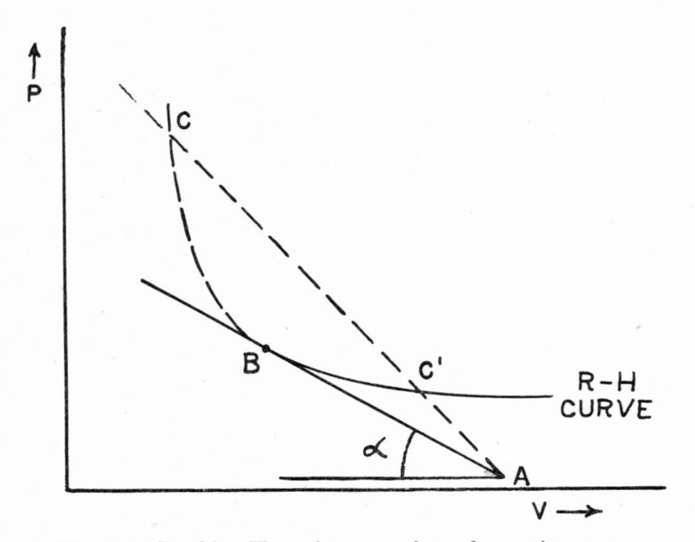

Fig. 3.1 Rankine-Hugoniot curve for a detonation wave.

Some such disturbance must therefore take place and unless the velocity of the detonation front is at least equal to $c + u$ it will be weakened by rarefaction waves. The existence and maintenance of a detonation wave therefore requires that D have a value equal to or greater than $c + u$.

The general picture is conveniently considered in terms of the pressure volume diagram required by the Rankine-Hugoniot equations. The actual computation of this P–V relation for practical explosives is a complicated chemical problem, discussed briefly later in this section and in section 3.3, which we assume here to give a result of the general form indicated in Fig. 3.1. The point A representing the initial condition of the explosive material must lie below the Rankine-Hugoniot curve for possible states of the products after detonation. From Eq. (2.29) the detonation velocity is related to the pressures and volumes P_o, V_o and P, V of the initial and final states by the equation

$$D = V_o \sqrt{(P - P_o)/(V_o - V)} \tag{3.2}$$

It is evident from the diagram that this relation may be written

$$D = V_o \sqrt{\tan \alpha}$$

where α is the angle a line such as AC, joining the initial point A and a point C on the Rankine-Hugoniot (R–H) curve, makes with the (negative) V axis.

The two possible states for a given value of D, corresponding to points C and C', will not have the same entropy, the value at C being higher. That this must be true is evident from the fact that, as Becker (7) points out, a transition from C' to C describes a shock wave of compression with initial state C' and final state C, and the velocity D. But, as discussed in section 2.3, the entropy is greater behind a shock front than ahead of it. The greater entropy at the upper point C of the two possible ones for a given D is therefore thermodynamically a more probable state, and we should expect the state of highest pressure and density compatible with other conditions to be realized. The other condition is of course that D must be equal to or greater than $c + u$.

It is readily shown, however, that D is equal to $c + u$ at the point B corresponding to the minimum possible value of D and is always less than $c + u$ for any point above B if, as is generally true, the adiabatic P–V relation for the products has positive curvature:

$$\left(\frac{\partial^2 P}{\partial V^2}\right)_s > 0$$

The more probable states above B are therefore destroyed by rarefaction waves running into the front, in spite of the higher detonation velocity. We therefore expect the state represented by B to be the one realized. The mathematical proof that D is equal to $c + u$ at B is a straightforward thermodynamical derivation.[2] From Eq. (2.29) the detonation velocity is related to the P–V conditions in two states (P_o, V_o) and (P, V) and the particle velocity u by

$$D = \frac{V_o}{V_o - V} u \tag{3.3}$$

and the velocity of sound behind the front is

$$c = \left(\frac{\partial P}{\partial \rho}\right)_s^{1/2} = V\left(-\frac{\partial P}{\partial V}\right)_s^{\frac{1}{2}} \tag{3.4}$$

[2] The proofs given here follow closely the derivations given by Kistiakowsky and Wilson (64). Thermodynamic considerations for detonation waves have also been discussed by Scorah (99).

In order to carry out the desired proof it is evidently necessary to consider the adiabatic P–V relation at points on the R–H curve. The relation between the adiabatic curve and the R–H curve at a given point will of course depend on the position of the point and it is useful to examine the possibility that for some point the two curves have the same slope. For an adiabatic change ($dS = 0$) we may write

$$dE = -PdV$$

and hence

$$\left(\frac{\partial E}{\partial V}\right)_s = -P \tag{3.5}$$

But if the slopes of the two curves are the same, the partial derivative at constant entropy must be the same as the corresponding derivative for the R–H curve. From Eq. (2.28), we have on the Rankine-Hugoniot curve

$$E - E_o = \tfrac{1}{2}(P + P_o)(V_o - V)$$

and hence

$$\left(\frac{\partial E}{\partial V}\right)_s = \tfrac{1}{2}(V_o - V)\left(\frac{\partial P}{\partial V}\right)_s - \tfrac{1}{2}(P + P_o) \tag{3.6}$$

Combining Eqs. (3.5) and (3.6) we obtain for the slope of the curves

$$\left(-\frac{\partial P}{\partial V}\right)_s = \frac{P - P_o}{V_o - V} \tag{3.7}$$

This is also the slope of the line AB in Fig. 3.1 passing through the initial point P_oV_o and this line is thus tangent to the R–H curve at B. The detonation velocity for this point is, from Eqs. (3.2) and (3.7), given by

$$D = V_o\left(-\frac{\partial P}{\partial V}\right)_s^{1/2} = \frac{V_o}{V}c$$

Combining this result with Eq. (3.3) we obtain

$$D = c + u$$

as was to be proved.

The proof that the detonation velocity is less than $c + u$ above the point B and greater than $c + u$ below B is somewhat more complicated. We note that the local sound velocity C is related to the slope of the adiabatic $P-V$ curves by

$$\left(-\frac{\partial P}{\partial V}\right)_s = V^2c^2$$

and from Eqs. (3.2) and (3.3)

$$D - u = V\sqrt{\frac{P - P_o}{V_o - V}}$$

P_o, V_o being evaluated at the initial point and P, V on the R–H curves. From these two equations, it is evident that investigation of the difference between D and $c + u$ is equivalent to determining the sign of the difference

$$V^2c^2 - V^2(D - u)^2 = \left(-\frac{\partial P}{\partial V}\right)_s - \frac{P - P_o}{V_o - V} \tag{3.8}$$

at various points on the R–H curve. The adiabatic slope and slope of the R–H curve at any point are related by the equation

$$\left(\frac{\partial P}{\partial V}\right)_R = \left(\frac{\partial P}{\partial V}\right)_s + \left(\frac{\partial P}{\partial S}\right)_v\left(\frac{\partial S}{\partial V}\right)_R$$

where the subscript R indicates differentiation along the R–H curve. Rearranging, we have

$$\left(-\frac{\partial P}{\partial V}\right)_s - \left(-\frac{\partial P}{\partial V}\right)_R = \left(\frac{\partial P}{\partial S}\right)_v\left(\frac{\partial S}{\partial V}\right)_R \tag{3.9}$$

In order to analyze the two factors on the right we note that from thermodynamics

$$\left(\frac{\partial P}{\partial S}\right)_v = -\left(\frac{\partial T}{\partial V}\right)_s$$

which is a positive quantity, i.e., an adiabatic decrease in volume is accompanied by a rise in temperature. (This fact is readily demonstrated from more familiar thermodynamic quantities: specific heat, thermal expansion coefficient, adiabatic compressibility.) From the first and second laws

$$TdS = dE + PdV$$

and hence

$$T\left(\frac{\partial S}{\partial V}\right)_R = \left(\frac{\partial E}{\partial V}\right)_R + P$$

From the energy equation (3.6) for points on the R–H curve we obtain

$$\left(\frac{\partial E}{\partial V}\right)_R = \tfrac{1}{2}(V_o - V)\left(\frac{\partial P}{\partial V}\right)_R - \tfrac{1}{2}(P + P_o)$$

Substituting for $(\partial E/\partial V)_R$ yields

$$\left(\frac{\partial S}{\partial V}\right)_R = \frac{V_o - V}{2T}\left[\frac{P - P_o}{V_o - V} - \left(-\frac{\partial P}{\partial V}\right)_R\right] \tag{3.10}$$

It is evident that the sign of $(\partial S/\partial V)_R$ depends on the relative slopes of the line joining the initial and final states on the P–V diagram and the tangent to the R–H curve at the point representing the final state. If the R–H curve is concave upward, as shown in Fig. 3.1 this difference in slopes is negative above the point B and positive below it. We must expect that this curve will have this form in general, as the pressure must become infinite as the volume approaches zero and must, for the same volume as in the initial state, lie above the point A. It is thus plausible that the R–H curve for any normal fluids has positive curvature much like an adiabatic curve of compression, and the fact that the sign of $(\partial S/\partial V)_R$ has the behavior stated can be demonstrated in terms of the curvature of the adiabatics.

Eqs. (3.9) and (3.10) permit us to write the velocity difference as given by Eq. (3.8) in terms of variables of known sign, with the result

$$V^2[c^2 - (D - u)^2] = \left(-\frac{\partial P}{\partial V}\right)_s - \left(\frac{P - P_o}{V_o - V}\right)$$
$$= -\left(\frac{\partial S}{\partial V}\right)_R \left[\frac{2T}{V_o - V} + \left(\frac{\partial T}{\partial V}\right)_s\right]$$

From what has been said, the right hand side of this equation is positive for points above B and negative for points below B. Hence, as was to be shown, D is less or greater than $c + u$ according as the final state lies above or below the point B on the R–H curve to which a line from the initial point A is tangent.

von Neumann (117) has made a more fundamental investigation of the Chapman-Jouget condition in which he considers that the conservation equations must hold not only for the complete reaction of the detonation front but also for intermediate stages of reaction in the detonation front. The transition from undisturbed material to the final product is therefore described by a series of states, each with its own R–H curve. The detonation velocity for each state is described by a line from the initial point A of Fig. 3.1 before reaction to the appropriate point on each curve. In the cases we have to consider the reaction zone is sufficiently narrow that all stages of reaction proceed with the same velocity and a single line intersecting all these curves must represent the detonation velocity. If the line representing the Chapman-Jouget condition does not intersect all the intermediate curves it cannot from this argument describe such states and is therefore incorrect. In such a case, the correct detonation velocity is, according to von Neumann, determined by the line which is tangent to the envelope of all possible R–H curves rather than to the final curve.

Investigations have been made by Lippmann, Brinkley, and Wilson (11) to determine the conditions under which the Chapman-Jouget condition is valid for various special cases of solid explosives, which indicate that the relation may break down for high densities of the initial material. The significance of the result is, however, impaired by the approximations necessary to permit the calculations. Perhaps the best justifications of the condition in practical cases are the facts that increased refinements of calculation lead to wider predicted ranges of validity for it, and the detonation velocities determined with its aid agree well with experimental values for conditions where calculations suggest that it may not be valid. As we shall see later, the exact form of the detonation wave is not, in most cases, of decisive importance for computation of shock waves and gas sphere motion. Therefore, a more detailed consideration of the possible failure of the Chapman-Jouget condition is not necessary.

With the assumption of the Chapman-Jouget condition and the Rankine-Hugoniot relations, the pressure, density, and velocity at the head of a detonation wave are determined in terms of the equation of state and thermochemical data for the medium. An explicit calculation can therefore be made and, once this has been done, the conditions behind the front can be computed by using these boundary conditions for integration of the equations of motion for the "fluid" behind the front. The pressure and velocity distribution behind the front will evidently depend on the shape of the front in space: plane, spherical, or other. It is important to realize, however, that the properties at the head of the wave do not depend on the shape as long as the thickness of the front is negligible, as this condition, which results in the Rankine-Hugoniot relations, makes any divergence of the wave immaterial.

If we consider a detonation wave established by the initial motion of a piston or some other cause which subsequently ceases, it is evident that gas at the point of initiation must come to rest, the agency by which this takes place being a wave of rarefaction moving outward. Immediately behind the detonation front the wave has a velocity $c + u$ equal to the detonation velocity D by the Chapman-Jouget condition, and at points behind the front its velocity falls off to a final value c_2 consistent with the adiabatic pressure density relation and the condition that the particle velocity u is zero. For a plane wave, the discussion in section 2.3 of the Riemann equations shows that the pressure and density are propagated forward with a velocity $c + u$ behind the front. The particle velocity u is given by

$$u - u_1 = \int_{\rho}^{\rho_1} c \frac{d\rho}{\rho}$$

where u_1 and ρ_1 are the particle velocity and density for some known state, in this case the state at the detonation front. The final state may thus be obtained by setting $u = 0$ and determining the density $\rho = \rho_2$ for which the integral equation is satisfied.

It will be observed that if the detonation wave has a constant velocity D, the front will travel a distance Dt in time t, the final state will have travelled a distance c_2t, and we expect that the whole scale of the wave profile is proportional to the time t. This is mathematically equivalent to the statement that the pressure, density and velocities behind the front are functions only of the ratio r/R, where R is the distance Dt travelled by the wave following its initiation at $r = 0$, $t = 0$. That this is true is readily seen from the fundamental equations of motion and of continuity, which are unaltered if t and r (or x) are replaced by βt and βr and the pressure, density, etc. are evaluated at βt, βr. This result is an example of the principle of similarity which will be discussed more completely for shock waves in Chapter 4. For the short periods of time before the steady state of the wave has been reached and the exciting cause has ceased, the condition will not be fulfilled, nor is it true in the front where dissipative processes cannot be neglected. In cases of interest, however, it is to be expected that the scaling law will be realized to a very good approximation before the disturbance has travelled any appreciable distance.

The detailed calculation of the properties of detonation waves is a relatively straightforward numerical integration once the conditions at the front and the adiabatic pressure-density relation for the products behind the front are known. These calculations have been made by G. I. Taylor for TNT, using as a starting point the conditions computed by H. Jones, and will be described in section 3.6.

3.3. The Equation of State for Explosives

A necessary preliminary to computation of conditions at and behind the detonation front in an explosive is, of course, knowledge of the equation of state and heat capacity of the products which, together with the heat of reaction and the Rankine-Hugoniot or adiabatic conditions, permit determination of the pressure-density relations. It is evident that the products of detonation, although to a large extent gaseous, are initially confined to the volume of the original solid or liquid. In addition, the products are at temperatures of the order 3,000–5,000 A. and pressures of 50,000 atm. or higher. These circumstances thus present two rather formidable problems: first, the choice of a suitable equation of state for products at much higher temperatures and densities than are ordinarily investigated, and second, determination of the appropriate distribution of the original atoms among the many possible molecular species in the product (i.e., the final composition).

A. *The ideal gas law.* Although we shall be interested in the reactions of solid explosives, in which the final state even of gaseous products is very poorly described by assuming ideal gas laws, it is of interest to consider the equations for ideal gases and some experimental results on explosions of gaseous mixtures, as these results give some indication of what composition products determine the detonation process. For an ideal gas the equation of state is

$$PV = \frac{nRT}{M}$$

where n is the number of moles of gas in M grams of the final gases (which we assume to be a mixture of products). The adiabatic expansion law is then

$$PV^\gamma = \text{constant}, \quad dS = 0$$

where γ is the ratio of specific heats and the velocity of sound c is given by

$$c^2 = V^2\left(-\frac{\partial P}{\partial V}\right)_s = \gamma PV$$

The increase in internal energy $E - E_o$ may be written

$$M(E - E_o) = Q + \bar{C}_v\,(T - T_o)$$

where Q is the heat absorbed in the reaction of M grams at the initial temperature T and final composition and $\bar{C}_v$ is the mean specific heat at constant volume of the final composition for the temperature range.

The Rankine-Hugoniot requirement for energy change at the shock front may be combined with these thermal data to give

$$(3.11) \quad E - E_o = \tfrac{1}{2}(P + P_o)\,(V_o - V) = \frac{Q}{M} + \frac{\bar{C}_v}{M}\,(T - T_o)$$

In order to solve for the final temperature it is necessary to eliminate P, V from this equation. The conditions of mass and momentum conservation at the shock front and Chapman-Jouget condition may be written

$$\frac{D}{V_o} = \frac{D - u}{V}$$

$$\frac{Du}{V_o} = P - P_o$$

$$D = c + u$$

Eliminating D, u by the first and third relations we obtain for the second

$$\frac{1}{V}\left(\frac{V_o}{V} - 1\right)c^2 = P - P_o, \quad \text{or} \quad V_o - V = (P - P_o)\,\frac{V^2}{c^2}$$

Substituting in the energy equation (3.11) and using $c^2 = \gamma PV$ we have

$$\tfrac{1}{2}M \frac{P^2 - P_o^2}{P\gamma} V = Q + \bar{C}_v(T - T_o)$$

Neglecting P_o^2 in comparison to P^2 and using the ideal gas law, this becomes

$$\tfrac{1}{2} \frac{nRT}{\gamma} = Q + \bar{C}_v(T - T_o) \tag{3.12}$$

If, in the explosion of a mixture of gases, the distribution of atoms in the molecular products were independent of temperature and pressure, both Q and n could be evaluated on the basis of quantitative reaction, and Eq. (3.12) could be solved for the final temperature T if the variation of γ and $\bar{C}_v$ with T is known. Actually, of course, the equilibrium constant of gaseous reactions depends on P and T. Furthermore, we have no a priori assurance that either quantitative reaction or the equilibrium condition is the one determining the state of the products at the head of a detonation wave. It is therefore of interest to compare results of both possible calculations with observed detonation velocities. Changes in equilibria affecting Q and n make the determination of T from Eq. (3.12) much more tedious than if the equilibrium were frozen, but Lewis and Friauf (67) have made such calculations for mixtures of oxygen and hydrogen with various excesses of both components and for the stoichiometric mixture ($2H_2 + O_2$) with added inert gases. The equilibria allowed for were $2H_2 + O_2 = H_2O$, $H_2 = 2H$, $2H_2O + O_2 = 4OH$, but not $O_2 = 2O$. Although the data used and assumptions made could be improved there is apparently no reason to believe that the results are seriously in error. Reference should be made to the original paper for further details. Once the final temperature T is determined calculation of the detonation velocity D is made from the equation

$$D = \frac{V_o}{V} c = \frac{V_o}{V}\sqrt{\frac{\gamma nRT}{M}} = (\gamma + 1)\sqrt{\frac{nRT}{\gamma M}} \tag{3.13}$$

obtained from the approximate relation that

$$V_o - V = (P - P_o)\frac{V^2}{c^2} \cong \frac{V}{\gamma}$$

The calculations of Lewis and Friauf compared with experimental data of Dixon (29) are copied in Table 3.1. It is seen that the equilibrium calculations in all cases agree better with experiment than the quantitative reaction. The agreement is, on the whole, extremely good and gives considerable support to a conclusion that detonation conditions are best calculated on the basis of equilibrium of the products. This result seems somewhat paradoxical when it is considered that the thickness of a shock or detonation front is estimated as less than 10^{-5} cm., which for a velocity of the order 3×10^5 cm./sec. allows a time of 10^{-9} sec. or less for reaction in the front. It is of course unreasonable to expect completion of the reaction in such a small time. The success of the equilibrium calculation indicates rather that the reaction is essentially complete before the conditions behind the shock front have

Composition	Calculated velocity (m./sec.)		Measured velocity (m./sec.)
	Quantitative reaction	Equilibrium	
$2H_2 + O_2$	3278	2806	2819
$+ N_2$	2712	2378	2407
$+ 3N_2$	2194	2033	2055
$+ 5N_2$	1927	1850	1822
$+ O_2$	2630	2302	2319
$+ 3O_2$	2092	1925	1922
$+ 5O_2$	1825	1735	1700
$+ 2H_2$	3650	3354	3273
$+ 4H_2$	3769	3627	3527
$+ 6H_2$	3802	3749	3532

Table 3.1. Calculated and measured detonation velocities in gaseous mixtures.

changed enough to invalidate the procedure. It should be mentioned that the detonation velocity is probably less sensitive to such errors than are the pressure and density immediately behind the front, and such values may be less accurate.

B. *The calculations of Jones.* The reactions of solid explosives are, of course, practically much more important than those in gaseous mixtures and present greater difficulties because of the higher temperatures and densities. A first approach to the question of a suitable equation of state for such conditions is to assume a modified Van der Waals equation of the form

$$P(V - b) = \frac{RT}{M}$$

in which b, the covolume constant, represents in elementary kinetic theory four times the total molecular volume from which molecular motion is excluded. This type of equation has been used by various workers in the past to represent explosion products of closed bomb experiments. As H. Jones points out, the values of b required to fit this type of data are not in very good agreement with other evidence and are furthermore totally unreasonable for the front of detonation waves as they lead to excluded molecular volumes greater than the total available volume. Jones (50) has therefore developed equations of state along somewhat different lines in order to calculate detonation conditions in solid explosives. At high pressure he fits an equation based on theoretical results for the solid state to data of Bridgman and at lower pressures employs a virial equation. The former equation Jones takes to be of the form

$$P = RTf\left(\frac{V}{N}, T\right) - \frac{dE_o\left(\frac{V}{N}\right)}{d\left(\frac{V}{N}\right)}$$

where $E_o(V/N)$ is assumed to be the potential energy of interaction of molecules in the gaseous product and N is the number of moles. The internal energy E_o is expressed in the form

$$E_o(V/N) = Ae^{-\alpha V/N} + B(V/N) + C$$

the first term representing repulsive forces, the second attractive forces.

In order to obtain suitable values of these constants, Jones considers first the energy of normal modes of vibration of molecules in a solid, assumed fully excited. With this interpretation, the function f can be evaluated from compressibility data for gases under high compression, and a constant value is found to be a good representation of Bridgman's results for nitrogen at 68° C., $6 < P < 15$ kilobars. The assumed equation of state is then of the form

$$P = \alpha Ae^{-\alpha V/N} - B + RTf$$

the constants found to fit Bridgman's data being

$$f = 0.313 \text{ cm.}^{-3} \qquad A = 855 \text{ kcal./mole}$$

$$\alpha = 0.263 \text{ cm.}^{-3} \qquad B = 0.139 \text{ kcal./mole}$$

The constant C in the energy equation is taken to be -5.80 kcal./mole to make the minimum energy equal to the heat of vaporization. At lower temperatures a virial equation is used:

$$\frac{PV}{N} = RT + bP + cP^2 + dP^3$$

the constants b, c, d being adjusted to give the same pressure and curvature as the equation for a higher pressure at an intermediate value of V.

These results are of course based on data for nitrogen at a low temperature. Jones makes the assumptions that the effect of interaction at high temperatures and comparable densities will not be appreciably different and that the same interaction can be applied to other gaseous products. With these assumptions the total energy of the products of an explosion becomes

$$E = \sum \left[N_a E_a + (N - N_s) \left(E_o + \frac{3}{2} RT \right) \right]$$

In this equation N_a represents the number of moles of each molecular species, whether solid or gas, and E_a is the energy per mole exclusive of interaction of molecules in the gas phase. $N = \sum N_a$ is the total number of moles; N_s is the number of moles of solid products. The last term thus represents the energy of interaction for the gas molecules plus vibrational energy $3/2\ RT$ per mole.

In addition to the interaction term it is therefore necessary to evaluate the energies E_a and the composition variables N_a. The energies E_a are evaluated from specific heat data at ordinary pressures, as interaction effects at high pressures are represented by E_o. The final problem is therefore determination of the equilibrium composition consistent with the Rankine-Hugoniot and Chapman-Jouget conditions. In order to determine the equilibrium of possible reactions as a function of temperature, the activities of the various products are introduced, which can be evaluated as a function of pressure from the equation of state. This procedure is made practical by the simplifying assumption that all molecules have the same field of force. The activity can however be expressed in terms of the composition variable N_a and the equilibrium constants for the assumed reactions as a function of temperature at normal pressures to which ideal gas conditions are applicable. Elimination between these equations and the decomposition equations for the particular explosive then permits determination of the N_a for various temperatures.

The increase in internal energy of the products must be the sum of the chemical energy Q released by the reaction and the work done by the pressure at the detonation front. A knowledge of E, the energy release Q, and the equation of state therefore provides sufficient information to solve the detonation front equations.

Although it is evident from this sketch that Jones' procedure involves several fairly drastic simplifying assumptions, the calculation is of importance because the results give an idea of the success obtainable in calculating detonation front conditions with an equation of state fitted to available physical and chemical data other than values from explosion measurements. The adiabatic relation for the burnt gases is also obtainable from these calculations. The two sets of results will be considered in more detail in sections 3.4, 3.5.

C. *The Wilson-Kistiakowsky equation of state.* A somewhat different approach to the problem of the equation of state was developed by Wilson and Kistiakowsky (64), who assumed an equation of the form

$$PV = \frac{NRT}{M}\left[1 + xe^{\beta x}\right] \tag{3.14}$$

$$\text{with } x = \sum(N_i k_i / T^{\alpha} V M)$$

where N_i is the number of moles of product i, k_i is an empirical parameter with dimensions of volume which may be called a covolume, and α and β are constants. In the original development, Wilson and Kistiakowsky took $\alpha = 0.33$, $\beta = 1.0$, $\sum N_i k_i = K$, but better results were later obtained using $\alpha = 0.25$, $\beta = 0.30$. This form of equation does not apply very well at low densities and it is therefore necessary to determine the k_i in some other manner than from the data at low temperatures. Instead of fitting a form suggested by theory of solids to compressibility data, as Jones does, a less fundamental but widely applicable procedure was adopted in which the k_i were determined to give agreement with detonation velocities at a number of initial (loading) densities for several explosives, a sufficient number being taken to include significant effects of the various possible product species. (It will be noted that the variation of D with the loading density should be a good measure of the k_i, as for perfect gas behavior with the k_i all zero the velocity is, from Eq. (3.13), independent of initial density.)

Actual calculations of the detonation velocity, and values P and V at the front, with this equation of state are carried out by numerical methods which are too involved to be more than very generally described here. The first step of the calculation is computation of a hypothetical detonation velocity D^* for products obeying the ideal gas law, in the manner already outlined for gaseous explosions. In the next step, the ratio D/D^* is computed with the ideal gas results as a starting point, using the equation of state (3.14) and corrected thermochemical data. The equilibrium composition to be used is computed from equilibrium constants versus temperature and stoichiometric conditions by successive approximations or by approximate rules for the reaction which give surprisingly good results. The pressure, density, and particle velocity of the products also result from these calculations and the properties of the front are thus all determined.

This method has been extensively applied to a number of organic explosives at various densities with results that agree with experimental detonation velocities to within experimental errors in nearly all cases. The other calculated quantities are not readily determined with any accuracy experimentally, and it is therefore difficult to judge how good the computed pressure, etc., are. The assumed reactions for specific explosives, comparison of experimental and calculated values for detonation, and the consequences of the results in determining the form of the detonation wave and the initial conditions of an underwater explosion will all be dealt with in more detail in section 3.4.

3.4. Calculated Conditions at the Shock Front

In section 3.2 it was shown that from the point of view of hydrodynamics the progress of a detonation wave in an explosive is described by the conditions of conservation of mass, momentum, and energy of the explosive material as it changes from its initial condition to the reacted products on passage of the detonation wave. The further Chapman-Jouget condition relating the velocity of the front to properties behind the front then fixes the detonation conditions at the shock front in terms of the pressure-density relation for the products and the chemical energies (heats of formation, specific heats) involved in the reaction. The detailed calculation of the velocity and initial state (pressure, volume, temperature) immediately behind the front is then possible for specific explosives if sufficient chemical data are available. An exact development from these principles is not possible because of two difficulties: inadequate knowledge of the properties of gases at the high pressures and temperatures existing behind the detonation front, and difficulty of accounting exactly for the variety of molecular combinations possible. The first of these difficulties is perhaps the more serious fundamentally, although the second presents the problem of determining with a tolerable amount of effort a reasonable final composition. It has already been suggested that the composition in chemical equilibrium at the final pressure and temperature is a plausible approximation on the basis of detonation velocity in gaseous mixtures, and for want of reaction rate data to suggest a better criterion, this condition has been assumed in the calculations to be described.

Two approaches to the determination of a suitable equation of state were outlined in section 3.3. The first is the approximation of Jones for TNT in which interaction effects of all molecules in dense phases were taken to be the same and represented by an equation suggested by the theory of solids fitted to experimental data for nitrogen, the pressure-density relation at lower pressures being represented by a virial equation. The second procedure, developed by Brinkley and Wilson (11), uses an empirical equation of state in which "covolume" constants for each type of molecule are adjusted to give best agreement for selected data on heats of formation and specific heats, and the detonation problem can then be solved for given equilibrium conditions. The second problem, of determining the equilibrium composition, is attacked in either approach by standard thermodynamic methods. The equation of state and energy conditions are of course not independent of the equilibrium and the final solution must thus be obtained by simultaneous solution of the two problems. The formal solution of the equilibrium problem is essentially the same in the two approaches but differs in detail because of different reaction constants assumed and the different equations of state employed. In order to indicate the nature of the equilibrium calculation a short summary of the steps in the procedure of Brinkley and Wilson is given here.

The work content A of a mixture of M grams of gases of sufficiently large volume V_o and temperature T is given by standard thermodynamics as

$$A(V_o, T) = \Sigma N_i \mu_{io} + RT \left[N \log \frac{P}{N} + \Sigma N_i \log N_i - N \right]$$

where N_i is the number of moles of the ith product gas and N the total number of moles. The chemical potential μ_i for a gas is defined in general as

$$\mu_i = \left(\frac{\partial A}{\partial N_i} \right)_{T,V,Nj \neq Ni}$$

and μ_{io} is its value for the pure gas at standard pressure and temperature T. At any other volume and the same temperature, the work content is given by

$$A(V, T) = A(V_o, T) - M \int_{Vo}^{V} P dV$$

Introduction of the modified Wilson-Kistiakowsky equation of state to eliminate V gives on integration

$$\mu_i = \left(\frac{\partial A}{\partial N_i} \right)_{T,V} = \mu_{io} + RT \left[\log_e N_i + \log_e \frac{P}{N} + \frac{e^{\beta x} - 1}{\beta} \right] + RT \left[\frac{N k_i}{\Sigma N_i k_i} x e^{\beta x} - \log_e F(x) \right]$$

In this equation, $F(x)$ is the function in the equation of state

$$PV = \frac{NRT}{M} F(x) = \frac{NRT}{M} [1 + x e^{\beta x}]$$

$$x = \frac{\Sigma N_i k_i}{T^{\alpha} M V}$$

and the k_i are the empirical covolume constants for the various species determined from detonation data.

If at equilibrium for the various reactions the number of moles of each species X_i is ν_i, counted as negative for reactants and positive for products, the reaction may be represented as $\Sigma \nu_i X_i = 0$, and for equilibrium we have the condition

$$\Sigma \nu_i \mu_i = 0$$

The thermodynamic constant K_p for reactions at various temperatures is defined as

$$RT \log K_p = - \Sigma \nu_i \mu_{oi}$$

Summing over the $\nu_i \mu_i$ and introducing the equilibrium constant gives

$$\log K_p = \Sigma \nu_i \log N_i + \Sigma \nu_i \left[\log_e \frac{P}{N} + \frac{e^{\beta x} - 1}{\beta} - \log_e F(x) \right] \frac{+ N x e^{\beta x} \Sigma \nu_i k_i}{\Sigma N_i K_i}$$

$$= \log K - \log G$$

where log K represents the first term on the right, log G the other terms. This equation represents the condition imposed on the reactions by their equilibrium constants. This expression may be written $K = K_p G$, which in turn may be expressed as

$$\Pi K_j = \Pi K_{pj} G_j, \quad \text{where} \quad K_j = \Pi_j (N_i)^{\nu_i}$$

the product being extended over the equilibrium considered.

The detailed calculation determining both the equilibrium constants and the final state is carried out in two steps: calculation of a hypothetical final state, using ideal gas laws, the heats of formation and specific heats of the products, and an estimated final composition and temperature in the Rankine-Hugoniot equations. This first estimate of the final state is improved using the actual equation of state. This revised estimate is employed to calculate approximate equilibrium constants K using the thermodynamic equilibrium conditions, this calculation involving solution of a set of simultaneous equations equal in number to the number of products. The whole process is then repeated with successive approximations giving the same equilibrium.

The complete process is evidently a long and somewhat complicated numerical procedure which will not be described in detail. It is, however, of interest to mention approximate composition rules developed by Brinkley and Wilson, which in most cases agree surprisingly well with the final result. For explosives described by the formula $C_q H_r O_s N_t$ the assumed decomposition equations are

$$C_qH_rO_sN_t \rightarrow \frac{r}{2}H_2O + \left(s - q - \frac{r}{2}\right)CO_2 + \left(2q - s + \frac{r}{2}\right)CO + \frac{t}{2}N_2$$

if $q + r/2 \leqslant s$, corresponding to an oxygen rich explosive, or

$$C_qH_rO_sN_t \rightarrow \frac{r}{2}H_2O + \left(s - \frac{r}{2}\right)CO + \left(q - s + \frac{r}{2}\right)C + \frac{t}{2}N_2$$

if $q + r/2 \geqslant s$ (oxygen deficient).

It will be seen from these equations that to a first approximation free atoms, even of hydrogen, are not significant, molecular hydrogen is not formed in quantity, nor do polyatomic molecules such as CH_4, C_2H_2, HCN, NH_3 occur in significant amounts. In this first approximation, the possible presence of OH, O_2, NO is also neglected. The reactions considered by Brinkley and Wilson for the equilibrium calculations are:

$$\begin{aligned}
H_2 + CO_2 &= CO + H_2O \\
2CO &= CO_2 + C \text{ (solid)} \\
H_2O &= \tfrac{1}{2}H_2 + OH \\
2H_2O &= 2H_2 + O_2 \\
H_2O + \tfrac{1}{2}N_2 &= NO + H_2
\end{aligned}$$

In later calculations, Brinkley and Wilson have also included the reactions

$$\begin{aligned}
3H_2 + CO &= CH_4 + H_2O \\
\frac{3}{2}H_2 + \tfrac{1}{2}N_2 &= NH_3
\end{aligned}$$

For oxygen deficient explosives, such as TNT, and aluminized mixtures the occurrence of solid products must be allowed for in the equation of state, which is done by excluding the volume V_s of solid products and using an effective volume $V - V_s$. In addition to equilibrium constants for the reactions, the theory requires a knowledge of the heats of formation of the various product molecules and the initial loading density.

Jones' calculations for TNT are made by assuming the reactions

$$\begin{aligned}
H_2 + CO_2 &= CO + H_2O \\
CO + \tfrac{1}{2}O_2 &= CO_2 \\
C(\text{solid}) + 2H_2 &= CH_4 \\
C(\text{solid}) + \tfrac{1}{2}O_2 &= CO
\end{aligned}$$

the equilibrium constants being taken from data given by Lewis and von Elbe (68).

The compositions found by Brinkley and Wilson for TNT of density 1.46 and by Jones for a slightly different value 1.5 are given in Table 3.2. Jones' results as compared to those of Brinkley and Wilson give larger quantities of CO_2 and CH_4 at the expense of H_2O and CO. The available experimental data of detonation products of course include equilibrium shifts on cooling of the product gases and so are not directly applicable for comparison. They do, however, indicate much smaller quantities of methane, and it is not unreasonable to believe that these may have been formed as a consequence of reactions during the cooling process. The reactions involving CO and CO_2 may also be expected to shift, and on the whole it is reasonable to suppose that the compositions obtained by Brinkley and Kirkwood approximate the actual conditions more closely. This is particularly true when the compositions are considered in connection with calculation of detonation properties, as the Brinkley-Wilson method is based in part on such data. As might be expected their results do give much better agreement on detonation rate than do those of Jones. The values are compared in Table 3.3.

The observed detonation velocity for $\rho_o = 1.50$ is 6,620 m./sec. from Messerly's data (72), and at $\rho_o = 1.46$ interpolation between values gives 6,470 m./sec. in excellent agreement with Brinkley and Wilson. Experimental data for pressures behind the detonation front indicate values of the order of 200 kilobars,[3] but are sufficiently inaccurate that they constitute only a qualitative indication.

Detonation velocities are measured by a number of methods, one of the most elegant employing a camera with a drum rotating at high speed on which is recorded the advance of the luminous detonation front along a stick of explosive initiated at one end. This same general technique has been employed to estimate the detonation pressure by an ingenious method in which the velocity of a shock wave started by the detonation in a lead strip is measured, the pressure being estimated from compressibility data for lead. The procedure involves an extrapolation which, together with experimental errors, makes the results only approximate. Shock wave velocities in water can also be measured and can similarly be used to estimate pressures near small charges, an experimental procedure described more fully in Chapter 6.

[3] This value was estimated from experiments made at the Explosives Research Laboratory, Bruceton, Pa., in which the shock wave velocity through a lead sheet in contact with a pentolite slab was measured. The measured velocity was 2,670 m./sec., corresponding to a pressure of 140 kilobars on the basis of extrapolated compressibility data. The detonation pressure was estimated to be from 40 to 80 kilobars higher, the difference being due to loss of pressure in expansion of the product gases against the lead. Because of the extrapolation of the pressure-velocity curve for lead and this correction, the figure obtained is only qualitatively reliable.

Product molecule		H_2	CO_2	CO	H_2O	N_2	OH	O_2	NO	CH_4	NH_3	C
Brinkley and Wilson	Approximate	0	0	15.77	11.12	6.67	0	0	0	0	0	15.77
	Equilibrium	0	1.92	11.64	10.96	6.60	0	0	0.04	0	0.04	17.32
Jones		0.013	11.6	0.27	4.85	6.60	0	0	0	3.08	0	16.9

Table 3.2. Composition of products in detonation of TNT, expressed as moles per kilogram of explosive.

The results on a large number of different explosives, and some at several loading densities, as computed by Brinkley and Wilson (11), show similarly good agreement with detonation rate data, the differences being in most cases within experimental error. The values for pressure and density are of unknown accuracy but are very probably not as good. The values obtained by Jones are less accurate, the calculated rate being seventeen per cent higher than the experimental value. This lack of agreement is hardly surprising when one considers the approximations involved and the fact that no experimental data from explosive measurements are used.

3.5. Adiabatic Pressure-Density Relation after Detonation

In order to compute the form of the detonation wave behind its front it is necessary to know the adiabatic law for the products. Jones has made calculations of the adiabatic relation for TNT (51), based on considerations similar to those in his calculations for the head of the wave,

	ρ_0 (gm./cm.^{3})	ρ (gm./cm.^{3})	T (°K)	P (kilobars)	D (m./sec.)	c (m./sec.)	u (m./sec.)
Brinkley, Wilson	1.46	1.86	3200	140	6640	5210	1430
Jones	1.50	1.94	3460	205	7720	5970	1750

Table 3.3. Calculated quantities for TNT at the detonation front.

as described in sections 3.2 and 3.3. As in such calculations, the problem is made difficult by the necessities of choosing an equation of state applicable to the high pressures and temperatures and determining appropriate compositions of the products. Both the density and composition are, for any very realistic representation, rather complicated functions of pressure and temperature.

The most fundamental procedure in fixing the initial conditions would, of course, be to employ a calculation of the type made by Jones which made no use of explosion data at all. Jones' calculations, however, led to a detonation velocity some fifteen per cent higher than experimental values, and the initial pressure and density are presumably liable to a similar error. In order to improve the adiabatic results, the pressure and density predicted by the theory for the observed detonation velocity, taken to be 6,790 m./sec. at an initial density of 1.50, are assumed. These data, together with the observed variation of detonation rate with initial density, are used to determine the constants of equation of state in the virial form

$$\frac{PV}{N} = RT + bP + cP^2 + dP^3$$

where N is the total number of moles of gaseous products per mole of TNT, and, unless the composition is "frozen," is a function of P and T.

In an adiabatic expansion, the work done by the gases must equal the sum of the energy liberated by the reaction and the loss of internal energy, giving the relation

$$PdV = d(H - E) \tag{3.15}$$

where H is the heat of formation of the products and E their total internal energy, both of which are functions of P and T. In order to make further progress it is evidently necessary to introduce conditions on the composition of the products. As the pressure and temperature fall during expansion, the equilibrium composition will change but at the same time the rates of reaction will decrease. The decreased rates make it reasonable to suppose that equilibrium is not attained in later stages of the expansion. Calculations based on the equilibrium assumed by Jones for TNT (see section 3.2) predict that at first the amount of CO increases rapidly at the expense of CO_2 and solid C, reaching a maximum at about 1,800° K. At this point, the reaction rates are enormously less than at 3,000° K., and Jones completes his adiabatic calculation with the assumption of this fixed composition for further expansion. As already mentioned, there is reason to question the equilibrium results of Jones, and the final adiabatic relation may be somewhat in error as a result. We shall see later, however, that the exact form of this relation is not of paramount importance for most of the calculations of pressures developed in underwater explosions.

In order to integrate the differential equation (3.15) for the adiabatic expansion, a rather complicated set of equations is developed in terms of the activity and composition variables. With the initial conditions of pressure and composition assigned, the initial values of activity and temperature fix a starting point for numerical step by step integration. The final P–V relation resulting from these calculations is plotted in Fig. 3.2. On the logarithmic plot a perfect gas condition would correspond to a straight line. It is seen that at the higher densities the pressures found are much higher, corresponding to the increased effect of repulsive forces on the internal energy. The decrease in slope at the smallest volumes is attributed to the smaller number of moles (decreased concentration of CO) at the higher pressures.

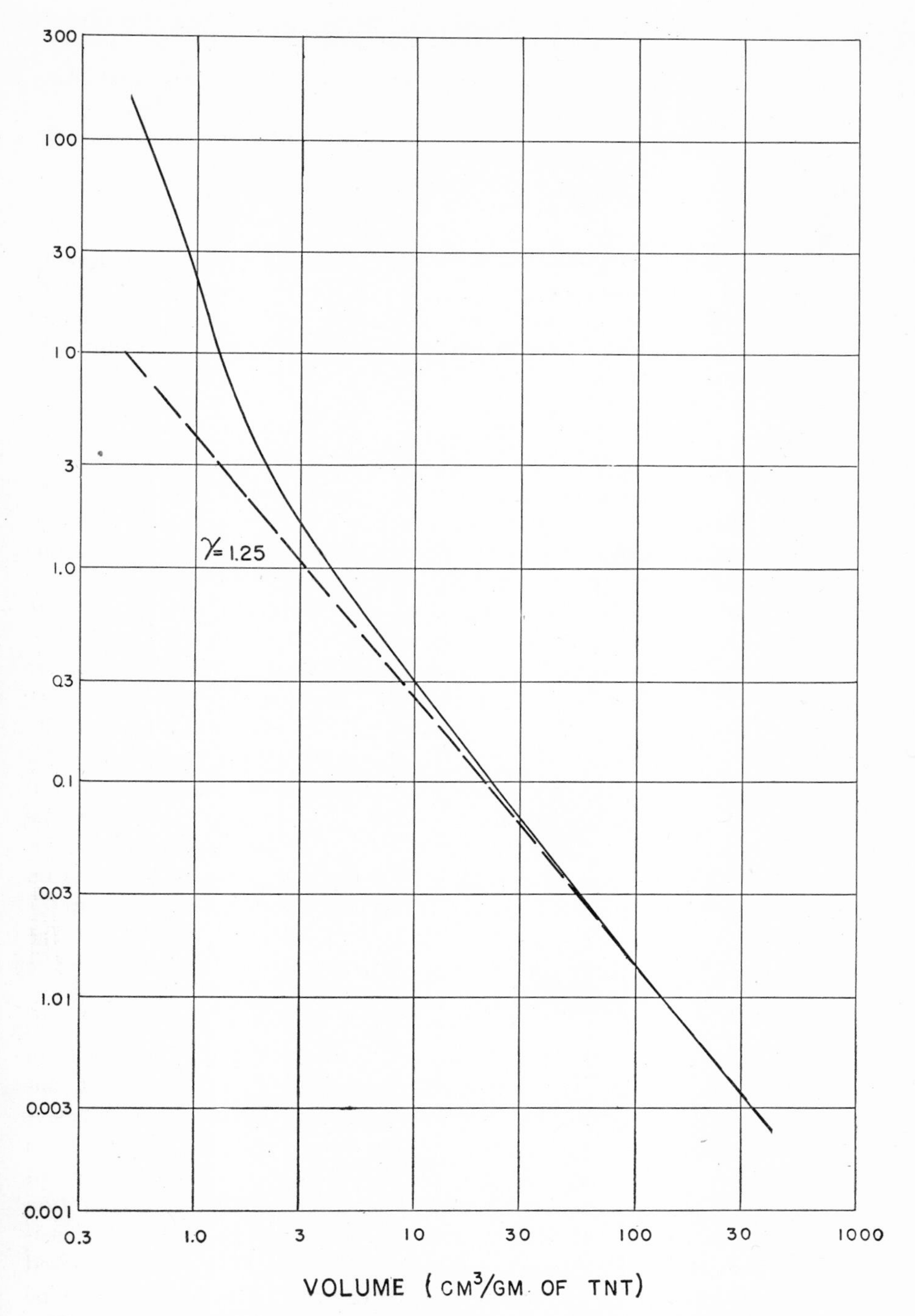

Fig. 3.2 Adiabatic pressure-volume relation for products of TNT, from calculations of Jones.

Jones has also made calculations (52), applicable to oxygen-rich explosives, by which the equation of state and adiabatic expansion are related to the variation of detonation velocity with loading density (as previously noted, this variation would not occur were the products ideal gases). An interesting result of these calculations, noted by Jones, is that the predicted compressibility of the gaseous products is quite similar to Bridgman's experimental data for nitrogen mentioned in section 3.3.

In addition to a knowledge of the pressure and density resulting from adiabatic explosion, the evaluation of conditions at the gas sphere boundary requires a knowledge of the sound velocity, Riemann function σ, and other parameters for the initial adiabatic in the gaseous products. The determination of these variables for any explosive is a quite straightforward derivation from the equation of state and heat capacity data. The explicit form of the necessary relations is, however, rather complicated and the description of them here is restricted to indicating the method of approach. The internal energy and entropy can be expressed as functions of density ρ and temperature T by means of the Wilson-Kistiakowsky equation of state in the form:

$$E(\rho, T) = E_o + N\overline{C}_v (T - 300) + \alpha NRTxe^{\beta x}$$

$$S(\rho, T) = S_o + N \int_{300}^{T} \frac{C_v}{T} dT - NR\left[\log_e \rho + \frac{1}{\beta}(e^{\beta x} - 1) - \alpha xe^{\beta x}\right]$$

where E_o, S_o are the ideal values for unit density and temperature 300° K. The sound velocity, given by $c^2 = (dP/d\rho)_s$, is then obtained from the equation of state and entropy expression by standard thermodynamics, it being assumed adequate for this purpose to use a constant heat capacity C_v for temperatures T not greatly different from T_o, the initial temperature of the products. The pressure-density relation can also be expressed as a function of temperature T and the equation of state variable x, as can the Riemann function σ and enthalpy ω. These latter are in the form of definite integrals, conveniently computed using the function $x = (\rho/M) KT^{-\alpha}$ as independent variable.

3.6. The Form of the Detonation Wave

When the Chapman-Jouget condition is satisfied, a specification of the adiabatic pressure-density relation in the products of explosion provides the necessary data for determining the form of the detonation wave behind its front. G. I. Taylor (106) has considered the general problem and obtained detailed results for TNT, using the detonation conditions calculated by H. Jones. Before describing these results it is of some interest to examine the consequences of the simplified case of a plane wave for which the products are assumed to be ideal gases.

A. *Plane wave, ideal gases.* If we assume ideal gases in which the ratio of specific heats is independent of temperature, the adiabatic relation at any point behind the front is $P = k\rho^\gamma$, where k and γ are taken to be constant. The detonation velocity, for such a mixture, was shown in section 3.3 to be

$$D = (1 + \gamma)\sqrt{\frac{P_1}{\gamma\rho_1}}$$

where P_1, ρ_1 are the pressure and density at the head of the wave. The velocity of sound c in these gases is given by

$$c^2 = \left(\frac{dP}{d\rho}\right)_s = k\gamma\rho^{\gamma-1} = \gamma\frac{P}{\rho}$$

and the velocity of sound c_1 at the detonation front is $c_1 = \sqrt{\gamma P_1/\rho_1}$. From this relation and the Chapman-Jouget condition, the particle velocity u_1 at the front is $u_1 = D - c_1 = c_1/\gamma$.

If the detonation is a plane wave, Riemann's analysis shows that behind the front constant values of particle velocity, density, and sound velocity travel with a speed $c + u$ and u is given by

$$u_1 - u = \int_\rho^{\rho_1} c\frac{d\rho}{\rho} \tag{3.16}$$

From the adiabatic law, $c = c_1(\rho/\rho_1)^{\frac{\gamma-1}{2}}$ and substituting for c in the integral gives

$$u_1 - u = \frac{2}{\gamma - 1}(c_1 - c) \tag{3.17}$$

If the detonation is initiated by any cause of duration short enough that the gases can be considered at rest for all times immediately after initiation, a given value of u will be propagated a distance r in time t such that $r = (c + u)t$. The detonation front will have travelled a distance $R = (c_1 + u_1)t$ and we have

$$\frac{R - r}{t} = (c_1 - c) - (u_1 - u)$$

Combining this relation with (3.17) gives

$$u_1 - u = \frac{2}{\gamma + 1}\left(\frac{R - r}{t}\right), \quad c_1 - c = \frac{\gamma - 1}{\gamma + 1}\left(\frac{R - r}{t}\right) \tag{3.18}$$

The particle velocity u will be zero for all values of r/t less than

$$\left(\frac{r}{t}\right) = \frac{R}{t} - \frac{\gamma + 1}{2} u_1 = \frac{R}{2t}$$

The particle velocity thus falls off linearly from its initial value, reaching zero halfway between the front and the starting point. The pressure distribution is also readily found from the adiabatic relation and Eq. (3.18), giving

$$\frac{P}{P_1} = \left[1 - \frac{1}{c_1}\frac{\gamma - 1}{\gamma + 1}\frac{R - r}{t}\right]^{\frac{2\gamma}{\gamma - 1}} = \left[1 - \frac{\gamma - 1}{\gamma}\frac{R - r}{R}\right]^{\frac{2\gamma}{\gamma - 1}} \tag{3.19}$$

The particle velocity and pressure distribution for $\gamma = 1.4$ are plotted in Fig. 3.3 (dashed curve). Initially $u_1 = D/(\gamma + 1)$, and an arbitrary

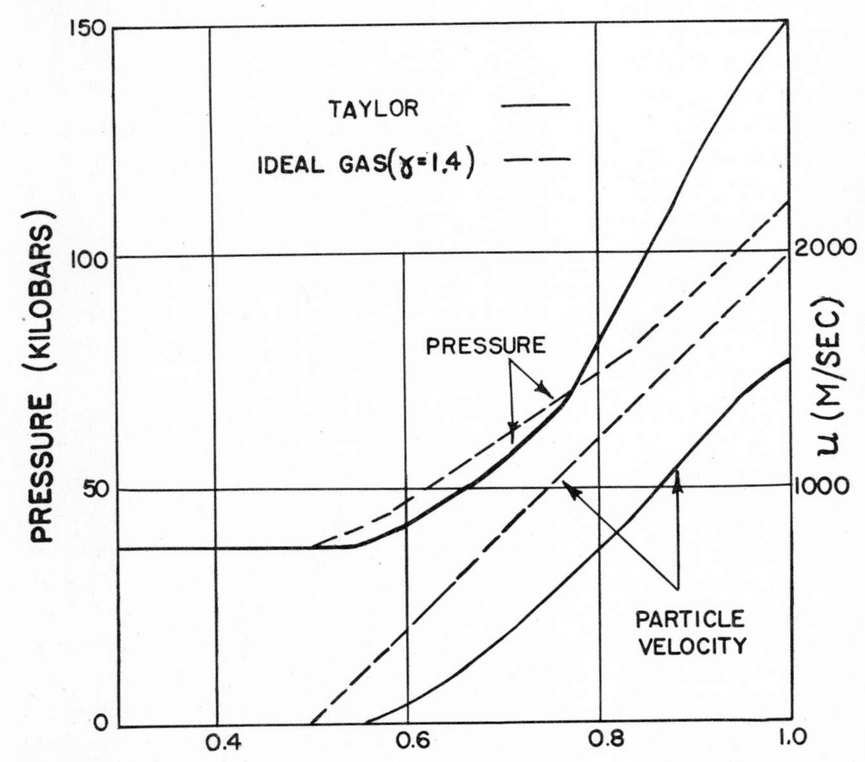

Fig. 3.3 Calculated pressure and particle velocity of a plane detonation wave.

initial pressure of 112 kilobars has been assumed. It is to be understood that this calculation is purely illustrative.

B. *Plane wave, TNT*. Determination of the profile of a plane wave for actual cases is formally much the same as for ideal gases; namely, if the values of pressure and density at the front are known, the Riemann analysis can be used to determine the particle velocity u as a function of density. The actual adiabatic law must replace the ideal gas relation and, if this has been determined in tabular form, the integration for u must be performed numerically. As a starting point of the calculation for TNT, Taylor chose $D = 6{,}380$ m./sec., $\rho_1 = 2.00$ gm./cm.^{3} The pressure and sound velocity for these values are from Jones' calculations $P_1 = 150$ kilobars, $c_1 = \sqrt{(dP)/(d\rho)_{\rho=\rho_1}} = 4{,}840$ m./sec. Numerical step by step integration of Eq. (3.16) using an adiabatic relation for TNT similar to Fig. 3.2 then yields a series of values of u for various values of ρ, and corresponding values of c and P are known from the adiabatic. The pressure and particle velocity are therefore determined as a function of $c + u$ or of $r/R = (c + u)/(c_1 + u_1)$, where as before r is the distance from the source at which the values c and u are realized after a time t and R is the distance for the head of the wave. Taylor's results are plotted as the solid curve in Fig. 3.3. The differences from an ideal gas law (dashed curve) are seen to be consistent with the difference of the adiabatics.

C. *Spherical wave, TNT*. Taylor (106) has also computed the profile of the wave behind a spherical detonation front, assuming initiation at the center which thereafter leaves the products at rest ($u = 0$). In the spherical case, the Riemann function is not directly applicable and a more complicated procedure is necessary. The fundamental equations describing the motion behind a spherical front are, from section 2.3,

$$\frac{\partial u}{\partial t} + u\frac{\partial u}{\partial r} = -\frac{c^2}{\rho}\frac{\partial \rho}{\partial r} \tag{3.20}$$

$$\frac{\partial \rho}{\partial t} + u\frac{\partial \rho}{\partial r} + \rho\frac{\partial u}{\partial r} = -\frac{2\rho u}{r}$$

If, as we assume, it is legitimate to suppose that a stationary detonation condition is attained before the disturbance has proceeded an appreciable distance, the conditions behind the front will scale in proportion to the distance the front has progressed. That this is true is evident on noting that if r and t are both changed by a factor β into βr and βt, the quantities ρ, u evaluated at βr and βt satisfy the same equations. The conditions at R and t, where R is the distance of the front from the initial point are, however, unchanged at βR and βt if the time to establish the detonation is small. Hence all the quantities behind the front can be functions only of the ratio r/t. The time t is given by $R/(c_1 + u_1) = R/D$, where, by assumption, D is constant. An equivalent statement is therefore that the properties behind the front are functions only of the ratio r/R, which may be considered a statement of the principle of similarity for a spherical detonation wave. This proposition, that the properties of the phenomenon are unchanged if both length and time scales are changed by the same factor, is of course not restricted to the spherical case, being equally true for any other symmetry, provided the time for the steady state condition to be established is negligible.

In order to integrate the equations for the spherical case, Taylor introduces the variables

$$\xi = \frac{u}{x}, \quad \psi = \frac{u}{c}, \quad z = \log x, \quad \text{where } x = \frac{r}{t}$$

in terms of which Eqs. (3.20) become

$$\frac{d\xi}{dz} = \xi \frac{3\xi^2 - (1 - \xi)^2\psi^2}{\psi^2(1 - \xi)^2 - \xi^2} \tag{3.21}$$

$$\frac{d\psi}{dz} = \psi\xi \frac{2\xi - \psi^2(1 - \xi)f}{\psi^2(1 - \xi)^2 - \xi^2}$$

where $f = (\rho/c^2)dc^2/d\rho$ and is a definite function of ρ from the adiabatic relation behind the front. These equations Taylor solves for TNT by step-by-step numerical integration back from the front, thus determining ξ and ψ in terms of z. A difficulty presents itself at the starting point in that both $d\xi/dz$ and $d\psi/dz$ become infinite for $r = R$, where $x = c_1 + u_1$. The values of ξ, ψ and $d\psi/d\xi$ are all finite, however, and the integration can be begun from a slightly smaller value z' obtained by a series expansion in the neighborhood of R:

$$z' = z_1 + \left(\frac{\partial z}{\partial \xi}\right)_{\xi_1} (\xi - \xi_1) + \frac{1}{2}\left(\frac{\partial^2 z}{\partial \xi^2}\right) (\xi - \xi_1)^2$$

Taking $z = \log r/R$, as is permissible since z appears in the equations only as a differential and noting that $(\partial z/\partial \xi)_{\xi_1} = 0$, this becomes

$$z' = \frac{1}{2}\left(\frac{\partial^2 z}{\partial \xi^2}\right)_{\xi_1} (\xi - \xi_1)$$

This expression can be evaluated from Eqs. (3.21). Noting that

$$\xi_1 = \frac{u_1}{c_1 + u_1} = 1 - \frac{\rho_1}{\rho_o}, \quad \psi_1 = \frac{u_1}{c_1} = \frac{\rho_1}{\rho_o} - 1$$

we obtain

$$z = -\frac{1}{2}\left(\frac{\rho_o}{\rho_1}\right)^2 \cdot \frac{1}{\left(\frac{\rho_1}{\rho_o}\right) - 1}\left[1 + \frac{f}{2}\right]_{\xi 1} \left(\xi - \frac{\rho_o - \rho_1}{\rho_o}\right)^2$$

This equation determines values of z near $r = R$ in terms of ξ, ψ, and with these initial values, successive increments of ξ, ψ are computed by stepwise integration of Eqs. (3.21) for $d\xi/dz$, $d\psi/dz$. The final results obtained by Taylor are shown in Fig. 3.4 and, although they are quite similar to the curves for a plane wave, the pressure and particle velocity fall off more rapidly behind the detonation front and the final pressure for $u = 0$ is smaller. The excess over the final value is negligible for $r < 3/5\ R$. This point is the crux of a basic assumption in some theories of shock-wave propagation and a numerical example may indicate more clearly the order of magnitude involved. If the detonation velocity D is 6,400 m./sec. the time required for the tail of the wave at $r = 3/5\ R$ to reach the point R is $2/3\ R/D$. For a change of radius 25 cm. (approx. 300 pounds TNT) this time is then roughly 25 microseconds, a comparatively small quantity relative to the time scale of the pressure wave developed in the water.

The calculations of Taylor for the spherical detonation wave in TNT suffer in accuracy from the fact that an approximate adiabatic relation for the explosion products was used. Appreciable errors in the calculated detonation wave result from small errors in this adiabatic because the function $f = (\rho/c^2)\ dc^2/d\rho$ is sensitive to small differences in values for $P(\rho)$. Dasgupta and Penney[4] have therefore carried out revised calculations, in which tabular values from Jones' calculations were used, and the initial conditions at the head of the wave redetermined to give the observed detonation velocity, taken as 6,780 m./sec. In order to satisfy this and the Chapman-Jouget condition, a higher pressure of 196 kilobars was necessary, as obtained by extrapolation of Jones' adiabatics, the corresponding sound velocity and particle velocity being $c = 4{,}870$ m./sec., $u_o = 1{,}920$ m./sec.

The detonation wave was then computed by stepwise numerical integration of Eqs. (3.21), which are solved for $d\psi/d\xi$ and $dz/d\xi$ to avoid the singularities at the detonation front encountered using z as an independent variable. The pressure and particle velocity values finally obtained are plotted as dashed curves in Fig. 3.4. It is seen that the pressure, while higher than Taylor's, decays more gradually and a constant value is reached for distances less than half the charge radius from the center.

Dasgupta and Penney have also computed, by the same method, the

[4] H. K. Dasgupta and W. G. Penney, British Report RC 373.

detonation wave for a 40/60 TNT/RDX mixture of density 1.5 from adiabatics calculated by Jones. The experimental velocity $U = 7{,}280$ m./sec. was used, and the pressure, particle and sound velocities were obtained as 177 kilobars, 5,660 m./sec. and 1,620 m./sec. The pressure,

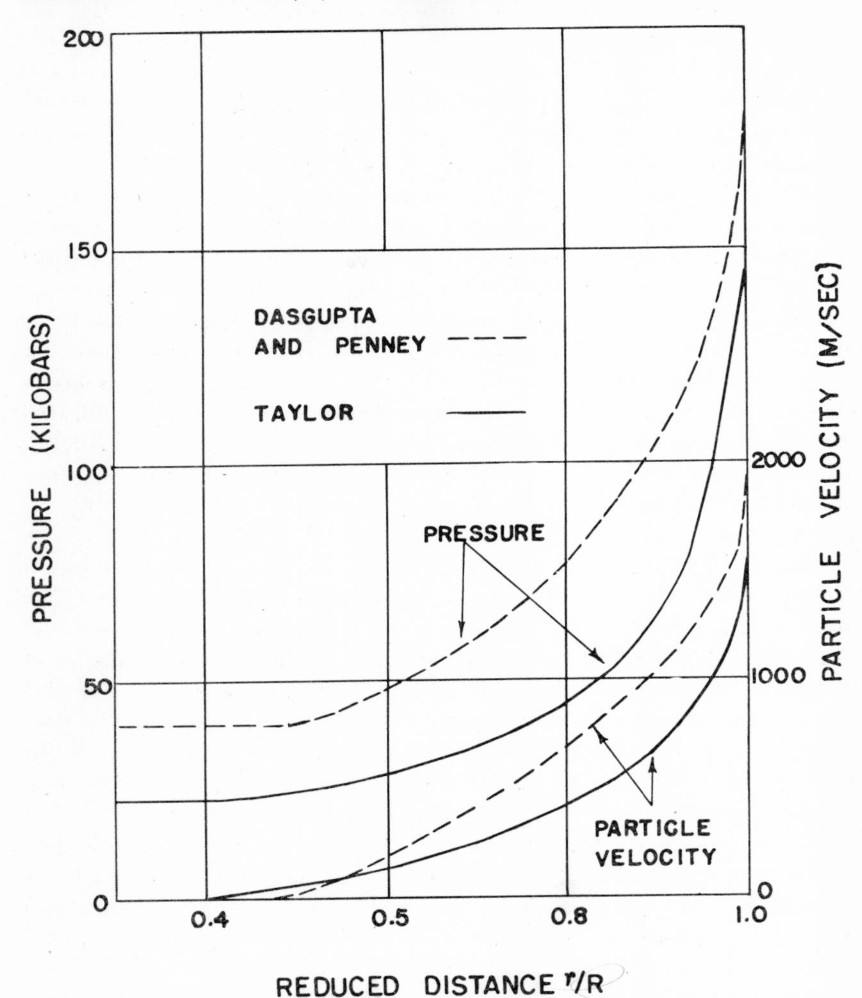

Fig. 3.4 Calculated pressures and particle velocities of a spherical detonation wave in TNT.

although found to be initially 10 per cent lower than for TNT, fell less rapidly to a final value of 44 kilobars, 10 per cent above that for TNT. Although these results are the best available for the actual form of the detonation front, no calculations of shock wave pressures from them have been reported at the time of writing.

3.7. The Approximation of Adiabatic Explosion at Constant Volume

An exact solution for the propagation of underwater shock waves should of course be based on a continuation of the solution for the detonation wave after it reaches the surface of the charge, applying the appropriate boundary conditions. These conditions require a shock wave in the water and a rarefaction wave travelling back toward the center of the charge, and a reasonably good solution taking account of both would be a matter of considerable difficulty. Kirkwood and Bethe (59, I) observe, however, that the exact form of the detonation wave near its front is in most cases of no great importance, for the reason that the initial portions of the underwater shock wave, determined by the head of the detonation wave, are rapidly destroyed by the overtaking effect as it progresses outward. In other words, even the front of the shock wave is, in their propagation theory, determined by conditions obtaining at times increasingly later than the time of complete detonation. It is therefore assumed in the theory that the actual initial conditions can be adequately approximated by assuming simplified conditions in which the explosion takes place adiabatically without change in volume, the explosion products being at uniform pressure throughout the volume.

Before describing the actual calculation of these simplified conditions it is of interest to make a rough estimate of the range in which neglect of the head of the detonation wave involves appreciable error. According to the Kirkwood-Bethe propagation theory the condition at the shock front is determined by the pressure on the gas sphere at a time τ which increases with the distance R the shock wave has travelled. The value of τ is related to R by the equation $x = e^{-\tau/\theta_1}$ in which, for any specific explosive, the dissipation factor x is a computed function of R/a_o (a_o is the original charge radius) and θ_1/a_o is characteristic of the explosive. Assuming this relation, the value of R/a_o corresponding to advance of the head of the detonation wave into the boundary may be computed. The time required is $2/3\ a_o/D$ from Taylor's calculation, and the value of x for this time is then $x = \exp(-2/3 \cdot a_o/\theta_1 \cdot 1/D)$. For TNT $\theta_1/a_o = 3.4 \cdot 10^{-3}$ m./sec. $D = 6{,}400$ m./sec. and so $x = 0.74$, which the theory predicts will be realized at $R/a_o = 2.0$. Although the calculation is a very crude one indeed, we might infer that the theory will be increasingly in error for distances of the order of two charge radii or less.

Calculations of the initial pressure and density following adiabatic reaction at constant volume have been made by Brinkley, Kirkwood, and coworkers (59, V, VII), using the modified Wilson-Kistiakowsky equation of state. The reaction must take place in such a way that the chemical energy released by the reaction is equal to the increase in internal energy of the products. The total energy change is of course independent of the path by which the transition is assumed to take place. It is convenient in calculation to evaluate the heat of reaction at the initial temperature and pressure (essentially zero), heat the products to the final temperature at constant volume which can be taken to be infinite and compress the products at the final temperature to the final volume. The expression of this process for M grams of explosive is

$$\sum N_i H_i - H_o + RT_o \sum N_i = \bar{C}_v \cdot (T - T\ ) + \int_\infty^{M/\rho} \left(\frac{\partial E}{\partial V}\right)_T dV \tag{3.22}$$

In this equation, N_i is the number of moles of product species i having a heat of formation H_i per mole, H_o is the heat of formation of M grams of the explosive, $\bar{C}_v$ is the total mean heat capacity of all products over the temperature range and M/ρ is the final volume. From the Wilson-Kistiakowsky equation

$$E(V, T) = E_o + N\bar{C}_v(T - 300) + NRT\alpha x e^{\beta x}$$

and so

$$\int_\infty^{M/\rho} \left(\frac{\partial E}{\partial V}\right)_T dV = NRT\alpha x e^{\beta x}$$

where x is a function of temperature and the final composition.

Knowing the specific heats and heats of formation, Eq. (3.22) gives an implicit solution for the final temperature which for a given composition can be solved by successive approximations. Use of the conditions $\Delta E = 0$ for the adiabatic conversion and $\Delta S = 0$ for the subsequent isentropic expansion, together with the thermochemical data, thus permits evaluation of the quantities necessary for approximate propagation theories. Numerical methods and auxiliary tables for computation are given in a report by Kirkwood, Brinkley, and Richardson (59, V).

3.8. Boundary Conditions and Initial Motion of the Gas Sphere

The propagation theory of Kirkwood and Bethe for the underwater shock wave reduces the determination of shock wave parameters to an evaluation of the "kinetic enthalpy" on the surface of the gas sphere at a variable retarded time τ after detonation is complete. As already discussed in section 3.5, Kirkwood and Bethe neglect the head of the detonation wave and use an initial condition of adiabatic explosion at constant volume. The basic problem is therefore evaluation of the pressure and particle velocity on the boundary as a function of time, in order to determine the kinetic enthalpy $G(a, t)$ defined by

$$G(a, t) = a\Omega_a(t) = a[\omega(a, t) + \tfrac{1}{2}u^2(a, t)]$$

where $\omega(a, t)$ is the enthalpy and a the radius of the gas sphere.

A. *The initial conditions.* The initial pressure $P(O)$ in the water must equal the initial pressure $P_g(O)$ in the gas sphere. (In what follows, quantities with subscripts g refer to the gas, without, to the water, and the suffix (O) indicates initial values.) This pressure is of course not the same as the adiabatic explosion pressure P_e calculated from thermochemical data, the equalization of pressure being achieved by generation of an outgoing compression wave in the water and an ingoing rarefaction wave in the gas products. Both of these waves leave the medium behind them moving with an outward particle velocity, and the second necessary boundary condition is that the particle velocities in the water and products be the same at the boundary. The rarefaction wave may be described as one in which the Riemann function $N_g = (\sigma_g + u_g)/2$ is initially zero.[5] The Riemann variable σ_g is, however, a calculable function of density, and hence pressure, of the explosion products.

A second necessary condition arises from the fact that the pressure and particle velocity in the water are not independent but are related by the Rankine-Hugoniot conditions and the derived tables (see section 2.6). Taken together, this condition and the continuity of velocity determine the initial pressure and particle velocity at the boundary, as indicated symbolically by the relations

$$(3.23) \qquad u = -\sigma_g(P_g), \text{ from the Riemann condition}$$

$$u = u(P), \text{ from the Rankine-Hugoniot condition}$$

The relation between σ_g and $P_g = P$ must be determined by calculations of the adiabatic expansion for the explosion products, as outlined in sections 3.5 and 3.7. With these results, the values of $P(O)$ and $u(O)$ consistent with Eq. (3.23) can be determined numerically or graphically.

B. *Pressure and particle velocity at the gas sphere.* The equality of pressure and particle velocity at the gas-water boundary must be true at all times after it is first established. The total time derivatives of

[5] It is important to note that $N_g = 0$ need not be, and in fact is not, true at later stages in a spherical rarefaction wave, nor does the propagation theory used by Kirkwood et al assume this.

these quantities, which include the effect of motion of the boundary, must therefore also be equal, giving

$$\frac{dP_g}{dt} = \frac{dP}{dt}, \quad \frac{du_g}{dt} = \frac{du}{dt}, \quad \text{at } r = a$$

The fundamental variable of the Kirkwood-Bethe theory is the function $G(r, t)$, constant values of which are propagated with a velocity $\bar{c}$ such that

$$\frac{\partial G}{\partial t} + \bar{c}\frac{\partial G}{\partial r} = 0$$

and it is therefore necessary to relate G to the boundary conditions. From its definition, G satisfies the differential equation

$$dG = d\left[r\left(\omega + \frac{u^2}{2}\right)\right] = \left(\omega + \frac{u^2}{2}\right) dr + rd\left(\omega + \frac{u^2}{2}\right)$$

where the enthalpy ω satisfies the fundamental thermodynamic relation $d\omega = TdS + dP/\rho$. The numerical calculations of section 2.7 show that the entropy term is small, and setting $dS = 0$ gives

$$dG = \left(\omega + \frac{u^2}{2}\right) dr + \frac{r}{\rho}\, dP + rudu$$

where $\omega = c^2 \int d\rho/\rho$. Substitution in the propagation equation for G gives

$$\left(\omega + \frac{u^2}{2}\right)(u + \bar{c}) + \frac{r}{\rho}\left(\frac{\partial P}{\partial t} + \bar{c}\frac{\partial P}{\partial r}\right) + ru\left(\frac{\partial u}{\partial t} + \bar{c}\frac{\partial u}{\partial r}\right) = 0$$

This equation is valid in both the water and gas products if the partial derivatives are evaluated in the medium considered.

The basic dynamical equations of motion and continuity must be true in either medium, and in the water they are

$$\frac{1}{\rho c^2}\frac{dP}{dt} = -\frac{\partial u}{\partial r} - \frac{2u}{a}, \quad \rho\frac{du}{dt} = -\frac{\partial P}{\partial r}$$

The equality of total derivatives at the boundary is of course not true of the partial derivatives if the boundary moves, and the next step is to solve for the desired total derivatives by elimination of the partial derivatives. For the water this gives

$$\frac{1}{\rho c}\frac{dP}{dt} = \frac{c(\bar{c} - 2u)}{c^2 - u\bar{c} + u^2}\cdot\frac{du}{dt} - \frac{c}{a}\,\frac{\bar{c}\left(\omega - \dfrac{3}{2}u^2\right) + 2u^3}{c^2 - u\bar{c} + u^2} \tag{3.24}$$

A corresponding equation for the gas products is obtained by replacing ρ, c, ω by ρ_g, c_g, ω_g (where $\omega_g = \int (c_g^2/\rho_g)\, d\rho_g$), and noting that the propagation velocity is $-\bar{c}_g$ for an ingoing wave, with the result

$$\frac{1}{\rho_g c_g}\frac{dP}{dt} = -\frac{c_g(\bar{c}_g + 2u)}{c_g^2 + u\bar{c}_g + u^2}\frac{du}{dt} - \frac{c_g}{a}\cdot\frac{\bar{c}_g\left(\omega_g - \dfrac{3}{2}u^2\right) - 2u^3}{c_g^2 + u\bar{c}_g + u^2} \tag{3.25}$$

The two equations, (3.24) and (3.25), can be solved for the derivatives dP/dt, du/dt at the boundary in terms of the various velocities and related functions in the two media. The resulting differential equations could then be integrated numerically, obtaining at each step new values of the variables. Rather than carry out these tedious calculations, Kirkwood and Bethe make use of the so-called peak approximation in which the falling off of pressure and enthalpy is assumed to be exponential, with a decay constant chosen to give the initial value of dP/dt correctly. Before describing this method, it is worthwhile to consider the approximations in the development so far outlined.

The motion of the gas products is assumed by Kirkwood and Bethe to be described sufficiently by a single ingoing rarefaction wave. After a finite time, this wave reaches the center of the charge and internal reflection must take place. At sufficiently great times, the boundary conditions in the gas will therefore be modified by successive compounding of internal reflections. The first such reflection will occur at a time greater than $2a_o/c_g(O)$, where a_o and $c_g(O)$ are the initial radius and sound velocity in the gas sphere. Inclusion of these reflections would require an elaborate analysis and is not attempted in the theory outlined. These effects occur, however, at sufficiently late times that the major portion of the shock wave is emitted, and the reflections make their appearance in the shock wave "tail." This conclusion can be inferred from the solution for dP/dt, which is readily shown to be always negative and decreasing in magnitude, the function $P(a, t)$ thus having a time variation similar to an exponential. The peak approximation already mentioned is therefore a natural one and should be a sufficiently accurate description of the pressure variation at times for which the theory is in any case reasonably reliable. The particle velocity u, however, is found first to increase to a maximum and then fall off. The peak approximation is therefore not suitable in this case, and the motion of the gas boundary requires a separate examination, as described in part (C).

A function $F(t)$ described by the peak approximation is assumed to have an exponential decay of the form $F(t) = F(O)e^{-t/\theta_1}$ and the decay constant θ_1 can be computed from the initial variation of $F(t)$ by the relation

$$\left(\frac{dF}{dt}\right)_{t=0} = -\frac{1}{\theta_1} F(O)$$

The time variation of $F(t)$ is thus determined in this approximation by the initial value and slope of the function. Kirkwood and Bethe apply this procedure to the enthalpy ω by writing $\omega = \omega(O)e^{-t/\theta_1}$. With the neglect of dissipation $d\omega = dP/\rho$, hence θ_1 is given by

$$\theta_1 = -\rho(O)\,\omega(O)\left(\frac{dP}{dt}\right)_{t=0}$$

and the initial value of dP/dt can be obtained from Eqs. (3.24) and (3.25). Assuming that initially $\bar{c} = c + u$, $\bar{c}_g = c_g - u$, this gives

$$\theta_1 = +\frac{\omega(O)\,a_o c(O)}{\rho_g(O)\,c_g(O)\,u(O)} \cdot \frac{\rho(O)\,c(O) + \rho_g(O)\,c_g(O)}{c(O)\,J(O) + c_g(O)\,J_g(O)} \tag{3.26}$$

where the functions J and J_g are given by

$$J = \frac{1}{cu}\left[\frac{c+u}{c-u}\,\omega - \frac{3}{2}u^2 + \frac{u^3}{c-u}\right]$$

$$J_g = \frac{1}{c_g u}\left[-\frac{c_g - u}{c_g + u}\,\omega_g + \frac{3}{2}u^2 - \frac{u^3}{c_g + u}\right]$$

With the values for $\omega(O)$ and θ_1, the variation of $\omega(a, t)$ is determined from the initial values and rates of change of quantities at the boundary. The leading term in the enthalpy function $G(a, t) = a(t)\,\omega(a, t) + \frac{1}{2}a(t)u^2$ is thus determined. In numerical applications of the theory, the radius $a(t)$ is taken to have its initial value a_o for the times at which the first term $a_o\omega(a, t)$ is important. At later times, the kinetic term $\frac{1}{2}a(t)u^2$ becomes increasingly important and must be considered.

C. *Motion of the gas sphere.* In order to investigate the expansion of the gas sphere boundary, Eq. (3.24) may be written

$$a\frac{du}{dt} + \frac{\bar{c}}{\bar{c} - 2u}\left(\frac{3}{2}u^2\right) = \frac{c^2 - u\bar{c} + u^2}{c(\bar{c} - 2u)}\,\frac{a}{\rho c}\,\frac{dP}{dt} + \frac{\bar{c}}{\bar{c} - 2u}\,\omega + \frac{2u^3}{\bar{c} - 2u}$$

The particle velocity u is much smaller than the velocity $\bar{c}$, and if u is neglected in comparison with $\bar{c}$, the equation becomes approximately

$$a \frac{du}{dt} + \frac{3}{2} u^2 = \frac{a}{\rho c} \frac{dP}{dt} + \omega$$

The peak approximation is appropriate for P and ω, which gives

$$a^{-1/2} \frac{d}{dt} (a^{3/2} u) = \omega(O) \left(1 - \frac{a_o}{c(O)\theta_1}\right) e^{-t/\theta_1}$$

If the factor $a^{-1/2}$ is approximated by $a_o^{-1/2}$ this equation is directly integrable with the result (noting that $u = da/dt$) that

$$(3.27) \qquad a^{3/2} u = a_o^{3/2} u(O) + a_o^{1/2} \omega(O) \theta_1 \left(1 - \frac{a_o}{c(O)\theta_1}\right)(1 - e^{-t/\theta_1})$$

A second quadrature gives

$$(3.28) \qquad a^{5/2} = a_o^{5/2} \left(1 - \frac{5u(O)\theta_1\alpha}{2a_o}\right) \left(1 + \frac{t}{\theta_2} + \frac{\dfrac{5u(O)\theta_1\alpha}{2a_o}}{1 - \dfrac{5u(O)\theta_1\alpha}{2a_o}} e^{-t/\theta_1}\right)$$

where

$$\alpha = \frac{\omega(O)\theta_1}{a_o u(O)} \left(1 - \frac{a_o}{c(O)\theta_1}\right)$$

$$\theta_2 = \frac{2a_o}{5u(O)} \cdot \frac{1 - \dfrac{5u(O)\theta_1\alpha}{2a_o}}{1 + \alpha}$$

The kinetic term $au^2/2$ then is given by combining Eqs. (3.27) and (3.28) with the result

$$(3.29) \qquad \frac{au^2}{2} = \frac{1}{2} \cdot \frac{(a^{3/2}u)^2}{(a^{5/2})^{4/5}}$$

$$= \frac{a_o u^2(O)}{2} \cdot \frac{((1 + \alpha) - \alpha e^{-t/\theta_1})^2}{\left(1 - \dfrac{5u(O)\theta_1\alpha}{2a_o}\right)^{4/5} \left(1 + \dfrac{t}{\theta_2} + \dfrac{\dfrac{5u(O)\theta_1\alpha}{2a_o}}{1 - \dfrac{5u(O)\theta_1\alpha}{2a_o}} e^{-t/\theta_1}\right)^{4/5}}$$

The rather complicated expression (3.29) for $au^2/2$ is not strictly consistent with the order of the peak approximation for P and ω used in its development, namely that these variables depend on time as

e^{-t/θ_1}, and terms of order $e^{-\alpha t/\theta_1}$ and te^{-t/θ_1} should properly be dropped. A further consideration is the fact that the value of ω used in the equation should be taken as the excess over the equilibrium value for hydrostatic pressure P_o in the undisturbed fluid. In the initial phases of the motion the pressures are so large that neglect of P_o is unimportant, but an investigation of the later motion of the gas boundary when the pressure is small must include the effect of hydrostatic pressure. As discussed in Chapter 8 by simpler methods adequate for the later motion, the hydrostatic term leads to long period radial pulsations of the gas products.

D. *The enthalpy function on the gas sphere.* The final step in formulating conditions at the gas sphere for use in the Kirkwood-Bethe theory is to construct a suitable approximation to the function $G(a, t) = a[\omega + \frac{1}{2}u^2]$ from the results of parts (b) and (c). To do this, Kirkwood and Bethe note that the kinetic term $au^2/2$ as given by Eq. (3.29) has for large t the asymptotic form

$$\frac{au^2}{2} \simeq \frac{(1+\alpha)^2}{\left(1 - \dfrac{5u(O)\theta_1\alpha}{2a_o}\right)^{4/5}} \left(1 + \frac{t}{\theta_2}\right)^{-4/5}$$

and for $t = O$ is of course $a_o u^2(O)/2$. In the initial phases of the motion the enthalpy term is dominant, and an approximate expression for $G(a, t)$ which reduces properly in the limits of short and long times is

$$G(a, t) = a_o\left[\omega_o + \left(1 - \frac{(1+\alpha)^2}{\left(1 - \dfrac{5u(O)\theta_1\alpha}{2a_o}\right)^{4/5}}\right)\frac{u^2(O)}{2}\right]e^{-t/\theta_1} + \frac{a_o u^2(O)}{2}\frac{(1+\alpha)^2}{\left(1 - \dfrac{5u(O)\theta_1\alpha}{2a_o}\right)^{4/5}}\left(1 + \frac{t}{\theta_2}\right)^{-4/5} \tag{3.30}$$

where θ_1, θ_2, and α have the values previously given.

The constant factor multiplying e^{-t/θ_1} in Eq. (3.30) is much larger than the factor multiplying $(1 + t/\theta_2)^{-4/5}$ and the two characteristic times θ_1 and θ_2 are comparable in value. As a result, an initially rapid reduction of enthalpy, and hence pressure, at the gas sphere is replaced at later times by a much more gradual decay, varying asymptotically at long times as $(t/\theta_2)^{-4/5}$. This later variation of pressure is essentially an effect characteristic of noncompressive flow, by which the motion is approximated for the small, slowly changing pressures at later times (see section 9.2). This asymptotic result is of somewhat dubious quantitative value because of the approximations involved. Experimental shock wave pressure-time curves do, however, show the initially rapid and later more gradual decay indicated by the analysis.

The major importance of the shock wave for many purposes is adequately described by the initial peak pressure and the rate of decay while the pressure is still an appreciable percentage of this value. If the contribution to the tail of the wave is neglected, the complete peak approximation of the Kirkwood-Bethe theory is obtained:

$$G(a, t) = a_o(\omega_o + \tfrac{1}{2}u^2(O))e^{-t/\theta_1} \tag{3.31}$$

This result adequately describes the prediction of the theory for the major effect on the shock wave and is the basis of numerical calculations based on the theory.

4. *Theory of the Shock Wave*

4.1. The Principle of Similarity

Before discussing detailed solutions of the hydrodynamical equations for shock wave propagation, it is worth-while to consider what can be inferred more generally about the solution from the form of the equations and the boundary conditions. As shown in Chapter 2, the basic equations describing the motion of a fluid are

$$\frac{d\mathbf{v}}{dt} = -\frac{c^2}{\rho}\,\mathrm{grad}\,\rho, \qquad \frac{d\rho}{dt} = -\rho\,\mathrm{div}\,\mathbf{v}, \qquad \text{where } c^2 = \left(\frac{dP}{d\rho}\right)_s \tag{4.1}$$

if viscosity and heat conduction can be neglected. The pressure P is a function only of the density ρ, the relation between the two being determined by the equation of state and heat capacity of the fluid. These equations have solutions of the form $P(\mathbf{r}, t)$, $u(\mathbf{r}, t)$, where $\mathbf{r}$ is the vector distance from an arbitrary origin.

Suppose that measurements of pressure have been made at a distance $\mathbf{r}$ from a charge of specified dimensions at a time t after it is initiated and that a new experiment is arranged in which all the linear dimensions of the charge are changed by a factor λ. The principle of similarity[1] asserts that the pressure and other properties of a shock wave will be unchanged if the scales of length and time by which it is measured are changed by the same factor λ as the dimensions of the charge. For example, the pressure and duration of the shock wave measured ten feet from a cubical charge one foot on an edge will be the same as the pressure and duration measured twenty feet from a charge two feet on an edge in units of time twice as large. The duration in absolute units is therefore doubled at the doubled distance for the charge of twice the linear dimensions (eight times larger weight). The principle does not state, nor is it true, that the pressure at any other distance and time than the scaled ones obeys any such scaling law regardless of other factors.

In order to examine the validity of this proposition, we observe first that the differential equations (4.1) are satisfied if the scales of measurement of both length and time are changed by a factor λ. For example, the first of Eqs. (4.1) becomes, on writing $\mathbf{r}' = \lambda\mathbf{r}$, $t' = \lambda t$,

[1] The earliest general statement of the principle of similarity for shock waves known to the writer is given in the classic paper of Hilliar (47).

$$\frac{d}{dt'}\mathbf{v}(\mathbf{r}', t') = -\frac{c^2(\mathbf{r}', t')}{\rho(\mathbf{r}', t')}\text{grad}' P(\mathbf{r}', t')$$

where grad′ indicates differentiation in the primed coordinates. We note that all the indicated derivatives are of the same order and we can therefore write

$$\frac{d}{dt}\mathbf{v}(\lambda\mathbf{r}, \lambda t) = -\frac{c^2}{\rho}\text{grad } P(\lambda\mathbf{r}, \lambda t)$$

the scale factor λ cancelling out. Hence the same differential equation is satisfied by $\mathbf{v}(\lambda\mathbf{r}, \lambda t)$ etc. as by $\mathbf{v}(\mathbf{r}, t)$. A similar result is obtained for the second of Eqs. (4.1) and we conclude that the equations are satisfied by values of P, ρ, $\mathbf{v}$ measured in scaled coordinates. At the shock front, Eqs. (4.1) are not valid, but the Rankine-Hugoniot equations (Eqs. (2.28) of Chapter 2) are easily seen to be satisfied by the same change of scale. We can therefore conclude that if the principle of similarity is true at any value $(\mathbf{r}_1, t_1)$, e.g., $P(\mathbf{r}_1, t_1) = P(\lambda\mathbf{r}_1, \lambda t_1)$, it is true for all values of $\mathbf{r}$ and t. The validity of the principle therefore depends on whether similarity is found to hold true in the initial stages of the explosion.

As was shown in Chapter 3, the process of detonation, once established, leads to a shock front advancing outward with constant speed and intensity. If the hydrodynamical equations (4.1) describe the situation behind the front then, by the argument just presented for the shock wave, the profile of the detonation wave is spread out in proportion to the amount the wave has advanced, but has the same form except for this change in scale. If the time required to establish the steady condition is negligible, the profile of the wave is the same for all geometrically similar charges, provided the scales of length and time used to specify it are proportional to the linear dimensions of the charge and the origins of time and distance are at the point of initiation. This is just the necessary condition for similarity to be established in the water shock wave and the remaining question is as to whether the boundary conditions at the interface of the explosion products and the water are compatible with similarity. In the absence of viscosity effects (shear), these conditions require continuity of pressure and normal components of particle velocity and hence of their total time derivatives, as described in section 3.8. We can easily convince ourselves that these conditions are satisfied if pressures and particle velocities are scaled geometrically, and the approximate relations so far developed to account for the shock wave are thus all consistent with the principle.

Deferring for the moment an examination of circumstances in which the principle of similarity fails, we consider what can be inferred with its aid about the form of the shock wave. The fact that the pressure and other properties are unchanged if the linear dimensions of the source and scales of length and time are all changed in the same ratio does not of course specify what the values are without other information. It is possible, however, to learn something about their functional dependence on charge size and distance. If the linear dimensions of the charge are specified in terms of a length, a_o, the principle can be satisfied only if the pressure depends on distance and time only as a function of the ratios r/a_o, t/a_o. The truth of this statement is evident from the fact that fixed values of these ratios correspond to the scaling which gives identical values of the pressure. The pressure P_m at the head of the shock wave (peak pressure) may therefore be expressed

$$P_m = f\left(\frac{a_o}{r}\right)$$

the form of the function f being undetermined. If the quantity θ is used to represent any measure of time duration of the wave, e.g., the time constant of an exponential decay, it is evident that θ/a_o can be a function only of the ratio a_o/r. Another important property of such a wave is the impulse associated with it, which measures the momentum imparted to the water by its passage. For unit area of the wave front the impulse I is given by

$$I(r, t') = \int_o^{t'} P(r, t)\, dt$$

where the origin of time is taken to be arrival of the shock front at r. The time t' to which the integration is carried should, for consistency, be taken proportional to the scale factor and we write $t' = Ka_o$, where K is a function only of a_o/r, and the pressure P depends on r and t only by the ratios a_o/r, t/a_o.

We may therefore write

$$I(r, t') = a_o \int_o^{K\left(\frac{a_o}{r}\right)} P\left(\frac{a_o}{r}, \frac{t}{a_o}\right) d\left(\frac{t}{a_o}\right)$$

and, the integral being a function only of a_o/r, we obtain

$$I(r, t) = a_o\, g\left(\frac{a_o}{r}, K\left(\frac{a_o}{r}\right)\right)$$

where g is an undetermined function. The impulse measured for proportional distance and time scales is therefore proportional to the linear dimension a_o.

The practical importance of the principle of similarity lies in the economy of effort it permits in determining the properties of shock waves and in the predictions it makes possible of the effect of a changed scale. For example, the numerical calculations of Penney (83, 85) for a spherical charge of TNT were explicitly made for a weight of 1,800 pounds of explosive. By a suitable change of scale, however, they also describe a charge of any other weight. These calculations were extended to a distance of 6 charge radii or 9.8 feet, and they therefore apply equally out to 6 charge radii for any other size charge, which for 100 pound weight is 3.8 feet. The more general theory of Kirkwood and Bethe, being based on the equations from which we deduced the validity of the principle, must and does automatically satisfy similarity and therefore predicts shock wave pressures for any size of spherical charge. Experimentally, the principle is economical because the form of the pressure and other functions can be determined by measurements over a range of either distance or charge size only, the effect of the other variable being determined by similarity. Once the form of the function is so determined, its value for other weights or distances is known, and not an independent result.

More detailed illustrations of the principle of similarity will be given later, but its importance and utility can be appreciated from the foregoing discussion. It is therefore important to examine its limitations and the conditions under which it is applicable. The principle will evidently fail in any circumstances for which forces not scaling geometrically are involved. One such force neglected in the equation of motion is the effect of viscosity which gives rise to terms of the form $\mu\partial^2u/\partial r^2$, where μ is the coefficient of viscosity. The derivative is, however, of the second order, and with its inclusion the equation of motion is not satisfied on substituting $P(\lambda\mathbf{r}, \lambda t)$, etc. At a shock front, processes at least similar in effect to macroscopic viscosity and thermal conduction must be significant and hence it is not to be expected that the principle of similarity applies to the rapid changes in the front itself. These changes do not appear to be of any practical significance and hence the applicability of the principle to this case is academic.

Another way in which similarity may be expected to fail is in the case that chemical reactions behind the detonation front are important. The rate of reaction does not of course change in proportion to the amount the front has advanced, and a significant effect of delayed reaction will make similarity invalid. No attempt is made in theoretical calculations to account for reaction rates, the chemical composition being assumed either in equilibrium or else "frozen" into a fixed relation.

There are some experimental departures from similarity which may be the result of delayed reaction or "afterburning," but for the most part this effect is probably insignificant.

The most important phenomenon in which the principle of similarity as here stated is not valid is in the later history of the explosion after emission of the shock wave. The laws of motion of the water and the gas sphere, as developed for calculation of the shock wave, neglect any external forces or boundaries. This is a perfectly legitimate procedure in view of the large internal forces acting in the times of significance, but is no longer so when the internal forces and accelerations have become weakened and mass flow of the water has established itself over large regions. The more important nonscaling factor under these conditions is gravity, because it is always present, but under many conditions boundaries, such as the surface and bottom, have an important effect on the mass flow and motion of the gas sphere.

The sphere of gaseous products expands very rapidly after its formation and soon occupies a volume much greater than that of an equal mass of water. As a result of this hydrostatic buoyancy the bubble can be thought of acquiring vertical upward momentum in addition to momentum associated with radial motion of the water. The energy associated with the upward rise must come at the expense of other forms of kinetic and potential energy and is more important for large charges in which the time scale of the motion is longer. If the vertical displacement of the bubble could be neglected and noncompressive flow assumed, the motion of the gas sphere would scale according to the principle of similarity, as shown in section 8.2. These assumptions are, however, not legitimate except as a crude first approximation, and inclusion of gravity effects makes the principle of similarity, as here stated, inapplicable. Other types of scaling law which permit a nondimensional approximation to the laws of motion for the gas sphere have been worked out by G. I. Taylor and will be described in section 8.5 (see also section 8.9).

4.2. The Detailed Evaluation of Shock Wave Propagation (Kirkwood-Bethe Theory)

In Chapter 2 it was shown that the propagation of a spherical shock wave could be described in terms of the kinetic enthalpy $\Omega(r, t)$ defined by

$$\Omega(r, t) = \frac{P}{\rho} + \frac{u^2}{2} = \omega + \frac{u^2}{2} \tag{4.2}$$

the contributions to Ω from entropy changes behind the shock front being neglected. The utility of this approach lies in the fact that Ω can be simply related to its value $\Omega[a(\tau), \tau]$ on the gas sphere boundary at an earlier retarded time τ given by

$$\tau = t - \int_{a(\tau)}^{r} \frac{dr}{\bar{c}(r, \tau)} \tag{4.3}$$

where $a(\tau)$ is the radius of the gas sphere, and $\bar{c}(r, \tau)$ is the velocity with which the function $r\Omega$ is propagated. This relation was shown in section 2.4 to be

$$\Omega(r, t) = \frac{a(\tau)}{r} \Omega(a, \tau)$$

As an illustration of how the pressure-time curve is related to these quantities, consider the asymptotic case of pressures at a point R sufficiently far from the charge that $u^2 \ll P/\rho$. In such cases $P = \rho_o \Omega(R, t)$, where ρ_o is the density of undisturbed water, and from Eq. (4.2) is given by

$$P(R, t) = \left(\frac{a_o}{R}\right) P(\tau) \tag{4.4}$$

where a_o is the initial radius of the gas sphere, and

$$P(\tau) = \rho_o \frac{a(\tau)}{a_o} \Omega(a, \tau)$$

If the particle velocity term is not negligible, the pressure must be obtained from the kinetic enthalpy using the equation of state for water. In either case, it is evidently necessary to evaluate the enthalpy $\Omega(a, t)$ at the gas sphere, determined in the manner described in section 3.8, at the retarded time τ. In order to do this it is necessary to make a number of approximations, the first of which is to represent the propagation velocity $\bar{c}$ of Ω by the velocity $c + \sigma$ and represent Ω by $(\omega + \sigma^2/2)$. It might seem more logical to use the particle velocity u rather than the Riemann function σ, but as shown in the Appendix the latter represents a better approximation for the difference $\sigma - u$ in the water behind the shock front. Furthermore, as shown in section 2.6, σ is also a convenient variable for expression of the properties behind the front. The question of importance with regard to this approximation is, of course, what error results from its use in evaluating $\Omega(a, \tau)$, and the fact that the error proves on investigation to be small is the justification for the approximation. The detailed analysis, described in the Appendix, is based on consideration of the error in terms of $(\partial/\partial r\ (r\Omega))_\tau$, which is not zero for the approximate τ, and can be written as a function of known or small quantities behind the shock front. Numerical estimates by Kirkwood of the error in $r\Omega$ lead to the conclusion that the approximate value will give too large values of pressure, the error not exceeding twenty per cent and in most cases probably much less.

A further approximation in the evaluation of $\Omega(a, \tau)$ lies in the assumption that its value for points not far behind the shock front may be obtained sufficiently accurately from a Taylor's series expansion about the value τ_o corresponding to the shock front. If the shock front has reached a point R at time t_o, we have

$$\tau = \tau_o + \left(\frac{\partial \tau}{\partial t}\right)_R (t - t_o) = \tau_o + \frac{1}{\gamma}(t - t_o)$$

where $1/\gamma = (\partial \tau/\partial t)_R$, and from Eq. (4.3)

$$\tau_o = \int_{a_o}^{R} \frac{dr}{U(r, \tau)} - \int_{a(\tau)}^{R} \frac{dr}{c + \sigma} \tag{4.5}$$

Differentiating Eq. (4.3), we have for γ

$$\gamma = \left(\frac{\partial t}{\partial \tau}\right)_R = 1 - \frac{1}{c + \sigma}\frac{\partial a}{\partial \tau} - \int_{a(\tau)}^{R} \frac{1}{(c + \sigma)^2}\left[\frac{\partial}{\partial \tau}(c + \sigma)\right] dr \tag{4.6}$$

The parameter γ, which is a measure of the time scale behind the shock front relative to that on the gas sphere, increases rapidly as the shock front travels outward. This is because $[\partial/\partial\tau\ (c + \sigma)]_r$ is negative at all times following passage of the shock wave if $\Omega(a, \tau)$ decreases with time during emission of the shock wave from the gas sphere boundary. This increase of γ with distance from the source means that the profile of the shock wave broadens as it is propagated outward. The increasing breadth made plausible by the Riemann formulation of the propagation equations thus appears naturally in this approach also. The limiting pressure-time relation, Eq. (4.4), may be written

$$P(R, t - t_o) = \left(\frac{a_o}{R}\right) P\left[\tau_o + \frac{1}{\gamma}(t - t_o)\right]$$

where

$$P(t) = \rho_o \frac{a(t)}{a(o)} \Omega(a, t)$$

With these approximations for $\bar{c}$ and the retarded time τ, it becomes practical to compute $\Omega(a, \tau)$ and γ from the equation of state for water and numerical values of the enthalpy on the gas sphere. The actual reduction of the expressions for τ_o and γ to a convenient form for calculation is somewhat involved and some of the details will be omitted in this presentation of the argument. As was shown in section 2.6, the various quantities associated with the propagation are conveniently written in terms of the Riemann function σ, defined as

$$\sigma = \int_{\rho_o}^{\rho} \frac{c}{\rho} d\rho, \text{ where } c = \sqrt{\frac{dP}{d\rho}}$$

In the approach to be outlined, we are interested in regions at or near the shock front. In this case, σ is not greatly different from the particle velocity u, and we may use the relations obtained in section 2.6 from the equation of state and the Rankine-Hugoniot conditions, which are

$$\begin{aligned} \bar{c} &= c + \sigma = c_o\,[1 + 2\beta\sigma] \\ u &= c_o\,[1 + \beta\sigma], \text{ where } \beta = \frac{n+1}{4c_o} \qquad (4.7) \\ \Omega &= c_o\sigma\,[1 + \beta\sigma] \end{aligned}$$

It will be noted that the propagation velocity $\bar{c}$ is approximated by $c + \sigma$ rather than $c + u$ and it is therefore not assumed that $\sigma = u$. In terms of σ, the basic propagation equation for $G(r, t) = r\Omega(r, t)$ becomes

$$G(r, t) = c_o r\sigma(1 + \beta\sigma) = a(\tau)\Omega(a, \tau) = G(a, \tau) \qquad (4.8)$$

A. *Evaluation of* $\Omega[a, \tau_o]$. From Eq. (4.5), the retarded time τ_o at the shock front is expressed as the sum of two integrals, the first integration being performed at the shock front, propagated from the initial radius $a(o)$ of the gas sphere to the point R, and the second being carried out at constant τ_o. These integrals can be expressed in terms of σ and $G[a, \tau]$ by differentiating Eq. (4.8) which gives

$$dr = -\frac{G(a, \tau)}{c_o^2}\cdot\frac{1 + 2\beta\sigma}{\sigma^2\,[1 + \beta\sigma]^2} d\sigma + \frac{1}{c_o}\cdot\frac{dG(a, \tau)}{\sigma\,(1 + \beta\sigma)}$$

Substituting this value of dr and noting that $dG[a, \tau] = 0$ for the second integration, we find after some rearrangement

$$\tau_o = \frac{\beta}{c_o^2}\left[\int_{\sigma}^{\sigma(a_o)} \frac{2G(a, \tau_o) - G(a, \tau(\sigma))}{\sigma[1 + \beta\sigma]^3} d\sigma + \frac{G(a, \tau_o) - G(a_o)}{\beta\sigma(a_o)\,[1 + \beta\sigma(a_o)]^2} + \frac{G(a, \tau_o)}{\beta}\int_{\sigma(a)}^{\sigma(a_o)} \frac{d\sigma}{\sigma^2(1 + \beta\sigma)^2}\right] \qquad (4.9)$$

In order to obtain $G(a, \tau_o)$ (which equals $a(\tau_o)\ \Omega(a, \tau_o)$) from this equation, it is convenient to define a new parameter $x(\tau_o)$ by the relation

$$x(\tau_o) = \frac{G(a, \tau_o)}{G(a_o)}$$

where $G(a_o)$ is the initial value at time $t = 0$ on the gas sphere.

We note from Eq. (4.8) that the quantity $\beta\sigma$ can be expressed in terms of (xa_o/R) by the relation

$$\beta\sigma = \frac{1}{2}\left[\sqrt{1 + \frac{4\beta\Omega(a_o)}{c_o}\left(\frac{xa_o}{R}\right)} - 1\right] \tag{4.10}$$

and hence solution for x in terms of (xa_o/R) determines x as a function of a_o/R. Substitution of x in Eq. (4.9) and a fairly tedious reduction leads to the equation[2]

$$K_o x^2 + x - K_1 = 0 \tag{4.11}$$

The functions K_o and K_1 are expressed for compactness in terms of an auxiliary variable Z defined by

$$Z(\beta\sigma) = \frac{1 + \beta\sigma}{\beta\sigma}$$

and this function with a subscript a refers to values on the gas sphere, so that

$$Z_a(\beta\sigma) = \frac{1 + \beta\sigma(a, \tau_o)}{\beta\sigma(a, \tau_o)}, \quad Z_a(O) = \frac{1 + \beta\sigma(a_o)}{\beta\sigma(a_o)} \tag{4.12}$$

It is obvious from Eq. (4.10) that Z and Z_a are functions of xa_o/R and $xa_o/a(\tau_o)$. With this notation, the functions K_o and K_1 can be shown to be

$$K_o = \frac{a_o}{c_o}\cdot\frac{G(a_o) - G(a, \tau_o)}{\int_o^{\tau_o} G(a, t)\, dt}\left\{\frac{Z_a(O) - 1}{2Z_a(O)} + \frac{\Omega(a_o)}{c_o}\int_{\sigma(a)}^{\sigma(a_o)} \frac{d\sigma}{(1 + \beta\sigma)^2}\right. \tag{4.13}$$
$$\left. + \frac{\beta\Omega(a_o)}{c_o}\left[\log\frac{Z}{Z_a(O)} + 2\left(\frac{1}{Z} - \frac{1}{Z_a(O)}\right) - \frac{1}{2}\left(\frac{1}{Z^2} - \frac{1}{Z_a^2(O)}\right)\right]\right\}$$

$$K_1 = 1 + \frac{a_o}{c_o}\cdot\frac{G(a_o) - G(a, \tau_o)}{\int_o^{\tau_o} G(a, t)\, dt}\left\{\frac{Z_a(O) - 1}{2Z_a(O)} + \frac{c_o}{2\Omega(a_o)}\int_{\sigma(a)}^{\sigma(a_o)} \frac{a^2(\tau)}{a_o^2}\, d\sigma_a\right\}$$

[2] For the derivation of this relation and the defining equations (4.13), the original report of Kirkwood and Bethe (59, I) should be consulted.

It might appear that the manipulations and reductions involved in expressing $G(a, \tau_o)$ as a function of $G(a_o)$ and x have not accomplished much toward a practical solution, as K_o and K_1 both involve $G(a, \tau_o)$, and the integrals in K_o and K_1 depend on τ_o through the value of σa on the gas sphere. Equation (4.11) is, as a result, strictly a transcendental equation for $x(\tau_o)$ which would have to be solved by approximation or other means. It turns out, however, that the integrals involving σa are unimportant compared to the other terms. Furthermore, it was shown in section 3.8 that the variation of the function $G(a, t)$ on the gas sphere can be approximated by an equation of the form

$$G(a, t) = G(a_o)e^{-t/\theta_1}$$

which is accurate for $t = 0$, but is increasingly in error for larger values of t. With this approximation, Eq. (4.11) may be solved explicitly for x as a function of xa_o/R, and hence a_o/R, given values of $\Omega(a_o)$ and θ_1. We shall have to expect, however, that the later portions of the shock wave will be somewhat in error from the peak approximation.

It is also to be remembered that the boundary conditions at the gas sphere, which led to the "peak" approximation, were also somewhat simplified by neglecting reflections at the boundary of the internal rarefaction wave. These reflections must also introduce perturbations of the later portions of the shock wave, and so too detailed an analysis neglecting this effect is not justified. The form of the correction for error in the peak approximation is, however, described in section 3.8. With the approximation and neglecting the integrals over σ on the gas sphere, Eq. (4.11) becomes

$$x = \frac{2K_1}{1 + (1 + 4K_oK_1)^{1/2}} \tag{4.14}$$

where K_o and K_1 are defined by Eqs. (4.13). If the integrals over σ are neglected, K_o and K_1 are functions of Z and $Z_a(O)$ only, which are expressed in terms of xa_o/R by the relations

$$Z = 1 + \frac{1}{2\Omega(a_o)}\left(\frac{R}{a_ox}\right) + \sqrt{\frac{1}{\Omega(a_o)}\cdot\left(\frac{R}{a_ox}\right) + \left[\frac{1}{2\Omega(a_o)}\cdot\left(\frac{R}{a_ox}\right)\right]^2} \tag{4.15}$$

$$Z_a(O) = 1 + \frac{1}{2\Omega(a_o)} + \sqrt{\frac{1}{\Omega(a_o)} + \left(\frac{1}{2\Omega(a_o)}\right)^2}$$

B. *Evaluation of* γ. In order to evaluate the time spread parameter γ, much the same procedure is used as in determining x. From Eqs.

(4.7) and (4.8), the particle velocity u and velocity $(c + \sigma)$ can be expressed in terms of σ and $G(a, \tau)$, and substitution in Eq. (4.6) gives

$$\gamma = 1 - \frac{\sigma_a}{c_o(1 + 2\beta\sigma_a)} - \frac{2\beta}{c_o^2}\cdot\frac{dG(a, \tau_o)}{d\tau_o}\int_{\sigma}^{\sigma_a}\frac{d\sigma}{\sigma(1 + \beta\sigma)\,(1 + 2\beta\sigma)^2}$$

Making the substitution $Z = (1 + \beta\sigma)/\beta\sigma$ and carrying out the integration yields

(4.16)

$$\gamma = 1 - \frac{1}{\beta c_o(Z_a + 1)} - \frac{2\beta}{c_o^2}\frac{dG(a, \tau_o)}{d\tau_o}\left\{\log\frac{Z}{Z_a} - \frac{4(Z - Z_a)}{(Z + 1)\,(Z_a + 1)}\right\}$$

With the peak approximation $G(a, t) = G(a_o)e^{-t/\theta_1}$ this becomes

$$\gamma = 1 - \frac{1}{\beta c_o(Z_a + 1)} + \frac{2a_o}{c_o\theta_1}\frac{\beta\Omega(a_o)}{c_o}\left\{\log\frac{Z}{Z_a} - \frac{4(Z - Z_a)}{(Z + 1)\,(Z_a + 1)}\right\}$$

As before, Z is a function of xa_o/R given by Eq. (4.15). The quantity Z_a is strictly a function of $xa_o/a(\tau_o)$, but it usually suffices to neglect the change in radius $a(\tau_o)$ for increasing τ_o and evaluate Z_a by the relation

$$Z_a = 1 + \frac{1}{2\Omega(a_o)x} + \sqrt{\frac{1}{\Omega(a_o)x} + \left[\frac{1}{2\Omega(a_o)x}\right]^2}$$

obtained by letting $R = a_o$ in Eq. (4.15). With this approximation, γ is determined as a function of xa_o/R and x for given values of $\Omega(a_o)$ and θ_1.

C. *Calculation of the pressure-time curve.* From the results of the preceding paragraphs, it is readily possible to compute the shock wave pressure-time curve. Given the values $\Omega(a_o)$ and θ_1, an assumed value of xa_o/R may be used to calculate the corresponding value of x. This value of x determines the point a_o/R which has the value and, knowing x and (xa_o/R), the corresponding value of γ may be computed. From the basic equation of the propagation theory, the kinetic enthalpy at a distance R as a function of time $t - t_o$ after arrival of the shock wave at time t_o is given by

$$\Omega(R, t - t_o) = \frac{G(a, \tau - \tau_o)}{R}$$

With the peak approximation $G(a, t) = G(a_o)e^{-t/\theta_1}$, this becomes

$$\Omega(R, t - t_o) = \frac{G(a, \tau_o)}{R} e^{-\frac{\tau - \tau_o}{\theta_1}}$$

Remembering the definitions

$$\tau - \tau_o = \frac{t - t_o}{\gamma}$$

$$G(a, \tau_o) = xG(a_o) = xa_o\Omega(a_o)$$

we obtain

$$\Omega(r, t - t_o) = x\left(\frac{a_o}{R}\right)e^{-\frac{t - t_o}{\gamma\theta_1}} \tag{4.17}$$

At the surface of the charge ($R/a_o = 1$), the pressure and particle velocity are known directly from the boundary conditions; this analysis in the approximation of Kirkwood and Bethe is described in section 3.8. At later times the pressure and other variables are determined from the kinetic enthalpy Ω by the equation of state and shock front conditions. The necessary relations, obtained numerically by Kirkwood and others, are discussed in section 2.6 and reproduced in part in Tables 2.1 and 2.2. These relations become increasingly uncertain near the surface of the charge, owing to lack of knowledge of the equation of state, but the initial approximations of the boundary value theory are such as to make the theory inaccurate in this region.

At sufficiently large distances from the charge, the particle velocity term $u^2/2$ in the expression $\Omega = P/\rho + u^2/2$ becomes unimportant, and the density ρ can be approximated by ρ_o, the initial value in undisturbed water. With this approximation, sufficiently accurate at distances greater than twenty-five charge radii for ordinary high explosives, the pressure is given by

$$P(r, t - t_o) = P(O)x\left(\frac{a_o}{R}\right)e^{-\frac{t - t_o}{\theta}} \tag{4.18}$$

where $P(O) = \rho(O)\Omega(a_o)$ and $\theta = \gamma\theta_1$

The theory therefore predicts that the pressure-time curve at a point R decays exponentially after the initial peak. The parameters x and θ both depend on (R/a_o), however, as does the velocity of propagation U.

The quantities $\Omega(a_o)$ and θ_1/a_o depend only on the explosive materials, and when their values are assigned, the values of x and γ are determined only by the ratio R/a_o and not by either R or a_o separately.

As a result, the pressure at any point for a given explosive depends only on the ratio (R/a_o) and the time scale at this point is proportional to the charge radius a_o, which is thus the linear scale factor required by the principle of similarity.

The calculation of x and γ as functions of (R/a_o) is straightforward, once the necessary values of $\Omega(a_o)$ and θ_1/a_o have been found from the equation of state, as described in section 3.8. Values of x and γ for three explosives at several values of R/a_o are given in Table 4.1, together with values of peak pressure P_m and reduced time constant θ/a_o.

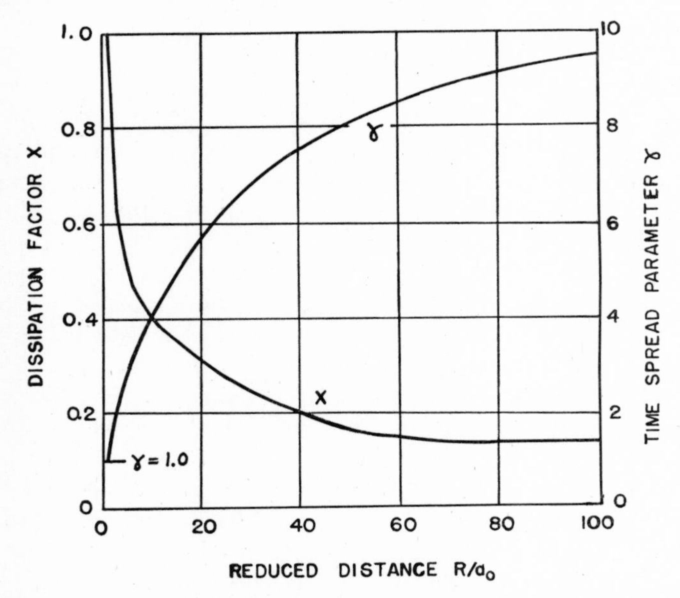

Fig. 4.1 Calculated dissipation and time spread parameters for TNT.

The values of x and γ for TNT plotted in Fig. 4.1 illustrate the variation of these quantities with distance.

It is often convenient to have an approximate functional relationship for these theoretical results in terms of charge weight and distance rather than a set of tabular values. It has been found that for limited ranges of the argument $W^{1/3}/R$ the peak pressure P_m can be approximated by power laws of the form

$$P_m = k\left(\frac{W^{1/3}}{R}\right)^\alpha$$

where W is in pounds and R in feet. This formula is a fairly accurate expression of theoretical results between 10 and 100 charge radii, as will be seen from Fig. 4.2, in which the tabular values of theory for TNT are plotted on double logarithmic scales. The empirically fitted straight line of slope $\alpha = 1.16$ gives a good fit except for $W^{1/3}/R > 0.3$ ($R/a_o < 25$). Similar curves for other explosives can be drawn to the same order of accuracy.

The variation of the reduced time constant $\theta/W^{1/3}$, also plotted in Fig. 4.2 for TNT, is not well represented by a power function. The derived quantities, impulse and energy, equal to $P_m\theta$ and $P_m^2\theta/2\rho c$ in the peak approximation, can be fitted quite well by power laws. A presentation of these laws and their agreement with experiment is given

(a) TNT, density 1.59

R/a_o	1	10	25	50	100
x	1.00	0.399	0.300	0.258	0.231
γ	1.00	4.04	6.43	8.11	9.55
θ/a_o (10^{-5} sec./cm.)	0.344	1.39	2.21	2.79	3.29
P_m (lb./in.^{2})	538,000	20,200	6,100	2,620	1,170

(b) Pentolite (PETN/TNT 50/50), density 1.60

R/a_o	1	10	25	50	100
x	1.00	0.372	0.277	0.237	0.211
γ	1.00	4.37	7.09	8.96	10.6
θ/a_o	0.310	1.37	2.22	2.81	3.33
P_m	606,000	21,200	6,310	2,710	1,205

(c) Tetryl, density 1.00

R/a_o	1	10	25	50	100
x	1.00	0.512	0.410	0.361	0.328
γ	1.00	2.96	4.35	5.29	6.19
θ/a_o	0.490	1.45	2.13	2.59	3.03
P_m	304,000	15,000	4,800	2,110	960

Table 4.1. Shock wave parameters predicted by Kirkwood-Bethe theory (from calculations of Kirkwood, Brinkley, and Richardson).

in section 7.4. It is interesting to note here that the variation of peak pressure with distance at a rate greater than $1/R$, and the spreading of the wave as it is propagated outward, are both confirmed by experiment. Although these departures may not appear large except at points close to the charge, they persist to great distances. As an example, the peak pressure at 100 charge radii is less than 60 per cent of what it would be for acoustic decay, and the time constant is more than twice as great.

The theoretical predictions as given in Table 4.1 represent only a small fraction of the calculations, which have been made for forty-eight different explosive compositions in all, comprising thirteen distinct combinations of components. For complete tabulations of the various results and variables of the theory, reference should be made to the original reports.[3] These reports also include formulas for calculation of the asymptotic variation of peak pressure and time constant at large distances, which are discussed in the next section.

4.3. The Asymptotic Behavior of Spherical Shock Waves

Although the detailed evaluation of the Kirkwood-Bethe propagation theory provides numerical values of spherical shock wave param-

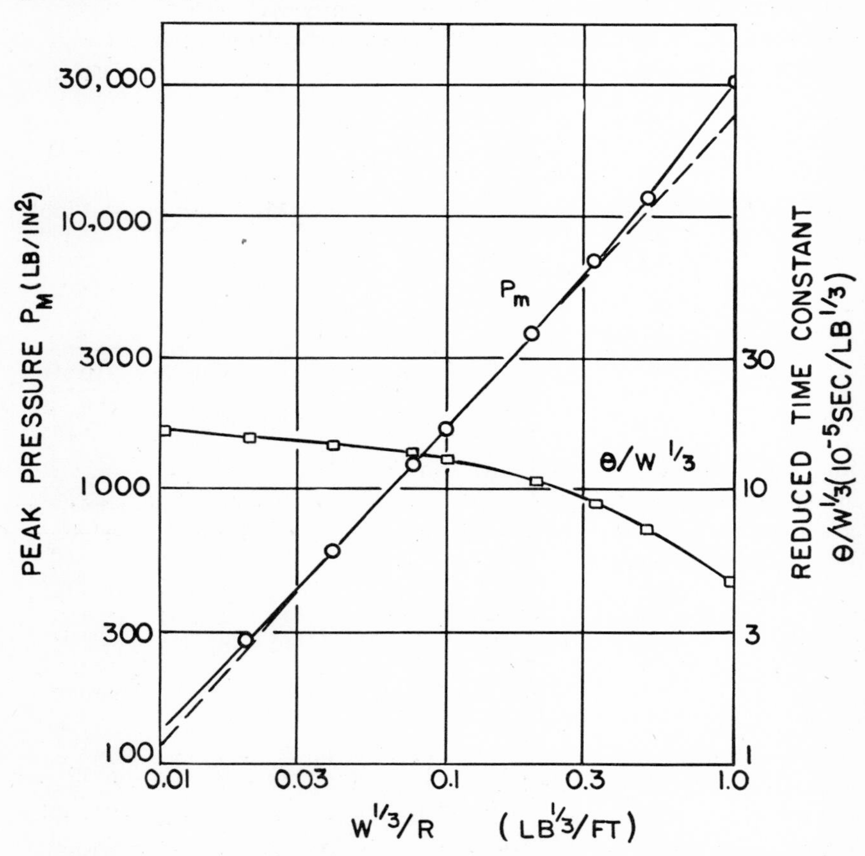

Fig. 4.2 Calculated peak pressure and time constant for TNT (Kirkwood-Bethe theory).

eters for selected values of distance and initial conditions, it does not permit a simple indication of the manner in which these parameters may be expected to depend on distance. Fortunately, it is a relatively simple matter to determine the limiting laws which hold at large dis-

[3] The final compilations are given in two reports by Kirkwood, Brinkley, and Richardson (59, V and OSRD 3949). It is to be noted that other calculations in earlier reports by Kirkwood et al are superseded by these results.

tances.[4] In order to carry out this derivation, we note that the quantity $\beta\sigma$ evaluated at the shock front decreases monotonically toward zero as a limit, as the shock front progresses outward, and the derived quantity Z (which equals $1 + (\beta\sigma)^{-1}$) therefore increases with increasing R/a_o. For large values of R/a_o, the behavior of the dissipation factor x and time spread factor γ will be determined primarily by the terms involving $\log Z/Z_a$ in Eqs. (4.13) and (4.14) for x and Eq. (4.16) for γ, the other terms varying only slowly or becoming insignificant.

From Eq. (4.10), the limiting behavior of Z with increasing R/a_ox is seen to be given by

$$Z = \frac{c_o}{\beta\Omega(a_o)} \cdot \frac{R}{a_ox} \quad \text{as} \quad \frac{R}{a_o} \longrightarrow \infty$$

and $Z_a(O)$ is, from the basic equation of propagation, determined by

$$\frac{Z_a(O)}{[Z_a(O) - 1]^2} = \frac{\beta\Omega(a_o)}{c_o}$$

Hence for large R/a_o we have

$$\log \frac{Z}{Z_a(O)} = \log \left(\frac{R}{a_ox}\right) \left[\frac{Z_a(O)}{Z_a(O) - 1}\right]^2$$

and the dominant term is $\log (R/a_ox)$ as $(R/a_o) \to \infty$. The quantity K_o, in terms of which Eq. (4.14) for x is expressed, becomes increasingly great compared to unity at large distances, and the asymptotic behavior of x is therefore given by

$$x = \sqrt{\frac{K_1}{K_o}} = \sqrt{K_1}\sqrt{\frac{c_o}{a_o} \cdot \frac{c_o}{\beta\Omega(a_o)} \cdot \frac{\int_o^{\tau_o} G(a, t)\, dt}{G(a_o) - G(a, \tau_o)}} \left(\log \frac{R}{a_ox}\right)^{-1/2} \tag{4.19}$$

The quantity K_1 defined by Eq. (4.13), which depends on R/a_o only through $G(a_o) - G(a, \tau_o)$ and the integral over $G(a, t)$,[5] is a slowly varying quantity as τ_o increases. The dissipation factor x itself therefore varies slowly for large R/a_o, the dominant factor being the logarithmic decay. Strictly, of course, Eq. (4.19) is transcendental, but the variation of x at large R/a_o is so gradual that its effect in the logarithmic term can be neglected. As a result, the factor x decreases slowly as $(\log R/a_o)^{-1/2}$ for large values of R/a_o, and even in this limit the peak pressure at the shock front decays more rapidly than in the acoustical

[4] These limiting laws are derived in the original report of Kirkwood and Bethe (59, I), which the discussion given here follows closely.

[5] The function $G(a,t)$ is a monotonically decreasing function, approaching zero for large t, and the integral accordingly varies slowly for sufficiently large t.

approximation (for which x would be independent of (R/a_o)). Neglecting $G(a, \tau_o)$ in comparison to $G(a_o)$, the asymptotic peak pressure variation is, from Eq. (4.18),

$$P_m = \rho_o c_o \sqrt{\frac{K_1}{\beta}} \left(\frac{1}{a_o^2} \int_o^{\tau_o} G(a, t)\, dt \right)^{1/2} \left(\frac{a_o}{R} \right) \left(\log \frac{R}{a_o} \right)^{-1/2} \tag{4.20}$$

and is therefore determined by the time integral of $Ga(t)$ on the gas sphere rather than its initial value.

The asymptotic behavior of the time scale factor γ is similarly found from Eq. (4.16) to be

$$\gamma = 1 - \frac{2\beta}{c_o^2} \frac{dG(a, \tau)}{d\tau} \log \left(\frac{R}{a_o} \right) \tag{4.21}$$

If the peak approximation $G(a, t) = G(a_o)e^{-t/\theta_1}$ is an adequate description we find that the time constant $\theta = \gamma\theta_1$ for the initial part of the pulse ($\tau = \tau_o$) is given by

$$\theta = 2 \frac{a_o}{c_o} \frac{\beta\Omega(a_o)}{c_o} \log \left(\frac{R}{a_o} \right) \tag{4.22}$$

and therefore increases gradually with R/a_o. This result is much more approximate than the expression for peak pressure. In the later portions of the pulse, the derivative $dGa(\tau)/d\tau$ is not accurately determined by the peak approximation and furthermore its value will become increasingly small. As a result, the second term in Eq. (4.21) for γ will become increasingly smaller compared to unity. At sufficiently late portions of the pulse, the broadening of the time scale represented by this term will become unimportant and the later profile of the shock wave is represented by a time scale increasingly similar to that on the gas sphere (except of course for the time required for the wave to be propagated from the gas sphere to the point considered).

4.4. The Shock Wave for Cylindrical Symmetry

The theories of the shock wave so far described have dealt with the case of spherical symmetry, which is the simplest symmetry corresponding to an experimentally realizable situation. In many practical problems it is neither necessary nor desirable to have the explosive charge in the form of a sphere detonated at its center. Any tractable theory for some other shape of charge is therefore desirable, if only to reveal the nature of the resulting differences. The simplest geometry for this purpose is evidently the one-dimensional case of an infinite cylinder detonated simultaneously at all points on its axis. This is a mathematical idealization not realized experimentally because no charge is strictly of infinite length and the detonation condition even for a finite cylinder is not easily realized. The best approximation to the relatively simple theoretical assumption is obtained for cylindrical sticks detonated at one end. In this case the detonation velocity, although not infinite, is sufficiently high that the shock front makes an angle of less than thirty degrees with the surface of the charge. For points not too far from the charge it is reasonable to expect that a theory based on cylindrical rather than axial symmetry would have a rough correspondence to observations under such conditions.

The basic differences between spherical and cylindrical symmetry are in the propagation equations for the water and explosion products, the equations of state and the shock front conditions remaining unchanged. Rice and Ginell[6] have extended the Kirkwood-Bethe propagation theory to the cylindrical case, and as in the application of the Kirkwood-Bethe theory the initial conditions are approximated by the pressure in adiabatic conversion of the explosive to its products at the same volume. This condition is unaffected by the symmetry and the essential differences are then in the disturbances propagated away from the discontinuity.

A basic difficulty in analysis of the propagation lies in the fact that cylindrical waves even in the acoustic approximation necessarily undergo a change of type as they are propagated.[7] There is no exact comparison with propagation as a function $F(t - r/c)$ as in plane or spherical waves, and only asymptotically is it found that pressure, for example, varies as $r^{-1/2} F(t - r/c_o)$ where F is an undetermined function (as compared with $r^{-1} F(t - r/c_o)$, valid at any distance for acoustic spherical waves). The development of a finite amplitude theory based on this dependence will not therefore be as simply related to the actual state of affairs, and errors incurred in approximations suggested by the relation will be larger than for spherical waves.

Rice and Ginell develop a propagation theory for the kinetic enthalpy Ω assuming a propagation function G, level values of which advance outward with a velocity $c + \sigma$, where c is the local sound velocity and σ the Riemann function. The discussion of the preceding paragraphs suggests as a first approximation, analogous to that for spherical symmetry, taking $G = r^{1/2}\Omega$, where $\Omega = E + P/\rho + \frac{1}{2}u^2$. Detailed calculations showed, however, that the assumption failed rather badly near the charge, despite its asymptotic validity. A better approximation,

[6] Rice and Ginell's treatment is presented in two reports (59, VI and VIII), the second being a revision and extension of the first.

[7] For a discussion of the propagation of cylindrical acoustic waves, see for example the book by Morse (75).

which was found to reduce the changes in G for a fixed value of $(t - r/(c + \sigma))$, was obtained by taking $G = r^{\alpha}\Omega$ with α equal to about 0.4 rather than 0.5.

The peak pressure is computed by procedures equivalent to those of Kirkwood and Bethe but differing in detail because of the changed sym-

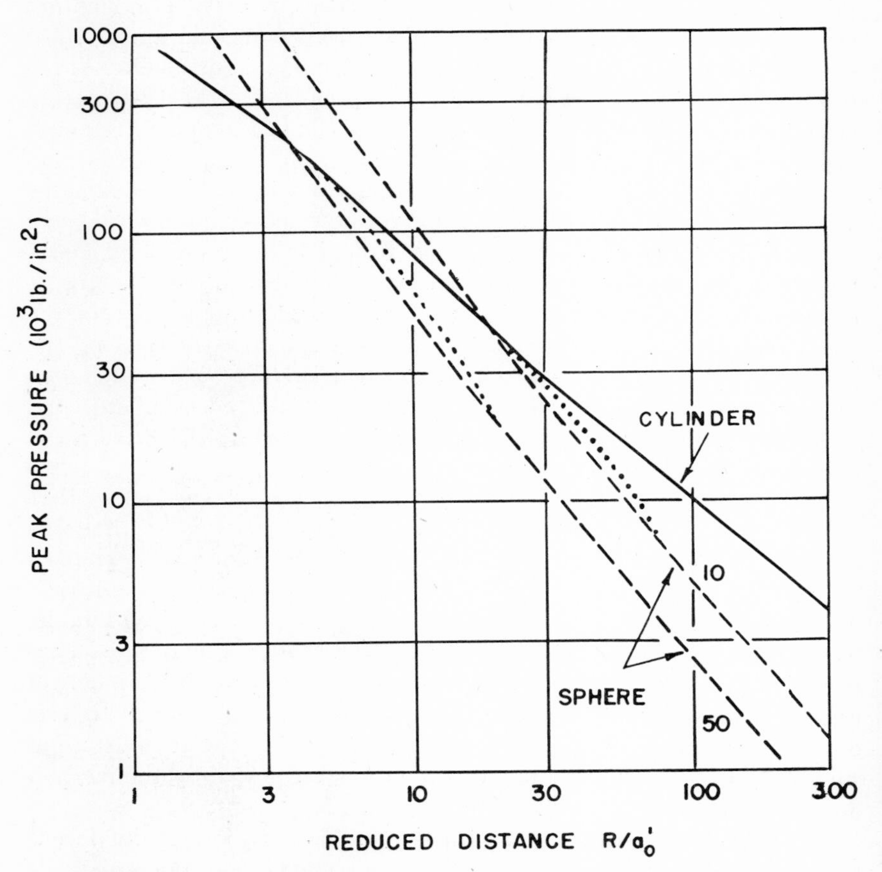

Fig. 4.3 Calculated peak pressure at the side of a cylindrical TNT charge. The dashed lines are for spherical charges and the dotted lines indicate transition to the dashed lines for charges of finite length.

metry. The pressure decays much more slowly with distance, owing to the fact that the energy of a given weight of explosive is concentrated in a cylindrical slice rather than a spherical shell. The initial rate of decay decreases as the shock front advances, and the rate of decrease remains pronounced at larger distances from the charge than in the spherical case. The decay of pressure at points behind the shock front strictly should be computed separately for each point, but this complication is perhaps not justified, as discussed by Rice and Ginell. The details of their calculations are sufficiently similar to those of Kirkwood and Bethe that reproduction of them here would not be worth-while, and the discussion here will be confined to the results obtained.

The calculated peak pressure P_m for cast TNT of density 1.59 is plotted in Fig. 4.3 as a function of R/a_o', where a_o' is the radius of the cylinder and R the distance from the axis. Near the charge the pressure falls off roughly as $(a_o'/R)^{0.6}$, and for greater values of R/a_o' changes in slope of the logarithmically plotted curve occur. These changes at large values of R/a_o' are, however, not likely to be realized in actual experiments because cylindrical charges of length many times the distance R, which itself is much greater than a_o', would be necessary to approximate the assumed symmetry.

As an estimate of peak pressures in the equatorial plane perpendicular to the axis of a charge of finite length, Rice and Ginell suggest that comparison be made with a spherical charge of the same weight. At small distances, the peak pressure should be essentially that for an infinite cylinder, and at larger distances the pressure should approach values for the sphere, the deviations from this symmetry becoming increasingly unimportant. The spherical charge of volume equal to a cylinder of radius a_o' and length L has a radius a_o given by

$$\frac{4}{3}\pi a_o^3 = \pi a_o'^2 L \quad \text{and hence} \quad a_o = (3L/4a_o')^{1/3}a_o'$$

The pressures for spherical charges of the same weight as cylinders with length/radius ratios of 10 and 50 are plotted in Fig. 4.3, as calculated from the Kirkwood-Bethe theory. The transitions from the cylindrical case to the spherical ones should occur in the region $R \approx L/2$, and the dotted lines suggest a reasonable transition. The decay of peak pressure with distance on this basis would change rather smoothly from a decay roughly as $(a_o'/R)^{0.6}$ near the charge to a decay as $(a_o'/R)^{1.15}$ at distances greater than the length of the charge. Similar estimates for the time constant θ as defined by $P(t) = P_m e^{-t/\theta}$ are plotted in Fig. 4.4, and the differences in the changes of θ/a_o with increasing distance are evident from this figure.

The theory of Rice and Ginell affords an at least semiquantitative prediction of the pressure variation off the side of a cylindrical charge, but the equally interesting question of what happens off the ends of such a charge has not been treated. The observed pressure-time curves at points along the axis reveal a number of interesting features, some of which can be understood qualitatively on theoretical grounds, and the analysis of such results is discussed in section 7.6. A comparison with the calculations of Rice and Ginell is furnished by measurements by

MacDougall and Messerly of the shock front velocity near cylindrical sticks of 50/50 pentolite[8] which agree within 3–6 per cent with values predicted from calculated peak pressures over the range 1–6 charge radii. The agreement is not necessarily this good, as the measured velocities may have been somewhat low.

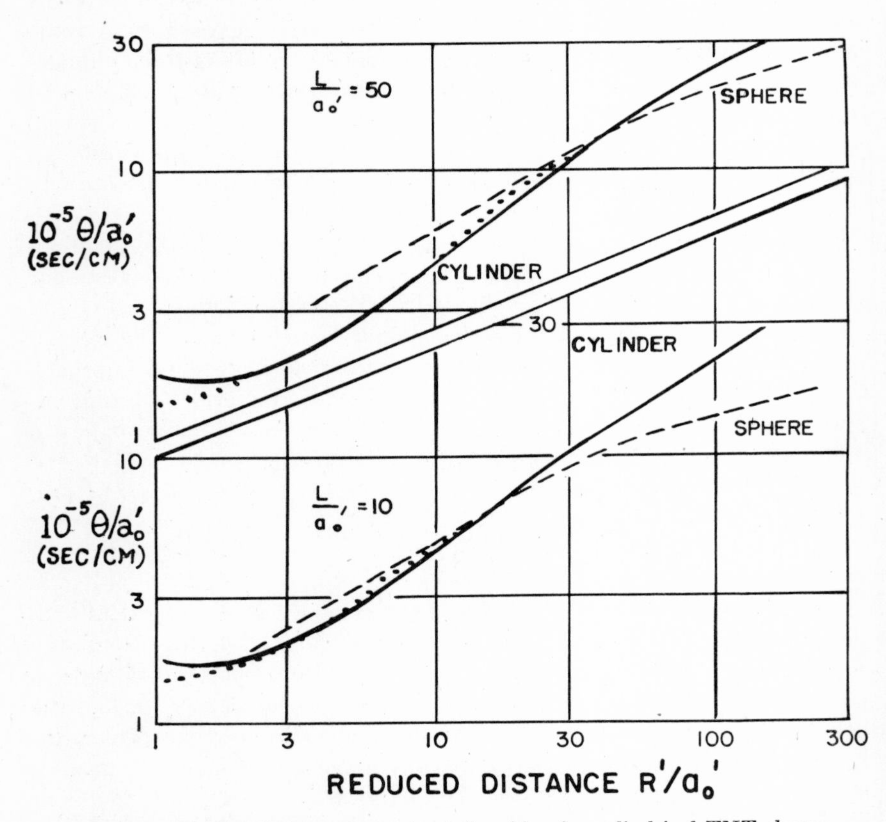

Fig. 4.4 Calculated time constants at the side of a cylindrical TNT charge. The dashed lines and dotted lines have the same significance as in Fig. 4.3.

4.5. Penney's Theory Based on the Riemann Equations

The differential equations for finite amplitude spherical waves have been solved numerically by Penney (83) and later Penney and Dasgupta (85) in the case of TNT. As was shown in section 2.3(c), these equations can be reduced to a form convenient for step-by-step integration by use of the Riemann functions N and Q, defined as

[8] These experiments, carried out at the NDRC Explosives Research Laboratory, are reported in UE Interim Report #13 (114).

$$N = \tfrac{1}{2}(\sigma + u) \tag{4.23}$$

$$Q = \tfrac{1}{2}(\sigma - u), \quad \text{where} \quad \sigma = \int_{\rho_0}^{\rho} \frac{c}{\rho}\, d\rho$$

The usefulness of these functions lies in the fact that the increments dN and dQ in a short interval dt and displacements dr can be found from the values of N and Q throughout the gas products and surrounding water at a time t. For the displacements dr indicated these increments are, from Eqs. (2.18),

$$dN = -\frac{cu}{r}\, dt, \quad \text{if} \quad dr = (c + u)\, dt \tag{4.24}$$

$$dQ = -\frac{cu}{r}\, dt, \quad \text{if} \quad dr = -(c - u)\, dt$$

Calculation of dN and dQ for a series of values of u and r at time t gives values at a time dt later. These values are of course only approximate, their accuracy increasing with the smallness of the interval chosen. Successive applications of the method then permit, with sufficient labor, building up the solution to any desired time. At each stage, the necessary values of u for the next stage are recovered from the relation $u = N - Q$, and the pressure is determined from $\sigma = N + Q$, which must be a known function of density and hence pressure in the water and gaseous products.

The starting point of the step-by-step calculations is determined by the conditions in the gas products after detonation is just complete. The position of the shock wave front is determined by the Rankine-Hogoniot conditions, as $R(t + dt) = R(t) + Udt$ and the shock front velocity U is a known function of pressure (see section 2.5). Although Eqs. (4.24) determine N and Q through most of the fluid, the displacements dr at which they are known after time dt leaves small shells in which they are not. The function Q, for instance, is an inward moving wave and its values at and near the shock front are progressively lost as t increases. Values at the shock front can, however, be recovered from the value of N and u, both of which are determined because the Rankine-Hugoniot conditions must be satisfied. The gap left in values of P and u on either side of the initial gas sphere boundary can be filled in by interpolation, and the solution built up.

In Penney's original calculation, this step-by-step process was started from an assumed initial pressure and density of the products corresponding to adiabatic conversion of the explosive at constant volume to its final state, the adiabatics used being those computed by H. Jones (see section 3.5), for which the initial pressure was 90 kilobars.

The solution after 27 steps for a charge of initial radius 30 cm. gave the pressure distribution up to a time 700 μsec. after detonation was complete, at which time the shock front in the water had advanced to 174 cm. or approximately 6 charge radii. The shock wave at later times was then estimated by approximate calculation of the "overtaking effect," (see section 2.4), as a result of which the front of the wave is

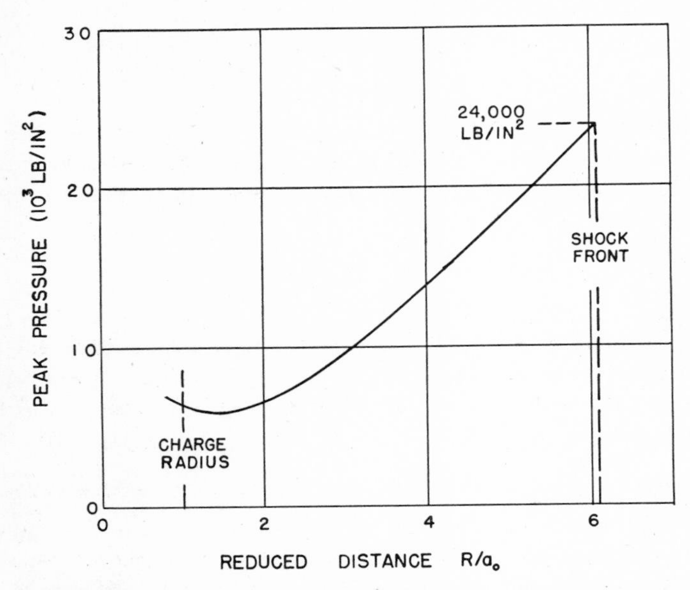

Fig. 4.5 Pressure distribution around an 1,800 pound TNT charge, from calculations of Penney and Dasgupta.

determined by pressures behind the front at earlier times, the portions nearer the front being lost by dissipation. The balance of the curve was then estimated by assuming an acoustic decay with distance, and the spreading of the wave predicted by the Kirkwood-Bethe theory is therefore not taken into account.

Penney's original results were inaccurate because of the assumption of adiabatic conversion for the initial conditions instead of the detonation wave, and because the equation of state used for water was based only on the available data at low pressures. An improved equation of state was made possible by later data of Bridgman (see section 2.6). Revised calculations were made by Penney and Dasgupta (85) using this equation of state and initial conditions based on Taylor's calculation of the spherical detonation wave in TNT (see section 3.6). The initial pressure and particle velocity at the front of the shock wave shortly after its generation were taken to be 60 kilobars and 1,420 m./sec., and corresponded to a chemical energy of 800 cal./gm. of explosive. The step-by-step process was then carried out on the basis of Eqs. (4.24) to approximately 6 charge radii, at which point the shock wave pressure was found to have a nearly linear decay behind the front, as shown in Fig. 4.5.

The calculated values of peak pressure P_m as a function of shock front radius R agreed closely with the formula

$$P_m \text{ (lb./in.}^2) = 103{,}000 \frac{a_o}{R} e^{2a_o/R} \tag{4.25}$$

over the range $a_o < R < 6a_o$, where a_o is the charge radius. The initial density of the explosive was taken as 1.5 gm./cm.³ and converting Eq. (4.25) to charge weight W in pounds and distance R in feet gives

$$P_m \text{ (lb./in.}^2) = 14{,}000 \frac{W^{1/3}}{R} e^{0.274\, W^{1/3}/R}$$

A detailed comparison with experiment is given in section 4.7, but it may be mentioned here that the predicted pressure of 12,600 lb./in.² at 10 charge radii is 20 per cent lower than recent experimental values at this distance.

4.6. The Propagation Theory of Kirkwood and Brinkley

An approach to the solution for shock waves of the basic hydrodynamic equations has been developed by Kirkwood and Brinkley (60), which is particularly convenient for extending experimentally determined shock wave data obtained at a single distance, or over a limited range, to other distances. As has been shown, the equation of motion and equation of continuity provide two partial differential equations for the pressure P and particle velocity u as functions of distance r and time t. In the case of spherical symmetry these equations are (see section 2.3)

$$\rho\left(\frac{\partial u}{\partial t}\right)_r + \rho u\left(\frac{\partial u}{\partial r}\right)_t + \left(\frac{\partial P}{\partial r}\right)_t = 0 \tag{4.26}$$

$$\frac{1}{\rho c^2}\left(\frac{\partial P}{\partial r}\right)_t + \left(\frac{\partial}{\partial r}\rho u\right)_t + \frac{2\rho u}{r} = 0, \quad \text{where} \quad c^2 = \left(\frac{dP}{d\rho}\right)_s$$

If these equations are specialized to the shock front, they provide two relations among the four partial derivatives of P and u. A third such relation can be obtained from the Rankine-Hugoniot condition for conservation of mass at the shock front, and the other shock front conditions permit evaluation of the shock front density and velocity as functions only of the pressure P. If a fourth relation among the partial derivatives and the shock front pressures can be obtained from other information about the shock wave, the set of equations can be formulated as a pair of ordinary or total differential equations adapted to numerical solution.

Before considering how this fourth relation is obtained, the transformation of the first three equations into more convenient form will be outlined. Instead of using r, the position in space of a point at which the derivatives are evaluated, Kirkwood and Brinkley choose as a variable R, the position in the undisturbed fluid (density ρ_o) of a volume element which has at time t the position r. To effect the transformation of Eqs. (4.26), it will be noted that by definition $(\partial r/\partial t)_R = u$, and a spherical shell of fluid with thickness dR and mass $4\pi\rho_o R^2 dR$ has at time t a thickness dr and mass $4\pi\rho r^2 dr$, hence $(\partial r/\partial R)_t = \rho_o R^2/\rho r^2$. Using these relations, Eqs. (4.26) become

$$\frac{R^2}{r^2}\left(\frac{\partial u}{\partial t}\right)_R + \frac{1}{\rho_o}\left(\frac{\partial P}{\partial R}\right)_t = 0 \tag{4.27}$$

$$\frac{\rho r^2}{\rho_o R^2}\left(\frac{\partial u}{\partial R}\right)_t + \frac{2u}{r} = -\frac{1}{\rho c^2}\left(\frac{\partial P}{\partial t}\right)_R$$

valid at any point behind the shock front. At the front itself $r = R$ and the equations are further simplified. The conservation of mass at the shock front requires that $P_m = \rho_o U u_m$ where U is the shock front velocity and P_m, u_m are the values of P, u at the shock front. This can be written as a relation between partial derivatives by noting that the derivative of P_m or u_m for a displacement dR of the shock front from a point R_o is given by

$$\left(\frac{\partial}{\partial R}\right)_{R-Ro} = \left(\frac{\partial}{\partial R}\right)_t + \left(\frac{\partial}{\partial t}\right)_R\left(\frac{\partial t}{\partial R}\right)_{Ro}$$

$$= \left(\frac{\partial}{\partial R}\right)_t + \frac{1}{U}\left(\frac{\partial}{\partial t}\right)_R$$

where U is the shock front velocity, and is determined as a function of peak pressure P_m for any fluid by the other two Rankine-Hugoniot conditions and the equation of state. Applying this operator to the equation $P_m = \rho_o U u_m$ gives

$$\frac{\partial u_m}{\partial t} + U\frac{\partial u_m}{\partial R} - \frac{g}{\rho_o U}\frac{\partial P_m}{\partial t} - \frac{g}{\rho_o}\frac{\partial P_m}{\partial R} = 0 \tag{4.28}$$

where $g = 1 - \rho_o u \dfrac{dU}{dP}$, and is determined by the value of P_m.

The fourth relation necessary to obtain the desired ordinary differential equations is determined by Kirkwood and Brinkley from the observed energy flux-time curve of the shock wave at a suitable distance from the source. The reason for choosing this property of the pressure wave lies in the fact that the work done by the shock wave after it passes a point R is ultimately dissipated as heat increasing the internal energy. This nonacoustic decay of a spherical shock wave and the loss of available mechanical energy are thus closely related.

From energy considerations the energy-time integral $F(R)$ at a distance R from the source can be expressed in terms of the peak pressure-distance curve integrated over distances greater than R, the relation being

$$F(R) = \int_{t(R)}^{\infty} r^2 P u dt = \int_{R}^{\infty} r_o^2 \rho_o h[P(r_o)]\, dr_o \tag{4.29}$$

The first integral is an expression of the work-energy principle, P being the pressure in excess of hydrostatic pressure and udt the displacement of a fluid element in time dt, and the integration is performed at constant R. In the second integral $h(P)$ is the increment in enthalpy, defined in section 2.4, for a fluid element through which the shock wave has passed.

The result Eq. (4.29) is obtained by considering the work W_o done by the source of initial radius a_o. The passage of a shock front through a fluid element leaves it in a state of higher entropy and internal energy, from which it returns adiabatically to hydrostatic pressure P_o. The total work done may be resolved into the sum of two terms: the increased internal energy of the fluid at pressure P_o within a sphere of radius R, and the work done on this spherical surface. Thus we have

$$W_o = 4\pi \int_{a_o}^{R} \rho_o r_o^2 E[P_m(r_o)]\, dr_o + 4\pi \int_{t(R)}^{\infty} r^2 \cdot (P + P_o)\, udt \tag{4.30}$$

where $E[P_m(r_o)]$ is the increase in internal energy per unit mass of fluid and the time integral is carried out for the volume element initially at R. The term involving P_o in the time integral gives the product of P_o and the volume displacement of the fluid element initially at R. This displacement is evidently the sum of the outward volume displacement ΔV_g of the inner boundary of the fluid (i.e., the expansion of the gas sphere) and the displacement of the volume of fluid initially between the shells of radii a_o and R to shells of radii a', R'. It can be written

$$4\pi \int_{t(R)}^{\infty} P_o r^2 udt = P_o \Delta V_g + 4\pi P_o \left[\int_{a'}^{R'} r^2 dr - \int_{a_o}^{R} r_o^2 dr_o \right]$$

$$= P_o \Delta V_g + 4\pi P_o \int_{a_o}^{R} \left(\frac{\rho}{\rho_o} - 1 \right) r_o^2 dr_o$$

the second step following from the relation $\rho r^2 dr = \rho_o r_o^2 dr_o$. Substitution in (4.30) gives

$$(4.31) \quad W_o = 4\pi \int_{a_o}^{R} \rho_o r_o^2 h[P_m(r_o)]\, dr_o + P_o \Delta V_g + 4\pi \int_{t(R)}^{\infty} r^2 P u dt$$

after introducing the enthalpy increment $h(P)$, which is $h(P) = E(P) + P_o d(1/\rho)$. The time integral is assumed to vanish as the shock front passes to infinite distance and hence in this limit we have

$$(4.32) \qquad W_o = 4\pi \int_{a_o}^{\infty} \rho_o r_o^2 h[P_m(r_o)]\, dr_o + P_o \Delta V_g$$

In this equation, the term $P_o \Delta V_g$ represents energy stored in the water by expansion of the gas sphere which, on later oscillations, may be dissipated by turbulence of the flow or by radiation of later "bubble pulses."

Subtracting Eq. (4.32) from (4.31) gives the final result. The second relation for $F(R)$ permits its expression in terms of P_m and R, and is conveniently written

$$(4.33) \qquad \frac{dF}{dR} = -\rho_o R^2 h(P_m)$$

The usefulness of considering the energy flux-time curve lies in the fact that, for a given weight of any particular explosive, its shape changes very little with increasing R. Kirkwood and Brinkley make use of this characteristic of F in formulating the desired fourth relation among the partial derivatives of P_m, u_m. As written, $F(R)$ does depend, however, on the strength of the source (weight of charge and its composition) and it is desirable to normalize the time integral to a value which is independent of this factor. This is done by expressing the integrand as a fraction of its initial value $R^2 P_m u_m$ and choosing a reduced time scale for which the initial slope of the integrand has unit value. Inasmuch as the observed energy-time curves are approximately of the form $R^2 P_m u_m e^{-t/\mu}$ it is convenient to use as time unit the initial logarithmic slope of the curve, defined by

$$\frac{1}{\mu} = -\left(\frac{\partial}{\partial t} \log r^2 P u \right)_{t=o}$$

Carrying out the indicated differentiation gives

$$\frac{1}{P_m}\frac{\partial P_m}{\partial t} + \frac{1}{u_m}\frac{\partial u_m}{\partial t} + \frac{2u_m}{R} = -\frac{1}{\mu} \tag{4.34}$$

and the energy function $F(R)$ can be written

$$F(R) = R^2 P_m u_m \nu(R)\mu \tag{4.35}$$

In this equation, $\nu(R)$ is the normalized integral over reduced time τ expressed by

$$\nu(R) = \int_o^\infty f(R, \tau)\, d\tau$$

where $$f(R, \tau) = \frac{r^2 P u}{R^2 P_m u_m}, \quad \text{and} \quad \tau = \frac{t - t_o(R)}{\mu}$$

The value of the integral $\nu(R)$ thus normalized is an expression of the shape of the shock wave, having the value one for an exponential decay and two-thirds for a linear decay (sawtooth wave). In the case of water, the shock wave is observed to be nearly exponential in form except for values of R greater than several hundred charge radii. Kirkwood and Brinkley therefore assign ν the constant value unity for calculations of underwater pressures, this being their "similarity restraint" on the energy flux-time curves. A different method of applying the restraint is necessary for blast waves in air; for details of this procedure the original report should be consulted.

The desired fourth partial differential equation is obtained from Eq. (4.34) by use of Eq. (4.35), the complete set being:

$$\frac{1}{u_m}\frac{\partial u_m}{\partial t} + \frac{1}{P_m}\frac{\partial P_m}{\partial t} + \frac{2u_m}{R} = -\frac{R^2 P_m u_m}{F(R)} \tag{4.36}$$

$$\frac{\partial u_m}{\partial t} + \frac{1}{\rho_o}\frac{\partial P_m}{\partial R} = 0$$

$$\frac{\rho}{\rho_o}\frac{\partial u_m}{\partial R} + \frac{1}{\rho c^2}\frac{\partial P_m}{\partial t} + \frac{2u_m}{R} = 0$$

$$\frac{\partial u_m}{\partial t} + U\frac{\partial u_m}{\partial R} - \frac{g}{\rho_o}\frac{\partial P_m}{\partial R} - \frac{g}{\rho_o U}\frac{\partial P_m}{\partial t} = 0$$

With this set of equations, it is possible to solve for the derivatives $\partial P_m/\partial R$, $\partial P_m/\partial t$ in terms of coefficients depending only the shock front pressure P_m and radius R and hence obtain an ordinary differential equation for P_m in terms of R:

$$\frac{dP_m}{dR} = \frac{\partial P_m}{\partial R} + \frac{1}{U}\frac{\partial P_m}{\partial t} = \varphi[F, P_m, R]$$

The function F is expressed in terms of P_m by Eq. (4.33), which is

$$\frac{dF}{dR} = -\rho_o R^2 h(P_m)$$

and the coefficients g, shock front velocity U, and density ρ appearing in φ are known functions of P_m. This pair of ordinary differential equations for P_m and F can therefore be solved by numerical integration subject to a choice of boundary conditions for determination of the constants of integration. The time variation of pressure behind the shock front as measured experimentally at constant r is initially given by the derivative

$$\frac{1}{\theta} = -\left[\left(\frac{\partial}{\partial t}\log P\right)\right]_{t=o}$$

corresponding to the peak approximation to the pressure-time curve $P(t) = P_m e^{-t/\theta}$ (see section 3.8). In terms of the coordinate R used in the present treatment, this becomes

$$\frac{1}{\theta} = \left[\frac{\partial}{\partial t}\log P_m\right]_R - u\frac{\rho}{\rho_o}\left[\frac{\partial}{\partial R}\log P_m\right]_t$$

which can be evaluated with the aid of Eqs. (4.36).

The two constants of integration can be determined in either of two ways: by the initial conditions at the boundary of the gas spheres following detonation, or by the experimental pressure-time curve at a selected distance.[9] If the assumption of adiabatic conversion of the explosive to its products is assumed, the first method is the same in principle as that outlined in section 3.7 for application of the Kirkwood-Bethe theory. The two constants may conveniently be taken to be the initial pressure at the boundary, and either the time parameter μ or the total shock wave energy. The explicit formulation of expressions for these quantities has been made by Kirkwood and Brinkley but is omitted here. If the semiempirical approach of using an experimental pressure-time curve to determine the constants is adopted, the natural quantities to choose are the peak pressure P_m and energy flux integral. The energy integral, however, is not accurately calculable from experimental pressure-time curves because of contributions from noncompressive flow energy, and is subject to greater experimental error than

[9] Tables and graphs for numerical application of either type of boundary condition are given in a report by Brinkley and Kirkwood (11).

peak pressure. The peak pressure as a function of distance R can be used equally well for determination of initial conditions from the total differential equation for dP_m/dR. This procedure has been used by Kirkwood and Brinkley to extrapolate measured shock wave pressures and compute shock wave energies for TNT and a mixed explosive. These results are discussed in sections 4.7 and 4.8.

4.7. Comparison of Shock Wave Theories

Three different theories of shock wave propagation have been outlined in preceding sections. All of these theories are approximate developments from the basic equations of hydrodynamics, but the methods of formulation and nature of the approximations involved are quite different in the three cases. It is therefore worth-while to compare the results by the different approaches with experimental evidence and with each other, in order to bring out their relative accuracy, flexibility, and fundamental limitations.

An unambiguous comparison of the various theories with each other and experimentally measured shock wave pressures is unfortunately not possible at the present time. This is so because results of all three theories have been obtained for only one explosive, cast TNT, and even in this case they are not directly comparable because different loading densities and heats of explosion were assumed in the developments. Penney and Dasgupta calculated the shock wave pressures for TNT with initial conditions corresponding to an energy release of 800 cal./gm.; calculations from the Kirkwood-Bethe theory are based on calculations leading to an energy of 1,060 cal./gm.; and the Kirkwood-Brinkley theory has been used to extrapolate experimental peak pressure measurements at low pressures to points closer to the charge. Other differences are in the range of pressures for which calculations have been made or involve less serious approximations, and are discussed in more detail later in this section.

The predicted dependences of peak pressure with distance R from a spherical charge of radius a_o are shown in Fig. 4.6, the product P_mR/a_o being plotted against R/a_o, using logarithmic scales to keep the graph within bounds. In addition, experimental values from piezoelectric gauge measurements over the pressure range 5,000–16,000 lb./in.^{2} are shown, these values being obtained from a number of measurements with a number of sizes of charge at Woods Hole. The close agreement of the curve from the Kirkwood-Brinkley theory for $R/a_o > 10$ is the result of the fact that the parameters of the theory were obtained from the plotted points, and only the part of the curve for higher pressures represents a theoretical prediction. It is seen that this curve lies very nearly parallel to, but lower than, the result obtained from the Kirkwood-Bethe theory, which lies fifteen-twenty per cent higher than the experimental points. The curve predicted by Penney and Dasgupta lies approximately the same amount below the points, but rises much more rapidly near the charge. It should be mentioned that later experimental values of peak pressures are slightly lower than the ones plotted in Fig. 4.6 (see section 7.4 for exact values), but the differences are unimportant for the present purpose.

The qualitative difference between the curve from the theory developed by Penney and Dasgupta and that of Kirkwood and Bethe is consistent with the fact that the former starts from the spherical detonation wave calculated by G. I. Taylor, whereas the latter is based

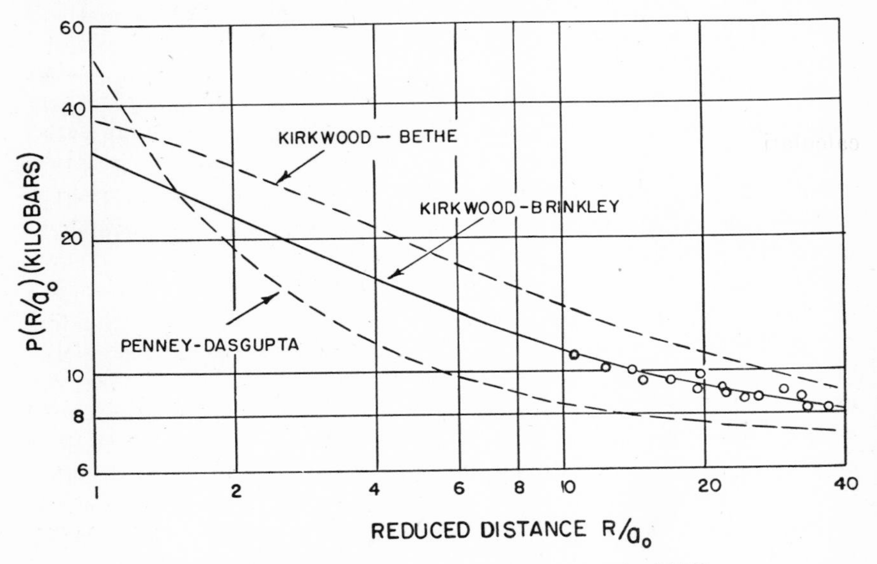

Fig. 4.6 Comparison of calculated pressures for TNT.

on an artificial detonation condition of adiabatic explosion at constant volume. The former is the preferable of the two and should lead to higher pressures very near the charge, because the front of the detonation wave is at higher pressure than exists behind it. This difference should, however, be decreased at greater distances because of dissipation in the shock front, as a result of which the initially higher pressures developed by the front of the detonation wave are not realized. It is, however, to be noted that the extrapolation by the Kirkwood-Brinkley theory, which involves no assumptions as to the detonation process because it works in from greater distances, resembles the Kirkwood-Bethe curve more closely than that of Penney and Dasgupta. Kirkwood and Brinkley have not presented this as a final conclusion, and it is possible that the approximations in the similarity restraint of their theory are responsible for the resemblance.

The low values of peak pressure predicted by Penney and Dasgupta at the distances of the experimental observations are to be attributed primarily to the low energy release assumed in the initial conditions. In addition, these values are the results of an extrapolation of the calculations, which were made from the surface of the charge out to six charge radii, and are subject to errors inherent in any such process. In view of these factors, it seems reasonable to suppose that an extension and revision of these calculations would give results in good agreement with experiment.

The assumption that the initial conditions are adequately represented by adiabatic conversion of an explosive into its equilibrium products is of course a rough one, and makes calculations of Kirkwood and co-workers rather inaccurate at distances of a few charge radii. Initial conditions similar to those employed by Penney and Dasgupta, which take account of the detonation wave front, are logically more satisfactory in this respect. Another approximation, in which the variation of conditions on the gas sphere boundary with time has been represented by the peak approximation, causes error in description of variations of pressure behind the shock front.

Despite the limitations of the calculations which have been based on the Kirkwood-Bethe theory, their analytical convenience has permitted the prediction and comparison of shock wave parameters for a number of explosives and conditions. These results are in good agreement with experiment in describing the nonacoustic variations of these parameters with distance from the charge, and further predict the relative order of explosives for pressure and duration with remarkable accuracy (see section 7.5). The methods of calculation developed by Kirkwood and collaborators therefore realize rather successfully the goals of any theory to predict and make understandable experimental results.

The more direct approach developed by Penney, based on numerical integration of the Riemann equations outward from the charge and using calculated detonation wave conditions, is undoubtedly inherently more satisfactory for calculation of the initial formation and development of the shock wave. Unfortunately its extension to greater distances involves a great deal of step-by-step numerical computation, with attendant danger of cumulative errors, unless precautions are taken of using very small intervals and providing adequate internal checks and controls. Penney (84) has suggested that calculations of shock waves could profitably be made by using the step-by-step calculations out a few charge radii to provide initial conditions for use of the Kirkwood-Bethe propagation theory at greater distances, but at the time of writing nothing of this kind has been done. Calculations along these lines would be of considerable value, particularly if compared with predictions from the alternative propagation theory of Kirkwood and Brinkley.

All of the theoretical developments are subject to some uncertainty at points near the charge for want of experimental data on the properties of water at the extremely high pressures involved. Any results for greater distances of course depend on these properties and on the exact conditions in the detonation wave. A more complete and accurate knowledge of underwater explosion pressures in general will therefore depend to a large extent on measurements and calculations for times during and immediately following detonation, and further work along these lines is highly desirable. Work of this kind has particular interest in determining the energy dissipated by the shock wave, as discussed in the next section.

4.8. Dissipation and Energy of Shock Waves

The energy in the shock wave from an underwater explosion is of importance from several points of view. The energy transport at a given distance from the source is a measure of the useful (or destructive) work which the shock wave can do. The total energy radiated from the source in the initial phases of an explosion determines how much of the available chemical energy remains for the later motion of the gas products and surrounding products.

Ultimately, of course, all the energy radiated in the shock wave is degraded into heat by dissipative processes as the wave is propagated outward. In an infinite medium, the transformation results in an increased temperature of the water, and the least ambiguous statement of the energy loss is in terms of this increased heat content. A formulation of this kind is used to advantage in the shock wave propagation theory of Kirkwood and Brinkley discussed in section 4.6. For purposes of analysis in terms of measured quantities this approach is undesirable, and it is necessary to examine the less clear-cut relation of the possible forms of energy to experimental shock wave pressures.

A. *Work and energy behind the shock front.* As discussed in section 4.6, the energy dissipated by the shock wave at points beyond a surface in the fluid defined by the position of the shock front at a given time must be equal to the work done in the displacement of this surface by the shock wave. Assuming spherical symmetry, the work $W(R)$ done on a surface having a radius R before arrival of the shock front at time $t(R)$ is given by

$$W(R) = 4\pi \int_{t(R)}^{T} r^2 P u dt \tag{4.37}$$

where u is the particle velocity for the shell of radius R and udt is therefore its displacement in time dt. This work is, however, not directly to be identified with the energy radiated by the shock wave and lost from further motion without some consideration of the time τ to which the integration is extended. If this time were taken as infinitely large, the integral in Eq. (4.37) would include the energy radiated in the later bubble pulses as well as in the primary wave. A smaller value of τ is therefore required to exclude these later contributions.

The proper value of τ and the interpretation of Eq. (4.37) are complicated by the fact that as the shock wave passes outward the work done on any boundary is increasingly that of noncompressive flow which gives the surrounding fluid kinetic and potential energy. This energy is returned to the gas products on later contraction of the bubble, and should not be included if one is interested in the energy radiated by the shock wave and lost for further motion. If the integration in Eq. (4.37) is extended to infinite time or over complete cycles of the pulsations this contribution is excluded because each outward flux of energy is cancelled by inward flux during the following contraction.

The energy radiated in later contractions of the gas sphere would be excluded by taking τ to be smaller than the time at which the gas sphere reaches its maximum radius. If this is done, however, the work calculated by Eq. (4.37) includes both the energy radiated in the shock wave and the recoverable kinetic and potential energies delivered to the fluid external to the surface of radius r. This combination of forms of energy is closely related to the afterflow term in the acoustic expression for the particle velocity u in spherical motion, which from section 2.2 is

$$u = \frac{P - P_o}{\rho_o c_o} + \frac{1}{\rho_o R} \int_{t(R)}^{t} (P - P_o)\, dt' \tag{4.38}$$

where P_o is the hydrostatic pressure. The second, or afterflow, term in this equation reduces, in the limiting case of noncompressive flow, to the velocity required by expansion of the gas sphere boundary, and the corresponding kinetic energy is returned to the gas sphere in its recompression.

If an evaluation of the work done at any point in the fluid from the measured pressure-time curves is desired, the particle velocity u can be eliminated from Eq. (4.37) by the use of Eq. (4.38), with the result

$$\frac{W(R)}{4\pi R^2} = \frac{1}{\rho_o c_o} \int_{t(R)}^{\tau} P^2 dt + \frac{1}{\rho_o R} \int_{t(R)}^{\tau} P(t) \int_{t(R)}^{t} P(t')\, dt'\, dt \tag{4.39}$$

if the hydrostatic pressure P_o is neglected. Strictly, this equation should be evaluated for a surface fixed in the water and hence moving with it, but except at points close to the gas sphere the resulting correction to values of P obtained at a fixed point is not important. If the total work done by both the shock wave and the noncompressive flow which the motion increasingly resembles is required, both terms in Eq. (4.39) should be employed. If, however, a measure of the radiated shock wave energy only is required, Eq. (4.39) is not correct as it stands, because of the recoverable changes in kinetic and potential energy included.

The second term in Eq. (4.39) is, in the noncompressive approximation, the change in kinetic and potential energy of the fluid flow. Therefore, the natural approximation to a value of the radiated energy is obtained by dropping this afterflow term. Actually, the two types of energy change are not distinct or mutually exclusive, and at no time can the pressure behind the shock front be regarded as the result of either compression or noncompressive flow alone. In cases of interest, however, the difference in variation with time of the two terms in Eq. (4.39) makes their contributions reasonably distinct. The first term increases rapidly during the initial portion of the shock wave for which the pressure P is large and later additions from the tail of the shock wave become less and less important. The afterflow term, on the other hand, increases much more slowly at first because of the integral $\int Pdt'$ in the integrand, but for the same reason continues to increase appreciably for pressures very much less than the initial peak pressure of the shock wave. The character of the afterflow term is also indicated by the factor $1/R$ as compared to the first term, thus being the result of the decrease of afterflow velocity with distance from the gas sphere boundary.

Returning to the question of the time τ to which the integrals in Eq. (4.37) or (4.39) should be evaluated, it is clear that no single answer is possible for all cases. A major part of the radiated shock wave energy is accounted for if τ is chosen to be several times the initial time constant of the wave, but the outward energy flux of noncompressive flow increases for much longer times (up to times of the order of one-tenth of the bubble period).

B. *Energy flux density.* The work done on a surface fixed in the fluid is not the only way in which energy behind a shock front can be expressed. Another measure which is sometimes useful is in terms of an energy current density, or rate of energy flow through a surface of unit area. A unit volume of fluid with particle velocity u and internal energy E per gram has energy of amount $\rho(E + \frac{1}{2}u^2)$, where ρ is its density, and the rate of transport through unit area is therefore $\rho u(E + \frac{1}{2}u^2)$. If we add to this the rate Pu at which pressure P in the fluid can do work on this surface, we obtain for the total energy transport as a result of fluid motion through unit area

$$E_f = \int \rho u \Delta \left(E + \tfrac{1}{2}u^2 + \frac{P}{\rho} \right) dt \tag{4.40}$$

where Δ indicates the increment over the value in the undisturbed fluid.

This expression for energy flux density is convenient in calculations based on the Kirkwood-Bethe theory, because the bracketed quantity in Eq. (4.40) is the kinetic enthalpy increment of the fluid, which is used as a fundamental variable in that propagation theory. The differences between Eq. (4.40) and the expression $\int Pudt$ is insignificant except for very large pressures (see section 7.2).

C. *Energy dissipation in shock waves.* The dissipation of energy in outward propagation of a shock wave as a result of pressure and velocity

Distance (charge radii)	Peak pressure (lb./in.²)		Shock wave energy (cal./gm.)		
	Calculated	Measured	Calculated total	Measured[1] to 6.7θ	Afterflow[2] to 6.7θ
1	460,000		561		
2	167,000		495		
3	91,400		449		
5	42,800		396		
7	26,600		366		
10	16,400	16,250	340	255	43
15	9,670	9,640	316	250	30
20	6,720	6,720	302	246	23
25	1,090	1,090	293	244	19

[1] Computed from the integral $\frac{4\pi R^2}{\rho_0 c_0} \int_0^{6.7\theta} P^2\, dt$

[2] Computed from the integral $\frac{4\pi R}{\rho_0} \int_0^{6.7\theta} P \int_0^t P\, dt'\, dt$

Table 4.2. Energy dissipation in the spherical shock wave from TNT.

gradients at the shock front is greatest near the charge where these gradients are largest. Experimental estimates in this region are difficult because of the damaging effects on measuring equipment, and at the present time theoretical calculations give the best estimate of the total energy radiated and the rapidity with which it is dissipated. Perhaps the best estimates of this kind are afforded by the propagation theory of Kirkwood and Brinkley discussed in section 4.5, in which the energy plays a central role. This theory has been applied to TNT using experimentally determined values of peak pressure for distances greater than ten charge radii to determine the initial conditions. The computed energies remaining in the shock wave at various distances are listed in Table 4.2. For comparison, values of energy flux computed from experimental pressure-time curves integrated to 6.7 times the initial time constant are also tabulated.

It is seen from Table 4.2 that the energy remaining in the shock wave falls off rapidly near the charge as the wave progresses outward, about thirty per cent of its initial energy being dissipated in the volume within five radii of the charge and forty-eight per cent in the volume within twenty-five charge radii. The energies computed from measured pressure-time curves are approximately twenty-five per cent lower than the theoretical values, but show corresponding decreases with distance over the range of measurement. The difference in absolute value is consistent with the fact that the experimental figures were obtained by numerical integration to an arbitrary, finite time after arrival of the shock front, in order to exclude significant contributions of noncompressive flow energy to the integral. For comparison, values of the afterflow integral to the same time are also tabulated.

The energy of 561 cal./gm. at 1 charge radius represents the total energy radiated by the shock wave, and is 53 percent of the estimated total energy of explosion of 1,060 cal./gm. of TNT. The remaining 47 per cent, or 500 cal./gm., is thus the remaining energy for later motion of the gas sphere and the surrounding water. This energy can, however, be computed rather accurately from the observed period of pulsation of the gas sphere (see section 8.3), and for TNT a value of 480 cal./gm. is obtained. This figure is in remarkably good agreement with the less directly estimated value of 500 cal./gm. Similar calculations for other explosives give correspondingly good agreement between calculated shock wave energies and indirect estimates from experimental measurements of the gas sphere motion. These agreements are to within the combined accuracy of the measurements and calculations, and thus give a very satisfactory accounting for the energy distribution in underwater explosions.

The rapid dissipation near an explosive charge represents a degrading or wastage of available energy to heat, this loss occurring at the steep gradient of the shock front. The fact that some twenty-five per cent of the total energy of explosion is lost for useful work within twenty-five charge radii of the explosion illustrates what might be termed the inefficiency of the shock wave for transmission of energy. This inefficiency has led to suggestions that greater effectiveness might be obtained by using slow burning propellant charges which did not develop steep pressure fronts. Whether or not an improvement would be realized depends of course on the necessary magnitude and duration of pressure for the purpose of interest, and whether the necessary energy could conveniently be released in the most suitable way.

5. *Measurement of Underwater Explosion Pressures*

The preceding chapters have dealt with the physical and chemical properties, physical laws, and mathematical developments necessary to a theoretical understanding of underwater explosion processes. Before proceeding to a detailed comparison of the theoretical results with experimental data, it is of importance to consider the techniques of measurements of underwater explosion phenomena, and this chapter is devoted to a brief discussion of some of the experimental methods. The more important and generally applicable of these methods fall into two broad classifications: mechanical and electromechanical devices for pressure measurements, and optical methods, which can be applied both to pressure measurements and to investigation of mass motion of the water as a function of time and other variables. This chapter is concerned with the first classification; optical methods are considered in Chapter 6.

5.1. Crusher Gauges

Early attempts at measurement of underwater explosion pressures were all based on devices in which the basic measured effect was plastic deformation of small lead or copper elements by the agency of the pressure. Measurements of this kind provide directly only an empirical comparison between different experimental conditions of charge weight, distance, etc., but interpretations in terms of fundamental quantities were attempted by comparing explosion effects with those of known hydrostatic pressures. The earliest investigations of this kind appear to be those of Abbot (1), who used what is commonly described as a crusher gauge, in which a steel piston acts on a small lead cylinder fixed on a massive support as sketched in Fig. 5.1(a). The device may be thought of as a mass (the piston) attached by a spring (the lead cylinder) to a fixed support. The action of pressure on the piston face causes an acceleration which is opposed by the plastic resistance of the lead to deformation. The amount of deformation suffered by the cylinder is then a measure of the force acting on the piston and hence of the pressure in the surrounding fluid. In the form used by Abbot and later by Schuyler, the crusher gauge is very similar to the crusher gauges used in measuring pressures developed in gun barrels by propellant charges.

The difficulty with use of such a gauge in measurement of shock wave pressures is the fact demonstrated by Hilliar (47) that the natural period of the large gauges with soft lead crushers used by Abbot must have been of the order 400 to 800 microseconds. As a result, the effective pressure acting on the crusher in the course of its deformation is a kind of average over comparable intervals (a more exact description of the actual mechanism of deformation is given later in this section). Times of this order are however comparable with shock wave durations of even larger charges. Exactly what the gauges used by Abbot measured is therefore highly uncertain, but it could not have been any kind of approximation to the initial pressure. That the measurements made

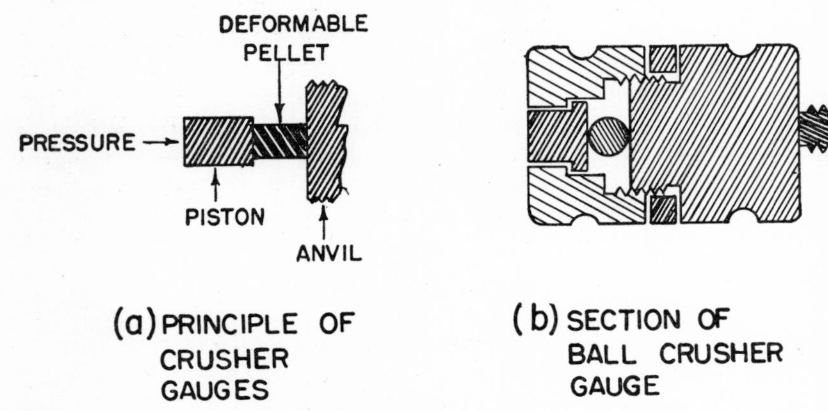

Fig. 5.1 Principle and construction of crusher gauges.

by Abbot have no fundamental significance of this kind is perhaps sufficiently well indicated by his finding pressures varying with weight W and distance R from an explosion roughly as $W^{0.7}/R^{1.4}$, results which are neither consistent with similarity nor in agreement with later work giving laws more nearly of the form $W^{0.38}/R^{1.14}$. The method is, however, of interest because it is the predecessor of fundamentally more useful gauges based on the same principle which will be considered in more detail.

Although cylindrical crusher pellets were used both by Abbot and Schuyler and by Hilliar in a classic series of measurements described later in more detail (see section 5.2), a more recent modification developed by the Naval Ordnance Laboratory, U. S. Navy (78), employs 5/32 inch or 3/8 inch diameter soft copper spheres as the "crusher" element, the construction of the gauge being indicated in Fig. 5.1(b). A considerable amount of work has been done to analyze the response of a gauge of this type to various types of applied pressures.[1] Calibration tests on deformation of the copper balls by impact of a piston travelling

[1] A number of such calculations are given by A. B. Arons, to be issued as a NavOrd Report by the U. S. Navy Bureau of Ordnance.

with known velocity show that the maximum depression of the ball is a function only of the energy of the piston before impact, a result suggesting that the resistance of the ball to deformation is primarily the result of work hardening. If the deformation S is measured, the average force acting during such a ballistic impact can be calculated from the initial energy of the striking piston, and with the aid of such experiments a calibration curve of force F versus plastic deformation S can be prepared. The use of copper spheres has the practical advantage of giving a very nearly linear calibration curve of the form $F = kS$ for deformations up to .05 and .08 inches for the 5/32 inch and 3/8 inch spheres, corresponding to static pressures of 3,000 and 10,000 lb./in.^{2}.

This is formally the same as a Hooke's law restoring force for a spring, with the difference that there is a permanent set of the sphere after the force is released rather than complete elastic recovery. The response of the entire gauge up to the time of maximum deformation is thus equivalent to that of a mass and spring subjected to a force applied to the mass and the familiar differential equation of a linear oscillator results for the motion of the piston:

$$m \frac{d^2S}{dt^2} + kS = AP(t) \tag{5.1}$$

where $AP(t)$ is the force on the piston of area A for an applied pressure $P(t)$ and m is the effective mass of the gauge. This effective mass is taken, in the theory developed by G. K. Hartmann (44), to be the sum of the piston mass plus ⅓ the mass of the copper ball (by analogy with the corresponding value for a spring of finite mass) plus a hydrodynamic mass. This last term represents the inertia of water near the face of the piston which moves with the piston, and is assumed by Hartmann to be $\frac{2}{3}\rho d^2$ where d is the diameter of the piston face, this being the value predicted by elementary hydrodynamics for the case of an infinite cylinder.[2]

The equation of motion (5.1) can be integrated for any assumed pressure variation. We consider the approximation to a shock wave of an initial peak followed by an exponential decay of the form $P(t) = P_m e^{-t/\theta}$. With this assumption, the solution of Eq. (5.1) is easily found by standard methods to be

$$S = \frac{AP_m}{k} \frac{(\omega\theta)^2}{1 + (\omega\theta)^2} \left[e^{-t/\theta} + \frac{1}{\omega\theta} \sin \omega t - \cos \omega t \right] \tag{5.2}$$

where $\omega = \sqrt{\frac{k}{m}}$

[2] This value is derived on the assumption that the motion of the water near the piston can be described as noncompressive flow. The formulation of the effective pressure at a moving boundary, from which the correction is obtained, is given in section 10.5.

with the understanding that this equation describes the motion only up to the first maximum of deformation of the crusher, after which time the ball retains the permanent plastic set. For an applied pressure of very long duration $(\theta,\ \omega\theta \to \infty)$ this maximum occurs at a time $\frac{1}{2} T_\infty = \pi/\omega = \pi\sqrt{m/k}$ and the deformation S_∞ is given by

$$S_\infty = \frac{2AP_m}{k} \tag{5.3}$$

which is twice the static value for a slowly applied pressure. The natural periods T_∞ of 5/32 inch and 3/8 inch balls in the NOL gauge for which the effective mass is about 15 grams come out to be 355 and 514

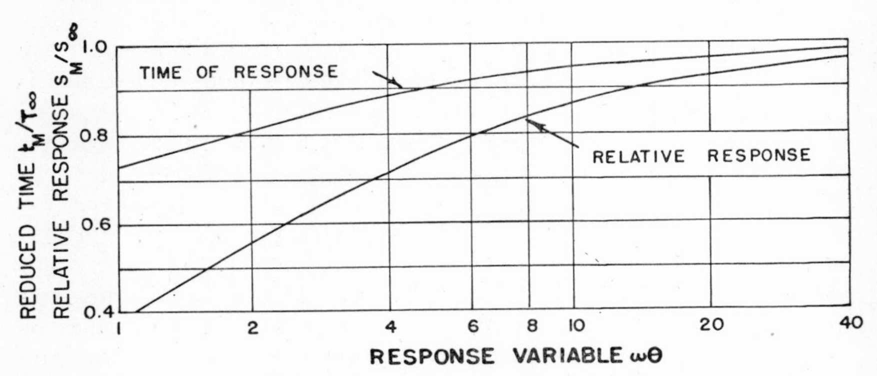

Fig. 5.2 Relative response and time of action of a ball crusher gauge.

microseconds, and Eq. (5.3) provides an accurate estimate of peak pressure P_m only if the time constant θ is much longer than these values. If this condition is not fulfilled, it is evident that the final deformation will be smaller, because of the smaller average pressure acting during the time of response of the gauge, and will be a function of the duration parameter θ. The actual deformation for a given θ can be found from Eq. (5.2) by determining the time t_m of maximum deformation for which $dS/dt = 0$, and solving Eq. (5.2) using this value of t_m. The time t_m is given by the condition

$$e^{-t_m/\theta} = \cos \omega t_m + \omega\theta \sin \omega t_m \tag{5.4}$$

and the deformation S_m is then

$$S_m = \frac{AP_m}{k} (\omega\theta) \sin \omega t_m \tag{5.5}$$

The time t_m can be computed from the transcendental equation (5.4) as a function of $\omega\theta$ and these values substituted in Eq. (5.5) to determine P_m as a function of kS_m and the product $\omega\theta$. In this application, the gauge gives a measure of peak pressure provided θ is known at least approximately. The ratio $t_m/T_\infty = \omega t_m/\pi$ and the deformation factor P_m/kS_m are plotted in Fig. 5.2 as a function of $\omega\theta$. It is seen that if $\omega\theta$ is not too small the deformation factor is not too sensitive to variations in $\omega\theta$. Hence, although the gauge does not, strictly speaking, measure peak pressure P_m without a priori information about θ, it performs this function reasonably well for sufficiently large charges if θ is known approximately.

Several complications have been tacitly ignored in the simple discussion of the ball crusher gauge which complicate the reduction of its readings to more fundamental quantities. It has been assumed that the pressure $P(t)$ acting on the gauge is the same as the free field pressure which would exist in the absence of the gauge and diffraction effects due to its presence are thus neglected. The magnitude of these effects has been estimated both theoretically and experimentally, with indications that they are not serious for conditions in which the gauge responds primarily to peak pressure. A second source of difficulty is the fact that the deformation of the copper crushers depends not only on the magnitude of the force but on the rate of its application, experimental tests by Seitz et al. (100) indicating that this "speed effect" may increase the constant k by as much as 15–20 per cent for strain rates $(1/S)\Delta S/\Delta t$ of the order 1,600 per sec. Corrections can be made on the basis of such tests but the existence of the effect makes the interpretation of observed depressions more uncertain. A third difficulty is the assumption of a specific form of pressure-time curve. While an exponential decay is certainly the best simple mathematical description of a shock wave in water, it is by no means true that it is always a good one. For example, the pressure-time curves in certain directions around cylindrical charges have multiple peaks (see Fig. 7.15, Chapter 7). The response of the gauge to such pressure could in principle be computed from the known form of the curve, but this approach obviously constitutes an investigation of the gauge rather than the pressure acting on it.

As a purely empirical device for comparison of different explosives or different conditions, the ball crusher gauge has very real value because of its reproducibility and simplicity in use and because one has at least a rough picture of what it is measuring. Comparison with the more detailed results obtained by electromechanical gauges which give a continuous record of pressure have confirmed the essential validity of the simple theory presented here. With this background the ball crusher gauge forms a useful tool in explosive investigations.[3]

[3] Detailed discussions of the use of ball crusher gauges and precision of measurements are given in reports by J. S. Coles (22), and by R. H. Brown (12).

5.2. The Hilliar Piston Gauge

As the discussion of section 5.1 shows, the ball crusher and other types of crusher gauge are most useful as indicators of peak pressure at a shock front. An extension of the basic idea of using plastic deforma-

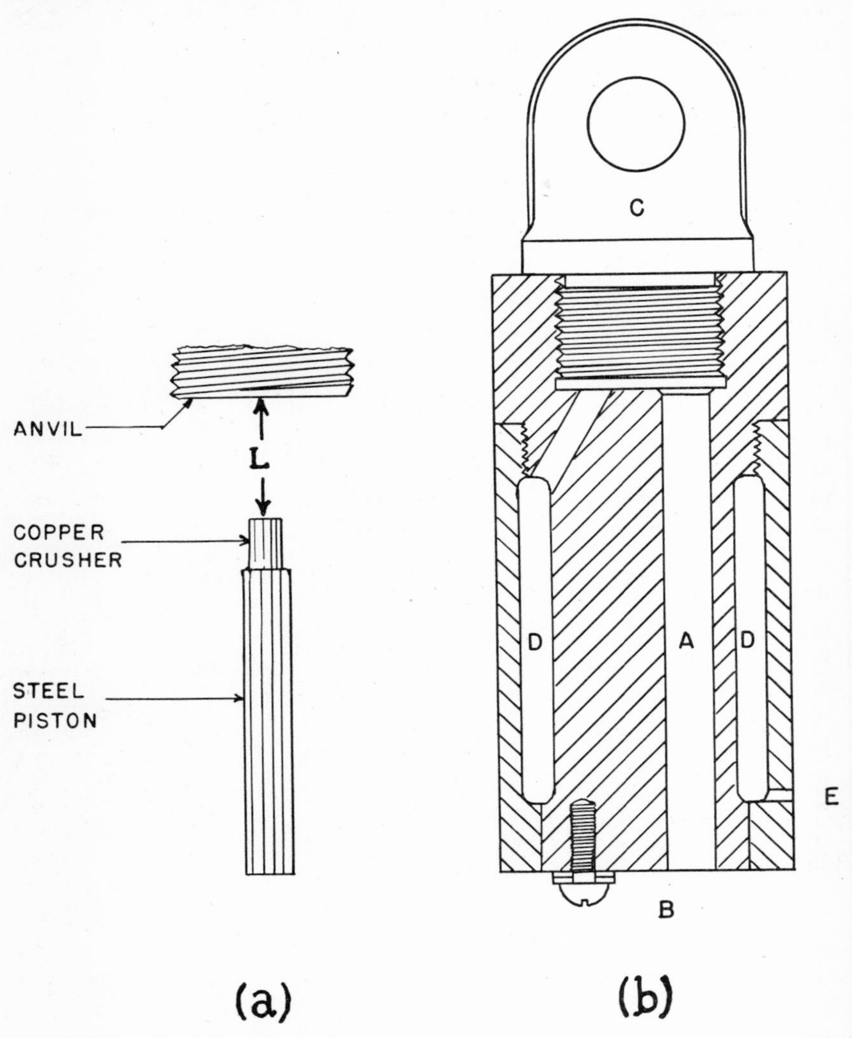

Fig. 5.3 Principle and construction of Hilliar piston gauge.

tions to make possible indications of pressures at later times was developed by Hilliar and applied by him to a classic investigation (47) of the properties of pressure waves from underwater explosions. The gauge consists of a piston and copper crusher as before, but with the difference that the piston and crusher are initially a known distance from the anvil rather than in contact with it, as indicated diagrammatically in Fig. 5.3(a). The piston thus has a length L of free travel before striking the crusher and in this time acquires a momentum proportional to the impulse or time integral of the pressure on its face. The impulse per unit area up to the time t at which the piston strikes the crusher will be denoted by $I(t)$, where

$$I(t) = \int_0^t P(t)\,dt \tag{5.6}$$

We may write, by conservation of momentum, that the velocity V of the piston on impact is given by

$$I(t) = \frac{mV(t)}{A} \tag{5.7}$$

where m is the effective mass of the piston and A the area exposed to pressure. Hilliar found, however, that the deformation S of the crusher as a result of impact by a piston is a function only of the energy of the piston. With the aid of calibration tests using pistons of known mass shot at measured velocities from a compressed air "gun," a calibration curve of kinetic energy E_c absorbed by the crusher versus crusher deformation can be prepared. If the deformation of a crusher struck by a piston accelerated by the pressure wave is then measured, its value gives a measure of the energy E_p of the piston and hence its velocity $V(t)$. Two corrections are necessary: the body of the gauge is also moving, although more slowly, and the piston does work on the copper before being brought to rest because of the pressure still acting. The first of these effects is accounted for by using a reduced mass $m' = m/(1 + m/M)$ where M is the mass of the gauge, and the second by subtracting a term $P \cdot S$ from the calculated energy absorbed, where P is the pressure during impact, with the result

$$E_p = \frac{1}{1 + m/M}\,(E_c - PS) \tag{5.8}$$

The kinetic energy of the piston is, however, given by $E_p = \frac{1}{2}mV^2$ and substituting in Eq. (5.7) gives

$$I(t) = \frac{1}{A}\sqrt{\frac{2m}{1 + m/M}\,(E_c - PS)} \tag{5.9}$$

The piston gauge in the form described thus measures the impulse of the wave up to an unknown time t. In order to make a more direct interpretation of the pressure-time curve it is evidently necessary to determine the time of piston motion. If this is known one could obtain an approximation to the pressure-time curve by writing $I(t) = P_{av} \cdot t$ where P_{av} is an average of the pressure over the time of action. The exact meaning of such an average depends of course on the form of the curve $P(t)$ but the difficulty can be resolved by using several piston-crusher units each acting over successively greater intervals. If each added interval is made small enough the average pressures over the added intervals represent good stepwise approximation to a continuous curve. If the pressure changes comparatively little over the first such interval the time of action of the piston can be computed assuming its acceleration to be uniform. The time t_1 for the first piston is then

$$t_1 = \frac{2L_1}{V_1}$$

where L_1 is the length of free travel and V_1 the final velocity computed from the crusher deformation. The average pressure over this interval is then

$$P_{av1} = \frac{m_1 V_1}{A t_1} = \frac{m_1 V_1^2}{2AL_1} = \frac{m_1}{1 + m_1/M} \cdot \frac{E_{c1} - P_1 S_1}{2AL_1} \tag{5.10}$$

All the quantities in this equation are known except for the final pressure P_1 which must be estimated. Its effect is not very large and can be estimated by judicious guessing, or if necessary a second approximation correcting an initially assumed value.

A second piston is designed to have a longer free travel L_2 and if necessary a larger mass m_2 in order to increase its travel time t_2. The extra travel time for this piston, again assuming an interval $t_2 - t_1$ short enough that the acceleration can be considered uniform, is

$$t_2 - t_1 = \frac{2(L_2 - L_1)}{V_2 + V_1}$$

The average pressure over the interval $t_2 - t_1$ is then given by

$$P_{av2} = \frac{m_2 V_2 - m_1 V_1}{A(t_2 - t_1)} = \frac{(m_2 V_2 - m_1 V_1)(V_2 + V_1)}{2(L_2 - L_1)A} \tag{5.11}$$

Again all the quantities can be calculated from known gauge constants and the average pressure over the interval $t_1 < t < t_2$ determined.

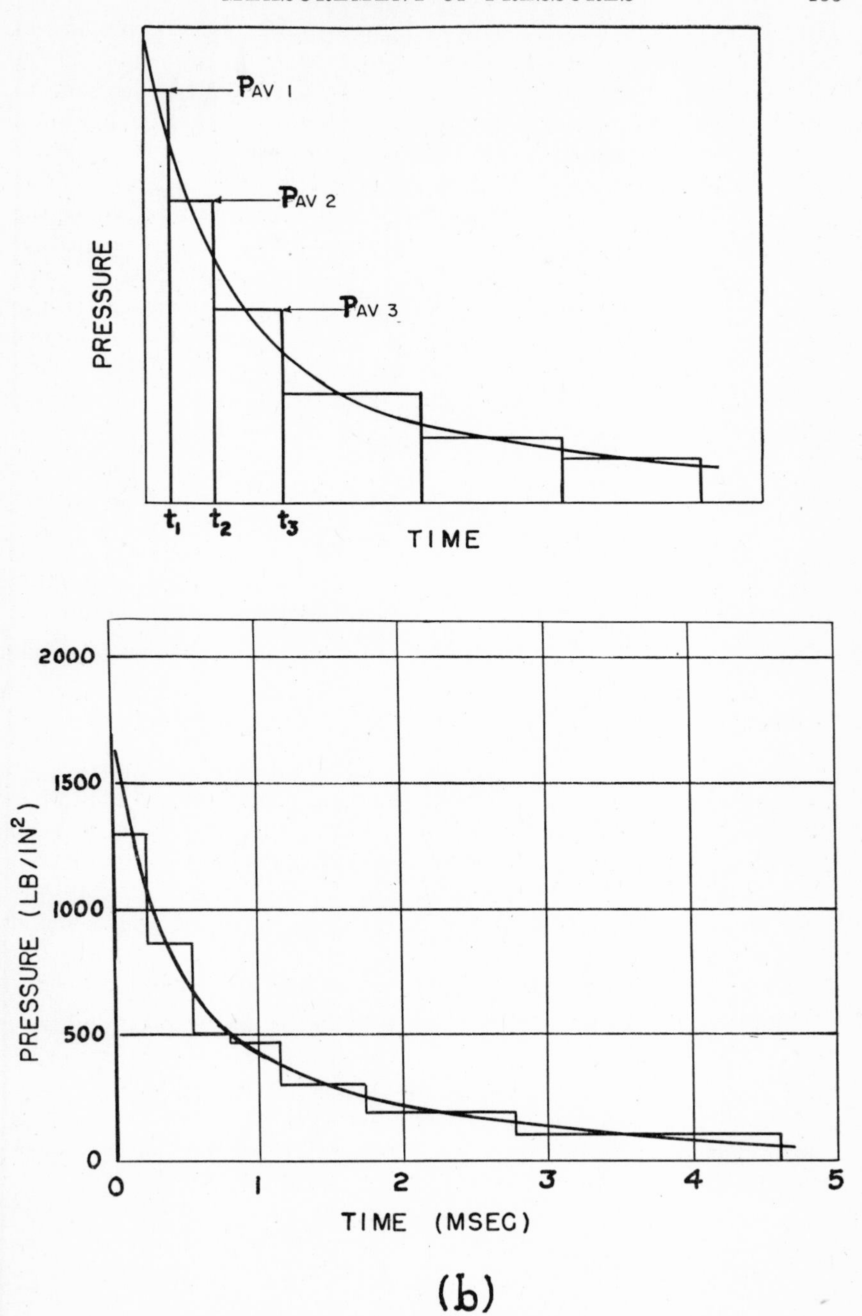

Fig. 5.4 Pressure-time curve obtained from Hilliar gauge.

With a sufficient number of piston-crusher units, a block diagram approximation to the actual pressure-time curves can be constructed, as sketched in Fig. 5.4(a), and a smooth curve drawn through the blocks, as indicated by the dashed curve, is taken to represent the actual pressure-time curve.

In Hilliar's experiments, as many as 10 sizes of multiple piston gauges were used, either 3 or 6 pistons and crushers being provided in gauges to give a better average. The pistons used were ½ inch in diameter and from ¾ inch to 8 inches long; the copper crushers, small cylinders 0.32 inch in diameter and 0.50 inch long, were found to give very consistent deformations in calibration tests; the free travels of the pistons varied from ½ inch to 2½ inches. A typical gauge with three units is shown in Fig. 5.3(b). The gauge was normally hung vertically and the pistons prevented from dropping out of the three chambers *A* (only one is shown in the drawing) by the washers *B*. The annular air chamber *D*, open to the water at *E* and to the chambers *A* below the eyebolt *C*, served to equalize hydrostatic pressure and reduce air compression by the moving piston. The crushers were attached to the pistons and crushed against the eyebolt.

With the use of piston and crusher gauges, Hilliar was able by a series of 109 shots in 1918–19 to determine the major features of shock waves, verify the principles of similarity, and determine weight-distance laws. A typical pressure-time record from Hilliar's report is shown in Fig. 5.4(b) for 300 pounds of TNT at a distance of 50 feet. Although the absolute pressures and durations obtained by Hilliar differ by ten-thirty per cent from more recent results, his determination of so many of the major features of shock waves with relatively simple equipment and facilities was a remarkable accomplishment.

Some of the limitations of Hilliar's techniques are fairly obvious: determination of a curve by what is essentially a series of time averages may smooth over or obliterate irregularities and details, and some care is necessary to pick the proper piston combinations. It has also been found in later work that good technique is something of an art and that considerable care and experience is necessary in order to obtain reproducible results. Nevertheless the method does determine the major features of the shock wave pressure-time curve and has been employed extensively up to the present time.[4] One interesting modification of the Hilliar gauge has been made by Hartmann (45), in which the working time is made sufficiently long, by increasing the piston inertia and decreasing its exposed surface, that the integrated pressure is measured over an interval longer than the shock wave duration; the device thus

[4] An improved design of Hilliar gauge has been developed by G. K. Hartmann and the Taylor Model Basin. The design of this multiple piston gauge, its precision, an analysis of errors, and results obtained are described in a comprehensive report by Hartmann (45).

constitutes an impulse or momentum gauge which gives a measure of the area under the pressure-time curve.

5.3. Diaphragm Gauges

Several types of gauges have been developed in which the plastic deformation of an air-backed plate or diaphragm is taken as a measure of explosive effectiveness. One such gauge, developed by the Italian Modugno about 1919,[5] consists of a ⅛ inch copper disk with an exposed surface 1 inch in diameter mounted on the lip of a cuplike gauge body,

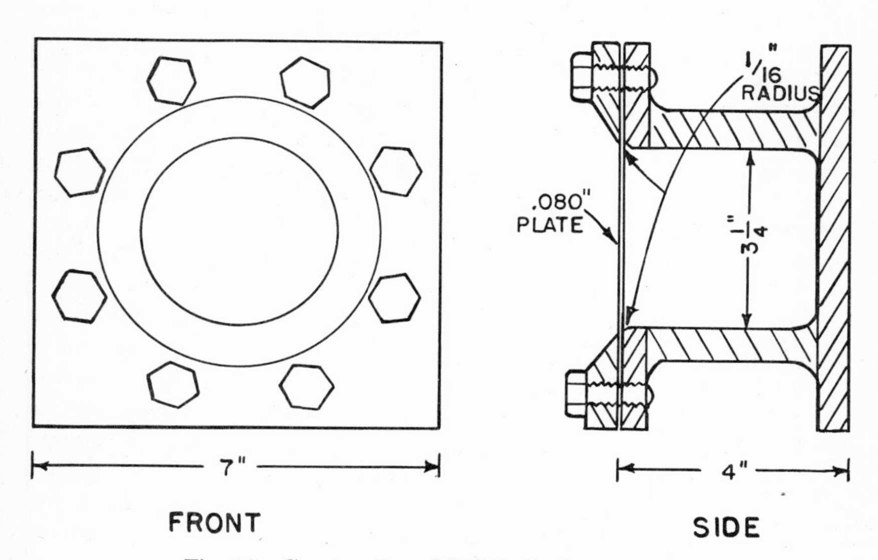

Fig. 5.5 Construction of UERL diaphragm gauge.

and fastened down by a screw cap which crimps the edge of the disk to constrain it. This device functions somewhat as does a ball crusher gauge, and the final deformation gives a measure of peak pressure from reasonably large charges. Any very realistic description of its operation is a sufficiently complicated function of the pressure and laws of plastic deformation that its utility, either as a tool of measurement or for investigation of mechanisms of damage, is somewhat limited.

Another type of diaphragm gauge is the British "pot" gauge and a modified form of this gauge used for explosive comparisons in this country. This latter gauge, developed at the Explosives Research Laboratory at Bruceton and the Underwater Explosives Research Labora-

[5] For a detailed description of this gauge, Reference (14) should be consulted (the original report by Dr. Francesco Modugno of the Royal Italian Navy is not accessible).

tory at Woods Hole,[6] is sketched in Fig. 5.5. The diaphragm *A* is ordinarily made of steel or copper .080 inch thick and is clamped to the gauge body by a face plate, the diameter of the area exposed to pressure being 3.4 inches. In practice, the gauge is placed a known distance from the explosion and the resultant roughly semi-spherical depression taken to be a measure of the explosive's effectiveness. As in the case of the Modugno gauge the law of its motion is not a simple function of

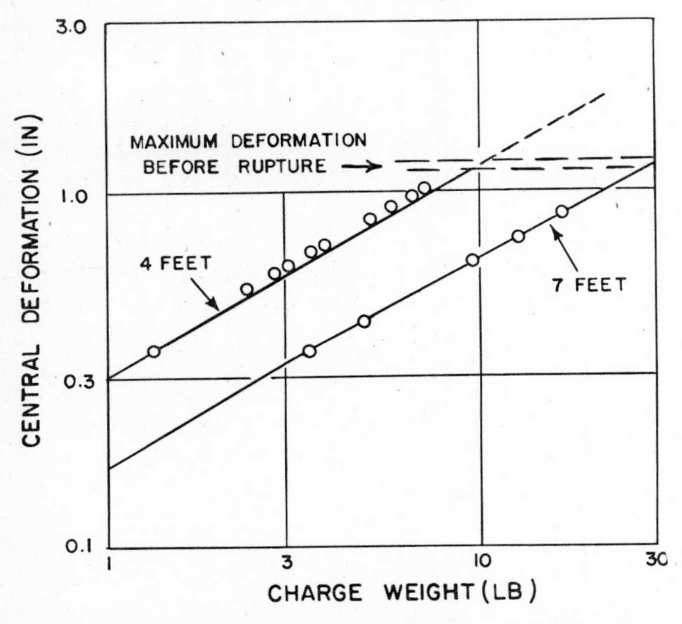

Fig. 5.6 Deformation of 3.4 inch steel diaphragms at 2 distances from small charges. The solid lines are calculated theoretical values.

any single property of the shock wave, and practical use of the gauge has been largely for empirical comparisons. The response of the gauge has, however, been used to test theoretical calculations of Kirkwood on the response of simple structures to shock pressures and has given a considerable amount of useful information. Half-scale gauges have also been used to demonstrate for special cases the validity of an extension of the similarity principle to structures, known as Hopkinson's rule. This rule states that if all the linear dimensions of a structure and the explosion to which it is subjected are scaled by a factor λ, the resultant damage will be increased in the same scale. Although the rule

[6] A detailed account of results obtained with this gauge at the Underwater Explosives Research Laboratory (Woods Hole) is given in a report by P. M. Fye and J. E. Eldridge (39).

is known not to be generally valid it was found true for this type of gauge.

The failure of diaphragm gauges to respond in a simple way to shock wave pressures is shown by the variation of maximum depression with charge weight and distance. These variations are illustrated for small charges by Fig. 5.6, in which central deformations as a function of charge weight are plotted to logarithmic scales for 2 distances. These and other results show that for charge weights up to 20 pounds, the deformation of steel diaphragms is proportional to the 0.6 power of weight and -1.2 power of distance. Other experiments with charges of 100 pounds or more (22, 23), give a weight exponent of 0.49 and distance exponent of -1.13. The differences of the exponents with range of charge weights and their failure to correspond exactly with simple shock wave parameters indicate the nonexistence of any simple correspondence with specific shock wave properties.

Despite the limitations of diaphragm gauges for direct measurement of fundamental quantities, they have continued to be very useful in explosive comparison. This usefulness is the result of comparative simplicity and excellent reproducibility when properly used. In addition, their use also furnishes valuable information for study of structural damage, as already mentioned and discussed in more detail in section 10.5. (Fig. 5.6, for example, shows the agreement with experiment of a theory developed by Kirkwood.)

5.4. Piezoelectric Gauges

Although the various mechanical gauges described in the last section make possible fairly detailed measurements of shock wave pressures, they evidently cannot practically give a continuous record of pressure versus time and may be quite inadequate for investigation of pressure waves which possess any appreciable irregularities of structure. They are also employed with increasing difficulty for very small scale measurements. Perhaps their most serious limitation is their inability to measure simply and directly the weaker but more sustained secondary pressure impulses following the shock wave as a result of later motion of the gas sphere. For such measurements, what is evidently wanted is a pressure sensitive device which can be arranged to give a continuous indication of pressure as a function of time. The natural type of development is then some kind of electromechanical gauge or transducer to convert pressure changes into a varying electrical signal which can be translated into a visible indication by some type of oscillograph and recorded photographically.

The simplest and most direct pressure-sensitive electrical phenomenon is that of piezoelectricity, the property exhibited by some classes of crystals of developing electrical charge on certain crystalline faces in response to applied mechanical stress. The effect is exhibited only by crystals of fairly low degree of symmetry, the practically most useful being quartz, tourmaline, and Rochelle salt. The first application of piezoelectric methods to measurement of underwater explosion pressures is apparently that of Keys (58) in 1921, as a result of a suggestion by Sir J. J. Thomson that the naturally occurring mineral tourmaline would be a suitable material. At about the same time, Langevin was developing the use of quartz crystals for underwater signalling and Nicolson had made pioneer investigations of Rochelle salt crystals as electromechanical transducers. The experiments of Keys demonstrated the feasibility of the method and tourmaline gauges have been extensively used by the Admiralty Research Laboratory in England from 1919 to the present.[7] Up to the outbreak of World War II these investigations were virtually the only sustained program of underwater explosion pressure measurements.

The original form of equipment used in these experiments was very simple. The gauge consisted of a mosaic of tourmaline crystals attached to a steel plate, which formed the low potential (grounded) electrode, a foil electrode connected the crystal faces of opposite polarity, and the whole assembly was mounted in a waterproof case. The total crystal area was some 30 square inches and the linear dimensions of the gauges were of the order 5–8 inches. With these large gauges, sufficient signal was developed by pressures of the order 200–2,000 lb./in.^{2} to permit direct connection of the gauge through cables several hundred feet long to the electrostatic deflecting plates of a cathode ray oscillograph. In the earlier experiments a time scale was provided by horizontal deflection of the trace from a condenser discharge circuit suitably synchronized with the arrival of the pressure wave at the gauge. The resulting deflection of the electron beam was recorded photographically on a film placed in the evacuated cathode ray tube chamber. The deflection sensitivity and time scales were calibrated by impressing a transient oscillation of known amplitude and frequency. Although this original technique has been refined and extended greatly as a result of advances in electronics, it contains virtually all the essential components found in later developments.

At the beginning of World War II, it became evident that piezoelectric or similar techniques were essential to obtain comparisons of military explosives, estimates of their effectiveness in various circumstances, and to make possible comprehensive studies of underwater explosion pressures. With these and similar objectives in mind, experimental programs were started at a number of laboratories in this country, the most extensive being at the Taylor Model Basin (U. S. Navy, Bureau of Ships) and, under contracts of Division 8 and Division 2 of the Na-

[7] Many of these investigations have been reported by Wood (123).

tional Defense Research Committee, at the Underwater Explosives Research Laboratory (Woods Hole Oceanographic Institution, Woods Hole, Mass.). The experimental techniques employed at these and other laboratories employed tourmaline pressure gauges perhaps more extensively than all other kinds put together. As such techniques were used to obtain most of the results on explosion pressures mentioned in this book, they will be described in some detail in this and following sections.

Of the various types of crystals which might be used for piezoelectric explosion pressure measurements, tourmaline stands virtually alone in its fortunate combination of properties. Mechanically, it is quite rugged and stable and can be so used as to withstand pressures of the order 40,000 lb./in.^{2} or more without damage. Although its absolute sensitivity, expressed as electrostatic charge per unit area developed by unit pressure, is much smaller than for Rochelle salt or similar crystals, it is comparable to quartz in this respect and has the great practical advantage over either quartz or Rochelle salt of responding to hydrostatic pressure.

The fundamental piezoelectrically active orientation of tourmaline is a cut perpendicular to the optic axis, charge being developed on faces perpendicular to the axis in response to pressure being applied both on these faces and on faces parallel to the axis. If the ratio of piezoelectric charge Q/A per unit area developed by hydrostatic pressure P is denoted by K, it is readily shown that the crystalline symmetry of tourmaline requires that $K = d_{33} + 2d_{31}$, where d_{33} and d_{31} are, in Voigt's notation, the piezoelectric moduli for stresses applied parallel and at right angles to the optic (Z) axis.

For X-cut quartz crystals, which have been used in pressure measurements, the faces of the crystal are perpendicular to the mechanical axis. The charge developed on these faces by unit stress parallel to the axis is represented by d_{11}, but stresses at right angles develop a charge $2d_{13}$ and it is easily shown that for equal principal stresses (hydrostatic pressure) the total charge, which is the sum of the effects, is zero. Hence if a quartz crystal is to be used, its edges must be protected from applied pressure. In some gauge designs, this has been done by mounting a circular quartz disk in a coaxial steel ring intended to relieve pressure on the edge of the disk.

It is well known that Rochelle salt and the more recently developed ADP (ammonium dihydrogen phosphate) crystals are piezoelectrically active for shearing stress. In X-cut crystals the charge is developed on faces parallel to the optic axis as a result of shearing stress applied to edges parallel and perpendicular to this axis. As for quartz, the crystal symmetry requires that the response to hydrostatic pressure be zero. As ordinarily used, Rochelle salt or ADP gauges employ crystals cut with edges at 45° and 135° to the optic axis and a normal stress applied only to one pair of edges results in a shearing strain, the other pair and the faces being protected by a rigid housing.

The advantage of sensitivity to hydrostatic pressure is, of course, the fact that no mechanical protection need be provided to permit only certain directions of strain being effective. Any such housing increases the effective size of the gauge, which is undesirable for gauges which must respond to pressures rapidly changing in space and time, and is also liable to shock resonances developed by the applied pressure wave which falsify the crystal response. From this viewpoint, there is no advantage but rather a definite difficulty in using quartz rather than tourmaline, and there is no advantage to quartz from the point of view of sensitivity, the two differing by only a few per cent. Although both Rochelle salt and ADP have much higher sensitivity, they also suffer from the disadvantage of having no hydrostatic response and must therefore be protected from unwanted stresses. The crystals are physically much less rugged, an important consideration when explosion pressures are involved, and Rochelle salt is very hygroscopic, likewise a serious problem when used underwater or in humid surroundings. Nevertheless, occasions may arise when the higher sensitivity attainable outweighs these drawbacks and both Rochelle salt and ADP gauges have been used to a limited extent for such purposes. (The statements made here refer of course to measurements of explosion pressures rather than underwater sound studies where the much lower pressure levels usually rule out tourmaline.)

As applied to explosion tests, Rochelle salt also suffers from its large temperature effect. This is due largely to ferroelectric properties which make its electrostatic capacitance vary by a factor of 10, becoming very large in the neighborhood of the Curie points at −10° C. and 25° C. The open circuit voltage changes by only about 20 per cent over the same range, and if the crystal is connected directly to a preamplifier or other high impedance circuit, the latter condition is quite closely realized. This method of use presents obvious difficulties when the amplifier must function properly during the passage of intense shock pressures. The other alternative, of connecting the crystal to a cable of considerable length, loads the crystal to such an extent that the output signal is determined by the coulomb sensitivity and its large temperature variations. These difficulties are present to a much smaller degree in the less sensitive ADP crystals which nevertheless have about ten times as large a response for a given size as tourmaline. ADP gauges have been used to a limited extent for explosion measurements and, for relatively large scale measurements at least, have showed considerable promise. A further difficulty with Rochelle salt gauges is transient distortion encountered with slowly changing pressures.

As a result of the considerations outlined and others, tourmaline pressure gauges have been the main reliance in determination of transient pressures. There are certain difficulties which must be considered in any use of tourmaline or other piezoelectric gauges. The first of these is the fact that, regarded as a circuit element, a piezoelectric gauge has the fundamental characteristics of an ideal generator in series with a small capacitance, of a few hundred micromicrofarads at most, which represents the electrostatic capacitance of the crystal.[8] The gauge is therefore inherently a high impedance device, a characteristic which introduces several complications in its use. Most devices suitable for recording fast transients are basically voltage sensitive and have a finite direct current resistance. The connection of such a resistance across the gauge thus provides a leakage path for the charge developed, and the voltage response to a suddenly applied and maintained pressure, for example, is an exponential decay, the rate decreasing as either leakage resistance or circuit capacitance is increased. The product of these two factors determines the rate of decay and must therefore be sufficiently large, a condition usually best achieved by making the leakage resistance large, as increased capacitance shunting the gauge reduces the voltage developed.

The necessity of essentially open circuit conditions for direct current or low frequencies not only presents requirements to be satisfied by the recording equipment to which the gauge is connected, but also presents problems in making this connection. If the electrical line is long enough, the time required for signals to be transmitted down its length is comparable with times of interest in transient pressure measurements and attention must be paid to the way in which such transmission takes place. It is a familiar fact that improper termination of electrical lines results in reflected electromagnetic waves and a distortion of the voltage at the end of the cable. This distortion is eliminated if the line is terminated by an impedance equal to the so-called surge impedance of the cable, which is quite closely equivalent to a resistance of the order 50–100 ohms. This value of resistance is very far from values of megohms necessary to prevent appreciable leakage of charge and special measures are therefore necessary to minimize distortion of this kind.

Another difficulty associated with high impedance gauges is the vulnerability of such circuits to unwanted signals. Pickup from disturbances in the neighborhood can be minimized by proper shielding and grounding of circuits, but another source of disturbance encountered in use of electrical cables, known as "cable signal," must be carefully considered. This phenomenon manifests itself as electrical charge de-

[8] This result is readily obtained from the basic piezoelectric equations (see Reference (3) for example), but is sufficiently evident not to justify a detailed proof here.

veloped as a result of stress on the cable and, in transient pressure measurements, results in an increasing potential difference between the cable conductors as more of the cable is struck by the pressure wave. The exact cause of the effect is not fully understood but in coaxial cables appears to be developed between the dielectric and outer shielding braid by some sort of frictional process. Its effect is evidently particularly serious in measurements of small pressures following an initially much greater pressure rise, as are encountered, for example, in an ordinary shock wave.

Another type of effect closely associated with piezoelectricity is the pyroelectric effect in crystals, in which electric charge is developed as a result of pressure changes. This effect may be a genuine one or a "false pyro charge," which is really piezoelectric charge induced by strains set up in the crystal by differential expansion from temperature changes. In tourmaline a temperature rise of 1° C. has been found to develop a charge corresponding to a decrease in pressure of about 200 lb./in.^{2} This effect is eliminated if both measurement and calibration are carried out under adiabatic conditions, but temperature changes in either situation resulting from conduction into the crystal of heat externally developed can cause appreciable errors. Fortunately, the effect is not appreciable for practicable conditions in underwater pressure measurements and calibration. It can, however, cause serious errors in air-blast measurements where proportionately much greater temperature changes are encountered and the durations are much longer.

Piezoelectricity is of course not the only pressure-sensitive electrical phenomenon which can be utilized as a measuring technique. Some of the more promising of other possible methods are use of pressure coefficient of resistance of metals (e.g., manganin) or electrolytic solutions; magnetostriction, in which the flux changes of magnetic alloys resulting from stress develop induced voltages in suitably arranged coils; the condenser microphone principle, in which mechanical displacements under stress cause capacity changes in an electrical circuit. None of these principles has as yet been developed into gauge designs as widely adaptable as the tourmaline or other piezoelectric gauges but some of them present attractive possibilities.

5.5. Response Characteristics of Gauges

An effect which must be considered in any gauge design is the relation of its size and mechanical properties to its response to rapidly changing pressures. Obviously, the ideal gauge for such measurements would be of vanishingly small size, so as to offer no obstruction to pressure waves or flow in the medium. This is hardly a practical idealization and the effect of finite size and specified mechanical properties of a gauge on its measurement of pressure and disturbance of the pressure to be measured must therefore be considered.

The simplest analysis bearing comparison with reality is that of the effect of finite gauge area.[9] Neglecting for this analysis any internal reflections or diffraction of the pressure wave by the gauge, we may consider the steady state response of a thin circular disk to a sinusoidal pressure acting on its face. If we let the response of unit area to a

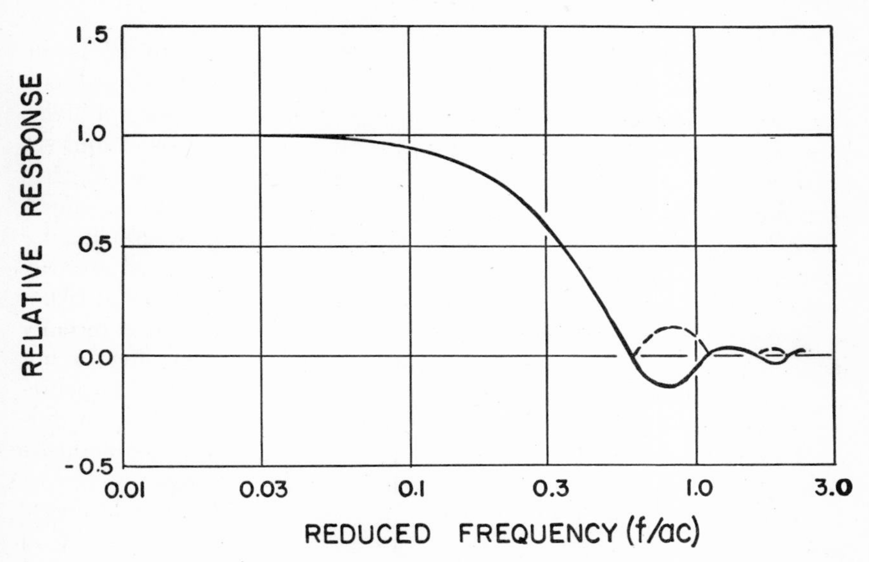

Fig. 5.7 Effect of finite gauge size on frequency response.

pressure P be KP and assume a pressure $P_m \cos 2\pi x/\lambda$ where λ is the wavelength, the response $R(\lambda)$ of a disk of radius a is found by direct integration to be

$$R(\lambda) = K(\pi a^2)P_m \frac{2J_1(2\pi a/\lambda)}{(2\pi a/\lambda)}$$

where $J_1(X)$ is Bessel's function of the first order. This result is for a standing wave. Letting $\lambda = c/f$, where c is the velocity of the wave and f its frequency, we obtain for the response to a wave $P(t) = P_m \cos 2\pi ft$, travelling parallel to the face of the disk,

$$R(f) = KAP_m \frac{2J_1(2\pi fa/c)}{(2\pi fa/c)} \tag{5.12}$$

[9] The analysis which follows is based on a report by Arons and Cole (3).

where $A = \pi a^2$ is the gauge area. The relative response $R(f)/KAP_m$ is plotted in Fig. 5.7 as a function of $(a/c)f$ and is seen to fall to zero for a frequency $f_o = 0.61c/a$ and to pass through a series of maxima and zeros of response for higher frequencies. The useful frequency range of such a disk, as limited by the transit time $2a/c$ across its face, thus extends to an upper limit somewhat less than the value f_o. For a gauge ½ inch in diameter in water this frequency is 150 kc./sec.

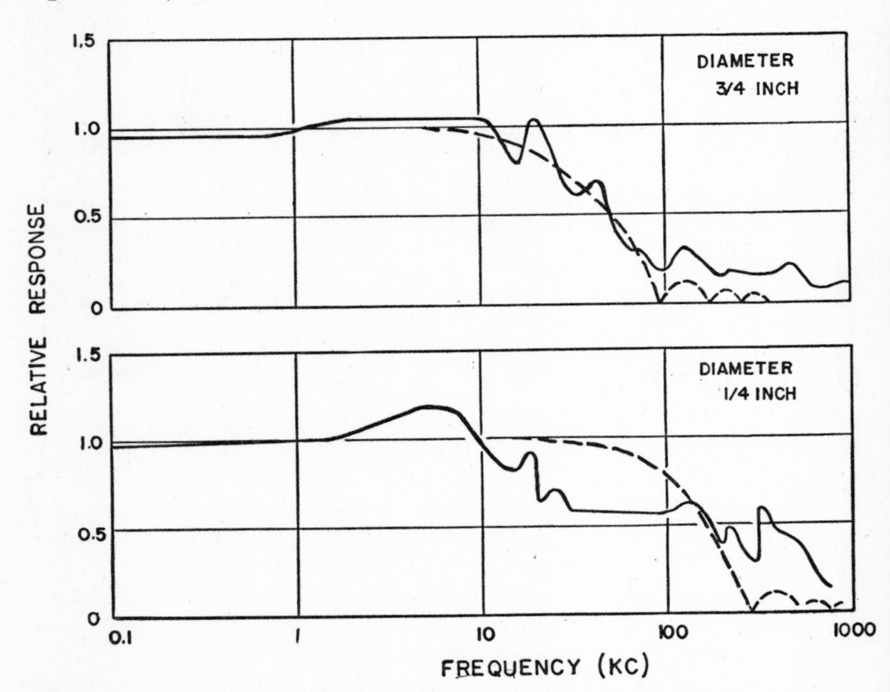

Fig. 5.8 Measured frequency responses of circular disk tourmaline gauges. The dashed curves are calculated from the geometrical size effect.

Steady state response measurements have been made on ¼ and ½ inch circular disk tourmaline gauges at the Mountain Lakes station of the Underwater Sound Reference Laboratory (Division 6, NDRC) by an experimental method developed for calibration of underwater sound hydrophones, which depends on the so-called reciprocity relation between the functioning of an acoustic transducer as a source and detector.[10] The results of these tests are shown in Fig. 5.8, where for comparison the responses predicted by Eq. (5.12) are indicated by dashed lines. The general agreement, apart from irregularities in the experimental curves, indicates that the very simple concept of transit

[10] The theory of reciprocity calibrations is given in a report by Dietze (27), and the measurements described here are from another report by the same author (28).

time of the wave across the gauge suffices to explain the general features of the response.

In measurements of explosion pressures one is interested of course in the response to a transient pressure rather than a sustained sinusoidal wave. The calculation for a circular disk with its face perpendicular to the wave front can be made by elementary methods for given forms of the transient. For mathematical convenience the response has been calculated for a suddenly applied step pressure P_m and for a saw-tooth wave for which

$$P(t) \begin{pmatrix} = 0 & , \quad t < x/c \\ = P_m \left[1 - \dfrac{t - x/c}{\theta}\right], & t > x/c \end{pmatrix}$$

This linear decay can be taken, to a sufficient approximation, to represent the initial portion of a shock wave. The response $S(t)$ for a step pressure has the form

$$\frac{S(t)}{KAP_m} \begin{cases} = 0 & , \quad t < -a/c \\ = \left[\dfrac{\varphi}{\pi} - \dfrac{\sin\varphi\cos\varphi}{\pi}\right], & -a/c < t < a/c \\ = 1 & , \quad t > a/c \end{cases}$$

where $\varphi = \cos^{-1}(-ct/a)$

The response $T(t)$ to the saw-tooth wave can be written

$$\frac{T(t)}{KAP_m} \begin{cases} = 0, & t < -a/c \\ = \left[\dfrac{\varphi}{\pi} - \dfrac{\sin\varphi\cos\varphi}{\pi} - \dfrac{a}{c\theta}\left(\dfrac{\sin\varphi}{\pi} - \dfrac{\varphi\cos\varphi}{\pi} - \dfrac{\sin^3\varphi}{3\pi}\right)\right], & -a/c < t < a/c \\ = 1 - t/\theta, & t > a/c \end{cases}$$

where the origin of time is taken to be the value for which the front of the wave is at $x = 0$. The relative response $T(t)/KAP_m$ is plotted in Fig. 5.9 as a function of the reduced time t/θ for various values of the ratio $a/c\theta$. The general feature of these curves is what one would intuitively expect: a rounding of the initial discontinuity followed by a closer approach to the true curve ($a/c\theta = 0$) as its rate of change decreases. The most serious differences occur for pressures at or near such shock fronts, which therefore determine the upper limits on gauge size for a tolerable error. For example, if one wishes to measure the peak pressure of a shock wave with a decay constant $\theta = 400$ microseconds to an accuracy of 2 per cent the ratio $a/c\theta$ should not, on the basis of the calculation, exceed 0.02. For water $c = 1.5 \times 10^5$ cm./sec. and the allowable radius is $a = 0.02 \times 1.5 \times 10^5 \times 4 \times 10^{-4} = 1.2$ cm.

Several objections can be made to this simple analysis: first, that the actual situation for a gauge of finite volume must be much more complex as a result of reflection, diffraction, and Bernoulli flow around the gauge; second, that orienting a thin disk with its face parallel to the pressure wave front is more efficient in reducing the transit time. Any very realistic attempt at a solution of the diffraction problem is of course rather complicated. The effect of reflections at the gauge surfaces can be seen qualitatively for a crystal gauge by considering an infinite slab

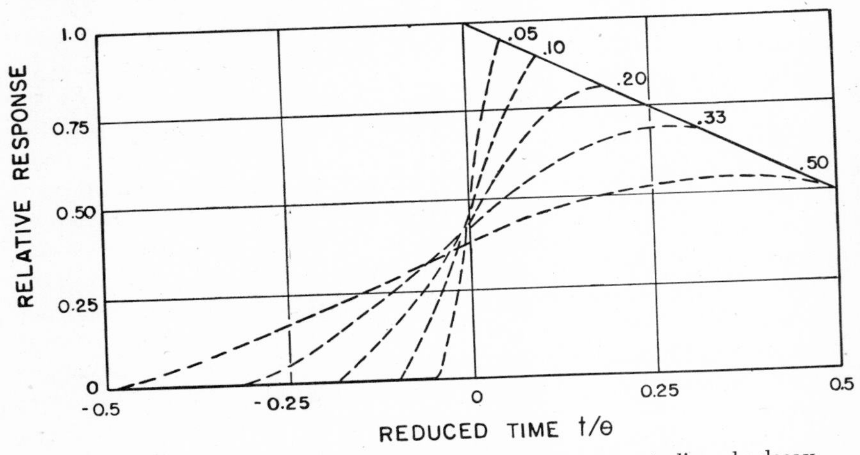

Fig. 5.9 Calculated response of a gauge of finite diameter to linearly decaying pressure.

of gauge material with faces parallel to the wave front. An incident step pressure wave is partially reflected at the first face because of the difference in density and sound velocity of the two media, the transmitted wave in the crystal is again partially reflected as a rarefaction at the second face (assuming a gauge medium denser than water), and a series of reflections of decreasing amplitude takes place.

The integrated pressure in the slab is easily seen to resemble a damped oscillation converging on the final equilibrium value, the period being the time for one round trip across the slab with the velocity of sound in the medium. This process is of course considerably modified by diffraction for a gauge of finite extent but one might qualitatively expect some roughly similar shock excitation to take place. Oscillations of this kind have been observed in the response of tourmaline disks subjected to explosion pressures, the effect being most marked if the face of the disk is parallel to the wave front. This larger effect is a factor in determining the best design and orientation of tourmaline gauges. It has been found empirically that the most satisfactory arrangement for general purposes consists of a stack of two or four disks in parallel, the faces of the disks being set parallel to the direction of advance of pressure. With this design the effects of internal reflections and the resultant oscillations of response are reduced, and the transit time for a required sensitivity is kept within reasonable limits. Some representative tourmaline gauge designs of this and other types are described in the next section.

5.6. Piezoelectric Gauge Design and Performance

A considerable variety of designs for piezoelectric gauges has been evolved for various types of applications and as a result of experience acquired in their use. The more detailed descriptions in this section are confined primarily to gauge designs which have been developed by the Woods Hole Laboratory and other cooperating groups. The material presented is not, therefore, a complete summary of the developments which have been made by all the various laboratories. The designs discussed, or similar ones, have, however, been the ones most extensively used and were employed in most of the explosion pressure measurements given elsewhere in this book.

As a result of the considerations outlined in sections 5.4 and 5.5, tourmaline has been the most extensively used piezoelectric material, both in the United States and in England. As mentioned in section 5.4, the pioneer development of tourmaline gauges was carried out in England after 1919 in a continuing program of research employing rather large units embodying a mosaic of crystals. Until quite recently, the basic design employed by the Admiralty Research Laboratory was used largely in England without major changes for measurements and comparisons of large charges.[11] Private communications have indicated that a program of investigation on small charges employing smaller gauges has been undertaken at the Admiralty Undex Works at Rosyth, but no reports of this work were available at the time of this writing.

A considerable program of gauge development in the United States began late in 1941 which was aimed chiefly at extending the range of usefulness and precision obtainable with tourmaline gauges. The initial efforts along these lines were made by A. B. Arons and E. B. Wilson, Jr. at Harvard. Much later development work and construction of gauges on their basic design, which might be called the hourglass gauge, was carried out at the Stanolind Oil and Gas Co. Development Laboratory under the direction of Daniel Silverman (104) and this group supplied

[11] A standardized design is described by Bebb (6).

gauges to a number of laboratories carrying out research programs. Further improvements were carried out in a later design, which will be called the Type B gauge, and was developed at the Reeves-Ely Laboratories under the direction of A. B. Arons and Clifford Frondel (38). In addition to these two basic designs, a number of special-purpose gauges have been evolved, some of which are described in part (C). It is beyond the scope of this section to give even a reasonably complete

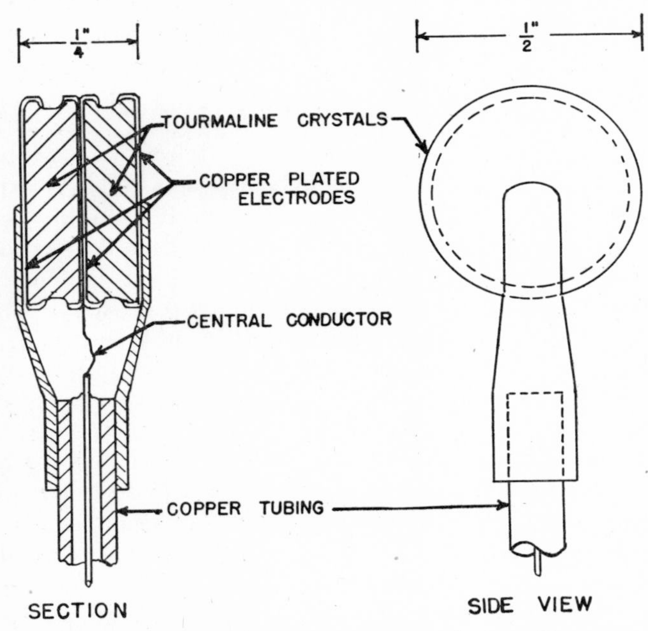

Fig. 5.10 Construction of hourglass crystal tourmaline gauge.

list of gauge developments, and reference should be made to more exhaustive discussions for further information.[12]

A. *The hourglass gauge.* This gauge design is based on the use of one or more circular tourmaline disks, the faces of which are chemically silvered and then electrolytically copper-plated for greater strength. It was found that the electrodes frequently peeled from the crystal if they made contact only with the faces. To remedy this, the edges were undercut to give the crystal a "waist" or hourglass appearance, which gave rise to the name. Electrodes are then applied by silvering and plating the entire crystal after which the metal in the undercut is ground away, as shown in Fig. 5.10, thus leaving electrodes plated over the edges to provide better mechanical adhesion. The crystal element

[12] See, for example, the reports by Greenfield and Shapiro (42), and by Arons and Cole (3).

is attached to the commonly used copper-tubing cable as shown, and the assembly is covered with a waterproofed coating.

The coating operation is of great importance, as the completed gauge must have high electrical resistance when in salt water and must remain watertight after being subjected to explosive shocks. The necessary qualities for a coating material are: toughness and strength without brittleness; freedom from air bubbles which collapse under transient pressures or tensions; good adhesion to the gauge assembly; and freedom from spurious electrical signals. It is also highly desirable that a thin coating be adequate, as the transient response characteristics are affected by the overall size, and ease in application and removal of the coating materially simplifies both the original construction and the repair of a damaged unit. Many types of coating, such as rubber molding, rubber tape and rubber cement layers, various lacquers and plastics, have been tried with varying degrees of success, but the best results have been obtained at Woods Hole using Zophar wax C276, a compounded mineral wax. Two or more thin layers of this wax are applied by dipping the gauge assembly into the molten wax and finally a thin film of tough lacquer (Tygon TP21) is applied. Although gauges so coated are not immune to failure, they have by comparison with other methods tried proved extremely rugged and easy to use and repair.

The hourglass gauge constructed as outlined, either with a single crystal or with 2 or 4 crystals, has been evolved and used in great numbers since 1941 and has proved quite successful. Most of the piezoelectric measurements made in the United States have involved the use of this or very similar designs. Gauges of this type have been used to record pressures from values of the order of 100 lb./in.^{2} up to 30,000 lb./in.^{2} or higher and have been employed both for shock and bubble pulse measurement. In the interests of a more symmetrical construction and to obtain greater sensitivity, 2 crystals mounted face to face have usually been used. The smallest practical crystal size with the hourglass construction is about 1/4 inch diameter by 1/8 inch thick, due to the necessity for undercutting the edge. Crystals much larger than 1/2 inch in diameter gave increasingly frequent gauge-to-gauge discrepancies and changes in sensitivity, the cause of which was not definitely determined but probably resulted from lifting of the electrodes from the crystal near its center. Difficulties with systematic discrepancies between the smallest gauges on small charge work (1/10 to 5 pound charge weights) also indicated that the basic design, although it led to gauges successfully used in accumulating a large amount of experimental data, still left much room for improvement.

B. *The Type B gauge.* The basic principles of this gauge design were originally developed for construction of large air-blast gauges, and the primary consideration was elimination of the hourglass feature, so as to permit the use of thinner crystals without the wasted area of the faces resulting from undercutting. Electrodes exhibiting much greater adhesion to the crystal than is provided by chemical silvering and copper plating can be formed by baking on silver-spray solutions. It was found, however, that much subsequent soldering to the electrodes set up stresses and weakened the adhesion considerably, thus resulting in electrodes which frequently loosened or otherwise made poor contact.

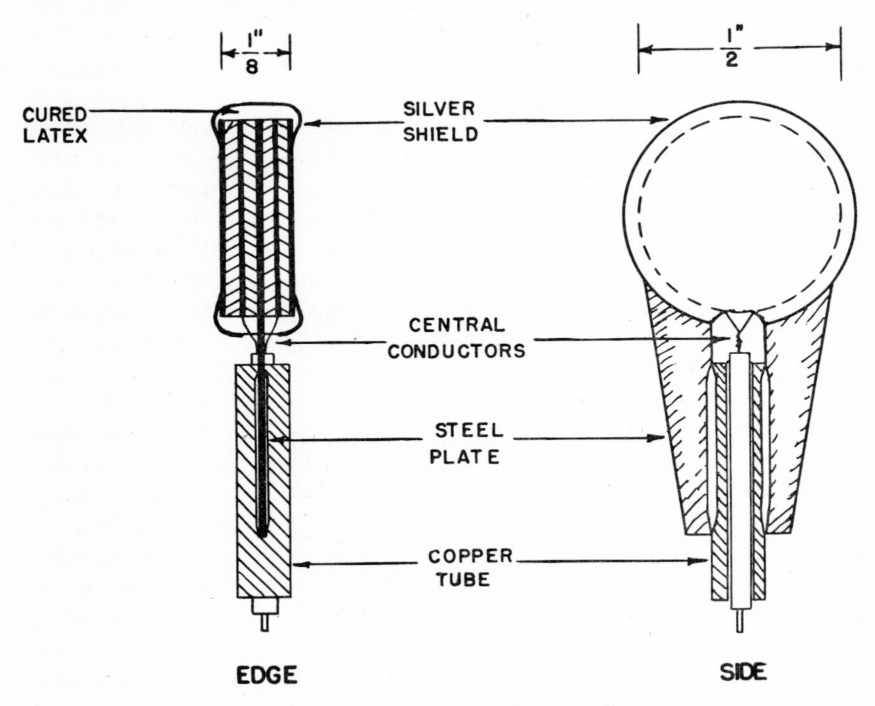

Fig. 5.11 Construction of Type B tourmaline gauge.

Gauges constructed by simply attaching electrical leads by local soldering were found to give unreliable and erratic results, and the Type B construction was devised to prevent local stresses and minimize handling of the electrodes. The construction is shown schematically in Fig. 5.11, and consists basically in sweating two or four circular disks to a central steel tab, to which a copper-tubing shielded cable (see section 5.8) is attached. Small leads are soldered to the other (ungrounded) faces of the crystals in the sweating operation, thus eliminating the need for subsequent local heating of the crystal by soldering. These leads are then connected to the insulated central conductor of the copper-tubing cable. The design also readily lends itself to balanced, or push-pull, gauge constructions. Either 2- or 4-pile gauges with a total thickness of $\frac{1}{8}$ inch to $1\frac{1}{8}$ inches have been successfully used, the gauge constants ranging from 2.5 (for $\frac{1}{4}$ inch doublet) to 90 (for $1\frac{1}{8}$ inch 8 pile) $\mu\mu$ coulombs per lb./in.^{2}

A detailed discussion of the performance characteristics of the Type B gauge is given in part (D), but it may be said here that the design fulfills the major requirements of a satisfactory gauge for recording explosion pressures in the majority of cases. Although it has not yet been used as extensively as earlier designs, sufficient field work and tests have been carried out to show the usefulness and reliability of the design for both shock wave and bubble pulse measurements over a wide range of pressures.

C. *Other piezoelectric gauge designs.* In addition to the large British gauges already mentioned, several interesting gauge designs have been developed by Osborne (80) at the Naval Research Laboratory. The most promising of these consisted of a small tourmaline element enclosed by an oil-filled Neoprene bag, or sheets, supported by rings several inches in diameter, leads from the electrodes being led out through small plastic tubing. This construction was intended to eliminate plastic coatings, which in earlier work of Osborne appeared to introduce nonlinearity of response at low pressure levels (< 100 lb./in.^{2}), and very small gauges of this type were used in measurements of small charges, for example, detonators. Although the design has considerable advantages in such cases, the relatively fragile construction makes it unlikely that such gauges would be practical for much larger charges.

In measurement of small explosion pressure changes, tourmaline gauges become difficult to use because of their relatively low sensitivity, and as a consequence some work has been done on development of piezoelectric gauges using the more sensitive Rochelle salt or ADP crystals. In tests made at Woods Hole, 45° X-cut crystals of these materials with cemented foil electrodes have been mounted in a lucite housing, as shown in Fig. 5.12, to prevent applications of pressure except on one pair of edges. The coulomb sensitivities of such gauges were of the order 400–500 and 115 $\mu\mu$ coulomb per lb./in.^{2} for Rochelle salt and ADP respectively. The Rochelle salt gauges were used only for qualitative measurements or timing purposes because of their large temperature coefficients of capacity and variation in response to a transient pressure with the duration of the pressure. Tests with ADP gauges were more successful and showed excellent quantitative agreement with tourmaline gauges in recording shock wave pressure-time curves; and calibration tests also indicated the absence of any significant low-frequency transient distortion as was encountered with Rochelle salt crystals. The primary difficulty with the ADP gauges which has limited their usefulness is the fragile nature of the crystals, which became gradually pulverized when used in large charge measurements. Gauges of this type should, however, be very useful if subjected only to low pressures or short durations of high pressure or if short life is not important.

D. *Performance of piezoelectric gauges.* Although the piezoelectric gauges which have been developed and used for underwater explosion pressure measurements are by no means foolproof or free of objectionable errors and difficulties in use, they are perhaps the most versatile single tool yet employed, and when properly used are capable of giving

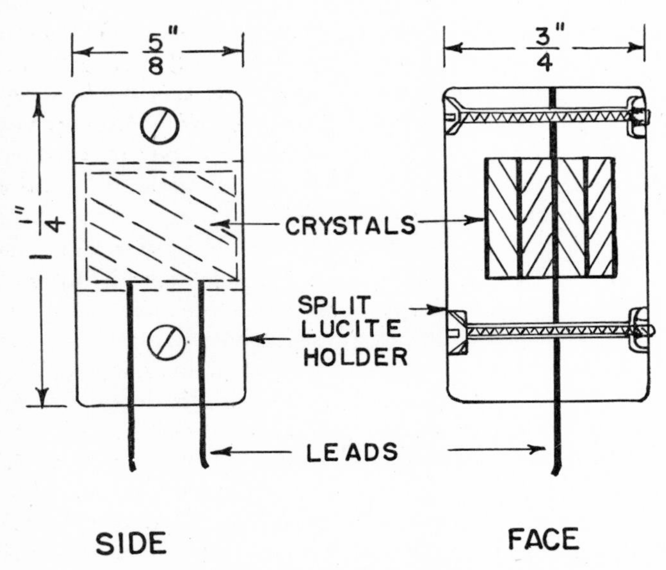

Fig. 5.12 Construction of gauge with Rochelle salt or ADP crystals.

very satisfactory quantitative results in a variety of applications. The pressure levels which have been measured with at least apparent reliability range from less than 100 to more than 30,000 lb./in.^{2}. Small $\frac{1}{4}$ inch Type B gauges have withstood pressures of the order of 90,000 lb./in.^{2} (3 inches from the center of a $\frac{1}{2}$ pound charge), failure under these conditions resulting from destruction of the copper tubing connecting cable rather than shattering of the crystal. Not all gauges will last for a large number of explosions at high pressures, but shock wave pressures up to at least 20,000 lb./in.^{2} from several hundred pounds of explosive can be practically measured as a routine matter. Tourmaline gauges of various sizes have been used to investigate secondary or bubble pulses of charges ranging from $\frac{1}{2}$ pound or less to 300 pounds in weight, and a variety of gauges have been used for providing synchronizing signals and measuring time intervals in underwater sound ranging. The tourmaline gauges used for shock wave measurements of service weapons at Woods Hole have for the most part been either the hourglass or Type B construction.

The majority of shock wave data on larger charges (50 to 700 pounds of explosive) cited in Chapter 7 were obtained with such gauges. In the work at Woods Hole, gauges were ordinarily used in pairs to provide two records of pressure under given experimental conditions, and four such pairs were routinely used in much of the work. The average deviation between two gauges of a pair in measurements of shock wave pressures was of the order of three-five per cent in routine work, and somewhat better precision has been obtained under more favorable experimental conditions. It has been possible in test work to distinguish between explosives differing by as little as two per cent in shock wave properties. It is fair to say that piezoelectric gauges constitute a proved and practical method for routine measurements on service weapons, both for comparison of effectiveness and for investigation of fundamental properties of shock waves.

The basic tourmaline gauge designs described have also been used to a considerable extent in investigation of secondary or bubble pulses. The earlier hourglass construction did not prove very satisfactory for this purpose because of difficulty in obtaining dependable gauges of sufficient sensitivity, but the Type B gauge design has been found quite satisfactory. The agreement between two gauges of a pair is not as good as for shock wave measurements, but a more serious difficulty limiting the accuracy of secondary pressure measurements, particularly for large charges, is lack of reproducibility of successive shots under presumably similar conditions. This question is more fully discussed in section 9.5, but it may be said here that the frequently disappointing accuracy of such measurements is only in part attributable to the instrumental technique. Even with the difficulties and errors of such measurements, piezoelectric gauges have provided much valuable information not otherwise easily obtainable, if at all, and their use has been fully justified.

In experimental investigations of underwater shock wave and other phenomena, it is frequently very desirable to work on a model scale with small charges, one pound or less of explosive, for reasons of simplicity in experimental arrangements and economy of time and money. As far as shock wave measurements are concerned, the main problem is one of obtaining sufficiently small gauges (in order to minimize distortion) of adequate and reliable sensitivity. Although the hourglass design of gauge has been used to a considerable extent in such investigations, both at the Taylor Model Basin and at Woods Hole, it never was wholly satisfactory. The major difficulties encountered were systematic differences in indicated pressures of supposedly similar gauges, sometimes amounting to ten per cent or more, which made absolute measurements open to considerable uncertainty in some cases, even if corrections were made for these differences. Comparative measurements, being unaffected by these errors if reproducible, could be made with more assurance, but even in this case the errors were larger than in work with larger charges and occasionally changes of gauge sensitivity were encountered. As a result of such difficulties, adequate small charge data with hourglass gauges were usually obtainable only by statistical procedures using a number of gauges in a series of shots. Individual gauge

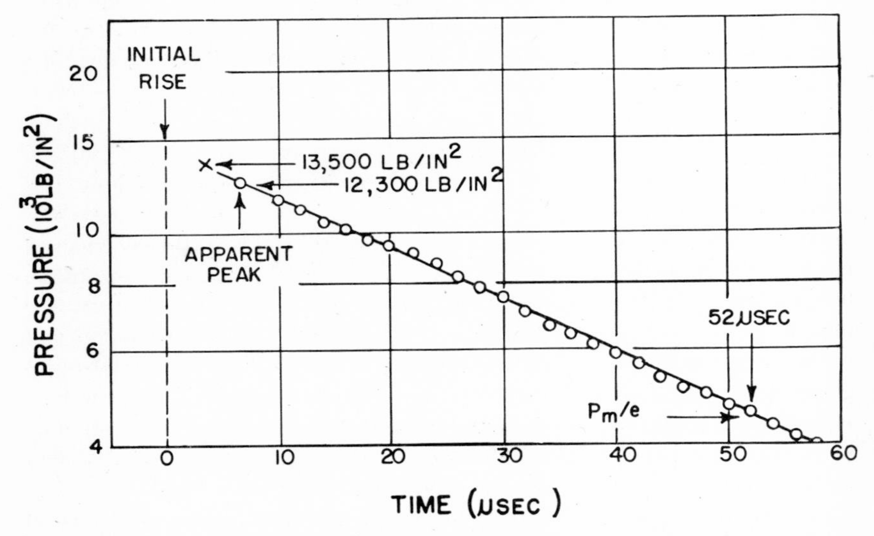

Fig. 5.13 Semilogarithmic plot of shock wave pressure against time to illustrate correction for gauge response (1.36 feet from a 0.55 pound spherical 50/50 pentolite charge).

records also often showed peculiarities of spurious bumps and oscillations not attributable to the pressure wave itself. (It is to be noted, however, that such irregularities may be real, and are more often encountered in small charge work, as discussed in section 7.6.)

The difficulties with small hourglass gauges were one factor leading to the development of the improved Type B gauge already described. Two-pile gauges of this later design using ¼ inch diameter disks have been fairly thoroughly tested in preliminary experiments with charges from 25 grams to 250 grams weight (5). These tests have indicated no systematic errors in gauge-to-gauge sensitivity exceeding the standard deviations of single gauge readings on successive shots. This latter error was found to be of the order of five per cent for the peak pressure of the shock wave, and is sufficiently small to make the use of Type B gauges for small charge work very practical and attractive. A fundamental difficulty which must be realized in such use is the distortion of the indicated pressure-time curve as a result of the finite gauge size. For peak pressure from a twenty-five gram charge, for example, the error predicted from the overall dimension of the gauge is seventeen per cent. If, however, the record is extrapolated back to the time at which the shock front has reached the gauge center, as shown in Fig. 5.13, a corrected peak pressure can be obtained. In the gauge tests mentioned, the correction found for twenty-five gram charges amounted to about fifteen per cent, in excellent agreement with the value predicted by consideration of gauge size. Other tests with larger charges gave similarly good results and hence it appears practicable to use the correction technique with confidence in its essential validity, provided the limitations of accuracy are realized.

5.7. Piezoelectric Gauge Calibration

The ideal pressure calibration of gauges, piezoelectric or otherwise, would be one in which a known transient pressure wave could be applied to the gauge under conditions similar to those in measurements made with the gauge. For this purpose, a suitable source should establish a pressure wave of simple form, accurately known by direct experimental proof or calculation, and amplitude comparable with pressures to be measured. This pressure wave should then be applied to the gauge and recording equipment as they would be used in measurement. Up to the present, no single method has been developed which satisfies all of these requirements, but the various techniques employed have led to fairly satisfactory determination of gauge characteristics.

A. *Static methods.* The simplest method of calibration is of course a static one in which a known hydrostatic pressure, read by a calibrated Bourdon or other pressure gauge, is applied to the gauge under test mounted in a liquid-filled pressure cylinder. This method in its simplest form is not applicable to piezoelectric gauges, the reason being that such gauges have no static response. For slowly changing pressures, a piezoelectric gauge is equivalent to an EMF V_o of value KAP/C_o in series with a capacitance C_o, the quantity KA being the gauge constant for applied hydrostatic pressure P, and C_o its electrostatic capacitance. Any external circuit to which the gauge is connected must have associated with it leakage resistance R and shunt capacitance C. (The leakage in general includes surface and volume leakage of the gauge crystal which, though never infinite, are very high for quartz and tourmaline.) With these parameters included, the schematic circuit to be considered is as drawn in Fig. 5.14(a). It is evident that the terminal voltage V developed by an EMF V_o resulting from an applied hydrostatic pressure decays exponentially with time to zero as indicated in Fig. 5.14(b). The time required for V to fall to $1/e = 0.368$ of its original maximum is given by the value of $R(C + C_o)$, the time constant of the circuit. For tourmaline gauge designs discussed in the preceding section, C_o has values of the order 10 to 200 $\mu\mu f$. Values of R greater than 10^{10} ohms are realized only by very special precautions and design of the complete circuit, and with $R = 10^{10}$ ohms, $C_o = 2 \times 10^{-10}$ farad, the value of RC_o is two seconds. This time can, of course, be increased by using padding capacitance C at the expense of decreased terminal voltage. In practical cases, leakage resistance and circuit sensitivity requirements limit the time constants conveniently obtainable to times measured in seconds rather than minutes. If the indicated voltage V is to be a direct measure of the EMF V_o it is evident that the times re-

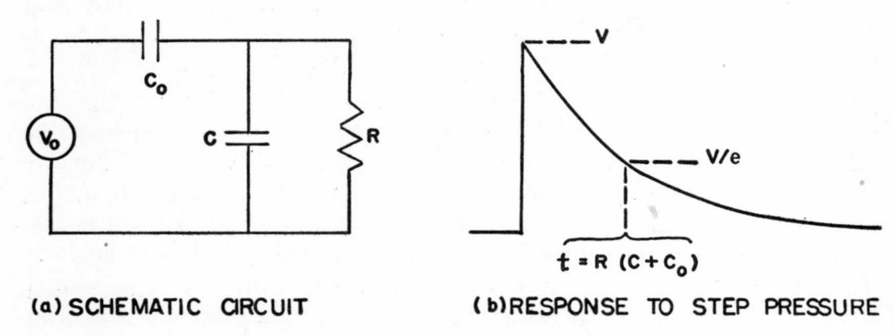

Fig. 5.14 Piezoelectric gauge circuit and long time transient response.

quired to establish the pressure and measure V must both be a small fraction of the time constant, say one per cent; as a result, slow pumping up of pressure and determination of V by a slow period meter are both ruled out.

The impractical requirements of static methods can be avoided by use of what may be called a quasi-static method, in which the gauge is allowed to come to equilibrium with an applied hydrostatic pressure; this pressure is then rapidly released by manual opening of a valve or by bursting of a diaphragm.[13] The resulting voltage is then measured by a high impedance voltmeter circuit, such as an electrometer type circuit, or by a recording oscillograph. Methods of this type permit pressure changes to be produced and their effects indicated in times of a few seconds or less; the time constant requirements can be conveniently realized, and the pressure level can be determined by standard methods.

Although these quasi-static methods might appear to satisfy adequately the primary requirement for a calibration, they suffer from two sources of error which impair their usefulness. The first of these is the phenomenon of cable signal (discussed in more detail in section 5.8), which manifests itself here as an electrical signal developed by electrical

[13] See, for example, the reports by Borden (9), and by Arons and Cole (3).

cables subjected to stresses in packing glands through which they are led to external circuits from the pressure cylinder. Difficulty with these signals can be avoided by using rigid leads through bushings in the pressure chamber, but if this is done, the crystal elements of a gauge must be calibrated prior to their assembly in the final gauge plus connecting cable, and the possibility enters of changes resulting from such mechanical operations.

A second source of error is the pyroelectric effect, or generation of electrical charge in a piezoelectric material as a result of temperature changes of the crystal. For tourmaline the magnitude of the effect is such that a temperature increase of 1° C. gives the same signal as a negative pressure of 200 lb./in.^{2}. Compression of either a gauge crystal or its surrounding medium always leads to such temperature changes, which can be calculated from the thermodynamic properties of the substance (specific heat, compressibility). Such temperature changes must occur both in the calibration and use of the gauge. Transient underwater pressure waves are in nearly all cases of such short durations that the crystal is heated adiabatically, and if the calibration conditions are also adiabatic there will be no error from the pyroelectric charge. If, however, the time of measurement is long enough, there may be an appreciable heat exchange with the external medium. In general, suitable pressure fluids exhibit much greater temperature rises than tourmaline. The apparent piezoelectric charge therefore decreases with time, owing to the negative sign of the pyroelectric effect for tourmaline, as has been experimentally observed (9). The presence of this error has the same effect as electrical leakage and sets a further condition on the maximum time of measurement.

If the sources of error described are realized and their effect can be made small, the quasi-static method furnishes a fairly satisfactory means for routine calibration which is widely used with good success. Both manual and diaphragm pressure release are practicable, the former being more convenient if many readings are wanted and the time necessary is not too long. Pressures up to about 5,000 lb./in.^{2} are readily obtained and measured with simple equipment and the electrical measurement presents no formidable difficulties. The method falls short of the ideal in that only relatively slowly changing pressures are involved which are not transient pressures in the sense of a travelling pressure wave.

B. *Steady-state calibrations.* Several methods are available which permit a better approximation to dynamic conditions than static or quasi-state calibrations. One of these, on which the Stanolind Oil and Gas Co. Research Laboratory has done preliminary tests (104), utilizes a reciprocating pump which produces approximately sinusoidal pressure variations up to 30 lb./in.^{2} at a frequency of 30 cps. This speed is sufficient to eliminate or greatly reduce pyroelectric and time constant errors. The method does not conveniently provide an absolute calibration of known pressure amplitude, and has so far only been used as a null method for comparing an unknown crystal against a standard one the constant for which has been otherwise determined. The method has shown considerable promise and should be a very convenient one, but has not as yet been fully tested.

An absolute method employing travelling acoustic waves has been developed by the Underwater Sound Reference Laboratory, operated by

Gauge	1	2	3	4	5	6
Static	0.85	3.45	3.55	5.35	11.2	95
Reciprocity	0.81	3.2	3.6	5.2	12.0	85

(a). Comparison of high pressure static with acoustic reciprocity calibrations of tourmaline gauges (sensitivities expressed in $\mu\mu$coul./lb./in.^{2}).

Gauge element	1	2	3	4	5	6
UERL, Woods Hole	1.35	3.53	3.83	3.82	11.1	23.2
Taylor Model Basin	1.34	3.53	3.84	3.91	—	—
Stanolind Lab.	—	3.64	3.86	3.87	11.2	23.4

(b). Interlaboratory comparison of gauge crystal calibrations (sensitivities expressed in $\mu\mu$coul./lb./in.^{2}).

Table 5.1. Calibration of tourmaline piezoelectric gauges.

Division 6, NDRC during the war. This method provides valuable information on the response characteristics of gauges, as discussed in section 5.5, but is not satisfactory for obtaining primary calibrations of explosion pressure gauges because of its limited accuracy and the very low pressure level. The limited accuracy is partly due to experimental difficulties in calibrating a steady state sound field (interference effects from reflections at boundaries of the finite medium), and partly results from the low sensitivity of gauges intended and suitable to measure much higher pressures than are obtained with acoustic sources. A pressure of 100 lb./in.^{2} is very low on the scale of underwater explosions, but sound projectors rarely develop pressures of more than a small fraction of an atmosphere under calibration conditions. This great difference is reflected in the very low signal developed by a tourmaline gauge tested by such pressures, and has the more fundamental disadvantage that any nonlinearity of gauge response with increasing pressure will not be detected. It is interesting that, in spite of these sources of error, the agreement between the steady state absolute calibrations and static calibrations at much higher pressures is within their combined experimental error, the results being shown in Table 5.1. Calibrations made at three different laboratories by similar, essentially static methods are included in the tabulation and indicate the satisfactory agreement obtained by different observers and equipment.

C. *Use of transient waves.* A fundamentally more desirable method than either static or steady state calibration would be one in which a shock wave, or discontinuity of pressure, with known amplitude and long duration passes the gauge to be calibrated. One possible method consists in releasing the pressure of a closed chamber, by bursting of a diaphragm for example, and initiating a compression wave in fluid at atmospheric pressure external to the chamber. Experiments of this kind attempted with thin copper diaphragms and oil as the pressure fluid met with only limited success, the difficulties being that, with the arrangement employed, the pressure wave developed had a rounded outline and was not very reproducible. This difficulty was the result of the time of about 100 μsec. required for rupture of the diaphragm and the failure of the wave so initiated to develop into a shock front in the rather unfavorable geometrical arrangement improvised for the tests. A further objection was the impossibility of obtaining an absolute calibration of the pressure; the method as tried was therefore useful only for comparison purposes.

A more promising line of approach is use of a plane shock wave in air formed by rupture of a plastic diaphragm separating a region at known gauge pressure from a tube containing air at atmospheric pressure. This method has been extensively employed in air-blast work[14] and has the advantage that pressure waves so initiated in air develop into shock fronts much more rapidly than in water. A further advantage of the method is the fact that the increase in shock wave velocity with pressure difference is very considerable in air even at moderate pressures; for example, an excess pressure of 20 lb./in.^{2} in air initially at atmospheric pressure and a temperature of 20° C. travels with a speed of 1,620 ft./sec., a value 47 per cent greater than the speed of an acoustic wave. This increase in velocity is, moreover, directly related to the pressure difference by the Rankine-Hugoniot equations (section 2.5) and velocity measurements of the progressive wave therefore permit absolute pressure determinations.[15] A gauge to be used in underwater measurements should of course be subjected to a pressure wave in water or a liquid, and this can be done by allowing the air shock wave to be reflected from

[14] A comprehensive report, with references to earlier work, is that of Fletcher et al (36).

[15] Experimentally determined pressures from velocity data are found to lie systematically below values predicted from the pressure released, indicating imperfect formation of a plane shock wave in the experimental tubes employed. See Reference (36).

a water surface and placing the gauge to be calibrated in the water. The pressure in the water will be double that in the air, except for divergence or attenuation of the wave and nonacoustic reflection (negligible for pressures less than 10,000 lb./in.^{2}). The duration of the wave so formed is determined by the lengths of direct and reflection paths, but is easily made adequate. A necessary precaution is careful determination of the initial pressure and temperature of the air in the tube, to both of which factors the velocity is sensitive. With suitable experimental modification, the "shock tube" methods used in air-blast

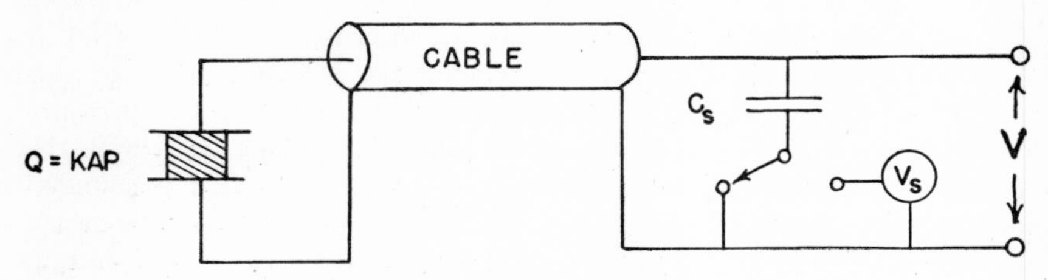

Fig. 5.15 "*Q*-step" calibration circuit.

calibrations may thus very well provide an excellent dynamic means of calibration, the chief disadvantage being the low pressure levels conveniently obtainable.

D. *Calibration standards.* The simplest absolute pressure standard is the free floating piston, or dead weight tester. For routine measurements to which it is applicable, a good quality Bourdon gauge, periodically calibrated against a dead weight tester, has been found very satisfactory for pressures up to 10,000 lb./in.^{2}, a sufficiently large value for most work. Another method for low pressures in air, based on the pressure-shock velocity relation, has already been mentioned (part (C) above). If much calibration of gauges is done, it is extremely convenient to use calibrated tourmaline crystals as a secondary standard, ideally in conjunction with a null comparison method. Carefully calibrated crystals are dimensionally stable and extensive checks have shown no significant aging effects.

Electrical calibration of the piezoelectric gauge circuit and associated recording equipment is also necessary and a very useful technique for this purpose and for calibration of pressure measurements is the so-called "*Q*-step" calibration circuit, shown schematically in Fig. 5.15. When the calibration response is to be recorded, the condenser C_s is connected in parallel with the gauge circuit, and the voltage V_P developed by pressure P is

$$V_P = \frac{KAP}{C + C_o + C_s}$$

where C is the capacitance of the cable. To calibrate the system, a known voltage V_s is applied to C_s in series with the gauge circuit and the resultant voltage V_c is given by

$$V_c = \frac{C_s V_s}{C + C_o + C_s}$$

The ratio of voltages V_P and V_c, and hence ratio of deflections of a linear indicating instrument, is then

$$\frac{V_P}{V_c} = \frac{KAP}{C_s V_s}$$

This ratio is, however, independent of both the circuit capacitances C and C_o and of the deflection sensitivity of the instrument. The gauge constant KA is thus determined in terms of known pressure P, voltage V_s, and capacitance C_s. In effect, the result is equivalent to impressing a charge $Q_s = C_s V_s$ on the circuit, hence the name "Q-step."

In actual use, the circuit also determines the transient response characteristics of the system for an equivalent charge Q_s, provided only that the condenser C_s has negligible shunt leakage currents and dielectric absorption current. The necessary step voltage V_s to determine a step response characteristic can be supplied by a calibrated source, either by a simple switch for static methods or by a more elaborate step generator for more rapid transient measurements (see section 5.9). A null method of gauge calibration has been described by Savic (97), in which the piezoelectric charge from pressure change is balanced against charge of opposite sign simultaneously applied from a known variable voltage source and capacitance, the circuit arrangement being quite similar to the Q-step technique. The piezoelectric constants originally obtained for tourmaline by this method are several per cent lower than accepted values, but later refinements in the method are reported to have given results in good agreement.[16]

5.8. Transmission of Piezoelectric Gauge Signals

As suggested in section 5.4, the internal impedance characteristics of piezoelectric gauges present several electrical problems in their use for measurement of underwater pressures. For greatest efficiency a piezoelectric gauge should be operated under essentially open circuit conditions, that is, with the minimum possible shunting effect by the input impedance of the electrical circuit connected to it.

Two effects of a low value of shunt impedance were discussed in section 5.7: neutralization of charge on the crystal faces by current flow

[16] Private communication from J. H. Powell of the Admiralty Undex Works.

through the shunting circuit, and reduction of terminal voltage by the crystal sharing charge with any shunt capacity. From this point of view, it is logical to think of using the crystal with a preamplifier or other impedance matching device close to it, which would minimize shunting effects and permit connection to the recording circuits with greater freedom of choice as to the electrical properties of the connecting elements. This solution, however, is a matter of considerable practical difficulty, as the impedance circuit must be sufficiently stable and rugged not only to survive the explosion and withstand rough usage, but also to operate properly while subjected to explosion pressures.

Although these difficulties can probably be overcome, the more profitable line of development has been found to be use of direct cable connection from the gauge to the recording equipment situated at a safe distance from the explosion. The necessary distance for service weapons is several hundred feet or more and use of cables this long presents a number of problems, both in handling and maintenance of the rigging and in achieving suitable electrical properties.

A. *Cable termination.* Although for slowly changing signals an electrical cable can be regarded as simply a lumped capacity, it is evident that for sufficiently rapid changes the time required for the signal to traverse the cable will no longer be negligible. The speed of propagation of a signal along an electrical cable is somewhat less than that of light or other electromagnetic radiation in free space, the value depending on the dielectric constant of the insulating material in the cable. For a cable of total distributed inductance L and total distributed capacitance C (these "total" values refer to values per unit length multiplied by the length), the transit time can be shown to be $\sqrt{LC}$ or, in terms of the surge impedance R_o of the cable defined as $R_o = \sqrt{L/C}$, transit time $= R_oC$.

If the surge impedance is expressed in ohms and the capacitance in microfarads the transit time is in microseconds. For a 1,000 foot cable with a capacitance of 30 $\mu\mu f$ per foot and 50 ohms impedance the time required for a signal to travel the length of the cable is 1.5 μsec. In this case, the cable can legitimately be regarded as a simple lumped capacity only for signals which change insignificantly over such time intervals. Even for such signals it is often necessary to examine the question of how this capacity is properly measured, as discussed in part (B).

If the cable cannot be treated as a lumped capacity, a more detailed investigation is necessary. It is a familiar fact that the propagation of signals along a cable generally results in electrical reflections at the ends of the cable unless the cable is properly terminated. The voltage at the receiving end of a cable is then an initial signal related to the original signal and the properties of the entire circuit, followed by a series of signals resulting from multiple reflections and traversals of the cable. These reflections are eliminated if the cable is terminated in a value of impedance characteristic of the cable and equal, for cables with negligible series resistance and leakage conductance, to the surge impedance $R_o = \sqrt{L/C}$. For practical cables, this surge impedance is a resistance of the order 50 to 100 ohms. It is evident that this requirement for no distortion by reflections is not directly compatible with the necessity of

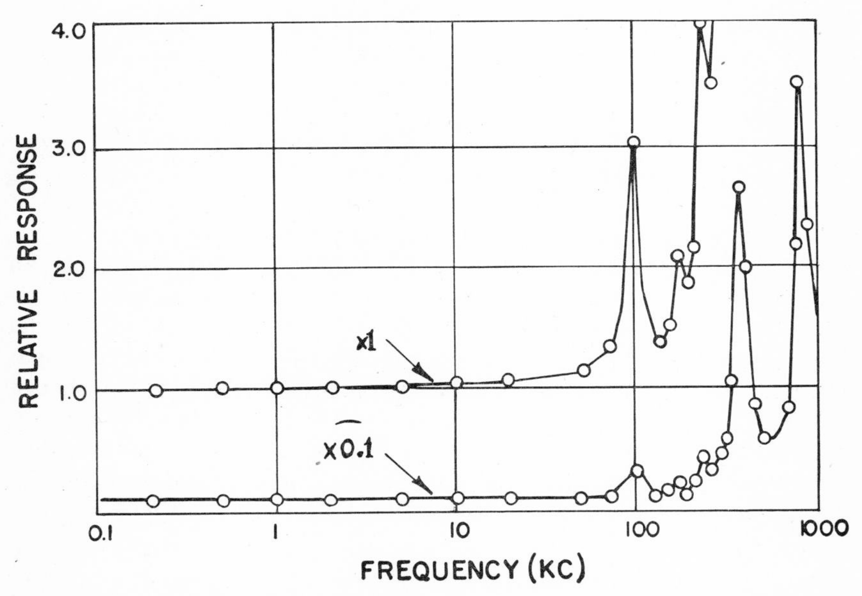

Fig. 5.16 Simulated frequency response of a piezoelectric gauge and unterminated cable.

high *d-c* resistance of the order of megohms across the piezoelectric gauge.

The type of behavior when a piezoelectric gauge is connected to a long cable without any termination at the receiving end is shown in Fig. 5.16. The cable used was 660 feet long with a capacity of 17×10^{-4} μf at a frequency of 10 kc. The characteristics of the gauge were simulated by an oscillator as source of constant voltage (20 volts) connected in series with a condenser of 100 $\mu\mu f$ representing the capacity of a gauge, and the output voltage was measured by a vacuum tube voltmeter. It is evident that for frequencies below about 20 kc. the output is independent of frequency and it has a value determined by the capacity of the cables. At higher frequencies, however, the response has a series of voltage maxima, characteristic of open-circuit resonances, for which the reflected signals reinforce each other. (These maxima would be infinite were it not for losses in the cable. It is also to be noted that the absolute value of response is plotted without regard to phase.) At these higher frequencies, therefore, the output signal bears no simple relation to the applied voltage.

In transient pressure measurements one is more interested in the response to a changing transient voltage, and the sketch in Fig. 5.17 illustrates the response observed when a fixed potential is instantaneously applied to the same network. The series of oscillations correspond to the arrival of successive reflections at intervals of $2R_oC$ (approximately 1.5 μsec. for the 660 foot cable already considered) and the damping is the result of cable and other circuit losses. These oscillations will of course obscure any variations of the applied signal in comparable time intervals, and if such variations are to be observed, a compromise between the condition for elimination of reflection and the condition for negligible circuit leakage at long times must be sought. This compromise is achieved by the use of terminating networks at the ends of the cable which, roughly speaking, are equivalent to capacities for slowly changing signals and approximate the surge impedance characteristics desired for rapidly changing signals.

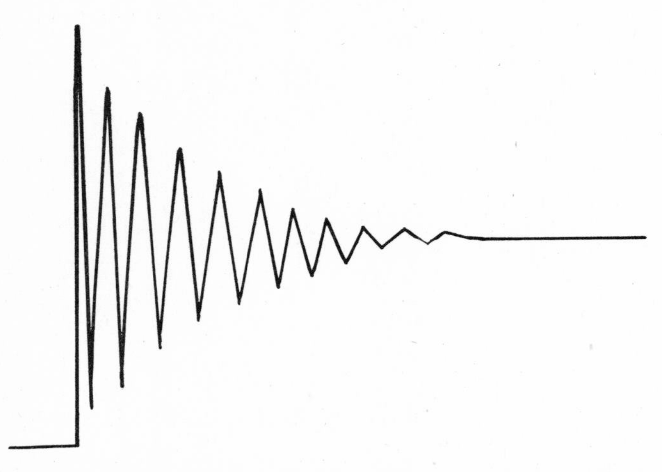

Fig. 5.17 Observed transient response of piezoelectric gauge and unterminated cable.

A sufficiently general network, similar to networks discussed by Burch (13), to provide adequate compensation is indicated schematically in Fig. 5.18. In this network, the source of voltage V_o and the capacitance C_o represent the piezoelectric gauge. The coaxial cable can, to a sufficient approximation, be regarded as loss free and hence characterized by series inductance L/d per unit length and capacitance C/d per unit length, d being the length of the cable. The input and output networks evidently reduce to shunt capacitances at low frequencies and approximate resistances for sufficiently high frequency of applied voltage. The steady state terminal voltage V_t for an applied voltage $V = V_o e^{pt}$, where $p = j\omega = j2\pi f$, and f = frequency, is found by standard methods[17] to be

$$V_t = C_o V_o \frac{pH(p)e^{-R_oCp}}{1 - J(p)e^{-2R_oCp}} \tag{5.13a}$$

In this equation, R_o is the surge impedance given by $R_o = \sqrt{L/C}$, and the quantity C_oV_o represents the charge developed by the equiv-

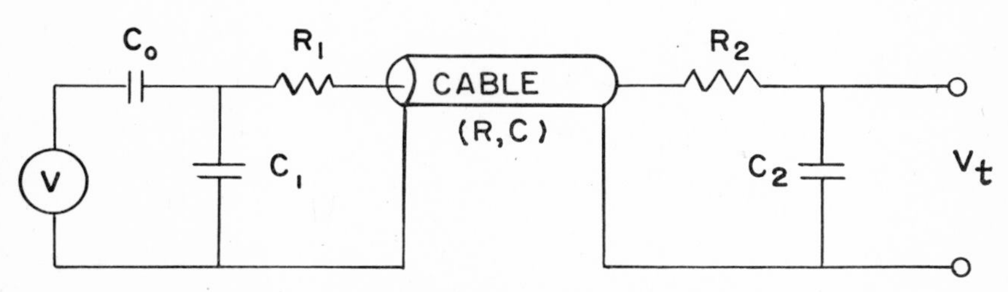

Fig. 5.18. Cable termination network for piezoelectric gauge.

alent piezoelectric gauge. The capacitance C_o can be neglected in comparison with the cable capacitance C, and with this approximation the functions $H(p)$ and $J(p)$, which depend on the parameters of the cable and terminal networks, are given by

$$H(p) = \frac{2R_o}{[1 + (R_2 + R_o)C_2p]\,[1 + (R_1 + R_o)C_1p]} \tag{5.13b}$$

$$J(p) = \frac{[1 + (R_2 - R_o)C_2p]\,[1 + (R_1 - R_o)C_1p]}{[1 + (R_2 + R_o)C_2p]\,[1 + (R_1 + R_o)C_1p]}$$

The equation (5.13a) and the defining relations (5.13b) are in convenient form for calculating transient response characteristics, for the reason that the exponential e^{-R_oCp} represents the phase shift corresponding to the transit time R_oC of a signal along the cable. An expansion of Eq. (5.13a) in powers of e^{-RCp} therefore corresponds term-by-term to the original signal and successive reflections. The transient effect of these terms can be obtained by use of Laplace transform theory or other operational methods. It is often desirable to analyze per-

[17] See, for example, the report by R. H. Cole (20), in which the material of this section is discussed in detail.

formance of terminating networks by a-c measurements, and for comparison with such measurements it is more convenient to rewrite Eqs. (5.13) in the form

(5.14)

$$\frac{1}{V_t(\omega)} = \frac{1}{C_oV_o}\left\{C\,\frac{\sin R_oC\omega}{R_oC\omega}\left[1 + j\omega(R_1C_1 + R_2C_2) - \omega^2(R_o^2 + R_1R_2)C_1C_2\right]\right.$$
$$\left. + (C_1 + C_2)\cos R_oC\omega\left[1 + j\omega(R_1 + R_2)\,\frac{C_1C_2}{C_1 + C_2}\right]\right\}$$

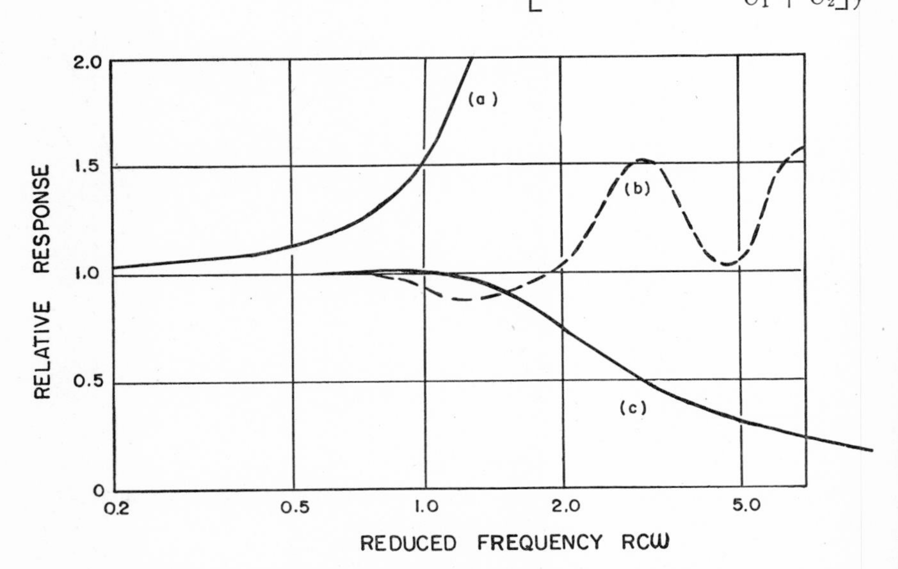

Fig. 5.19 Frequency response of piezoelectric gauge circuit for various cable terminations.

It is easily seen that this expression reduces, for low frequencies, to the value obtained by considering only the capacitances in the circuit:

$$V_t = \frac{C_oV_o}{C + C_1 + C_2}$$

Some special values of terminating network parameters will now be considered to illustrate more specifically the circumstances under which compensation becomes important and the improvement obtainable with two types of compensation.

(1). Capacitance only at receiving end ($R_1 = R_2 = C_1 = 0$). The steady response predicted by Eq. (5.14) reduces to

$$V_t(\omega) = \frac{C_oV_o}{C\dfrac{\sin R_oC\omega}{R_oC\omega} + C_2 \cos R_oC\omega}$$

It is evident that the denominator has zeros whenever $(\tan R_oC\omega)/R_oC\omega = -C_2/C$, corresponding to voltage resonances which are not in practice infinite because of cable losses. The response for $C_2 = 2C$ is plotted in Fig. 5.19 (curve (a)) as a ratio of the value for a given $R_oC\omega$ to the value for $R_oC\omega = 0$. The transient response to an applied step

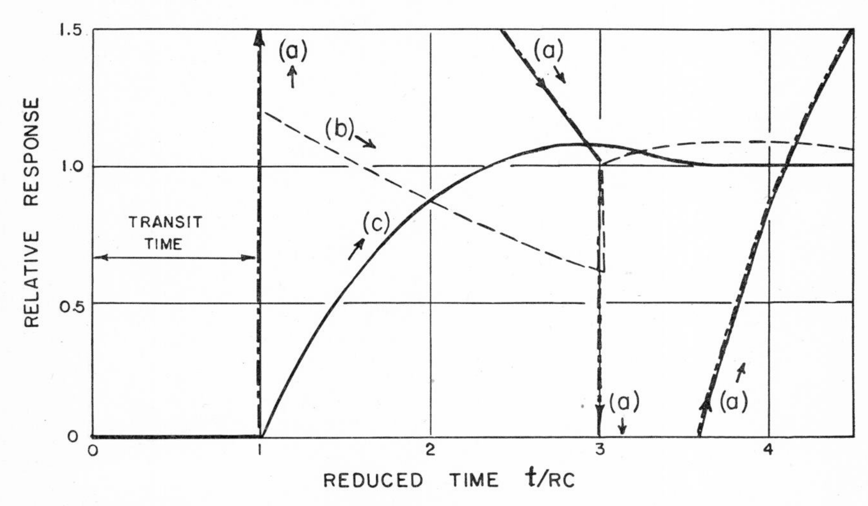

Fig. 5.20 Step transient response of piezoelectric gauge circuit for various cable terminations.

voltage V_o at time $t = 0$ can be computed by series expansion of Eq. (5.13) in powers of e^{-RCp} and term-by-term transformation. The result for the first signal and the first reflected signal is

$$V_t(t) = C_oV_o \left\{ \frac{2}{C_2} e^{-\frac{t-R_oC}{R_oC_2}} - \frac{2}{C_2} e^{-\frac{t-3R_oC}{R_oC_2}} \left[1 - 2\left(\frac{t - 3R_oC}{R_oC_2} \right) \right] \right\}$$

$$t \geq R_oC \qquad\qquad t \geq 3R_oC$$

The response for $C_2 = 2C$, relative to the response as $t \to \infty$, is plotted in Fig. 5.20 (curve (a)). The time interval $t = R_oC$ before any response develops is of course the transit time, and the second interval of $2R_oC$ represents the time for the first reflection at the receiving end to travel the length of the cable and back. It is evident that, while the infinite response for $C_2 = 0$ has disappeared, the approximation to a step response over these intervals is very poor.

(2). Resistance and capacitance at receiving end. A better compensation developed by Lampson (66) adds a resistance R_2 in series with C_2, the terminal voltage being taken across C_2. Specializing Eq. (5.14) to this case gives for the steady state response

$$V_t(\omega) = \frac{C_oV_o}{C\frac{\sin R_oC\omega}{R_oC\omega}(1 + j\omega R_oC_2) + C_2 \cos R_oC\omega}$$

Empirical tests and calculations show that if $C_2 = 2C$ the optimum value of R_2 is about $1.5R_o$ and the response for these values is shown in Fig. 5.19 (curve (b)). The resonances are reduced but analysis shows that no combinations of R_2 and C_2 can eliminate them. The first two terms in the corresponding step response are

$$V_t(t) = C_oV_o\left\{\underset{t \geq R_oC}{\frac{2}{5C}e^{-\frac{t-R_oC}{3R_oC}}} + \underset{t \geq 3R_oC}{\frac{2}{15C}e^{-\frac{t-3R_oC}{3R_oC}}\left[1 + \frac{18}{5}\left(\frac{t-3R_oC}{3R_oC}\right)\right]}\right\}$$

This expression, plotted in Fig. 5.20 (curve (b)), shows an improvement over use of capacitance alone but retains the resonant characteristic. It should be noted here that all these results are based on the assumption that cable losses are negligible. Bancroft has shown that Lampson's network can be adjusted to match the surge impedance of a cable with finite ohmic resistance of the conductors. The appropriate value of capacitance C_2 which accomplishes this turns out to be extremely large (of the order 0.25 μf), and would give very greatly attenuated voltage V_t. Tests at the Taylor Model Basin have indicated satisfactory performance with smaller values,[18] but where considerations of sensitivity are a factor it is desirable to have a more efficient network.

(3). Resistance and capacitance at both ends of cable. A consideration of the behavior of Lampson's network, and of cable properties more generally, suggests that the incipient resonances of the single network can be reduced by terminating both ends of the cable, in other words using all the elements in the schematic circuit of Fig. 5.18. The steady state characteristics of this network have been found to give satisfactory results for resistances R_1, R_2 about equal to the cable surge impedance R_o, and capacitances C_1, C_2 equal to the cable capacitance C. Comparison of calculated frequency response with experimental measurements on rubber dielectric cables with surge impedance of approximately 50 ohms shows satisfactory agreement (20), considering the approximations of the theory and failure of the cable to have exactly the ideal characteristics ascribed to it.

[18] These are described in a report by Greenfield and Shapiro (42).

The transient responses for this network are readily obtained by the same methods as before. For $R_1 = R_2 = R_o$, the result is

$$V_t(t) = \frac{C_oV_o}{C}\left[\left(\frac{t - R_oC}{2R_oC}\right)e^{-\frac{t-R_oC}{2R_oC}} + \frac{1}{6}\left(\frac{t - 3R_oC}{2R_oC}\right)^3 e^{-\frac{t-3R_oC}{2R_oC}} + \cdots\right]$$

$$t \geq R_oC \qquad\qquad t \geq 3R_oC$$

and for $R_1 = R_2 = kR_o$ one obtains

$$V_t(t) = \frac{2C_oV_o}{(k + 1)^2C}\left(\frac{t - R_oC}{R_oC}\right)e^{-\frac{t-R_oC}{(k+1)R_oC}} + \cdots$$

$$t \geq R_oC$$

This expression is plotted in Fig. 5.20 (curve (c)) for the value $k = 1.12$ (i.e., $R_1 = R_2 = 1.12\,R_o$). It is seen that the saw tooth characteristics of single-ended terminations are replaced by a smooth rise similar to critical damping characteristics in simple R–L–C circuits. The time to rise to the final value is of the order 1.5 R_oC, which for a 600 foot cable with $R_o = 50$ ohms, $C = .015\ \mu f$ is about 1 microsecond. This time is short enough to permit faithful transmission of exponential pulses with a time constant of 100 microseconds or more, and so is adequate for shock wave measurements of charges larger than about 10 pounds. The necessary or obtainable performance under other conditions is easily predicted by obvious proportions.

The double-ended compensation described has been used for many shock wave measurements on service weapons (22), and has proved very satisfactory. A practical problem in its use is the necessity of incorporating an electrical network at the gauge end of the cable. Tests have shown, however, that the network can be placed fifty to seventy-five feet from the gauge if necessary without significant loss in performance, and the network can be spliced into the cable as a waterproof assembly.

From the point of view of explosion pressure measurement, it is evident that compensation for cable transmission characteristics is important chiefly for short duration shock waves, and is so much less critical in measurement of the more sustained "bubble pulses" as ordinarily not to be necessary. The compensations discussed by no means represent the ultimate in performance readily attainable; they have, however, proved adequate without further elaboration.

B. *Dielectric properties of cables.* The discussion of cable characteristics so far given has been based on the assumption that, except for the distributed nature of the parameters, the cable capacitance could be regarded as a constant independent of the nature of signal applied to the cable by a piezoelectric or other high impedance gauge. This assumption requires a perfect dielectric material for cable insulation, in which the capacitance is independent of frequency and the dielectric loss is negligible. All solid dielectrics exhibit some departures from this ideal state, although the effects are sufficiently small as to be negligible for polystyrene and polyethylene cables used in gauge circuits. Cables with these dielectrics are desirable for small charge work where they can be handled fairly gently.

Unfortunately, the standard varieties of low-loss cable often do not survive the rough treatment encountered in work from vessels with large charges, and also exhibit spurious effects known as "cable signal," as discussed in part C. The cause of failure is breakdown of the dielectric by handling and salt water absorption. To avoid these difficulties more durable rubber dielectric cables have been used almost exclusively in the past for field work. Cables of this type all show appreciable variations of capacitance with frequency, some being very poor indeed. Although the low-loss rubber varieties are much better, it is still not safe to assume that over any appreciable range of frequencies or time intervals they can be represented as a simple capacitance and it is therefore necessary to determine the errors involved and devise means for their correction.

Over the frequency range fifty cps. to fifty kc./sec., the capacity of a good low-loss rubber dielectric cable varies by ten per cent or more, and the response to a gauge signal impressed on the cable will vary by about the same amount over this range of frequency or the corresponding range of times. If the cable response characteristics are determined as a function of frequency, by bridge measurements or otherwise, it is formally possible to correct any observed transient response for the effect by use of the superposition theorem and operational methods. As a practical matter, this procedure would be prohibitively difficult because of the necessity for a series of response measurements, repeated for every use of the cable, and tedious calculations. An alternative solution is introduction of a compensating network, which has been done, but with only limited success because no simple network can be successfully used over the wide range of frequencies necessary. A simpler procedure, which is adequate for reasonably good dielectrics, is to calibrate the transient characteristics directly and use this calibration record for correction of the observed records, the procedure for correction being quite simple if the effects are not too large.

A suitable and very convenient procedure for calibration of the cable characteristics is afforded by the Q-step procedure described in detail in section 5.7, by which in effect a known quantity of charge is impressed on the circuit from a step voltage source (see section 5.9(D)) and a standard condenser of known capacitance. The standard condenser is conveniently made the terminating condenser in the terminating networks described in part (A) of this section. A step voltage applied to this condenser in series with the cable then gives a response curve similar to the sketch in Fig. 5.21(a), the gradual falling off with time being the result both of dielectric effects (change in capacitance with frequency and dielectric loss) and of leakage resistance of the cable circuit. The transient characteristic thus obtained is actually a complete calibration of the charge sensitivity of the complete electrical circuit (including the recording circuit), except for times so short that transient effects in long cables are of importance. That this is true can be seen from the equations derived in section 5.7, as the capaci-

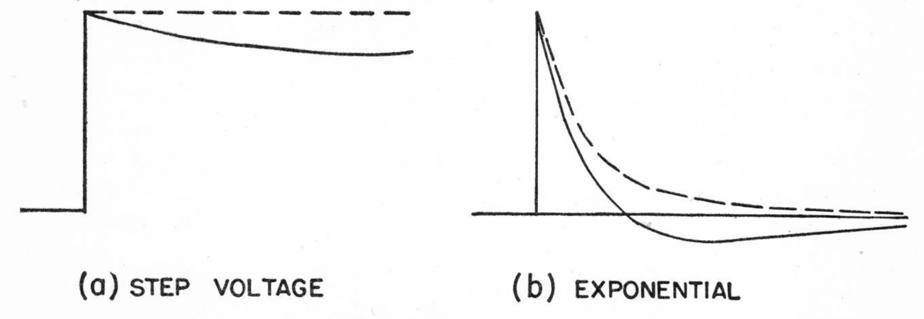

Fig. 5.21 Effect of dielectric loss and leakage resistance of cables on transient response.

tance C representing the circuit capacitance can actually be a fictitious equivalent capacitance (real or complex) of any two-terminal network.

For rapidly changing signals, such as the initial peak pressure in a shock wave, the initial response is evidently the correct one to use, and for more slowly changing signals correspondingly later values of response. The exact formulation of the relations can be made by the superposition theorem, which relates the response of a linear system for any applied function to its value for a unit function or "step voltage." The nature of the error for an exponential pulse is indicated in Fig. 5.21(b), drawn roughly to correspond with the step response indicated in Fig. 5.21(a). It is seen that the deviations from the ideal response become increasingly great with time, hence errors in area measurement (to give the impulse of a pressure-time curve) can be serious for relatively small loss in step response. If, however, this loss in step response is small, quite adequate correction for it can be simply made.

Consider the case of an applied voltage $V(t) = V_o e^{-t/\theta}$ and a step response of the form $S(t) = S_\infty + (S_o - S_\infty)e^{-t/\tau}$, corresponding to an initial response S_o followed by an exponential decay of time constant τ

to a lower value S_∞. By the superposition theorem,[19] the observed response $R(t)$ is found to be

$$R(t) = V_o e^{-t/\theta} - (S_o - S_\infty)\left(\frac{\theta}{\tau - \theta}\right)\left[e^{-t/\tau} - e^{-t/\theta}\right]$$

If the decay of the step response is slow, and hence τ much greater than θ, one finds to a first approximation that

$$V(t) = \frac{R(t) - (1 - S_\infty/S_o)\,[\theta/(\tau - \theta)]R(O)\,(1 - t/\tau)}{S_o - (S_o - S_\infty)\,[\theta/(\tau - \theta)]}$$

where $R(O)$ is the value of response at time $t = 0$, and terms of order $(t/\tau)^2$ and higher have been neglected. The true time constant θ can be evaluated from the $R(t)$ curves with sufficient accuracy. If the ratios t/τ, θ/τ can also be neglected the even simpler formula results that

$$V(t) = \frac{R(t) - (1 - S_\infty/S_o)\,(\theta/\tau)R(O)}{S_o - (S_o - S_\infty)\,(\theta/\tau)}$$

If the area under the $V(t)$ curve to a time t' is required it can be shown that for the same assumptions

$$\int_o^{t'} V(t)\,dt = \frac{1}{S(t')\,\dfrac{\tau}{\tau - \theta} - S_\infty\,\dfrac{\theta}{\tau - \theta}} \int_o^{t'} R(t)\,dt$$

If the time t' is sufficiently greater than τ that $S(t') \approx S_\infty$ this equation becomes

$$\int_o^{t'} V(t)\,dt = \frac{1}{S_\infty}\int_o^{t'} R(t)\,dt$$

and the proper calibration value is the limiting value S_∞ for long times. Equations of this type have been found useful and practical in correcting shock wave records for leakage and dielectric effects in cable circuits. If these effects are not too large, similar correction formulas are readily worked out for other types of applied signal and response. The important conclusion to these arguments is that the presence of such effects may result in serious errors if precautions are not taken to minimize and evaluate them.

[19] See, for example, the book by Gardner and Barnes (40).

C. *Cable signal.* In addition to the purely electrical properties of cables, an effect of applied stress on electrical cables must be considered in their use with piezoelectric or other high impedance gauges. This phenomenon, described as cable signal, manifests itself as an electrical charge developed on the conductors of a coaxial cable as a result of shock pressures on the cable. The exact origin and nature of the effect are somewhat obscure, but the evidence indicates that static charge is developed on the outer surface of the insulating dielectric, where it is in contact with the outer braided shield, as a result of friction between the dielectric and shield. In a low impedance circuit, charge of this kind is quickly neutralized, but a piezoelectric gauge circuit is necessarily one of high direct current resistance which is as effective in retaining such spurious charge as it is in retaining piezoelectric charge.

The phenomenon is not very reproducible under conditions of use. It has been found, for example, that a given cable exhibits progressively decreasing signal on repeated applications of shock pressures at fairly short intervals, being, so to speak, beaten into submission, but if left unused for several days again becomes as bad as it was at first. Cables with different dielectrics vary greatly as regards cable signal, polyethylene cables being particularly bad, and presumably identical pieces of cable frequently differ greatly. Insofar as anything very simple can be said about such a relaxation phenomenon, it is true that the amount of signal is proportional to the length of cable subjected to stress, and that the voltage developed varies inversely as the total circuit capacitance (thus leading to the conclusion that a "false piezoelectric" charge is developed).

The fact that the charge developed increases with length of cable exposed to pressure means that cable signal is most serious when one is interested in measuring slowly changing pressures, particularly when these are preceded by much higher initial pressures. For example, a shock wave with a large initial peak pressure or followed by a gradual decay develops a signal when it encounters the connecting cable from the gauge which increases until the whole of the wave has passed the gauge and met the cable. Release of pressure from sections near the gauge reduces the signal, but this is balanced by the signal from the cable further along. A nearly constant signal is thus developed shortly after the cable is first struck which persists until drained off by circuit leakage or by attenuation of the pressure wave with distance. The cable thus acts to give a sort of integration of the pressure-time curve, and the effect obviously becomes increasingly serious if one attempts to measure relatively weak pressures following a much stronger pressure which has had full effect on the gauge cable.

The errors resulting from cable signal may be so serious as to spoil completely quantitative measurement, even though tests seem to indicate its unimportance. As an extreme example, suppose that the maximum signal developed by an exponential pressure wave is fifteen per cent of the peak signal from the gauge. This signal will decrease very gradually and the relative gauge and cable signals as a function of time will look somewhat as sketched in Fig. 5.22. It is evident that the indicated signal will be double the true value when the latter is fifteen per cent of the original peak signal. The error in indicated area under the pressure-time curve also becomes increasingly great; for the final time in Fig. 5.22, it is already forty-five per cent of the true value.

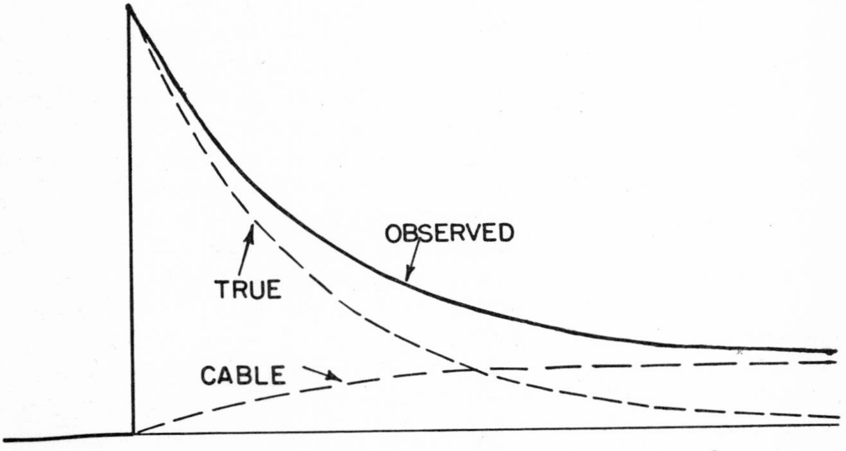

Fig. 5.22 Effect of cable signal on piezoelectric gauge records.

The indicated value of rapidly changing pressures, such as peak pressure in the example, will not be in error because of the integrated, or "low-frequency," nature of cable signal.

With a given cable, there are several expedients which can be used to minimize the signal. The most obvious is that of leading the cable directly away from the source of pressure in order to minimize the length of cable exposed and realize attenuation of pressure as rapidly as possible. A second, which has been used in measurement of bubble pulse pressures, is to lead the cables to the water surface as directly as possible and run them just below the surface, a region in which the pressure wave is almost immediately cancelled by the reflected pressure wave from the surface. A third expedient makes use of the "seasoning" effect already mentioned, whereby cable signal is reduced after application of pressure. Although this technique has been used with some success, it is clearly makeshift and, it is to be hoped, is of historical interest only.

Inasmuch as none of the standard types of coaxial cable has been found satisfactory as regards cable signal, efforts have been made, both in this country and in England to devise better special cables. One such design, devised at the Taylor Model Basin (42), employes a $\frac{1}{8}$ inch outside diameter soft copper tubing with 1/16 inch bore. A central copper wire (#22 or smaller) is used as central conductor, and the tube is filled with ceresin wax for dielectric. This filling may be accomplished in a variety of ways, one of the most effective being to heat the tube by an electric current, pumping fluid wax into the cavity, and cooling gradually to prevent formation of voids. Cables of this type have been used with considerable success in measurement of shock wave and bubble pulse pressures. No two cables are exactly alike, and tests of all sections of tubing used are necessary if a minimum of cable signal is to be realized. Good cables of this construction develop one-fifth to one-tenth the signal of the best of a number of rubber dielectric cables tested at Woods Hole, and these cables have been found adequate to reduce cable signal error on shock wave measurements to the order of one or two per cent. Similar cables used with larger gauges in bubble pulse measurements give rise to errors of the order of five per cent. It might appear that use of these cables would be hopelessly difficult in large charge work where the distance from gauge to recording equipment is hundreds of feet, but only the fraction of this length corresponding to the distance travelled by the pressure wave in the desired time interval need be composed of low signal cable; for the rest, standard types of cable may be used. A somewhat similar type of cable, devised by the Stanolind Oil and Gas Co. Research Laboratory, consists of a conventional coaxial cable except for an outside lead sheath replacing the usual copper braid. This cable has been found to be considerably better than the standard construction, but is usually somewhat inferior to the copper tubing design.

A British cable known as "Telconax," developed by the Telegraph Construction and Maintenance Co. for air blast work, has been found to have very low cable signal. This cable is much like a coaxial cable except for the use of a relatively soft and porous dielectric. Because of its inferior mechanical properties, use of this cable in marine work seems doomed to failure except under ideal conditions.

The probable origin of cable signal at the interface between the dielectric and outer shield of conventional cables suggests the possibility of reducing the signal by making the dielectric surface conducting, either by use of colloidal graphite or by use of conducting rubber near the outside surface. Several types of cable along these lines have been or are being developed. At the time of writing, preliminary underwater tests have been made indicating that such cables are equal or superior to copper tubing cable but further tests are needed.

One final possibility remains to be considered, that of using a balanced, or push-pull, line. If the two sides of such a balanced line develop equal cable signals, these signals will cancel each other exactly if the rest of the system is arranged to respond only to potential differences between the two sides. Hence the use of push-pull gauges and amplifiers with a perfectly balanced line represents the ideal solution in principle. In practice push-pull operation has usually been found to reduce the signal considerably but never completely. The improvement was found only if the two high potential conductors were surrounded by a common dielectric and shield. The use of two separate single conductor cables is, on the other hand, as likely as not to make matters worse, an example of the nonreproducibility of the phenomenon. The failure of two conductor cables ever to be perfectly symmetrical means that a cable should be used in which the separate signals are small. If this condition is satisfied, as it appears it would be with graphite or conducting rubber layers between dielectric and shield, the use of a balanced line should be the best ultimate solution for the cable signal problem. That it has not been used more extensively up to the present is explained largely by limitations of available equipment and cable.

5.9. Electrical Recording of Underwater Explosion Pressures

The most powerful tool for recording of transient pressures is undoubtedly the cathode ray oscillograph, and in many applications where high time resolution is necessary it has no satisfactory substitute. As a result, nearly all measurements of underwater pressure-time curves involve its use. In such transient measurements, a number of factors are involved in addition to those encountered in the more qualitative uses of oscillographs. These added requirements are sufficiently severe that, for the most part, present-day commercial equipment is inadequate and one must either modify such equipment considerably or design instruments specifically for the purpose. This section outlines the more important requirements and some of the solutions which have been developed.[20]

The necessary elements of a transient recording system can be grouped in five broad categories: (a) signal amplifiers, (b) cathode ray tube indicator, (c) time resolution of the record and synchronization, (d) photographic or other means of obtaining a permanent record, (e) calibration of the sensitivity and time resolution. The relation of these functions is indicated schematically in Fig. 5.23. The signal from the piezoelectric or other pickup is amplified by whatever factor is necessary and applied to one pair of deflection plates of the cathode

[20] Designs of electronic equipment for such transient measurements are described in reports by Greenfield and Shapiro (42) and by Cole, Stacey, and Brown (21).

ray tube. The screen of the tube is photographed by a suitable camera, a time scale being provided by motion of the film or by deflection of the cathode ray tube trace. In either case, it is usually necessary to brighten the cathode ray tube trace at the time of the event to be recorded, and if a linear time base is used its operation must also be synchronized. Suitable triggering signals can be obtained by a gauge or other device operated by the pressure wave, or from an electrical circuit

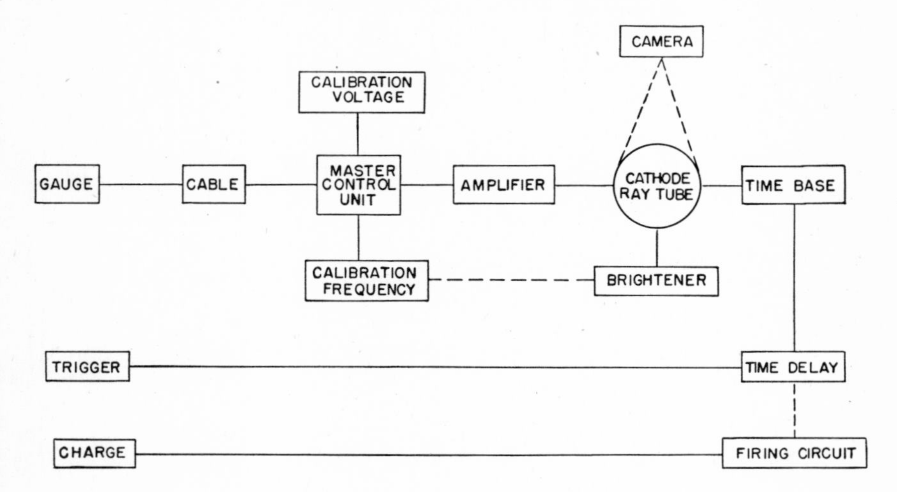

Fig. 5.23 Block diagram of equipment for transient recording.

firing the explosive charge; in some cases it is convenient to include an electrical delay circuit. Both the time resolution and voltage sensitivity must be calibrated in terms of suitable standards. In order to insure efficient and dependable operation of the entire equipment various auxiliary items of equipment are desirable, such as partly or fully automatic control circuits and regulated voltage supplies. The various components of a complete recording system are described more fully in what follows.

A. *Amplifier circuits.* The specific requirements of signal amplifiers for recording explosion transients depend, of course, both on the transient and the gauge or transducer used. In any case, the amplifier system must be stable and rugged, possess sufficient gain and frequency response, and be capable of providing an adequate linear output signal to the cathode ray tube. An amplifier used with piezoelectric or other high impedance sources must also have a high input impedance in order not to distort the input signal.

The frequency response requirements are frequently rather severe, both at low and high frequencies, and can best be made evident by simple examples. Faithful reproduction of the initial discontinuous rise and rapid decay of shock wave pressures makes necessary sufficiently rapid response that the peak value will not be rounded off appreciably. In many cases the initial behavior of shock wave pressure $P(t)$ is satisfactorily represented by a negative exponential of the form $P(t) = P_m e^{-t/\theta}$, where P_m is the peak pressure and the time constant θ

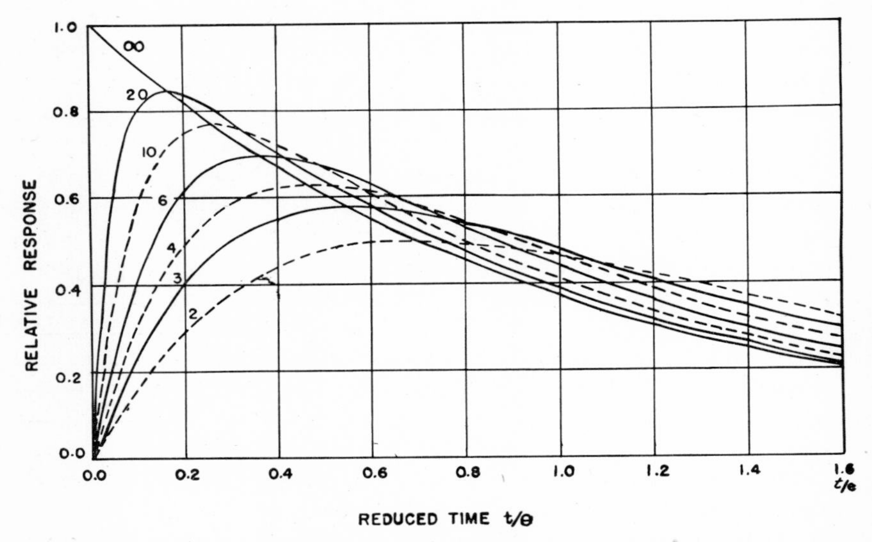

Fig. 5.24 Effect of amplifier high-frequency cutoff on an exponential pulse.

has values from a few microseconds to a millisecond. In a simplified analysis of amplifier characteristics, the loss in high frequency response arises from shunting of a load resistance R, across which the voltage response is developed, by circuit capacitance C. The relative response $F(j\omega)$ of such a circuit at frequency $f = \omega/2\pi$ is given by

$$F(j\omega) = (1 + j\omega\tau)^{-1}$$

where the time constant $\tau = RC$.

The transient response $R(t)$ to the exponential $P(t)$ is readily shown to be

$$R(t) = \frac{P_m}{1 - \tau/\theta} [e^{-t/\theta} - e^{-t/\tau}]$$

This result is plotted in Fig. 5.24 as a function of the reduced time t/θ for various values of the ratio θ/τ. The response for the limiting case that $\theta \rightarrow \infty$ is an exponential rise to the final value P_m as given by $R(t) = P_m(1 - e^{-t/\tau})$, with time constant τ determined by the circuit. It is evident from Fig. 5.24 that unless the value of τ is much smaller than the time constant θ of the applied signal there will be considerable distortion of the true curve (represented by $\theta/\tau \rightarrow \infty$) and loss of indicated peak value. It is easily shown that the indicated peak height always lies on the true curve, and the time of rise to this value must be small compared to the time constant θ. For example, if a pulse of duration $\theta = 50$ μsec. corresponding to the shock wave 3 feet from a ½ pound charge is to be recorded with a loss of less than 2 per cent in peak value, the value of θ/τ must exceed 300 and τ must be less than 0.17 μsec.

Expressed in terms of frequency response, the requirement on τ means that the simple RC stage should have 70 per cent of its midband response at a frequency $f_c = 1/2\pi RC = 950$ kc./sec. For larger charges the demands will be less severe, and in the case of service weapons (300–500 pounds of explosive) a high frequency response flat to 100 kc./sec. is usually adequate. The analysis presented here is of course oversimplified and in practice multistage compensated amplifiers must be considered. Nevertheless, the analysis does show the nature and order of magnitude of the requirements which any amplifier must satisfy.

The low frequency requirements on amplifier circuits can be made evident in a similar manner. As already discussed in sections 5.7 and 5.8, the input impedance of an indicating circuit used with a piezoelectric gauge must be high in order to prevent loss of pressure-developed charge, the exact requirements depending of course on the transient. A similar problem which must be considered in a-c amplifiers is the loss of low-frequency response as a result of capacitative interstage couplings. Either of these problems can be analyzed in terms of a series resistance-capacity circuit, the input voltage being applied across the combination of resistance R and capacitance C in series, and the output voltage developed across the resistance. The relative response in this case is given by

$$F(j\omega) = \frac{j\omega\lambda}{1 + j\omega\lambda}$$

where the time constant $\lambda = RC$ is the time required for the response to an applied step voltage to fall to $1/e$ of its original value.

If the input signal is again a negative exponential $P(t) = P_m e^{-t/\theta}$, the transient response $R(t)$ is given by

$$R(t) = \frac{P_m e^{-t/\theta}}{1 - \theta/\lambda} [1 - (\theta/\lambda)e^{(1-\frac{\theta}{\lambda})\frac{t}{\theta}}] \tag{5.15}$$

This expression is plotted in Fig. 5.25 as a function of t/θ for the case $\theta = \lambda/10$ (circuit time constant a factor of ten greater than the pulse time constant), and also for $\theta/\lambda = 0$, corresponding to perfect low-

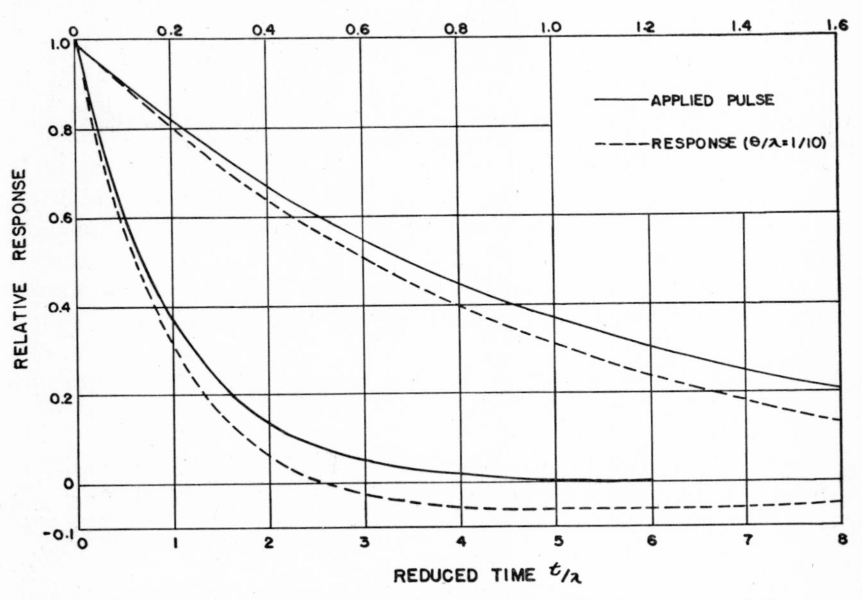

Fig. 5.25 Effect of amplifier low-frequency cutoff on an exponential.

frequency response. It is seen that the response curve has the same general form as the true curve but lies increasingly below it at increasingly longer times. The values of indicated response thus became increasingly in error, and are determined as much or more by the amplifier decay constant λ as they are by the time constant θ of the applied signal. The magnitude of the error in any specific case is readily calculated from Eq. (5.15). If, for example, $\lambda/\theta = 50$, the response measured at time $t = 3\theta$ will be in error by 36 per cent, and the response for $t = 5\theta$ is negative and greater in magnitude than the true signal. The value of λ for $\theta = 1$ msec. is 50 msec. in this example, and corresponds to a circuit response 30 per cent below the mid band value at a frequency of 3 cycles/sec. This response is by usual standards very good indeed, but from the example is seen to give rather poor simulation of the true curve.

In measurements of explosion pressures, the area under the pressure-time curve, or impulse, is of considerable importance and it is evident that the low-frequency response will modify measured areas considerably. One can, in fact, show very easily that the integrated response to infinite time of a pulse of finite area is zero for any a-c amplifier or other circuit with no d-c response, a conclusion made plausible by the negative values of response at long times indicated in the example of Fig. 5.25. (It is interesting to note that, if the circuit considered does have a non-zero response to a fixed voltage, it is quite generally true that the integrated response is independent of the high frequency response characteristics, being determined primarily by the low-frequency response.) If the integration of the response curve is carried out to a finite time t', the fractional error ΔA in the area $A(t')$ for an exponential pulse is readily shown from Eq. (5.15) to be approximately given by

$$\frac{\Delta A}{A} = \frac{t' - \theta}{\lambda}$$

if the time t' is less than 10λ. Thus, if the error is to be less than 5 per cent, the amplifier low-frequency time constant must be at least 20 times that of the pulse.

The low-frequency demands of pulses encountered in explosion measurements are so severe that, except in a very few cases, commercial standards of response (flat to say 20 cycles/sec.) are grossly inadequate. Frequently the simplest solution, as far as amplifier characteristics are concerned, is to eliminate the problem entirely by the use of d-c amplifiers. This solution cannot be applied to the input circuit for a piezoelectric gauge which has inherently no static or d-c response because of the equivalent capacitative internal impedance of the gauge. Although low-frequency compensation could be applied to improve the characteristics of such input circuits, it has been found adequate, and simpler in practice, to use uncompensated circuits with input impedances as high as 100 megohms. (Higher values than this are usually difficult to obtain in field work with cables in salt water, and the value quoted is sometimes inconveniently large.)

The requirements on amplifiers, in addition to those of frequency response, are conventional except that the linearity and gain stability need to be rather better than ordinarily found in commercial amplifiers. The more important elaborations for this purpose are regulated supply voltages and liberal use of feedback. In addition, instruments which are to perform reliably in the field under frequently trying conditions need to be developed with attention to such things as ruggedness, accessibility and ease of repair; only the existence of these factors in design and use can be indicated here.

B. *Cathode ray tube circuits.* The most important requirements of the cathode ray tube and its associated circuits are that the accelerating voltages be high enough to provide adequate brilliance of the trace for photography, and that these voltages be sufficiently well regulated to prevent changes in sensitivity and undesired modulations of the light intensity. For trials with service weapons, overall accelerating potentials of 2,000 volts are usually adequate, while for small charge work on shock waves potentials as high as 5,000 volts may be necessary.

A factor in quantitative use of cathode ray tubes often not appreciated is the fact that the deflection sensitivity of the tube varies inversely as the applied accelerating voltage. This voltage must, therefore, be kept constant to at least the same degree as the desired precision of measurement; this regulation is accomplished by use of sufficiently steady a-c supply voltage or regulation of the high voltage supply. High voltage regulation is of advantage also in preventing supply voltage modulation of the light intensity, which otherwise frequently results when the cathode ray tube grid is used to switch the beam on and off in transient recording. This switching, almost always necessary if an electrical time base is used and often so with moving film recording, is described below.

C. *Time resolution and synchronization.* In some respects, the simplest way of obtaining time resolution of a transient signal is simply to use a high-speed continuous film movie camera started manually just before the occurrence of the transient. Electrical problems of synchronization disappear and a virtually unlimited time duration is obtainable, and these advantages may be sufficiently great to outweigh the problem of handling and wasting large amounts of film. Usually, however, this is not the case, and other means are preferable. One method is deflection of the cathode ray tube trace by a linear time base, or single sweep, circuit initiated electrically just before the occurrence of the transient. The requirements of the circuit are linear deflection with time, reliable triggering by readily available synchronizing pulses, and a means of turning the tube beam on and off at the beginning and end of the deflection. The circuits used for the purpose are, for the most part, evolutions from standard designs. This method of obtaining a time scale is best for relatively fast transients where most efficient use must be made of the available light output from the cathode ray tube, but it gives only a limited recording time for given time resolution.

A useful compromise between use of a movie camera and of electrical trace displacements is the rotating drum camera, in which, as the name implies, a short length of 35 mm. film is attached to a drum rotating on an axis parallel to the displacement of the cathode tube trace by the signal. Drums with diameters of 3 or 4 inches driven at speeds up to

3,600 RPM or more are commonly used, giving time resolutions on the film up to 1 or 2 inches per millisecond.

Either the electrical time base with fixed film or the drum camera with moving film requires synchronization of the recording interval with the events to be observed. One method is to use changes in the firing current of the detonator cap initiating the explosion. Unless special caps are used or large firing currents employed, the detonation of the charge is usually delayed by several milliseconds after the circuit is closed and the interval is often not sufficiently reproducible if accuracy to a fraction of a millisecond is needed. The breaking of the firing circuit by the force of the explosion is also quite undependable, particularly for large charges. A number of special firing circuits have been devised which permit satisfactory operation, but none has been found foolproof. Another expedient is to use a pressure operated switch suitably placed near the explosion. The most satisfactory switch of this kind is probably simply a piezoelectric "pilot" gauge designed primarily for ruggedness, as little maintenance is required with a good gauge and its operation is both simple and readily predictable. In these methods or others, it is often desirable to delay the operation of the various circuits to be synchronized by controllable intervals, and relatively simple trigger circuits accomplish this purpose satisfactorily.

D. *Photographic recording.* In photographic recording of transient oscillograph traces there is seldom an excess of light and it is usually desirable to use the fastest available lenses of reasonably good quality, the most generally employed type being an f/2.0, 50 mm. focal length lens. The trace of either a single tube or a combination of several tubes mounted near one another is recorded on the film. The type of film used is not critical, although the best results have been obtained with Eastman Fluorographic, and adequate densities for most purposes are obtained without use of prefogging or intensification. A trace 1 millisecond long on a 5 inch tube screen with a record of an exponential pulse of time constant 50 microseconds can be photographed with good readability using an f/2.0 lens and 2,200 volts overall accelerating potential on the cathode ray tube. For fixed film recording using a time base either 35 mm. film and a rebuilt miniature camera, or cut film with a special holder to permit several exposures on a single film, have been used.

E. *Calibration circuits.* Any system for recording transients must include accurate means of calibrating voltage sensitivity and the time scale, and it is highly desirable to extend the meaning of sensitivity to include the complete transient characteristics of the electrical systems. A common practice in much oscillograph work is to calibrate both voltage and time scales at once by recording a sine wave of known amplitude and frequency. Although this method is economical of time, it has the disadvantages that a sine wave, being a smooth curve, is not ideally suited for precise interval measurements, and is rather difficult to calibrate with high accuracy. A further objection is that the record of a single frequency gives virtually no indication of transient response characteristics. A better procedure is to separate the two types of calibration and perform each in a more convenient and useful way.

The Q-step principle of calibrating piezoelectric gauge circuits has already been discussed. As was pointed out in section 5.7, this method of applying a known transient voltage, and in effect a known quantity of charge, provides a complete transient response calibration of the entire electrical system (cable, amplifier, cathode ray tube), except, as ordinarily used, for effects of distributed parameters in a long cable. The most useful type of transient voltage for purposes of analysis is usually a step, or Heaviside unit function, although sometimes other functions, such as a negative exponential, may be desirable. Various means of generating such transients have been developed, which have in common an accurately known source of d-c voltage and a conveniently controllable switch (either mechanical or electrical). A simple form of transient generator, consisting of a high-capacity dry-cell and precision voltage divider switched by a relay and associated control circuit, may be adequate for some purposes. For more critical applications, more elaborate circuits have been developed which include an electronically regulated supply switched by a vacuum-tube "gate" circuit, a means of delaying the operation for an adjustable interval after a synchronizing pulse, and an internal potentiometer circuit for checking the calibration.

For timing calibration, pulses of sharply peaked wave form can be provided from standard crystal or tuning fork frequencies and their submultiples obtained by multivibrator or scaling circuits. These pulses can be applied to the cathode ray tube as a signal or intensity modulation, or used to control a glow discharge tube mounted optically adjacent to the tube, which thus provides timing marks on a moving film.

Summary. To sum up the methods of transient recording, most of the requirements can be met by extension or adaptation of standard electronic and photographic techniques. The important consideration is that such equipment must be capable of giving quantitatively reliable records dependably under field conditions. Most existing commercial equipment falls short of these demands and, although some instruments can be modified to perform acceptably, it is usually more satisfactory to build the equipment specifically for its intended purposes.

5.10. Experimental Arrangements for Pressure Measurements

Although the arrangement and handling of gear for locating explosive charges and measuring equipment involves few points of fundamental scientific interest, it is a necessary part of the experimental work which must be properly done if trustworthy results are to be obtained efficiently. The details of such experimental layouts will of course depend very much on the specific problem and the available facilities, and as a result there are not many principles and procedures sufficiently general in application to merit a detailed discussion here. There are, however, several basic considerations which must always be taken into account and are discussed here, together with brief descriptions of experimental arrangements which have been used for some of the large and small charge work at Woods Hole.

The most obvious characteristics peculiar to underwater explosion work are the facts that high explosives by definition react violently and can cause serious damage, wanted or otherwise, in their vicinity; that a sufficiently large body of water must be available; and that an explosion happens only once. If work with small charges only is intended, it is not too difficult to provide a suitable volume of water, necessary safety features are readily obtained, and arrangements of gear can be made with small expense in manpower and materials. For example, tests using standard detonator caps and/or charges up to an ounce or so in weight (usually tetryl) have been performed at a number of laboratories in tanks of steel or concrete of moderate size, say three feet or more in each linear dimension, and the major precautions are those of proper explosives handling and prevention of interference with other activities in the immediate vicinity.

Charges up to a pound or more in weight can usually be fired in small ponds or off docks not too near buildings. At Woods Hole, for example, a great number of investigations using charges as large as one-half pound have been performed off a dock in twenty feet of water, an arrangement which was particularly convenient because booms and tackle at the edge of the dock could be used for easy handling of gear and recording equipment could conveniently be housed nearby. Artificial or natural ponds can be used which still permit shore installation of handling and recording equipment. At Woods Hole, a shallow natural pond was enlarged by excavating to provide water eight to ten feet deep in which charges up to five pounds were fired. Larger charges were impractical because of the small depth and short time interval before reflections occur from the boundaries, and because of filling up by loosening of the sides of the pond.

Damage to the walls and limited volume of water will probably be the limiting factors on the size of charge which can be fired in a small body of water, and work with larger charges, such as service weapons, has without exception been done either in lakes or coastal waters of the ocean. Many types of gear have been used for different purposes in the small charge work, but the most useful arrangement for many purposes has been to support charges and gauges or other equipment on a steel ring of suitable size, for example, rings ten or fifteen feet in diameter of one inch steel rod. The entire rig is then lowered from a boom or overhead cableway to a suitable depth for firing.

The successful firing of charges weighing hundreds of pounds under controlled conditions is at best a rather complex operation involving considerable manpower and equipment, including vessels to handle gear, in addition to the basic gauges and auxiliary instruments. Any such explosions must be fired in reasonably quiet water, usually with an available depth of eighty feet or more. Even with purely mechanical gauge instrumentation, rather large and cumbersome rigs are required to provide known positions of gauges and charge. Work of this kind done by the U. S. Navy Bureau of Ordnance at Solomon's Island, Maryland has involved the use of pipe frameworks supported from the surface by large logs or telegraph poles, the whole gear being set from large barges. In some British work, the gear has been set from a moving vessel and supported by floats, as has also been done in the work at Woods Hole. The latter work, employing the schooner "Reliance," has given the most comprehensive data for large charges on effects of explosive weight, shape and composition, variation of explosion pressures with distance, and so forth, the instrumentation including both piezoelectric and mechanical gauges. For most of the work, the gauges and charge have been disposed along a steel cable forty feet below the surface, the various elements being supported by floats, and the whole rig kept taut by towing.[21] Electrical firing line and cables from the piezoelectric gauges are led along the surface on floats to the Reliance, a seventy-six-foot Gloucester schooner fitted out with electrical recording equipment, laboratory and shop space, and sources of electrical power.

A typical rig of the type described is shown schematically in Fig. 5.26, the insets giving details of the various gauge mountings. On some shots, as many as 8 piezoelectric gauges, 25 ball crusher gauges, 8 diaphragm, 8 Modugno and 8 mechanical momentum gauges have been employed, and at the date of writing 450 large charges of various kinds have been fired. For further details of rigging and handling gear, detailed analysis of experimental techniques and measurements, and so forth, the research reports should be consulted (22, 23).

In all explosive work the hazardous nature of the work must constantly be kept in mind and adequate safeguards against misfires, improper handling, and so on, set up and observed. The characteristic of explosions that they happen only once adds to the difficulty of good measurements, because the instruments must be functioning properly at the time of firing if any results at all are to be obtained from that

[21] The methods used in measurements of large charges (50 to 1,000 pounds of explosive) at the Underwater Explosives Research Laboratory (Woods Hole) are fully described in a report by J. S. Coles (22).

shot, adjustments and tests while the event occurs being almost invariably difficult and often impossible. These difficulties are particularly great for large charge work, in which a single shot represents a large

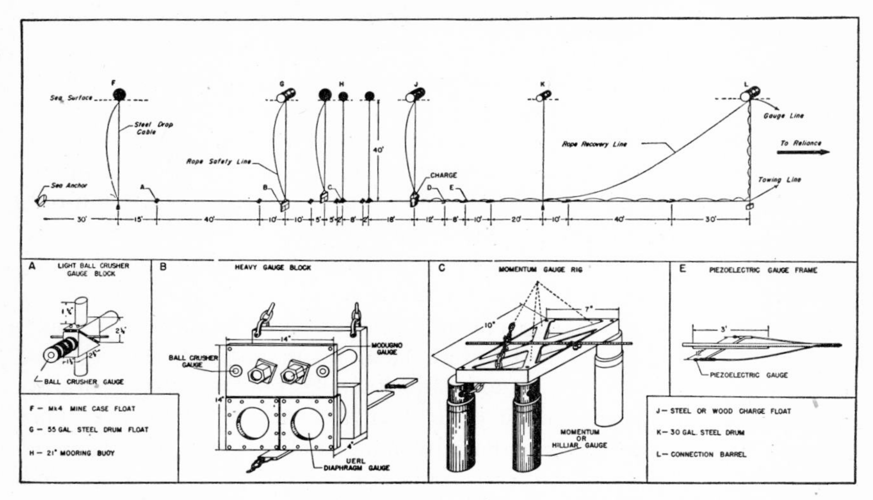

Fig. 5.26 An arrangement of gear for large scale explosion tests.

investment in time and money, and seemingly excessive precautions to insure reliable functioning of equipment usually justify themselves in the end.

6. Photography of Underwater Explosions

The optically visible effects of underwater explosions constitute valuable experimental tools in measurement. The advantage peculiar to techniques based on optical phenomena is the fact that for many purposes no distortion of the phenomenon results from the measurement, a state of affairs which it is often difficult and sometimes impossible to achieve by use of mechanical or electrical devices. A second advantage lies in the fact that photographic records can give information over a field of view, rather than being point measurements. Largely because of these characteristic features, photography of underwater explosions has supplied a great deal of both qualitative and quantitative information and is potentially capable of supplying much more. It is the purpose of this chapter to review some of the techniques which have been developed, suggest their advantages and limitations, and give illustrations of their applications.

6.1. Surface Phenomena

The simplest and most obvious photographic data obtainable from underwater explosions are records of the phenomena visible above the surface. Apart from their pictorial interests, such records can be made to yield quantitative information as to the propagation of shock waves and the motion of water around the gas bubble formed by the explosion products. The surface phenomena arising from these causes are discussed in detail in sections 10.1 and 10.2, but the types of measurement involved are mentioned here to indicate some of the technical requirements.

The initial arrival of the shock wave at the surface is made evident even at great depths by a region of darkened water, often described as the slick. The front of this region spreads out from a point above the explosion with a speed which is at first infinite (neglecting relativity effects!) and decreases asymptotically to the velocity of sound in water, or about 5,000 ft./sec. The higher speed at points closer to the charge is the result both of the geometry of the source and surface and of supersonic velocity of finite amplitude waves. Measurements of the disturbance are thus capable in principle of application to measurements of depths of explosions and to determination of shock front velocities. In either case, the large velocities involved require motion pictures taken at high speed if detailed records are to be obtained, as motion pictures taken at normal rates of 16–64 frames/second show the phenomenon on one or at most only a few frames before it leaves the field of view. For charges of a few hundred pounds or less, frame rates of 1,000 per second, such as are possible with Eastman high-speed or Fastax cameras, are needed.

A somewhat slower phenomenon, but one which still requires high film speeds, is the rise of the dome of spray thrown up from the surface by reflection of the incident shock wave. The initial velocity of the upper surface of this dome is simply related to the pressure in the shock wave producing it, and can therefore be used to compute shock wave pressures. The development and contour of the dome is easily seen to be affected by the depth of the explosion, and the phenomenon has been made the basis of several methods for measuring depths of explosion.

The gaseous products of explosion rise upward after their formation and, after a time determined by the depth and explosive, break through the surface, projecting the familiar plumes of spray if the charge is sufficiently close to the surface. The time of the process and the nature and extent of the upward water motion are clearly closely related to the motion of the buoyant gas volume, and can provide valuable evidence as to the nature of the motion when the breakthrough occurs.

The phenomena above the surface, as sketched briefly in the foregoing, are examined in more detail in Chapter 10, but the usefulness of photographic records suitable for quantitative measurement should be apparent. If such records are to be obtained, two general types of requirement must be met. The first of these is the obvious necessity that the photographic quality must be good. Proper exposure, good contrast, and in the case of motion pictures proper frame speed, must all be realized for good measurements. These necessities are both straightforward and obvious, but they are mentioned because of the large amount of otherwise useful data which have been lost as a result of such things as surface glare on the water and unsuitable or faulty equipment.

A second type of requirement is that of proper scales of length and time, incorporated directly in the record if possible. Although it is sometimes possible to make useful measurements without either type of scale, any record is more valuable for them and many measurements are virtually impossible otherwise. An example of the latter is photographic determination of underwater shock wave velocities and pressures, which can hardly be accomplished without accurate knowledge of a distance scale and precise indications of film speed in terms of a known standard.

6.2. Optical Considerations in Underwater Photography

Although, as suggested in the preceding section, phenomena visible above the water surface can provide much information about an explosion below it, these phenomena are to a considerable extent byproducts of the major physical events. Many of these events can thus be inferred only indirectly unless the photography is carried out beneath the surface. The substitution of water as the optical medium and the fact that the phenomena of interest are capable of destructive effects both present difficulties in securing a sharp, adequately illuminated image at the photographic plate, and methods of satisfying this basic requirement are discussed in this section.

A. *Transparency of the water.* Water is at best an imperfect medium for transmission of light waves, owing to a greater decrease in intensity with distance from a source than predicted by geometrical considerations, such as the inverse square law for a point source. This attenuation is largely the result of scattering by impurities, and varies tremendously with the amount and nature of such impurities.

If the laws of geometrical image formation hold, all the light which reaches the camera from a given point on an object to be photographed combines at a corresponding image point on the plane of the film. If scattering and absorption occur in the intervening medium, however, part of this light does not reach the camera and, what is worse, light scattered from other geometrical paths does arrive, thus producing a decrease in definition and contrast of the image.

This reduction in visibility is inherent in any given medium, and becomes rapidly more serious as the distance increases, as the attenuation and scattering increase exponentially with the path traversed.

Loss of intensity by absorption and scattering could of course be compensated by using brighter illumination, but the failure of the light to form a good image cannot. An upper limit to possible working distances is therefore set by the clarity of the water, and this limiting factor is an important one for measurements on any scale. In tests on a laboratory scale, such as might be performed in a tank several feet on a side with detonator caps, reasonably clean tap water is often satisfactory, but must be changed when contaminated by dirt, solid carbon explosion products, or growths. With larger charges, transparency is a more difficult problem, because larger working distances are necessary which in turn require larger volumes of clearer water. If an artificial basin is used, the water must be changed or cleaned periodically, which becomes a considerable expense for work with even a few pounds of explosive. For charges of more than a few pounds, work must almost always be carried out in natural bodies of water, and the clarity of this water is frequently the limiting factor which makes satisfactory pictures difficult or impossible to obtain.[1]

[1] Problems of underwater photography in general have been investigated by the Woods Hole Oceanographic Institution, as reported by Ewing, Vine, and Worzel

A very simple and effective measure of water transparency is obtained by using the Secchi disk. This is a circular disk, eight inches in diameter and painted white, which is lowered in the water until it becomes invisible from above the surface. The depth of water at which this happens is called the Secchi disk reading, and a rough but useful empirical rule is that reasonably good pictures can be taken up to half the Secchi disk reading. This reading in sea water varies greatly with nearby sources of contamination, nature of the bottom, etc., but generally is very much less in coastal waters than offshore. For example, harbor waters on the New England coast may give readings of the order of 20 feet, while readings of 120 feet are obtained 100 or 200 miles offshore. Tropical waters are much preferable, for this and other reasons, readings of 130–140 feet having been obtained within a mile of islands in the Bahamas for example. The readings quoted here are not intended as source data, but merely to give an indication of what order of distances can be used. In tests in Woods Hole harbor on the Massachusetts coast, object-camera distances of more than 5 feet are frequently impossible, but in tests off the Bahamas motion pictures of good quality have been taken at a distance of 60 feet.

A further inherent characteristic of water is of course the lower speed of light corresponding to an index of refraction of the order 1.34, the exact value depending upon the temperature, density, and salinity. The effective object distances and field of view for a camera with air between the lens and film are therefore reduced by this factor, as compared to air. The difference in optical path merely requires a changed lens setting for proper focus, but the greater working distance required for a given field, as compared to air, is unfortunate because of the scattering effects already discussed. The index of refraction is, however, a useful experimental tool in investigation of pressures and flow following explosions, as discussed in section 6.3.

B. *Explosive light sources.* An explosive is an appropriate light source for study of explosions, and some of the most successful sources for still photographs with extremely short exposures have employed explosive "flash charges." Similar charges have also been proved most satisfactory for fundamental studies of small scale explosions, and a description of their preparation is therefore merited.

The basic design, developed at the NDRC Explosives Research Laboratory in Bruceton, Pa. (34), consists of a centrally detonated spherical charge of cast 50/50 pentolite (an equal mixture by weight of PETN and TNT, see section 3.1). This charge is surrounded by a concentric layer of argon gas at atmospheric pressure. The spherical

(33); a number of techniques and applications to underwater explosions are described in reports by Eldridge, Fye, and Spitzer (31), and by Swift et al (105).

detonation wave strikes this layer and a brilliant flash of light results from emission following excitation of the atoms by the high temperature and pressure of the shock wave in the gas. The duration of intensity can be made very short by decreasing the thickness of the layer, which is excited at supersonic velocity, and the total illumination can be varied by using different charge sizes. A 0.55 pound charge 2.5 inches in

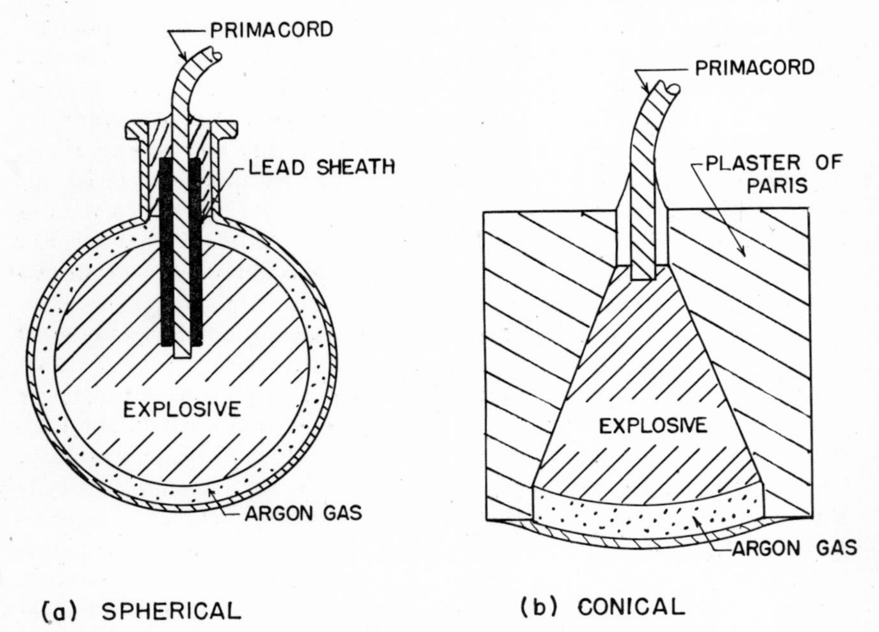

Fig. 6.1 Explosive flash charges for underwater photography.

diameter with an argon layer 0.19 inches thick has been estimated to give approximately one million candlepower with a duration of about one microsecond.

Two adaptations of this basic design are shown in Fig. 6.1.[2] The spherical charge in Fig. 6.1(a) is detonated at its center by a piece of primacord, a detonating fuse, which is wrapped in a lead sheath to prevent initiation of the main charge until the detonation wave reaches the end of the fuse. The charge is cast in a plaster of paris mold with the fuse in place, and mounted in a split round bottom flask of larger internal diameter; the two halves of the flask are then cemented together and the air space evacuated and filled with argon. This design from its symmetry gives nearly uniform illumination over a sphere. Charges without the glass flask and argon filling have been found to give remark-

[2] Methods of preparing these and other types of explosive charges are given in reports by Eldridge, Fye, and Spitzer (31), and by Newmark and Patterson (79).

ably uniform and reproducible shock wave pressures, and hence are very useful in fundamental investigations.

The spherical intensity distribution from the design of Fig. 6.1(a) is to a large extent wasted in many applications. This waste is accompanied by the obvious disadvantages of using unnecessarily large quantities of high explosives near equipment or gear. Reflectors can of course be used to concentrate more of the total light output in a given region, but a better solution in many cases is simply to use only a suitable section of the sphere in the form of a conical charge, as shown in Fig. 6.1(b). With an angular opening of 60° this charge has approximately $^1/_7$ the weight of a spherical charge giving the same intensity over this angle of field without reflectors. The charges described are not usable without modification at depths of more than about 20 feet below the surface, but protective cases of various types and other charges permit flash photography at depths of 600 feet or more. For details of these and other designs of flash charges and their preparation the original reports should be consulted.

Explosive flash charges have the advantages of very high intensity with extremely short duration and are thus best suited to single flash photographs of rapidly changing phenomena, especially underwater shock waves. Although two or more charges have been fired at preset intervals, they are not useful for sequence records of slowly changing phenomena. The use of flash charges has the further disadvantage of involving high explosives, which of course cause underwater effects in addition to those intended to be made visible, and experimental arrangements must be planned to prevent interference by these disturbances. The use of high explosives also involves hazards. These are already present in photography of an explosion, and it may well be that alternative sources, such as high voltage discharges, result in comparable hazards, as well as presenting difficulties in field operations.

C. *Other flash sources.* Electrically operated light sources can also be used in underwater flash photography. The General Radio microflash unit, which comprises a gas-filled lamp through which a condenser previously charged to 7,000 volts is discharged after ionization of the gas by a triggering impulse, is one electrical counterpart of an explosive flash charge, and the General Electric electronic photolight, employing a high voltage discharge through a capillary arc, is another.[3] Both these units give intense flashes with durations of the order of two to four microseconds, and can therefore essentially stop underwater shock waves. A number of investigations at the Taylor Model Basin on shock waves and intersecting shock waves from small charges have

[3] The General Electric "Electronic Photolight" is an example of the capillary arc; underwater flares for photographic purposes are described in a report by E. L. Patterson (81).

been reported by Campbell (16, 17), in which microflash units have been employed.

Another type of source employs discharge of a condenser through a spark gap in the water. A spark arranged to act as a very small source of high intensity and duration of a few microseconds has been developed by Libessart (69). Magnesium electrodes are employed, the light being emitted from a tiny hole in one electrode, and the main discharge circuit from a high voltage condenser is closed by auxiliary gaps excited from a spark coil. This type of spark is particularly simple to use in many applications, because it acts virtually as a point source, and sharp records can be made on photosensitive material without use of a lens.

As compared with explosive flash charges, electrical flash methods are much more convenient for tests which can be performed on a laboratory scale. This convenience is largely lost in field work with larger charges, because of the limited light output without multiple units and the increased electrical difficulties. For such work, the explosive flashes are more useful, particularly if facilities for charge preparation are available.

The use of flash sources similar to those described has the disadvantage that a single unit gives only a single flash or requires a considerable interval before it can be triggered again. This objection can be met by the use of a source of high intensity and longer duration if the film is moved rapidly past a slit and a "streak" photograph taken. Experimental methods of this kind (72) have been extensively used in studies of detonation waves, which are self luminous, and have been successfully applied to measurements of shock wave velocities in water, illumination being supplied by the self-luminous shock wave excited in an argon-filled tube, as shown in Fig. 6.2. The advantage of the continuous time record from this method is of course obtained at the price of limitation of the observable field by the slit.

D. *Sources of greater duration.* Flash sources or very high speeds of mechanical motion are essential for stopping the motion of shock or detonation waves which travel several feet in times of a millisecond or less. For slower but still, by ordinary standards, rapidly changing phenomena, such as the expansion of the gas sphere, a number of types of light source are possible. Stroboscopic illumination with continuous film motion and continuous high-intensity illumination using high-speed motion picture cameras have both been used. The stroboscopic methods are based on Edgerton's developments, and the lamps employed in the General Radio power stroboscope, for example, have been used to give flashes at rates up to 1,500 per second with exposure times of the order of 30 microseconds. High-speed motion picture cameras employing rotating prisms to provide the equivalent of intermittent film motion, such as the Eastman High-Speed and Western Electric Fastax cameras, can be used at even higher frame speeds but the duration of exposure is greater (e.g., 80 microseconds for the Eastman High-Speed running at 3,000 frames per second).

A number of varieties of continuous high intensity sources have been used, among which may be mentioned capillary high-pressure mercury arcs and underwater flares. For the relatively short times of 1 or 2 seconds at most, photoflash bulbs with durations of from 6 to 100

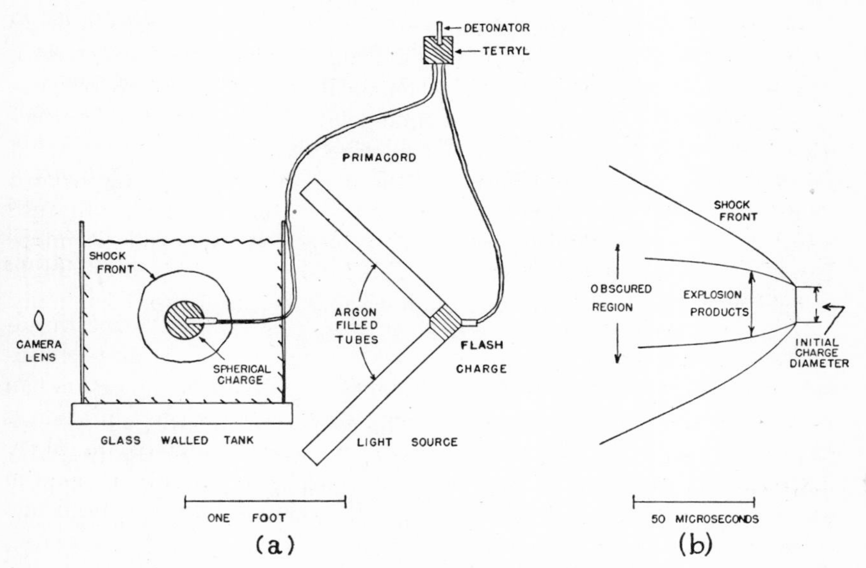

Fig. 6.2 Shock wave velocity measurements for small charges.

milliseconds can be simply employed, several bulbs being fired in sequence by mechanical or electrical means if it is necessary to prolong the illumination. A difficulty common to many such sources is their relative fragility, as a result of which they must be suitably protected from shock wave pressures if they are not to be extinguished or damaged. This consideration evidently also applies to cameras, which must be housed in cases strong enough to withstand hydrostatic and explosion pressures. In extremely clear open water natural sunlight may be adequate for photography of relatively slow motions. As an example, motion pictures of the bubble pulsation from 50 and 300 pound charges have been taken from 60 feet with natural sunlight in water off the Bahamas.

E. *Lighting and synchronization.* The most effective lighting arrangements evidently depend on the end to be achieved, and so many variations are possible that it would be futile to attempt any very general rules. Such things as suitable arrangements of source and camera positions and the need for introduction of reflecting or diffusing surfaces or backgrounds must be determined for the case of interest. For many purposes a silhouette is desirable, which can be obtained, for example, by a translucent diffusing screen in front of a source facing the camera. In small scale experiments, light sources and cameras can often be mounted in air above the water surface by use of suitable mirrors.

In high-speed simple flash records especially, accurate synchronization of lighting with the explosion event is necessary. This can be done in a number of ways, the more common involving changes in firing current on detonation, use of multiple firing circuits with flash charges, or action of the shock wave on a mechanical or electrical triggering device (see section 5.9). Controlled time delays can be obtained by electrical delay circuits, or by use of explosive time delays from unequal lengths of detonating fuse initiating the main and flash charge from a single detonator cap. The velocity of detonation in primacord fuse being 20.6 feet per millisecond, a delay of 4.0 microseconds is introduced by a 1-inch path difference. This method is particularly simple in field work, and with various modifications has been used for many purposes.[4]

6.3. Pressure Measurements by Optical Methods

A. *Shock velocity measurements.* The velocity of an advancing shock front is directly related to the pressure difference across the front by the Rankine-Hugoniot conditions if the equation of state for the fluid is known. In the case of water, the departures of the shock velocity from the limiting acoustic value are appreciable only for pressures very close to the charge, the increase being 30 per cent for a pressure difference of 70,000 lb./in.^{2} (see section 2.5). Pressure determinations from velocity measurements are evidently of limited accuracy unless this excess velocity is appreciable, and methods based on velocity are therefore inherently best suited (in the case of water) to measurements close to the charge. This region is one which is explored with difficulty using mechanical or electrical gauges, and the optical technique is therefore a valuable supplement to more direct pressure measurements. A drawback, as far as pressure determination is concerned, is the increasing uncertainty as to the pressure-velocity relation at higher pressures, owing to meagerness of data for determining the equation of state.

Measurements have been made at Bruceton[5] using a rotating drum camera and slit, the photographic arrangement being as shown in Fig. 6.2(a). Illumination was provided by the luminous shock waves in two

[4] A difficulty in such applications is the fact that primacord, being an explosive, develops a shock front which may interfere with the phenomenon of interest if not properly arranged.

[5] See DF Report Number 20 (35).

argon-filled tubes excited from an auxiliary charge, the angle and lengths of the tubes giving illumination for 70 microseconds. The resulting streak record, a tracing of which is shown in Fig. 6.2(b), can be measured to give displacement as a function of time, and the velocity then ob-

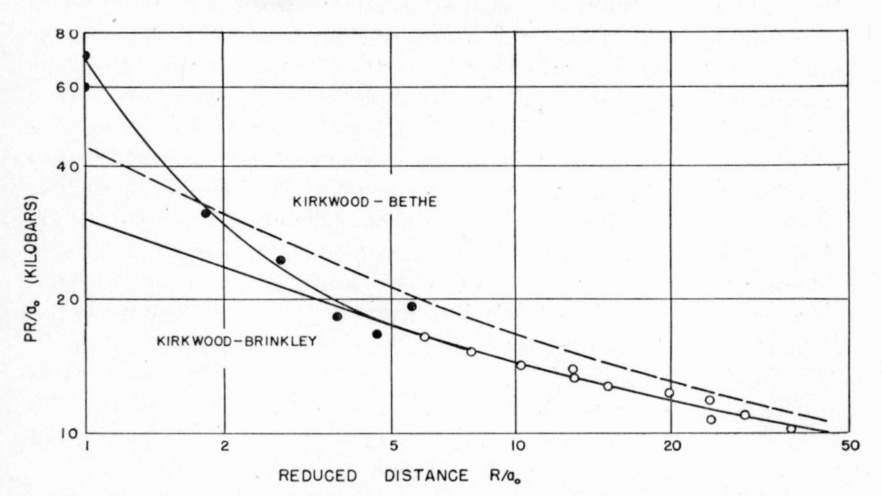

Fig. 6.3 Shock front pressures from velocity data compared with theory and piezoelectric gauge measurements.

tained by differentiation. Values obtained for spheres of a cast explosive two inches in diameter and of density 1.80 are listed in Table 6.1, together with shock front pressures computed from Kirkwood and Montroll's calculations (59, II) for fresh water. The values of both

Distance (charge radii)	Velocity U (m./sec.)	Peak pressure (kilobars)
1.00	5710[1]	>70
	4885	>70
1.86	2675	17.3
2.73	2290	9.5
3.60	1975	5.0
4.46	1875	3.8
5.33	1865	3.7

[1] Measured at a flat surface of explosive.

Table 6.1. Peak pressures for spherical pentolite charges, calculated from the shock front velocity.

velocity and pressure are uncertain at the surface of the charge, the former because of the rapid change in velocity at this time, and the latter both for this reason and because the shock front calculations have not been extended to these velocities. A minimum value is

16,000 ft./sec., corresponding to pressures in excess of 10^6 lb./in.^{2}, and measurements of the less rapidly changing plane wave velocities at a flat charge surface give 19,000 ft./sec.

The pressures determined from these measurements are compared with predictions of the Kirkwood-Bethe and Kirkwood-Brinkley theories in Fig. 6.3, in which distance R is expressed as a multiple of the charge radius a_o. The parameters of the latter theory were obtained from the experimental piezoelectric gauge data plotted as open circles. The velocity data (solid circles) are seen to agree within their scatter with gauge measurements where the two meet, but give much higher pressures close to the charge than either of the theoretical predictions. The difference from the Kirkwood-Bethe theory is probably explained by the assumed initial conditions of adiabatic conversion of the explosive to its products at the same volume. The similar disagreement with the Kirkwood-Brinkley theory cannot be attributed to this assumption, as it is not involved, but may be the result of inadequate treatment of the approximation methods employed. It is to be noted that the various data of Fig. 6.3 are not strictly comparable for several reasons (different explosive loading densities, velocity results for fresh water), but the necessary corrections are much smaller than the differences.

The data obtained by any method are very meager for pressures close to the charge and further investigations would be desirable. Although the velocity method is indirect and gives only shock front pressures, this line of attack is one of the most promising ones. The initial rate of expansion of the gaseous products is also indicated in Fig. 6.2(b), but the outline is distorted by refraction of light passing through the high pressure region (see part B).

B. *Optical distortion methods.* The velocity of light in water depends on the pressure as a result of the difference in density, the index of refraction changing from 1.344 at atmospheric pressure and 17° C. to 1.365 at 20,000 lb./in.^{2}, an increase of 1.5 per cent. This increase is great enough to cause marked refraction effects, particularly for oblique incidence at a discontinuity, and has been used as a basis for optical studies of pressures. The deviation in a light ray passing through a spherical front is indicated diagrammatically in Fig. 6.4(a). A ray originating at Q travels along the broken path through the points S and P. The curved path within the shock front is drawn for decreasing pressure and decreasing velocity at points behind the front.

The actual path between fixed points of entry and exit is determined by the requirement of geometrical optics that it be the one which minimizes the travel time between these points. This condition leads to Snell's law of refraction, and for a linear velocity gradient behind the front the path is easily shown to be an arc of a circle. Rays actually originating at S or P appear to an observer at Q to have originated at points S' or P' on the virtual ray. For clarity, the deviations indicated are very much exaggerated over ones actually occurring. Coordinate lines on screens placed as shown in Fig. 6.4(a) appear to be distorted outward, as indicated in Fig. 6.4(b).

The optical refraction effect just described has been used by Halverson (31) as the basis of a method for determination of shock wave pressures in water. In this method a ruled screen of lucite in the position of the points S, S' is photographed from point Q. The resultant image of the displacement SS' on the photographic record can then be used, together with the known radius of the shock front, distance of the

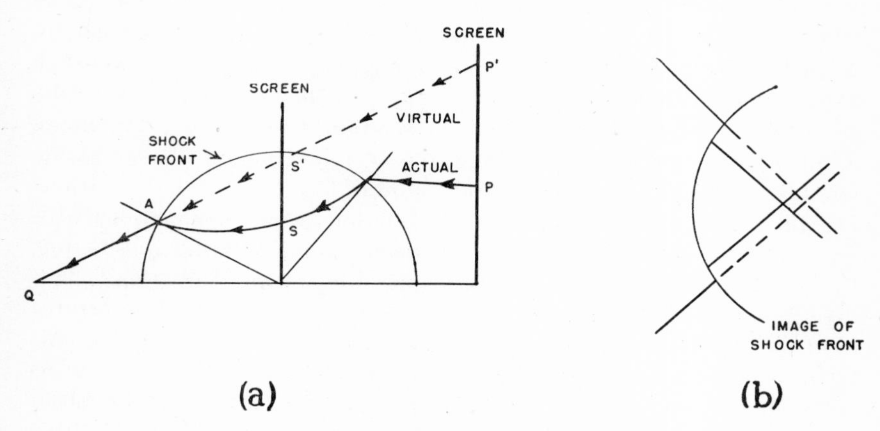

Fig. 6.4 Refraction by a spherical shock wave.

camera, and separation of coordinate lines, to determine the angles of emergence from the spherical front at point A. If the actual path AS is approximated by a straight line an average index of refraction can then be computed for the path by applying Snell's law at point A. This average value, and average pressure deduced from it, has no immediate correspondence with any point values, but the peak pressure at the shock front can be obtained by extrapolation to the limiting ray, the error in the straight line approximation vanishing in this limit if the true pressure-distance relation is a smooth curve behind the front.

The necessary geometrical relations for a spherical front have been obtained by Halverson and need not be repeated here. Computations made on the basis outlined for shock waves $14\frac{1}{2}$ inches from three 0.55 pound tetryl charges gave an average peak pressure of 17,500 lb./in.^{2} with a spread of ± 500 lb./in.^{2} This value is 15 per cent higher than the figure from piezoelectric gauge similarity curves. Approximately the same differences were obtained in another set of measurements with a different explosive.

The systematically higher values from refraction measurements have not been accounted for. It seems unlikely that the piezoelectric gauge data are ten-fifteen per cent in error, in view of their much better agreement with results from ball-crusher gauges, dome velocity calculations, and independent calibrations and measurements. The more likely possibility is therefore the one of inaccuracies in the method. Shock distortion of the screen by the pressure wave was found, by measurements on the records, to be small. Another source of error is uncertainty as to the refractive index-pressure relation for the salt water used in the experimental arrangement. No direct figures for salt water were available, and the relation employed was based on isothermal values at high pressures corrected to adiabatic conditions from data at lower pressures. Further uncertainties exist because of a necessary wavelength correction and possible temperature effects in second order terms. There are reasons to believe that none of these uncertainties has a large effect on the relation actually used, but either a direct determination for salt water or repetition of the explosion measurement in fresh water would be desirable.

Halverson has also extended the refraction measurement to determinations of the decay constant of the pressure wave with distance behind the shock front, the profile being assumed exponential in form. This determination evidently requires an examination of the actual path behind the shock front. Assuming that Snell's law applies, i.e., that the ray is described by geometrical optics, the differential equation of the path can be integrated numerically for any assumed value of decay constant $\theta(r)$ in the expression $P(r) = P_m(R)e^{-(R-r)/\theta(r)}$ where R is the shock front radius. This calculation is repeated for several possible values of $\theta(r)$ to find by interpolation the one consistent with the observed displacement of a coordinate intersection behind the shock front. The decay constant so obtained is of course for spatial variation, and a conversion must take into account the variation with R of peak pressure P_m, and strictly should include variation of $\theta(r)$ with R. The latter effect of spreading is unimportant, however, and can be neglected.

The calculations outlined have been carried out for the shots previously mentioned and gave an average time constant of 40.4 μsec. for 5 points behind the front, the average deviation from the mean being ± 1.7 μsec. with no indication of systematic variation for different points. The initial time constant of decay from piezoelectric data is 50 μsec. or about 20 per cent higher.

The measurement of scale distortions described in the foregoing is open to the basic objection that the results obtained are not purely analytic deductions from basic principles, but depend in part on assumptions as to the form of the refraction gradient. It can readily be shown that this is an essential limitation of the method, as there are finite regions near discontinuities which any possible ray must completely traverse, and hence cannot be explored in detail. In many cases this objection is not serious, however, and the exceedingly simple equipment needed for optical distortion studies recommends the general method as a potentially very useful one.

C. *Other refraction methods.* Although measurements of optical displacements are the most promising means of utilizing refraction effects in pressure waves having discontinuous fronts, two other possibilities should be mentioned. The first of these is an interferometer method devised by E. Mach and further developed by his son L. Mach (70), which is essentially an application of the principle of the Rayleigh or Jamin interferometer. The experimental arrangement consists in splitting light from a slit source into two geometrically separated beams by use of a half-silvered mirror. These two beams are recombined at a second half-silvered mirror after reflection from suitably placed reflecting surfaces, and interference fringes are formed in the combined beam. If the refractive index of a finite region in one beam is changed, the corresponding part of the fringe system is displaced in proportion to the total change in optical path in this part of the beam. The method is of course very sensitive to small changes in refractive index and has been extensively used for investigation of supersonic flow in air or gases. The differences in refractive index for liquids are so much greater, however, that even moderate pressures would cause impractical magnitudes of fringe shift. Deviations in the path by refraction would make elementary analysis based on undeviated rays inapplicable. The most useful application of interferometer methods is to small, slowly changing velocity gradients which are not characteristic of the more interesting underwater explosion phenomena.

A more promising possibility is the Schlieren method, in which the region of disturbance causes angular deviations of parallel light from a source bounded by a straight edge. The beam is brought to a focus by a lens in whose focal plane a second straight edge parallel to the first is placed to block off all but a narrow strip of light. The deviations in the beam result in more or less light passing this barrier into a camera focused on a suitable plane in the disturbance. The original intensity of the disturbance as recorded by the camera will then be changed in proportion to the angular deviation. The method is less sensitive to changes in refractive index than are interferometer techniques, but the limitation of infinitesimal deviation of the beam must be respected if elementary analysis is to be applied, and large discontinuities or gradients are handled with increasing difficulty. The method is nevertheless a potentially useful one, but has not to the writer's knowledge been applied to studies of pressure waves in water.[6]

[6] The mathematical developments and limitations of this and other refraction methods have been discussed in a report by Weyl (119).

6.4. Direct Photography

The applications of photographic techniques which have so far been described involve special optical or mechanical arrangements to obtain detailed information about pressures and velocities. The more conventional methods of flash and motion picture photography are also of great importance, a particular advantage being that information is obtained over a field of view rather than merely at points in the field. This advantage is particularly evident in studying shock wave fronts, the form and motion of the gas volume, and effects at targets. A number of examples of such "straight" photography and results obtained by its use are given in other sections, but a few representative illustrations are presented here to indicate their application and limitations.[7]

A. *Form of shock fronts.* Although shock fronts in water are not self-luminous, large pressure gradients are made visible when suitably illuminated as a result of the accompanying changes in refractive index which have a focusing lens effect. In the case of a spherical shock front, for example, the refraction effect acts to concentrate uniform illumination from behind the spherical front into paths nearer that of the tangential ray grazing the spherical surface than the ones in the absence of the front (see Fig. 6.4). The region near the grazing ray therefore appears more brightly illuminated on a photograph. The apparent radius of the shock front seen in this way is evidently somewhat greater than the true value, because the tangent rays subtend a larger angle at the camera than the secant lines from the great circle on the sphere in the plane normal to the optical axis. This geometrical error is small unless the camera is close to the shock front, but must be taken into account in measurement.

A number of photographs of shock waves in successive stages of development are reproduced in Plates I and II. The pictures were taken by light from flash charges behind the main charge and facing the camera. The desired time interval between detonation of the main and flash charges was obtained by use of primacord fuse, as described in section 6.2. The camera shutter was left open through the interval of both explosions, and the outline of the main charge is made visible in the center of the pictures by light from its detonation. Bare charges were intentionally used to permit this double exposure and to reduce the amount of stray material in the field of view to a minimum.

The development of the shock wave from a spherical charge of pentolite 2⅝ inches in diameter is shown in the 3 records of Plate I, taken at times of 13, 58, and 104 microseconds after detonation. The

[7] These illustrations are from photographs taken at the Underwater Explosives Research Laboratory, Woods Hole. These and many other similar photographs are reproduced in reports by Eldridge, Fye, and Spitzer (31), and by Swift et al (105).

scale marks in the first two cases are 7 inches apart, in the last, 13 inches. The black horizontal rods were introduced to determine the distortion of the dark gas sphere outline inside the shock front, and supporting wires and detonator fragments are also visible. The first picture shows that the shock front shortly after detonation is very nearly spherical and concentric with the charge. Fine structure of the shock wave just behind the front is evident in the second and, to a lesser extent, third records at later times. These secondary fronts are believed to be real rather than optical illusions, and become invisible at greater distances. Even in this highly symmetrical case, the shock wave close to the charge has a fine structure sufficiently pronounced to be registered by refraction, a result of importance in confirming shock wave pressure-time curves exhibiting similar irregularities. It is perhaps not amiss to point out that these and other photographs give little support to assumptions that the shock wave pressure has a significant time of rise, nor does the gas bubble outline exhibit large irregularities in its expanding stages.

Shock waves from a cylindrical stick of pentolite ($1^7/_{16}$ inches diameter, $5\frac{7}{8}$ inches long) are shown in Plate II. The reference marks are 7 inches apart in the first two pictures taken 10 and 54 microseconds after initiation, and 13 inches apart in the third picture taken at 101 microseconds. The first picture shows detonation about half completed, the shock front and gas products being visible although somewhat obscured by general illumination. The second picture is of particular interest, as it clearly shows multiple fronts in regions off both ends of the charge. Two secondary fronts are visible at the detonator end and a single one at the other end. These multiple discontinuities are also observed in piezoelectric gauge records at greater distances, and must have their explanation in the relief of pressure discontinuities at fronts intersecting shortly after their formation, as discussed in section 7.7. The multiple shocks are present but closer to the front and less evident at the later time of the last record. The shock wave front at this time is less elongated and pictures at later times show the approach to the spherical shape ultimately realized.

A clear cut example of the Mach effect in intersecting shock waves from two 0.11 pound loose tetryl charges placed 8.4 inches apart and fired simultaneously is shown in Plate III. The shock fronts have travelled approximately 12 inches from the charges (located below the margin of the picture) and the normals to the fronts at the intersection include an angle of 71°. The characteristic Mach region of intersection discussed in section 7.7 is clearly evident, although the exact details are subject to error from refraction effects.

B. *Photographs of the gas bubble.* The pictures in Plates I and II show the initial rate of expansion of the gaseous explosion products after detonation. The possibility of distortion by refraction in passage through the shock front must be recognized in analyzing any such pictures. This was tested by measurement of the known separations of steel rods visible in Plate I, and the apparent distance found to be from 2 to 8 per cent greater than the actual values. (The elastic compression of the rods by the pressure wave is computed to give displacements of less than 0.1 inch.) The apparent bubble diameters for the 2⅝ inch spheres were 3.7, 5.9 and 6.9 inches at times of 13, 58 and 104 microseconds after detonation. The initial radial velocity of the bubble is therefore of the order of 4,000 ft./sec., this figure being correct only as to order of magnitude because of uncertainties as to the exact times and distances.

The motion of the bubble over the greater part of its pulsations is very much less rapid than in the initial phase, and can be followed in sufficient detail with motion pictures. For small charges with bubble periods of several hundredths of a second or less, special high speed cameras are needed. The records reproduced in Plate IX and Fig. 8.1 for a 0.5 pound charge with a period of 28 microseconds were taken, for example, with a 35 mm. Fastax camera running at about 1,800 frames per second, illumination being provided by flash bulbs. For work with larger charges, lower film speeds and longer exposures are possible. As an example, Plate IV shows the gas products from a 56 pound TNT charge fired at a depth of 100 feet. The photographs were taken from 40 feet with natural light by a Jerome motion picture camera running at approximately 45 frames per second. The first frame at 0.18 seconds after detonation shows the spherical outline of the bubble near its maximum expansion, and the second at 0.34 seconds the spiked protuberances at the first contraction.

Many investigations of bubble motion have been made by photography, a number of which are described in Chapter 8. Comparatively few results have been obtained except on a laboratory or model scale, owing to obvious practical difficulties in obtaining adequate lighting conditions and clear water in field work. Some very beautiful small-scale photographs of both bubble motion and shock waves were obtained in German investigations during World War II, but at the time of this writing few technical data concerning them were available, and they cannot therefore be adequately described here.

C. *Effects at and near targets.* Photographic techniques have proved useful in understanding and analyzing the motion of water near boundaries, such as the free surface of the water and deformable plates. Some examples of this kind discussed in section 10.4 show very clearly the formation of cavitation in water under small negative pressures near such surfaces, and a number of investigations have been carried out to examine the mechanism of such cavitation.

Photographic studies of the damage process in structures are very useful in an understanding of the important phenomena. Measurements of deformation as a function of time have been carried out by several groups, one of the more important conclusions from such investigations being the demonstration of the importance of secondary bubble pulses in causing damage under favorable conditions. A time history of deformation is obviously a first requirement in such studies, and a motion picture record is the best single method of obtaining such information. As illustrations of such work, there may be mentioned studies at the Taylor Model Basin by Hudson (48) on ten inch circular diaphragms, in which displacements were measured by reflection of light into a moving film camera from small steel balls or aluminum spots on the diaphragm. A number of photographic studies on cylinders of various sizes and internal supporting structure have been made by Decius and Fye (26) at Woods Hole, which have shown very strikingly the importance of later bubble pulses in ultimate damage. For more detailed accounts of such investigations reference should be made to the original reports.

7. Shock Wave Measurements

7.1. The Form of the Shock Wave

The underwater shock wave from high explosives is found in most cases to be a highly reproducible phenomenon, which can often be represented at a given point to a first approximation by a discontinuous rise in pressure followed by an exponential decay in time. As the shock wave spreads out from a charge, its peak value decreases and its duration, as estimated for example from the time constant of exponential decay, increases gradually. These broad general features are shown by the reproductions in Plate V of pressure-time curves, obtained using tourmaline gauges, at distances of 20 and 500 feet from a cylindrical 300 pound charge of TNT.[1] The peak pressure of 6,000 lb./in.^{2} at 20 feet falls to 150 lb./in.^{2} at 500 feet, and the time constant of 500 μsec. at 20 feet, defined here as time for pressure to fall to $1/e = 0.37$ of its initial value, increases to 900 μsec. at 500 feet. (The irregularities at later times are the result of reflections from the surface and bottom, not characteristics of the pressure in an infinite body of water.)

It is evident from Plate V that the description of the shock wave as a simple exponential is only an approximation. For charges reasonably close to a sphere in symmetry, the approximation is a reasonably good one for pressures greater than about one-third the peak value, but two types of departure are observed at later times, even for charges of high symmetry. First, a hump or irregularity of variable magnitude and shape, depending on the shape of charge and explosive, and second, a very much more gradual decrease of pressure than the continuation of the initial exponential decay. The gradual decrease is generally observed, and has its explanation in the expansion of the gas sphere following detonation and, in the incompressive approximation, a falling off of pressure at points in the surrounding water with time as $t^{-4/5}$. This characteristic of the pressure-time curve has been considered in section 3.8 according to the theory of Kirkwood and Bethe, and is further discussed in section 9.2.

The "hump," or local region of pressure in excess of a smooth curve, which frequently occurs somewhere near or after the time constant, has been found for geometrically similar charges to be characteristic of the explosive, and it is often possible to distinguish pressure-time curves and identify the explosives responsible by the appearance of these humps.

[1] The experimental data quoted in this chapter were nearly all obtained at the Underwater Explosives Research Laboratory, Woods Hole, by methods described in Chapter 5. More complete compilations of data will be found in reports of the UE series (114), and formal reports by J. S. Coles (23) and by Arons and Smith (5).

Their character is also affected by the shape of the charge, as are features of other positions of the pressure-time curve. The origin of these humps has not been conclusively demonstrated, but it is reasonable to presume that they arise from internal reflections in the product gases of the rarefaction wave of pressure generated in these products when the initial detonation wave first reaches the surface of the charge. No detailed analysis of the complicated dynamics inside the products has been attempted, but the time of appearance in the shock wave in relation to the initial peak is at least of the order of magnitude to be expected from this explanation.

It is to be expected that charges which do not have spherical symmetry will give rise to a shock wave which is not symmetrical, and differences in form of the wave at different points around the charge are in fact observed. An example is the pressure-time curve, shown in Plate VI for the shock wave at one end of a cylindrical charge containing 300 pounds of TNT. In this case, two pressure peaks are observed, the second being higher than the first, and it is evident that approximation of the curve by an exponential is only a crude representation of it. Double peaks of this kind are usually observed at points on or near the extended axis of cylindrical charges, their relative magnitude and displacement depending on the relative dimensions of the charge and distance of the point of observation. These complications are accentuated for small charges of insensitive explosives, for which relatively large booster charges of more sensitive materials must be used to insure proper detonation. The explanation of such departures is considered in section 7.7, but can be considered roughly to be the result of a nonlinear combination of shock wave pressures from different parts of the charge.

The conclusion of importance for the present is that, while an exponential curve is a simple and convenient approximation to the form of an underwater shock wave, it is by no means a perfect representation and in some circumstances is a rather poor one.

7.2. Experimental Shock Wave Parameters

A complete, explicit representation of experimentally measured shock wave pressure-time curves for any type of charge would be an equation giving the pressure as a function of time after arrival of the initial peak in terms of the charge weight and position of the point of measurement. Even for the simplest forms of charge conveniently realizable, an exact result of this kind would be cumbersome and not readily visualized. On the other hand, a simple functional representation by means of a negative exponential curve is not a fully adequate representation, as the discussion of section 7.1 shows, and some compromise between the two possibilities is usually desirable.

A. *Peak pressure and time constant.* Despite the limitations of an exponential decay as a description of shock waves, its analytical convenience is so great that parameters based on it are widely used as a first approximation, and in many cases such parameters furnish a quantitatively reliable description of the earlier parts of the curve. In this

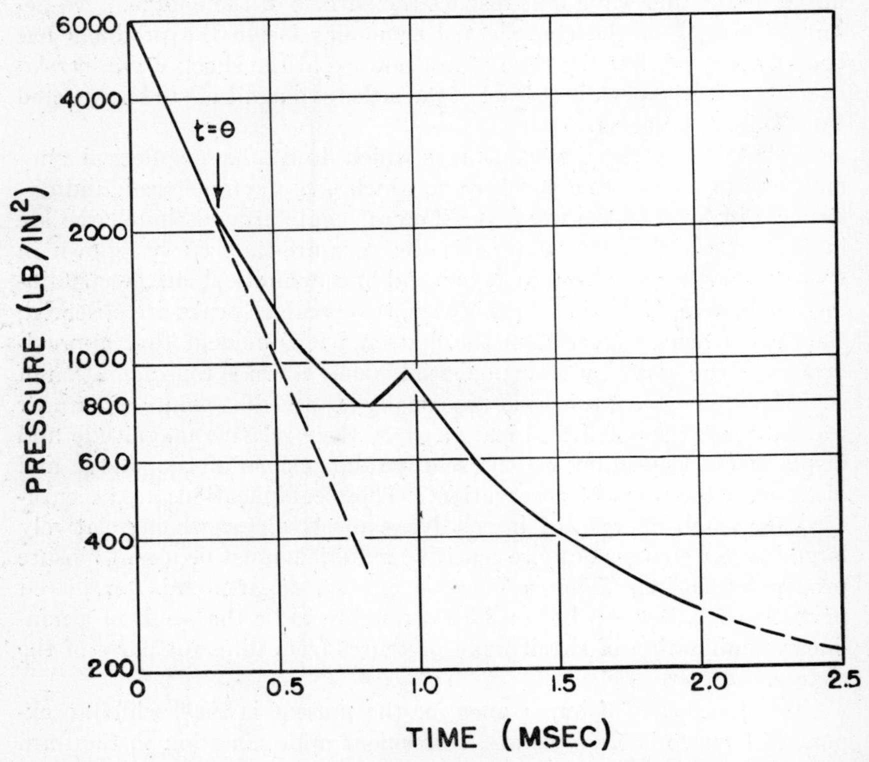

Fig. 7.1 Semilogarithmic plot of pressure against time (14 feet from an 80 pound 50/50 pentolite charge).

approximation, the pressure P as a function of time t after arrival of the shock front is expressed as

$$P = P_m e^{-t/\theta} \tag{7.1}$$

where P_m is the initial peak pressure and θ is the time constant of exponential decay. The accuracy of this approximation in any particular case can be tested by plotting log P as a function of t, which should give a straight line of negative slope $1/\theta$. An example of such a plot is given in Fig. 7.1 for the pressure 14 feet from an 80 pound charge of pentolite. It is seen that the form of the actual curve is well represented by a straight line out to about 0.3 msec., at which time the pressure is about

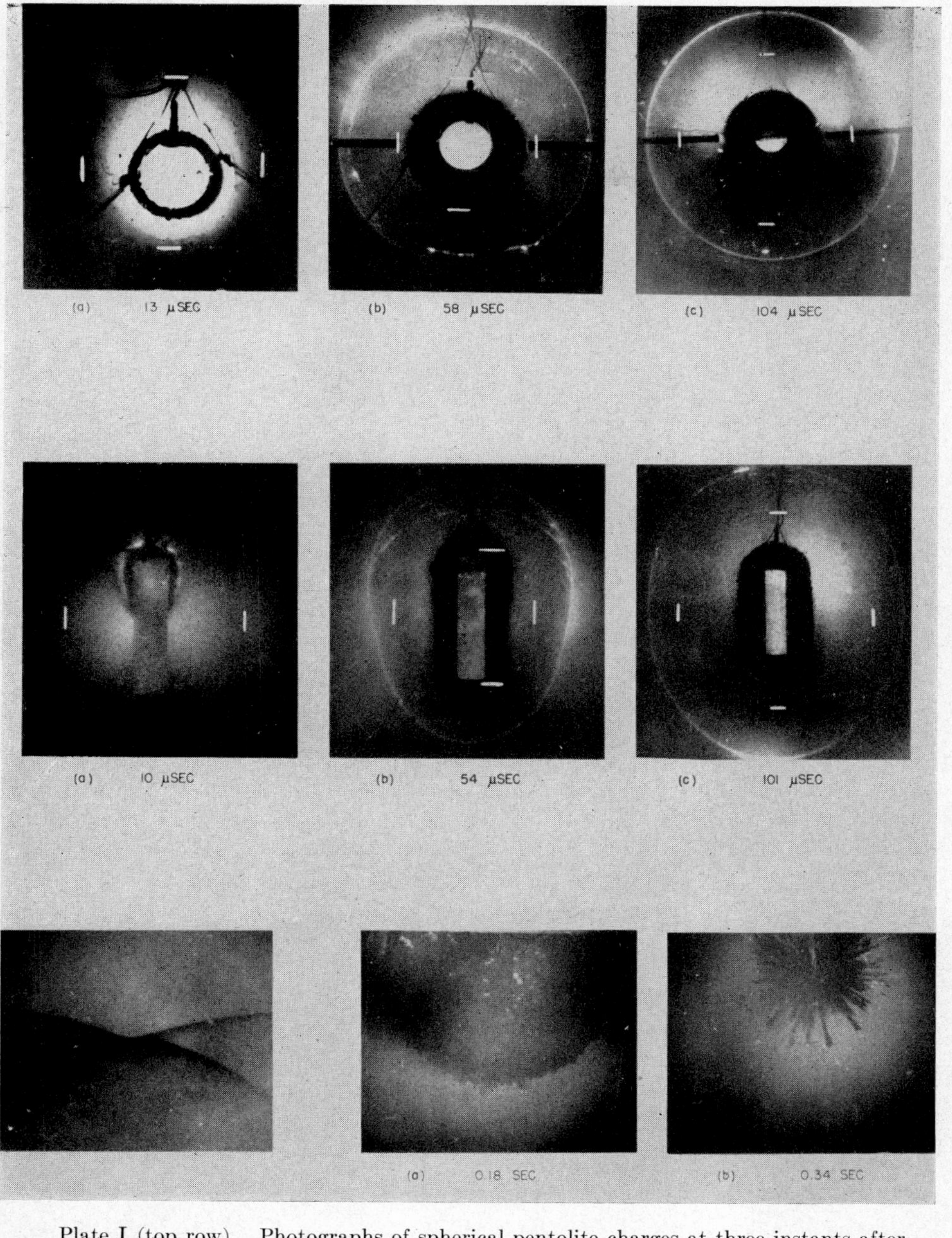

Plate I (top row). Photographs of spherical pentolite charges at three instants after detonation.

Plate II (second row). Photographs of cylindrical pentolite charges (4/1 length-diameter ratio) at three instants after detonation.

Plate III (lower left). Mach intersection of two shock waves at a point 12 inches from two 0.11-lb. tetryl charges 8.4 inches apart.

Plate IV (*a* and *b*, lower right). The gas products from a 56-lb. TNT charge, fired 100 feet below the surface, at two instants.

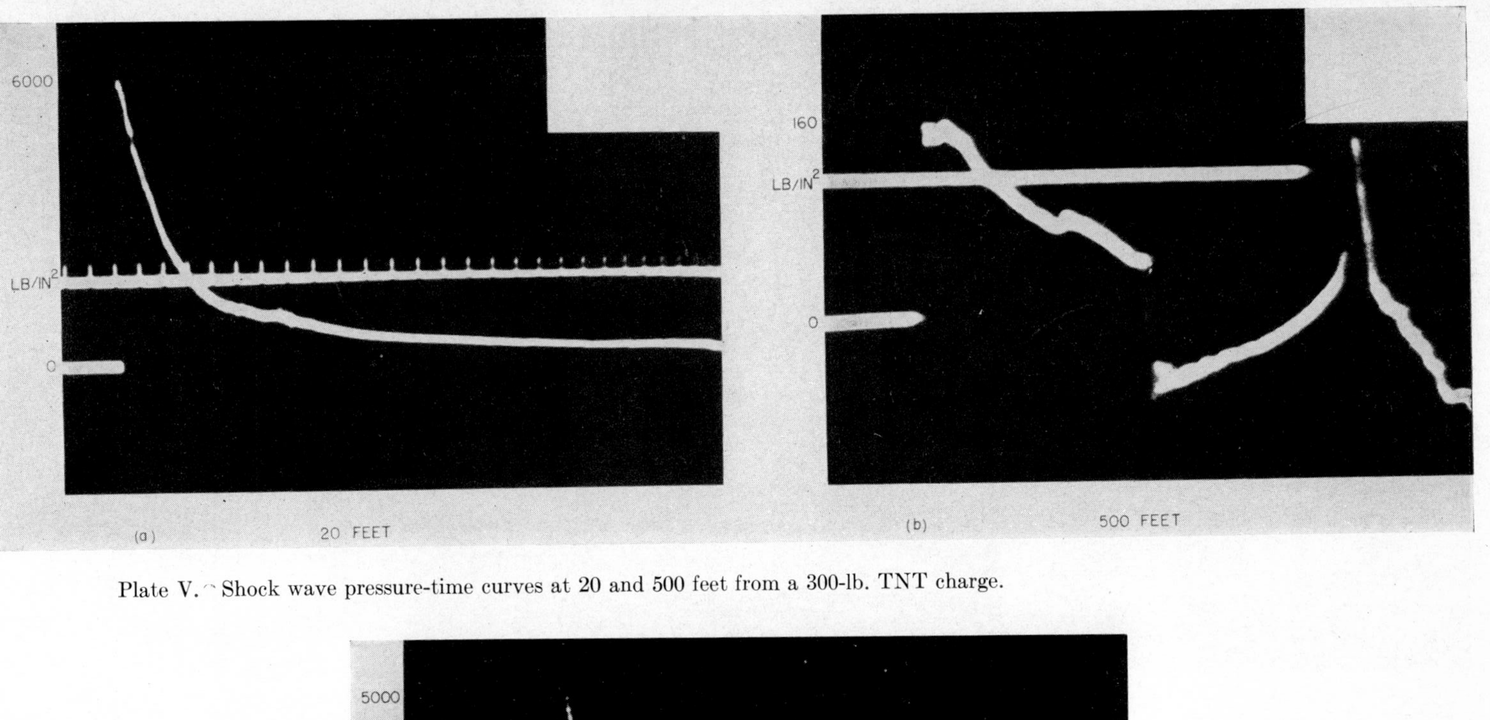

Plate V. Shock wave pressure-time curves at 20 and 500 feet from a 300-lb. TNT charge.

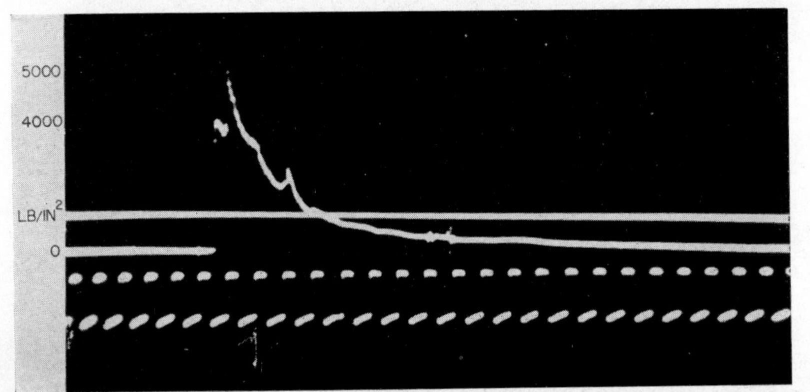

30 per cent of its initial value, but at later times the "tail" of the shock wave decays much more slowly than in the exponential approximation. This result is typical of charges which do not differ too much from spherical symmetry, and illustrates the extent to which the exponential approximations of Eq. (7.1) can be expected to apply in simple cases. Although it would be possible to devise auxiliary functions to fit the later portions of the curve with additional parameters, increasing experimental uncertainty as to the exact values of the smaller pressures ordi-

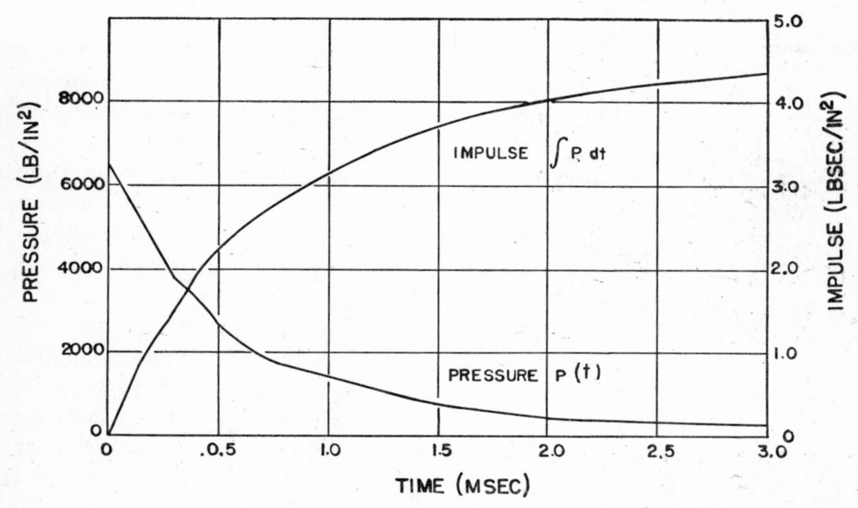

Fig. 7.2 Increase in impulse from later portions of the shock wave (pressure 20 feet from a 300 pound charge).

narily does not justify the necessary labor, and other measures of the pressure-time curve, which include these later portions less explicitly, are used in preference.

B. *Impulse.* For many purposes, the effectiveness of a shock wave depends on the time-integral of pressure, or impulse, more significantly than on the detailed form of the pressure versus time. This may be true, for example, in cases of structural damage if the time of deflection of the structure is much longer than the duration of pressure in the incident wave. By definition, the impulse of unit area of the shock wave front[2] up to a time t after its arrival is given by:

$$I(t) = \int_{o}^{t} P(t)\, dt \tag{7.2}$$

[2] The proper definition of impulse is the time integral of force, and the quantity referred to as impulse in the text is strictly a specific impulse for unit area. The term impulse for time integral of pressure is so general, however, that this usage is followed.

Strictly, the pressure $P(t) - P_o$ in excess of hydrostatic pressure should be used in this equation, but for most cases of interest the shock wave pressure $P(t)$ is so large that the difference is of no importance.

Unfortunately for simplicity, the form of the shock wave is such that the integral $I(t)$ does not converge rapidly to a limiting value with increasing values of t, and it is not practically possible or desirable to define an upper limit on time for which the impulse is obtained. This difficulty arises from the existence of the shock wave tail, or interval of slowly decaying pressures, the area under which makes a continually increasing contribution to the integral of Eq. (7.2), as shown in Fig. 7.2. The reason for the tail is the onset of essentially noncompressive motion exterior to the gas sphere, and the pressure associated with the motion becomes zero (strictly hydrostatic) only after the gas pressure has fallen to hydrostatic. This time is of the order of one-tenth the period of pulsation of the gas sphere (see section 9.3) which is many times the time constant θ of the initial high pressure region. For example, the time constant θ is approximately 0.5 msec. thirty feet from 300 pounds of TNT, and the duration of positive excess pressure is of the order of 80 msec. for this charge detonated 40 feet below the surface.

The determination of impulse out to some 200 times the initial time constant of the wave would, for many reasons, be a time consuming and difficult problem both in experimental arrangements and analysis. Such an integration in many cases would reduce the value of the figure finally obtained as a measure of a short-lived transient pressure. For these reasons, impulse values for shock waves are customarily obtained by integration over only the earlier high-pressure region to an arbitrarily assigned upper limit on time, which should be long enough to include any characteristic features of the pressure-time curve when the pressures are a significant fraction of the initial peak value. One way of assigning the upper limit on time is to specify it in terms of the initial time constant, for example as five or ten times the time constant. Such a measure presents difficulties if pressure-time curves from considerably different explosives or from charges of different shapes are to be compared, and in such a case a simpler measure is a specified value for a given charge weight or volume.

In order to arrive at a suitable criterion for the integration time, the effect of different choices in this time on impulse ratios for charges of different explosives and shapes has been examined.[3] The data used included pressure-time curves for charges of three explosives, in approximately spherical and elongated cylindrical shapes. The measured time constants differed by factors as great as three, but impulse ratios for times greater than five times the mean value showed little change with increasing values of the limit on time, although individual values in-

[3] Unpublished analysis by the writer.

creased continuously as indicated in Fig. 7.2. It was accordingly concluded that a significant measure of impulse for comparison purposes is obtained by integrating to five times the time constant of a spherical charge of the same weight. Values of time corresponding approximately to this rule have been used for most impulse calculations from data obtained at Woods Hole, although for special cases other values have been used. Whatever the criterion that is chosen, its arbitrariness must be recognized in any analysis of experimental data. (This point is discussed, for example, in section 9.3 in comparing shock wave impulse with values for the later secondary, or bubble, pulses.)

C. *Energy flux density.*[4] Another significant measure of the shock wave is in the energy flux across unit area of a fixed surface normal to the direction of propagation. The rate of energy transport across unit area in a fluid of density ρ and particle velocity u is, from section 4.8, given by

$$\rho u \Delta\left[E + \tfrac{1}{2}u^2 + \frac{P}{\rho}\right]$$

where $\Delta[E + \frac{1}{2}u^2]$ is the increase in kinetic and potential energy for unit mass of fluid, and P is the pressure. The energy flux density to time t after arrival of the shock wave is then

$$E_f = \int_0^t \rho u \Delta\left[E + \tfrac{1}{2}u^2 + \frac{P}{\rho}\right] dt \tag{7.3}$$

In the limit of small amplitudes, the internal and kinetic energies can be neglected in comparison with P/ρ, and for an outgoing spherical acoustic wave the particle velocity u is given by (see section 2.2)

$$u(R, t') = \frac{P - P_o}{\rho_o c_o} + \frac{1}{\rho_o R}\int_0^{t'} (P - P_o)\, dt$$

The energy flux density in terms of pressure then becomes

$$E_f = \frac{1}{\rho_o c_o}\int_0^t (P - P_o)^2\, dt + \frac{1}{\rho_o R}\int_0^t (P - P_o)\left[\int_0^{t'} (P - P_o) dt\right] dt' \tag{7.4}$$

The second term, representing the effect of the excess particle velocity or afterflow left by an outgoing spherical wave, rapidly becomes negli-

[4] This quantity is frequently described in the literature as energy flux or simply energy, when in reality a specific quantity for unit area is meant.

gible compared with the first at increasing distance R from the source. Its magnitude increases with the time t to which the integration is extended as a result of the slow decrease of P in the tail of the shock wave.

The effect of the afterflow is not ordinarily considered at distances exceeding ten-twenty charge radii, and energy flux density is estimated by the first term only in Eq. (7.4). This term in itself is not exact, because it is obtained in the acoustic approximation. An estimate of the correction necessary at the shock front can be made by expressing the exact result of Eq. (7.3) in terms of pressure by employing the Rankine-Hugoniot conditions of section 2.5, which are valid at the shock front. These give:

$$E_f = \frac{1}{\rho_o} \int \frac{P^2}{U - \dfrac{P}{\rho_o U}} dt$$

Expressing the shock front velocity U in terms of pressure by the relation $U = c_o(1 + \alpha P)$ and expanding in powers of P then gives correction terms to the integral $(1/\rho_o c_o) \int P^2 dt$. For an exponential shock wave of peak pressure P_m (lb./in.²) in sea water, these reduce to the result (4) that

$$E_f = (1 - 1.67 \times 10^{-6} P_m - 4.9 \times 10^{-12} P_m{}^2)\,(1/\rho_o c_o) \int_o^t P^2 dt \tag{7.5}$$

For a pressure $P_m = 20{,}000$ lb./in.², the correction amounts to 3.5 per cent and hence is insignificant at lower pressures.

D. *Units.* Of the many units which have been, and are, employed to express explosion pressures, the most common in English speaking countries is pounds per square inch (lb./in.²). In some of the theoretical developments in Chapters 3 and 4, the more convenient metric unit of kilobars (1 kilobar $= 10^9$ dyne/cm.² $= 14{,}513$ lb./in.²) is employed, but the use of (lb./in.²) is so general in experimental and engineering work that the experimental results discussed in this and later chapters are expressed in this unit. Lengths such as distance from the charge and depth of water are expressed in feet, also in conformance with the most usual practice, and times are given in seconds or decimal fractions of seconds (milliseconds, microseconds).

The derived quantity impulse is conveniently expressed in units of (lb. sec./in.²). The energy flux density, as defined in part (C), has really the nature of intensity, or energy per unit area, and the total energy flux through a surface is given by the integral of this quantity over the surface. The engineering unit of work or energy which is readily obtained from the experimental used pressure unit is inch-pounds, and the energy flux density function is conveniently expressed in units of (in. lb./in.^{2}). It is evident that the units chosen have little to recommend them from the point of view of logic or internal consistency and their adoption here is justified by the convenience of established usage.

The energy flux density in any approximation is somewhat arbitrary, as in the case of impulse, because of the necessity for some convention as to the time t to which the integration is carried, and this must be remembered in specifying or attaching significance to experimental values. The dependence on P^2 of the leading term in Eq. (7.4) makes this integral converge more rapidly than the impulse integral, however, and for most purposes the same upper limit as for impulse, of approximately five times the time constant, is fully adequate to give a fair measure of energy flux density.

7.3. Validity of the Principle of Similarity for Shock Waves

The most direct predictions concerning the behavior of shock waves are embodied in the principle of similarity, discussed in section 4.1, and conclusions drawn from it. According to this principle, if the linear dimensions of the charge and all other lengths are altered in the same ratio for two explosions, the shock waves formed will have the same pressures at corresponding distances scaled by this ratio, if the times at which pressure is measured are also scaled by the same ratio.

A. *Scale experiments for spherical pentolite charges.* The most direct experimental test of similarity is the obvious one of making two or more experiments with charges of different weights in which all linear dimensions have the same ratio. If the observed pressure-time curves, measured at distances in this ratio, are identical when corrected for the difference in time scale, the principle is verified. Since the linear dimensions of a charge are proportional to the cube root of the volume, and hence weight W, the distances in such experiments should be made proportional to $W^{1/3}$. The observed pressure-time curves at scaled distances R satisfying R = constant $\times W^{1/3}$, and only at these distances, should be the same at times t proportional to $W^{1/3}$, or what is the same thing, should superpose if plotted to new scales of time taken proportional to the reduced time $t/W^{1/3}$.

The simplest and most clear cut tests of this kind have been made using spherical cast pentolite charges of three different weights. Pentolite is an excellent explosive material for this and other types of fundamental investigation because it can readily be cast into charges of varied size and shape and is sensitive enough to be detonated with a minimum size of initiating, or "boostering," material. In small charge work, which is in many ways the more difficult, spherical pentolite charges have been found to give more reproducible results than any other explosive, presumably because of the symmetry and lack of booster material.

A series of measurements (23) have been made for 51 and 80 pound charges at a number of distances from 12 to 100 feet, which can be com-

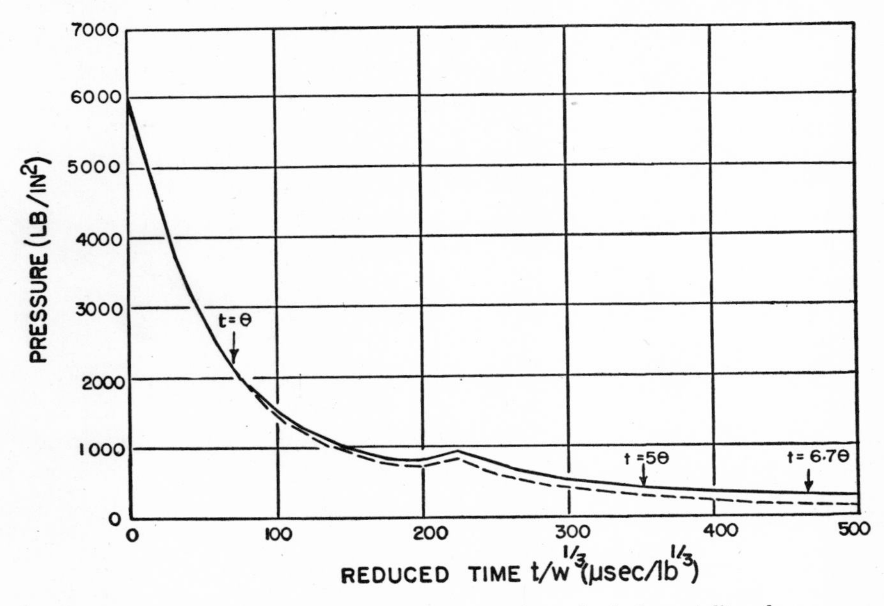

Fig. 7.3 Shock wave pressure-time curves for spherical pentolite charges, plotted against reduced time $t/W^{1/3}$.

pared with data for 3.8 pound charges at a distance of 5 feet. The pressure-time curve for the small charges should, if similarity holds, superimpose on the curve for 51 pound charges at a distance of $(51/3.8)^{1/3}\cdot 5 = 11.9$ feet, and the curve for 80 pound charges at 14

Charge weight W (lb.)	80	51	3.8
Distance R (ft.)	14	11.9	5.0
$W^{1/3}/R$ (lb.$^{1/3}$/ft.)	0.308	0.312	0.312
Peak pressure P_m (lb./in.^{2})	5910	6060	6040
Reduced time constant $\theta/W^{1/3}$ (μsec./lb.$^{1/3}$)	69.6	72.8	69.7
Reduced impulse $I/W^{1/3}$ (lb. sec./in.^{2}/lb.$^{1/3}$)	0.604	0.604	0.558
Reduced energy density $E/W^{1/3}$ (in. lb./in.^{2}/lb.$^{1/3}$)	287	263	273

Table. 7.1. Experimental verification of geometrical similarity for shock waves from spherical charges of pentolite (50/50 PETN-TNT).

feet, provided that the curves are plotted against a reduced time $t/W^{1/3}$. Averages of individual gauge measurements are plotted in Fig. 7.3, the solid curve being the average of 4 records for 51 and 80 pound charges and the dashed curve the average of 14 records for 3.8 pound charges.

The general agreement is excellent, even to the shape and position of the bump at $t/W^{1/3} = 225$ μsec./lb.$^{1/3}$. The differences in pressures at later times are to be attributed to experimental inaccuracies rather than to failure of similarity.

The agreement among results for the 3 charge weights can also be shown by comparison of values for peak pressure and reduced values of time constant, impulse, and energy density. The reduced values are obtained by dividing by $W^{1/3}$, this being necessary to make the values comparable as all depend linearly on time, which scales as $W^{1/3}$. The results in Table 7.1 are seen to be in excellent agreement except for the impulse (integrated to 6.7θ), the difference in this resulting from the discrepancies in the tails of the measured curves.

Parameter	Functional dependence	Acoustic law
Peak pressure P_m	$f(W^{1/3}/R)$	constant $\cdot (W^{1/3}/R)$
Characteristic time t' (e.g., time constant θ)	$W^{1/3} f(W^{1/3}/R)$	constant $\cdot (W^{1/3})$
Impulse $I(t'/W^{1/3})$	$W^{1/3} f(W^{1/3}/R)$	constant $\cdot (W^{2/3}/R)$
Energy density $E(t'/W^{1/3})$	$W^{1/3} f(W^{1/3}/R)$	constant $\cdot (W/R^2)$

The following parameters are functions only of the ratio $W^{1/3}/R$:
Peak pressure P_m, reduced time constant $\theta/W^{1/3}$, reduced impulse $I/W^{1/3}$, reduced energy density $E/W^{1/3}$.

Table 7.2. Functional dependence of shock wave parameters on charge weight W and distance R required by similarity. $f(x)$ represents an undetermined function of the argument x. The last column gives the limiting acoustic law for a spherical wave.

A natural question is whether the verification of similarity for spherical charges is also obtained for other charge shapes. There appear to be no direct experiments establishing similarity in such cases, but indirect evidence gives no reason to doubt the validity of the principle, nor does the hydrodynamical theory. The results of Fig. 7.3 and Table 7.1 thus provide a good experimental basis for accepting the principle as generally valid for shock waves, and this acceptance is further justified by less direct tests described in the following.

B. *Similarity curves for spherical charges.* The principle of similarity, as developed in section 4.1, leads directly to simple predictions of the functional dependence of shock wave parameters on distances from the charge and its dimensions. These predictions are summarized in Table 7.2, and follow simply from the basic proposition which can be stated: pressures for the same value of $W^{1/3}/R$ are the same at equal values of $t/W^{1/3}$. To show how this is fulfilled in practice, values of the various reduced parameters listed in Table 7.2 can be plotted as functions of $W^{1/3}/R$ for different values of both $W^{1/3}$ and R. If similarity is in fact realized, all the points for each reduced parameter should lie on a single curve, and the equation of this curve then specifies the parameter completely for the particular explosive.

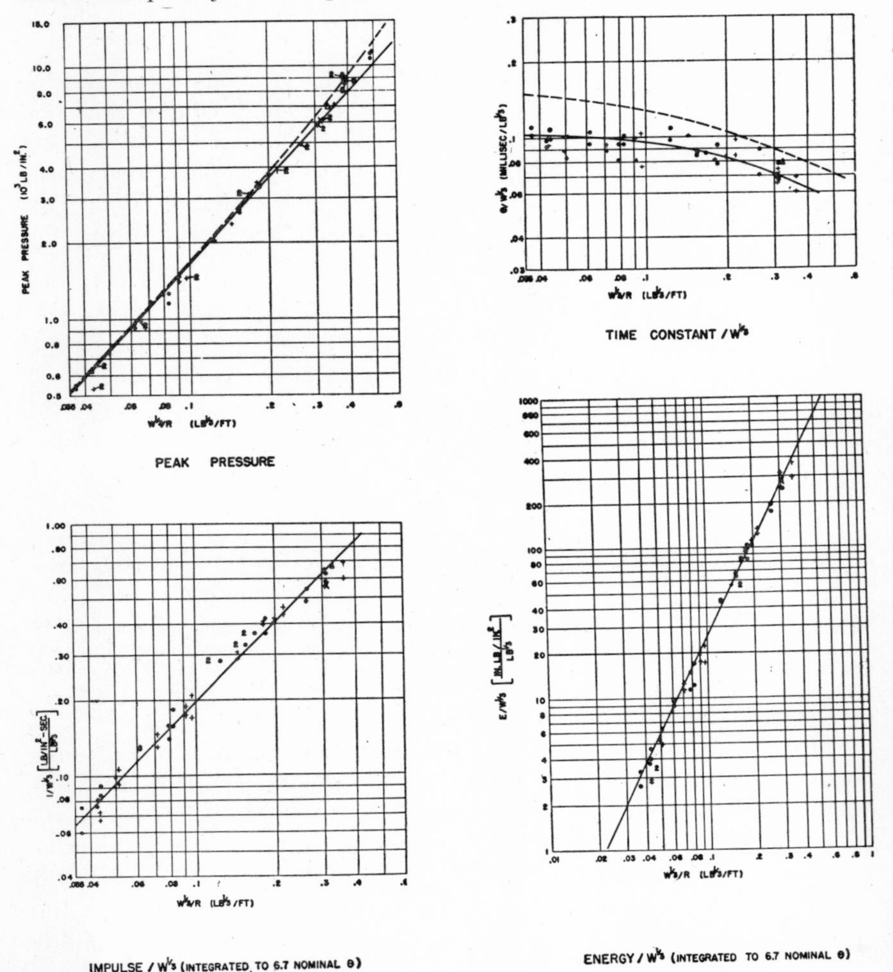

Similarity curves for spherical pentolite charges plotted against $W^{1/3}/R$.
Fig. 7.4 (upper left), Peak pressure P_m.
Fig. 7.5 (upper right), Reduced time constant $\theta/W^{1/3}$.
Fig. 7.6 (lower left), Reduced impulse $I/W^{1/3}$ to time 6.7θ.
Fig. 7.7 (lower right). Reduced energy flux density $E_f/W^{1/3}$ to time 6.7θ.

Extensive studies of shock wave similarity have been made at Woods Hole with spherical pentolite charges. Data were obtained from 47 records at 12 distances from 51 and 80-pound charges, and 75 records at 4 distances from 0.5, 3.8, and 7.5-pound charges. Peak pressures P_m are plotted against $W^{1/3}/R$ with logarithmic scales in Fig. 7.4 and are seen to follow a smooth curve. The solid straight line of slope 1.13 represents the empirical equation

$$P_m(\text{lb./in.}^2) = 2.25 \times 10^4 \left(\frac{W^{1/3}(\text{lb.})}{R(\text{ft.})}\right)^{1.13}$$

Theoretical values from calculations based on the theory of Kirkwood and Bethe are plotted as the dashed curve and are seen to be in excellent agreement with the observations. This agreement is rather better than is observed for most explosives (more detailed comparisons with theory are presented in section 7.4).

The reduced time constant $\theta/W^{1/3}$ is plotted in Fig. 7.5 and is seen to increase slowly with distance from the charge. The experimental values show somewhat greater scatter than in the case of peak pressure but show no systematic departures from the estimated empirical curve. The dashed line predicted by the Kirkwood-Bethe theory gives systematically larger time constants, which are to be attributed to the exponential or peak approximation, which constrains the theoretical curve to exponential form. The reduced impulse, integrated to a time $t = 6.7\theta$, is plotted in Fig. 7.6. The points are fitted quite well, except close to the charge, by the formula:

$$I(6.7\theta) = 2.18W^{1/3}\left(\frac{W^{1/3}}{R}\right)^{1.05}$$

The theoretical results are an even better fit, as indicated by the dashed line, but the agreement in absolute value is fortuitous, as the experimental results were obtained by integration of the experimental curve to a finite time, and the theoretical result is for integration of an exponential formula to infinite time. It is interesting to note, however, that the shape of the curve is so well represented by the theory.

The energy flux density function E_f, integrated to 6.7θ, is plotted in Fig. 7.7 and is fitted over the entire range by the relation

$$E_f = 3.27W^{1/3}\left(\frac{W^{1/3}}{R}\right)^{2.12}$$

The Kirkwood-Bethe theory gives the same slope of 2.12 and a constant of 4.23. The excess of this slope over the value 2.00 is a result of departures from the acoustic inverse square law, as is the difference of the peak pressure slope of 1.13 from the acoustic variation as $1/R$.

The confirmation of the principle of similarity can be extended even further to the magnitudes and times of occurrence of the minimum and following maximum in the pressure-time curve. The plots of these pressures and reduced times are omitted here for reasons of space, but it is interesting to note that the pressures fall off with distance as $(W^{1/3}/R)^{1.05}$. The exponent 1.05 is not, nor does similarity require it to be, the same as the value 1.13 for peak pressure, and the difference is understandable in view of the lower pressure level and closer approach to acoustic conditions.

Although the data for pentolite give the most detailed verification of the principle of similarity, investigations with other explosives (TNT, tetryl, and mixed explosives) have established similarity curves for these explosives also, some of which are discussed in comparison with each other and with theory in the next section.

7.4. Similarity Curves for Various Explosives

Shock wave similarity curves for a considerable range of charge weights and distances from the charge have been obtained for several

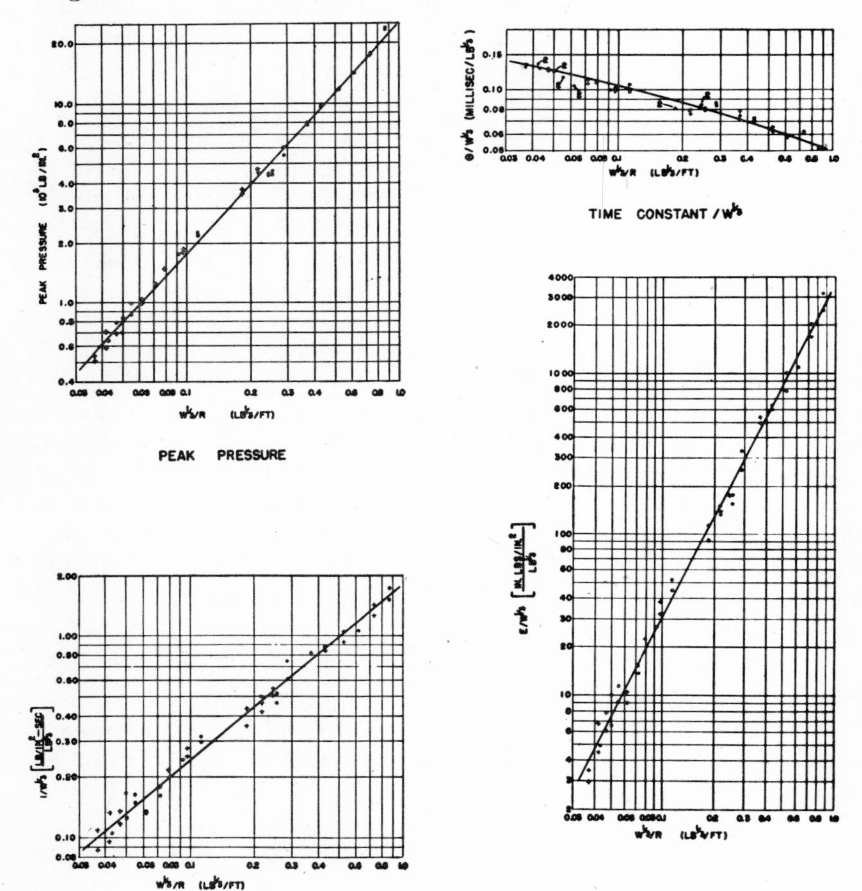

Similarity curves for spherical TNT charges plotted against $W^{1/3}/R$.
Fig. 7.8 (upper left), Peak pressure P_m.
Fig. 7.9 (upper right), Reduced time constant $\theta/W^{1/3}$.
Fig. 7.10 (lower left), Reduced impulse $I/W^{1/3}$ to time 6.7θ.
Fig. 7.11 (lower right), Reduced energy flux density $E_f/W^{1/3}$ to time 6.7θ.

explosives. These results are very similar in general form to the results obtained for pentolite, as discussed and illustrated in section 7.3. Experimental values have been obtained for the following conditions, among others:

(1). Spherical TNT charges, pressures from 500 lb./in.^{2} to 20,000 lb./in.^{2} for charge weights of 48 and 76 pounds, density 1.52 gm./cm.^{3}

(2). Cylindrical loose tetryl charges, pressures from 4,000 lb./in.^{2} to 17,000 lb./in.^{2} for charge weights of 4 and 5 pounds, density 0.93 gm./cm.^{3}

For many years, TNT has been the standard high explosive, more widely used than any other, although other materials are superior to it in effectiveness or for fundamental investigations. Because of its widespread use as a standard of reference, similarity curves for TNT are plotted in Figs. 7.8–7.11. These curves are seen to be qualitatively very similar to those for pentolite in section 7.3, and provide a means of estimating shock wave parameters over a rather wide range of charge weight and distance. It should be remembered that these curves apply strictly only to spherical charges of density 1.52 gm./cm.^{3}, and departures from spherical shape or differences in loading density will result in changes of the parameters.

It is generally characteristic of shock wave results for high explosives that peak pressure P_m and the reduced energy flux $E_f/W^{1/3}$ are accurately represented over the range of measurement by power laws of the form: constant $\times$ $(W^{1/3}/R)^\alpha$, where α is a constant. The reduced impulse $I/W^{1/3}$ and time constant $\theta/W^{1/3}$ are less accurately fitted by such a power law, but the agreement is fair for these over a range of 10 to 1 or more in $W^{1/3}/R$, except very close to the charge (distances less than 10 charge radii). The power law thus gives a compact, reasonably precise method of representing the data, and is also a reasonably good representation of results from the theory of Kirkwood and Bethe (see section 4.2). Accordingly, it is convenient to write the peak pressure, impulse, and energy flux in English units as

$$P_m = k\left(\frac{W^{1/3}}{R}\right)^\alpha \tag{7.6}$$

$$I(t/W^{1/3}) = lW^{1/3}\left(\frac{W^{1/3}}{R}\right)^\beta$$

$$E_f(t/W^{1/3}) = mW^{1/3}\left(\frac{W^{1/3}}{R}\right)^\gamma$$

In quoting values of the constants, charge weights are taken in pounds, and distance R from the charge in feet.

Experimental values of the constants (k, l, m) and exponents

(α, β, γ) for the series listed are summarized in Table 7.3. As has been remarked in section 7.2, impulse and, to a lesser extent, energy flux values depend on the time to which the integration of the pressure-time curve is carried, and this time for the different series is therefore also entered. For comparison with theory, values of the parameters fitted to calculations based on the Kirkwood-Bethe theory (59) are entered in parentheses.

Important questions from the purely empirical point of view are the accuracy of the experimental values upon which the curves are based and the accuracy with which the curves represent these points, as these determine the confidence with which pressure and derived quantities can be obtained from Eqs. (7.6). The reproducibility of experimental

Explosive	Peak pressure P_m		Impulse $I(t)$		Energy density $E(t)$		Time of integration
	$10^{-4}k$	α	l	β	$10^{-3}m$	γ	
TNT Density 1.52	2.16 (2.60)	1.13 (1.21)	1.46 (1.50)	0.89 (0.86)	2.41 (3.78)	2.05 (2.11)	6.7θ
Loose tetryl Density 0.93	2.14 (2.50)	1.15 (1.22)	1.73 (1.50)	0.98 (0.86)	3.00 (3.20)	2.10 (2.04)	5.0θ
Pentolite Density 1.60	2.25 (2.85)	1.13 (1.23)	2.18 (1.65)	1.05 (0.88)	3.27 (4.23)	2.12 (2.11)	6.7θ

Table 7.3. Parameters of shock wave similarity curve for several explosives. Values predicted from the Kirkwood-Bethe theory over the same pressure range are given in parentheses.

values differs for the various series quoted, and specific estimates should be obtained from the original reports, but standard deviations of mean values, where applicable, are of the order of ± 2 per cent for peak pressure, and ± 4 per cent for impulse and energy flux. These estimates of error are based entirely on internal consistency of the data and rather larger errors in absolute value are possible, but enough independent evidence for isolated pressure values exists to make it improbable that these errors are more than a few per cent. The standard deviation of single experimental points from the best straight lines are of the order of 4 per cent for peak pressure, and 8–12 per cent for impulse and energy. It is to be noted that these estimates of error include failure of the straight-line relation to fit the data over the range of measurement as well as scatter of the points, and are therefore somewhat pessimistic. It is therefore reasonable to conclude that over the experimental range of $W^{1/3}/R$, the formulas (7.6) give reasonably good values of the parameters.

The theoretical estimates listed in Table 7.3 give, on the whole, a very good account of the experimental observations.[5] The predicted peak pressures are of the order of 10 per cent higher than experimental, but the exponents of $W^{1/3}/R$ agree rather well, the poorest agreement being in the case of tetryl, for which the data were scantier and covered a smaller range of $W^{1/3}/R$. The comparison of absolute values of impulse is somewhat meaningless unless some decision is made as to when and how the shock wave ends (see section 4.8), but the agreement of observed and calculated values is as good as can be expected in view of this difficulty. The energy-flux parameters are somewhat less ambiguous and, although absolute theoretical values are systematically high, the weight-distance exponent is in good agreement. In nearly all cases, comparison of the Kirkwood-Bethe theory with experiment shows that the former leads to pressures and initial time constants of decay which are somewhat too large, the discrepancy being larger for time constants.

The variations with distance of all shock wave parameters are predicted rather well, and it can be concluded that the theory agrees with experiment at least as well as could be expected.

The upper limits of pressures to which the similarity curves discussed are fitted do not exceed 25,000 lb./in.^{2} corresponding to a distance of about 7 charge radii. At closer distances, the effects of dissipation at the shock front and the true form of the detonation wave in the explosive are more important. Rather limited data at higher pressures reflect this difference by showing that the rate of decay with distance becomes greater close to the charge. More extensive measurements in this region would be of great interest in determining the conditions before the shock wave has progressed very far, and in particular for estimating the total shock wave energy release (see section 4.8). The data so far available give only meager or rather inaccurate values of pressure and duration, although they do show the expected increase in rate of dissipation and rate of pressure decay (decreased time constant).

At the other extreme of low pressures, a number of complicating phenomena are present, which are associated with the long paths involved. Ultimately, in an ideal fluid of infinite extent, an ordinary acoustic wave of much lower pressure and longer duration than the shock wave near the charge would be expected from theory. Water is, however, not an ideal fluid, seemingly small viscosity effects are no longer negligible, differences in temperature with depth cause refraction of bending of sound rays, and the surface and bottom of the sea introduce reflected waves which interfere with the direct wave to a point of observation. A more detailed discussion of these complicating phe-

[5] The agreement is not immediately obvious from the constants and exponents listed in Table 7.3, as numerical values of the constants are sensitive to small changes in the exponent.

nomena is given in section 7.9, but it may be said here that similarity curves of the form discussed in this section and section 7.3 appear to describe the pressures until they have fallen below 100 lb./in.^{2}, corresponding to distances of the order of 900 charge radii.

7.5. Comparison of Shock Waves for Different Explosives

A. *Similarity curves.* Similarity curves giving peak pressure, impulse, and energy furnish the most comprehensive basis for the comparison of the shock waves. By their use, estimates can be made of ratios of any of these parameters at any distance. As the approximate formulas are written and curves are plotted, ratios can be most easily compared for equal weights of explosive. If the equations are used,

Explosive	Pentolite 50/50	Loose tetryl
Peak pressure	1.04	0.97
Impulse	1.29	1.09
Energy density	1.27	1.19

Table 7.4. Ratios of shock wave parameters for two explosives relative to an equal weight of TNT at the same distance.

it is to be remembered that they represent simple approximations which give a reasonably good overall fit for the experimental ranges of $W^{1/3}/R$, and hence are subject to some error.

The peak pressure, impulse, and energy density ratios for equal weights of explosive depend not only on the explosive but also on the distance considered (because of differences in the distance decay laws). A quantitative comparison for $W^{1/3}/R = 0.4$ (corresponding to a distance of 17 feet from a 300 pound charge) is given in Table 7.4, the values being obtained from the data used in obtaining the similarity curves and formulas. Compared to TNT both Pentolite and tetryl show rather small differences in peak pressure, but appreciable increases in impulse and energy. Both these quantities depend on the time integral of pressure, and the increases therefore reflect the increased duration of the shock wave for equal weights of these explosives. This effect is shown even more strikingly in more powerful explosives, peak pressure increases of 20 per cent being accompanied by energy increases of 50 per cent or more.

The differences in increase of different parameters illustrate the meaningless nature of a blanket statement without qualification that loose tetryl, for example, is 10 per cent better than TNT as an underwater explosive. If by better one means the impulse for the same value of $W^{1/3}/R$, the statement may be true, but the 10 per cent figure is clearly wrong if peak pressure or energy are more significant factors. It must also be remembered that ratios of this kind may depend significantly on the pressure, and statements true for low pressures at considerable distances from the charge may not be true at smaller distances.

More precise values of peak pressure and other ratios than are given by similarity curves can be obtained by direct comparisons in which charges of different explosive materials are fired under as nearly the same experimental conditions as possible. A vast amount of experimental data has been acquired in this way, not only for comparison of explosive materials, but also to determine the effects of differences in methods of preparing the materials, effects of charge shape and orientation, and so on.[6] All that can be given here is a brief discussion of the ways in which results of such comparisons can be presented, and their significance.

B. *Shock wave ratios for equal volumes of explosive.* In many cases, explosive comparison tests are performed by using equal volumes of explosives in cases of the same dimensions. The ratios of shock wave parameters at equal distances will evidently be the same as for equal weights only if the two explosives have the same density. On an equal volume basis, as compared to equal weight, the material of higher density is evidently favored because of the greater mass of material, and this difference further illustrates the necessity for caution in being sure what one really means or is concerned with when comparing explosives.

The equal volume basis is somewhat more convenient when comparing shock wave parameters with theoretical results, which are naturally obtained directly in terms of charge dimensions, and this representation will be considered first. Strictly, peak pressure and other ratios should be stated for a specific value of a_o/R, where a_o is the charge radius (or other characteristic dimension) and R the distance. It is almost always found to be the case that in the pressure range 1,000 to 15,000 lb./in.^{2} these ratios do not change significantly with pressure level, and for such pressures these results can usually be accepted without more specific limitation. The standard deviation of mean ratios obtained in tests at Woods Hole depend on the explosive and parameter, but are of the order 2 per cent for peak pressure, 4 per cent for impulse (integrated in most cases to approximately five times the time constant), and 6 per cent for the energy function.

As in the case of comparisons for equal weights of explosives, the ratio of one explosive to another depends on the shock wave parameter considered, the values being progressively larger for impulse and energy as compared to peak pressure, owing to increase in both peak pressure

[6] The most complete summary of such data is given in a report by J. S. Coles (23), in which results from tests at Woods Hole and from representative British and German work are given.

and duration of the shock wave. These increases suggest the possibility of alternative forms of expressing the comparison in terms of weight or volume ratios, which in many cases give simpler representations of the data.

C. *Weight and volume ratios.* A weight ratio for a given measurement of two explosives such as shock wave pressure may be defined as the ratio of weights of the two explosives which gives the same peak pressure (or other measure) at the same distance. On this basis, the ratio for explosives more powerful than TNT is less than unity, i.e., less weight is needed. In order to avoid this situation, the reciprocal of the figure just defined is more often used and referred to as an equivalent weight. Ordinarily, the comparison explosive is taken to be TNT and the equivalent weight of an explosive in terms of TNT is therefore equal to the weight of TNT in pounds required to give the same peak pressure at a given distance, for example, as one pound of this explosive. Similar considerations can be applied to volume of charge rather than weight in order to obtain equivalent volumes in terms of TNT or some other reference explosive.

The usefulness of equivalent weights or volumes lies in the fact that the values obtained, when different shock wave parameters or other measures of explosives are considered, are often very nearly the same. The reason for this lies in the fact that an explosive which develops a greater peak pressure than the same weight of another explosive, TNT say, often but not always has an increase in time scale of pressure in roughly the same proportions. If this is true, then a suitably larger weight of TNT will develop very nearly the same shock wave as the more powerful explosive, and this equivalent weight will be the same whether deduced from peak pressure, impulse, or energy measurements. Under such circumstances an equivalent weight is a more concise statement of experimental results than the totality of peak pressure and other ratios for equal weights or volumes.

The extent to which one may speak of the equivalent weight of an explosive, and the magnitudes of values obtained for various explosives, can only be shown by explicit comparisons of calculations based on shock wave peak pressure, impulse, and energy flux density. With a few exceptions, the differences between variously obtained values for a given explosive are less than or comparable with the experimental uncertainties of the ratio. It is therefore possible in many cases to speak of an explosive as equivalent to, say, 1.3 times its weight of TNT without serious ambiguity. Even in this example, however, one can run into difficulty for the reason that what is an equivalent weight at one distance from the charge is not at another. This behavior must be the result of different variations with distance of the shock wave parameters, and an example where this sort of thing occurs is for shock

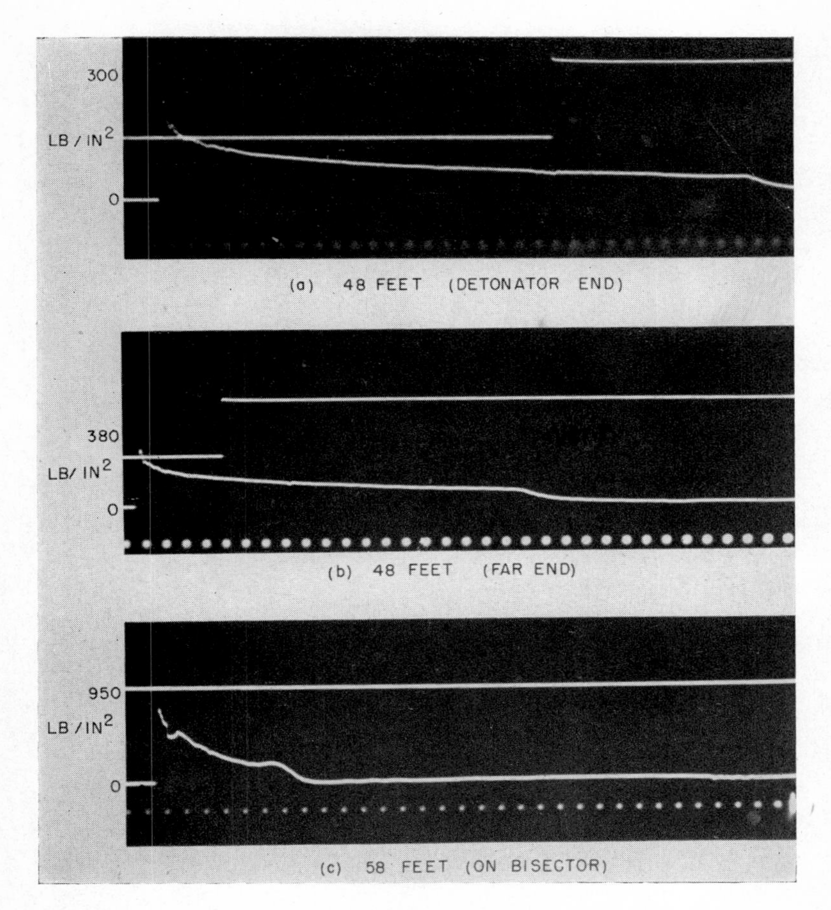

Plate VII. Shock wave pressure-time curves for line charges (50-lb. TNT, 25 foot length). The interval between timing dots is 200 μsec.

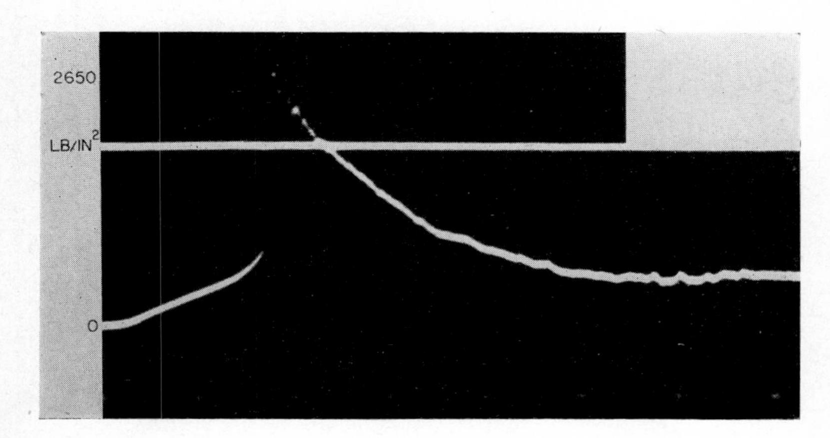

Plate VIII. Ground shock and water shock wave 50 feet from a 300-lb. TNT charge, charge and gauge 4 feet above bottom.

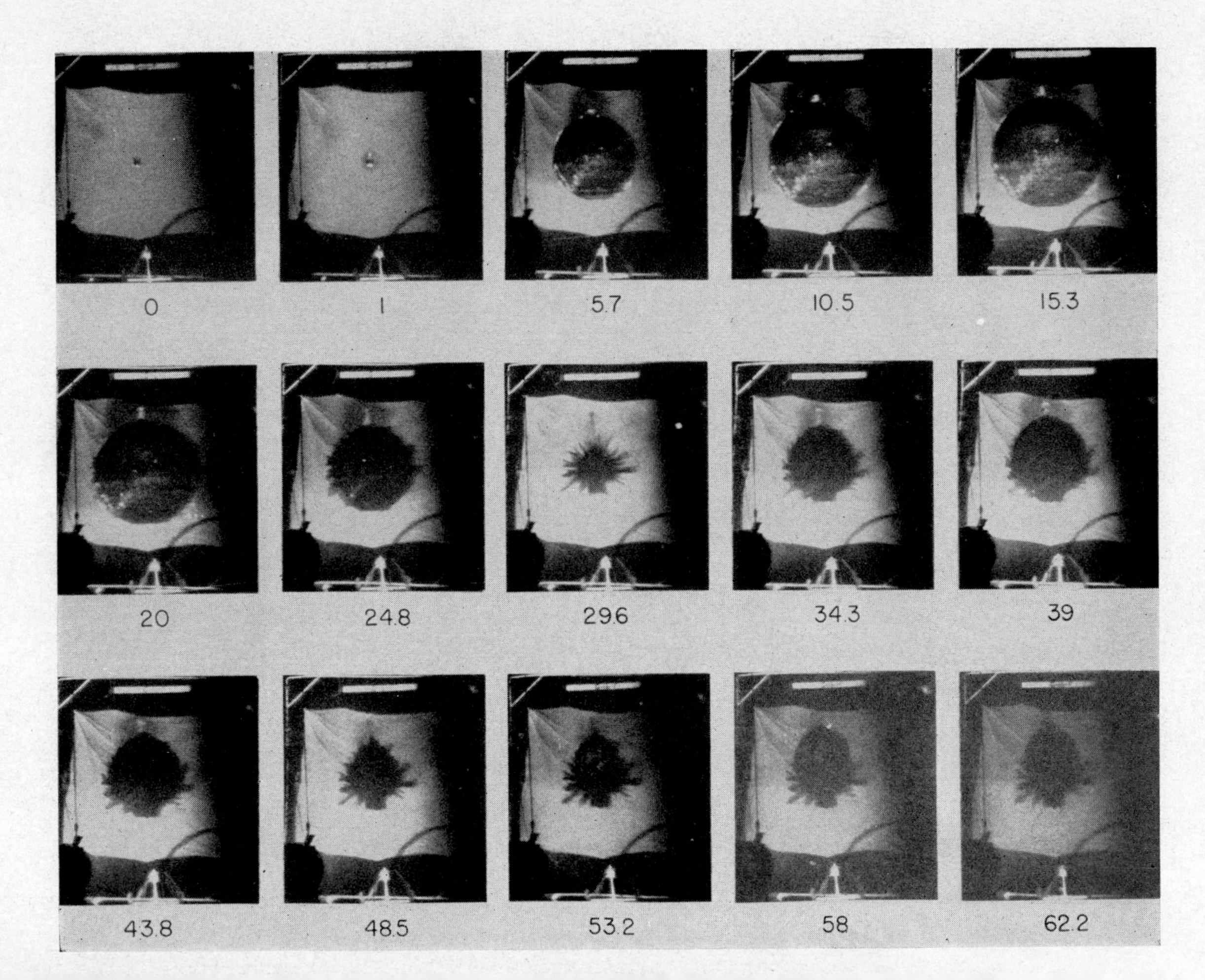
0
1
5.7
10.5
15.3
20
24.8
29.6
34.3
39
43.8
48.5
53.2
58
62.2

waves very close to the charge. More powerful explosives which develop higher initial pressures may be expected to produce shock waves which are dissipated more rapidly in their initial stages, the shock wave may well be relatively stronger near the charge, and the equivalent weight is therefore greater.

In a few exceptional cases, there is no single figure suitable as even an approximate equivalent weight, and the advantage of the concept is to some extent nullified for this or similar cases. Even in these circumstances, use of equivalent weights for specific parameters can be defended on the ground that it gives a somewhat fairer picture than do ratios of shock wave parameters at equal distances. This is particularly true for peak pressure: a 15 per cent increase in peak pressure obtained by using a more powerful explosive may seem relatively unimportant, but it is not if a 50 per cent greater weight of the inferior explosive would otherwise be required. This difference in viewpoint results from the fact that the peak pressure varies only as the 0.33–0.4 power of charge weight, and corresponding but smaller differences result for impulse, which varies with weight as $W^{0.6-0.7}$.

D. *Other measures for explosive comparison.* Shock wave ratios or equivalences are not the only relative measures of explosives which can be used in their comparison. Another useful type of ratio is a distance ratio: the ratio of distances from the charge at which equal volumes, or weights, of explosive produce the same value of peak pressure, impulse, or other parameter of the explosive. The distance ratios have as a corollary volume ratios, which are the ratios of volumes of water around the charge in which the parameters of interest exceed a specified value.

In addition to the various shock wave parameters, as measured by piezoelectric or mechanical pressure gauges, more empirical measures can be used, such as deformations of diaphragms or plates at known distances from the explosion. These empirical values are usually related to properties of the shock wave in rather complicated ways (see section 10.5), and are therefore to be related to specific shock wave parameters with caution. They are, however, often highly reproducible and comparatively easily obtained results and so have been widely used for the purpose of empirical comparison. From the present point of view of shock wave comparison, the difficulties in interpretation of such data make their detailed consideration here out of place, but it would be misleading not to mention the importance of such methods in practical testing. The results obtained, particularly if equivalent weights are determined, usually give values which are closely consistent with results obtained from more directly determined shock wave parameters. This is of course to be expected, and apparently anomalous results can usually be understood in terms of differences in character of the shock waves.

The various measures of shock wave comparison are clearly related to one another. For example, comparisons at equal weight and at equal volume differ from one another because of differences in density of the explosives and conversion from one basis to another can be made using known values of density. The conversion also involves a knowledge of the weight dependence of the parameter under consideration, which must therefore be known. In many experiments, equal volumes of explosives have been compared, and ratios for equal weights are derived quantities, but the reverse procedure may equally well be followed. Similar considerations apply to the other relative figures which have been mentioned, and one can clearly devise a large number of such figures all describing the same pair of explosives but from different points of view. Systematic procedures for deriving a desired figure in terms of another known one have been devised, for which reference should be made to original reports.

A decision as to what figure of explosive comparison is significant depends of course on the reason for making the comparison. If, for example, damage of structures of rapid time response by large explosions is of interest, the peak pressure only may be of concern, and at the opposite extreme the impulse may be the dominating factor. It may equally well be that one is interested in the total energy release both during emission of the shock wave and at later times, in which case shock wave energy parameters are significant but only a part of the story. Any discussion of the many possible applications is beyond the scope of the present remarks, which have for their purpose only to list some of the figures of comparison which have been obtained and to indicate their relation to one another.

7.6. Effects of Charge Shape and Orientation

The discussion of preceding sections has been limited to the simplest realistic case of shock waves possessing spherical, or nearly spherical, symmetry. In the majority of practical cases, this situation should ultimately be approached at sufficiently large distances from any charge of finite dimensions, but it does not by any means follow that initial differences from a spherical wave become insignificant within a few charge dimensions. The case is, in fact, quite the opposite, for example, differences in shock wave pressure-time curves around cylindrical depth charges observed at distances of 20 feet (roughly 15 times the average linear dimensions of the charge) are found at distances of 100 to 500 feet. These differences can clearly, therefore, be of practical as well as fundamental importance, and representative experimental measurements will be considered in this section.

The idealization of a spherical charge detonated at its center is one that is approached in practice only by special methods of preparation (see section 6.2); for military and practical purposes less symmetrical forms of charge and methods of detonation are therefore almost exclusively used, however unfortunate this is from the point of view of simplicity in theory and measurement. Other reasons than expediency may also dictate nonspherical shapes of charge, as one may be interested in concentrating explosive effects in one direction at the expense of others.

A. *Line charges.* Perhaps the simplest experimentally realizable form of charge which differs considerably from a sphere is a line charge.

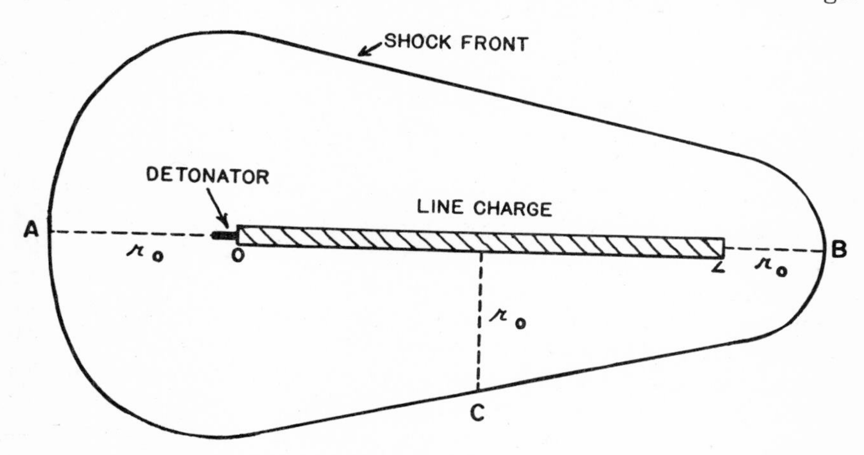

Fig. 7.12 Shock front around a line charged detonated at one end.

The line charge is an extremely elongated cylinder which is detonated at one end, the detonation wave progressing the length of the charge with finite velocity and generating a conical shock wave about the cylinder as axis. This situation does not correspond to one-dimensional cylindrical symmetry in which the shock wave is specified by the radius vector from the cylinder axis, but is rather one of axial symmetry in which the coordinate parallel to the axis must also be considered. The extension of the Kirkwood-Bethe theory to cylindrical symmetry by Rice and co-workers, as discussed in section 4.4, does not apply directly to line charges, both because of the finite length of charge and because of the finite detonation velocity. Simpler approximate considerations, however, suffice to explain qualitatively the general differences in pressure at different points around the charge.

In Fig. 7.12, a line charge of length L is detonated from point O. The detonation wave proceeds with a velocity D toward L, and a shock wave forms in the surrounding water. After detonation is complete, the shock will have an outline as indicated, if it is assumed that the cumulative effect of successive elements of the explosive is to be obtained by simple addition. In order to understand qualitatively the differences in pressures measured at points A, B, C, it is convenient to assume that the pressure at any point and time can be obtained by adding the pressures which would be developed independently by successive elements of the charge. In this approximation, which has been developed in a Road Research Laboratory report,[7] the pressure at a point at a distance r from an element dx is then the sum of pressures $P = (P_m e^{-t/\theta}/r)\,dx$, where P_m and θ are constants. If the difference in times of initiation are to be taken into account properly, these contri-

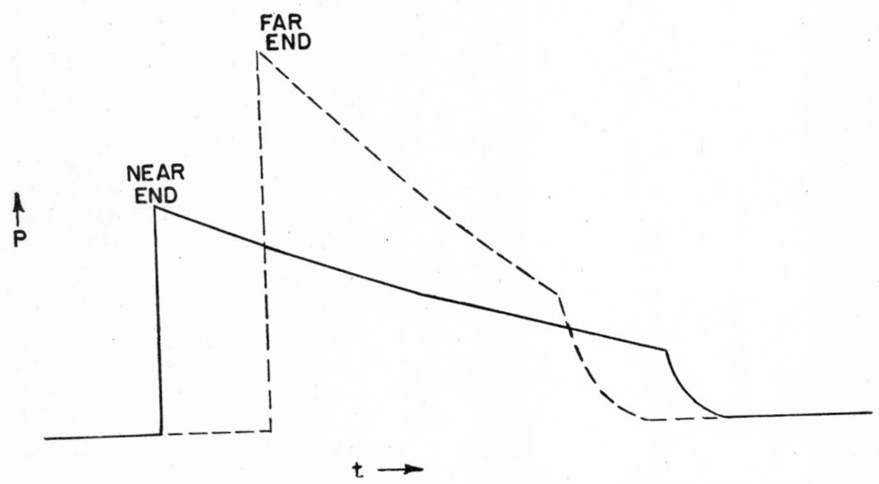

Fig. 7.13 Shock wave pressure-time curves at points on extended axis of a line charge.

butions must be evaluated for a retarded time $\tau = t - r/c - x/D$, where r is the path in water with velocity c, and x is the distance of the element from the point of detonation. For an element to contribute we must then have $\tau \geqslant 0$, and the pressure at any point is

$$P = P_m \int \frac{e^{-\tau/\theta}}{r}\,dx \tag{7.7}$$

the limits of the integral over x including all values of x from which a signal can reach the point of observation (i.e., for which $\tau \geqslant 0$).

Explicit evaluation of the integral Eq. (7.7) must be carried out numerically and this has been done for several specific cases, but the general nature of the solutions can be inferred quite simply. Considering point A of Fig. 7.12, it is evident that the first signal arrives from

[7] The details of this theory and its application to experimental results are given in three reports listed as Reference (92).

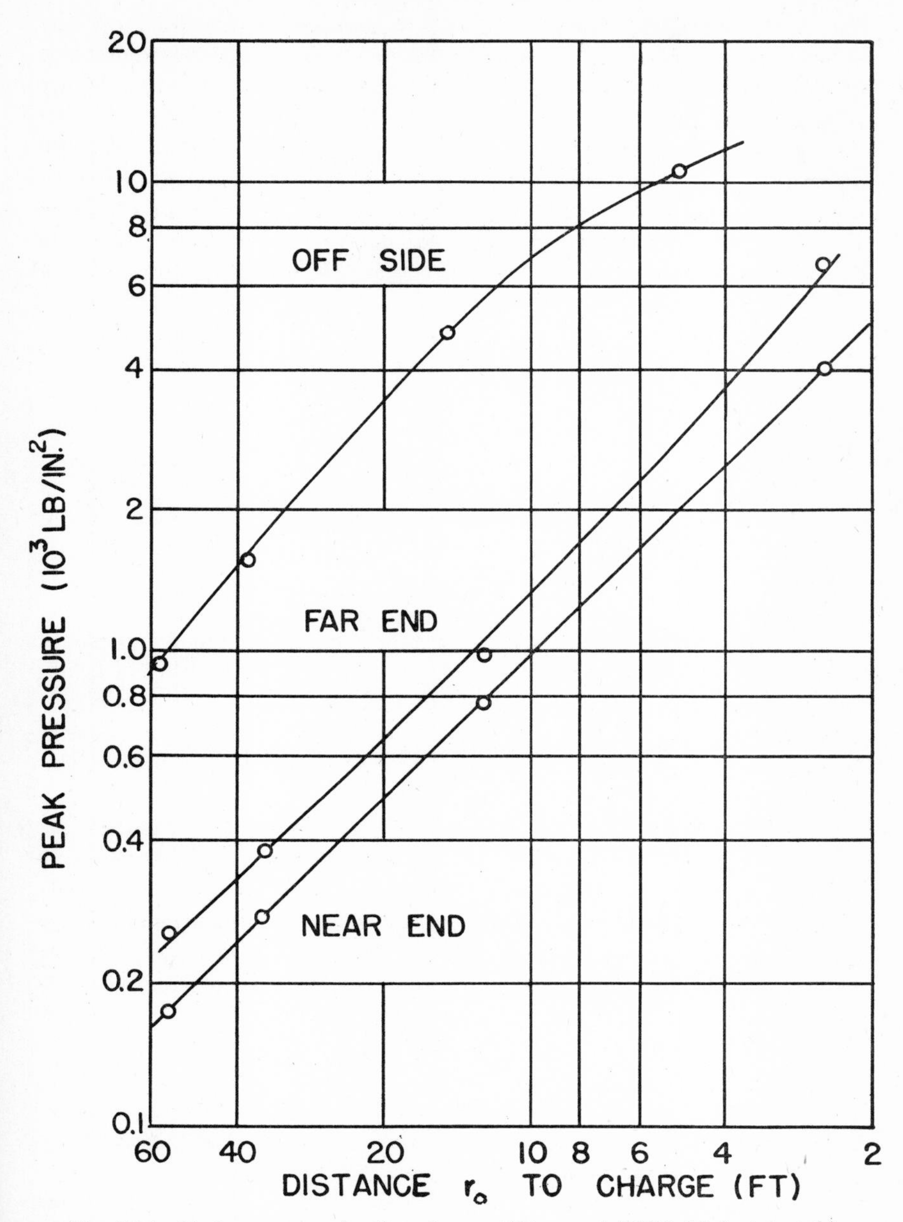

Fig. 7.14 Peak pressures for line charges (50 pound TNT, 25 foot length) as a function of distance r_o from the nearest point of the charge.

0, and the pressure is rapidly built up to a maximum, actually by formation of a shock front. The pressure then falls off, until after a time interval $t_1 = L(1/c + 1/D)$ the last, much weaker, signal arrives from point L. The duration of the wave is then very nearly t_1. At point B, the first signal arrives from point L, and at a time $t_1 = L(1/c - 1/D)$ later the last weaker signal from 0 arrives. The duration at point B, in the direction of detonation, is thus shorter than at the other end, and the pressure-time curve is, so to speak, compressed at point B. Calculations based on Eq. (7.7) show that, approximately, the pressures at B would be increased over those at A in the ratio that the time scale is reduced. This ratio is $(D + c)/(D - c)$, which for $D = 20$ ft./msec., $c = 5$ ft./msec. is 5/3. The pressure-time curves at opposite ends should therefore look roughly as sketched in Fig. 7.13.

The qualitative predictions represented by Fig. 7.13 are confirmed surprisingly well by experiment. Calculations based on Eq. (7.7) have been applied to experimental data for charges of cordtex (similar to primacord, 3 grams of explosive per foot) and found to represent the general characteristics of the observed pressures quite well with $P_m = 5{,}000$ lb./in.^{2} and $\theta = 17$ μsec. The experimental records unfortunately did not show the full duration of the waves, the later portions of which were cut off by surface reflections. Similar experiments with larger charges (50 pounds of TNT in a 25 foot length) in much deeper water[8] gave durations of 6.5 and 3.8 msec. for positions corresponding to A and B in Fig. 7.12, as compared to values of 6.25 and 3.75 msec. calculated using $D = 20$ ft./msec., and $c = 5$ ft./msec. The peak pressures were roughly 40 per cent higher at points A, as compared with the estimated increase of 67 per cent. The measured impulses, however, differ much less, being of the order of 15–20 per cent higher at B, as expected from the approximate estimates.

For points on the perpendicular bisector to the charge, such as point C in Fig. 7.12, the situation is quite different. The duration of pressure for continuing detonation depends on the distance from the axis, but must always be shorter than at points off either end, and decreases as the distance increases. At points sufficiently far from the charge, the peak pressures are much higher than off the ends, as shown in reproductions of the observed pressure-time curves in Plate VII. Near the charge, however, the peak pressures increase less rapidly than they would for a spherical charge, and the peak pressures off the end increase more rapidly. This difference in behavior is shown in the plots, on logarithmic scales in Fig. 7.14, of peak pressure P_m as a function of $1/r_o$ for 50 pound, 25 foot TNT charges. At distances greater than about 30 feet (from the nearest point of the charge), all the curves have

[8] These results are given in more detail in a report by J. S. Coles and R. H. Cole, to be issued as a NavOrd Report by the U.S. Navy Bureau of Ordnance.

nearly the same shape, the slope being somewhat larger than the value of one for a spherical acoustic wave. Near the charge, the pressure off the side increases much less rapidly and gives indications of varying more nearly as $(r_o)^{-1/2}$, which would correspond to a cylindrical acoustic wave. This is qualitatively reasonable, for at nearby points on the side the charge appears to be more nearly of infinite extent and the shock front more nearly plane (see also Fig. 7.12).

The pressure-time curves obtained off the side are very much more like those for spherical charges in shape and duration than are the ones obtained on the extended axis. The impulse at points off the side is

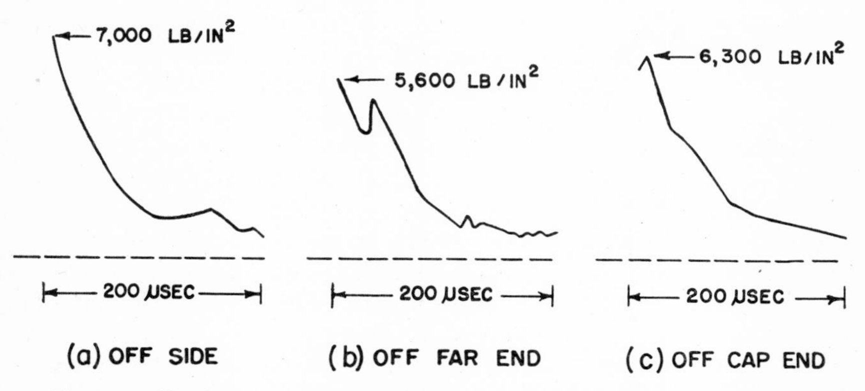

Fig. 7.15 Shock wave pressure-time curves for cylindrical charge (3.5 feet from 0.55 pound loose tetryl, length-diameter ratio 1.4).

some 30 per cent higher than values off the end. At a distance of 15 feet from the charge the initial rate of pressure decay corresponds to a time constant of 0.32 msec., which increases to about 0.70 msec. at 58 feet, although the total duration is less than 1.4 msec. A nonacoustic spreading of the wave profile thus occurs even though the total duration, corresponding to the end of detonation, decreases.

It is also evident from the observed pressure variations plotted in Fig. 7.14 that at no point can the pressure decay with distance be considered acoustic. Any calculations based on this assumption are therefore very approximate, and it is perhaps surprising that so much of the general behavior can be simply understood on this basis.

B. *Cylindrical charges.* The asymmetry of charge shape of which line charges are the extreme case is that of a cylindrical charge detonated at one end. Many studies have been made of such charges for various ratios of diameter to length. The general features of the shock wave pressure-time curves at different points around the charge are broadly similar in kind to those found for line charges, and can be qualitatively understood in the same way.

The qualitative correspondence of cylindrical to line charges is an imperfect one, however, as illustrated by pressure-time curves of Fig. 7.15.[9] These results were obtained at a distance of 3.5 feet from 0.55 pound loose tetryl charges with a length to diameter ratio of 1.4. As for line charges, the peak pressure is highest off the side, but is higher off the cap end than off the far end and the order of pressures for points on the axis is thus reversed for the greater symmetry.

The pressure-time curves off the side of the charge are found to be very similar to those from spherical charges in that the initial peak is followed by a smooth, roughly exponential decay. At points on the axis, however, the curves are quite different. Not only are the peak pressures somewhat lower, but they are followed by second peaks and rather irregular decay, initially more nearly linear than exponential in form. Off the detonated end, the separation of the two peaks is smaller and the difference in shape less marked, but these effects are observable. These multiple peaks of course make meaningless a single description of the curves by peak pressure and time constant, and explicit description of such curves is much less simple.

The simplest quantitative comparisons of curves such as those of Fig. 7.15 are in terms of initial peak pressure (remembering its less clear significance off the ends), impulse, and energy. Analysis of records for various length to diameter ratios show that the impulses change very much less with relative position than the peak pressures, the total difference in impulse being less than five per cent for the case of Fig. 7.15 while the difference in pressure is twenty per cent. The rate of decay of pressure is thus affected in the opposite way as peak pressure, so as very nearly to neutralize changes in impulse. This action is of course reasonable from considerations of conservation of momentum, as the vector momentum of the entire system must be zero (neglecting the external force of gravity). The component of momentum parallel to the charge axis, to be obtained by integration over the entire disturbance, must also be zero, which implies the approximate equality of measured impulse at opposite points. The energy density varies roughly as the product of peak pressure and impulse and accordingly is greatest for the directions of maximum initial pressure, being determined by the integrated square of pressure versus time.

Rather complex pressure-time curves are observed for small cast charges containing significant amounts of booster material and for large charges with noncentral boosters. The geometry and manner of detonation is not simple in most service weapons and in most cases they correspond only roughly to spherical or cylindrical symmetry. Measurements on large cylindrical charges detonated near one end do, how-

[9] From measurements by W. G. Schneider, reported in UE 15 (114).

ever, show the same general features of multiple peaks and irregular decay (Plate VI is an example).

The presence of multiple peaks is not accounted for by simple acoustic theory, but is rather the result of nonlinear interference of pressure fronts from different parts of the charge. The theory of such effects, as discussed in section 2.10, has been only partially developed, but does give a qualitative description of the phenomenon. Because this explanation involves the consideration of multiple sources, of which asymmetrical charges are hardly a simple example, further discussion is postponed to section 7.7.

It is interesting to note, in conclusion, that the propagation of the double peaked shock wave outward from the charge offers an illustration of the overtaking effect in shock waves. At distances of a few

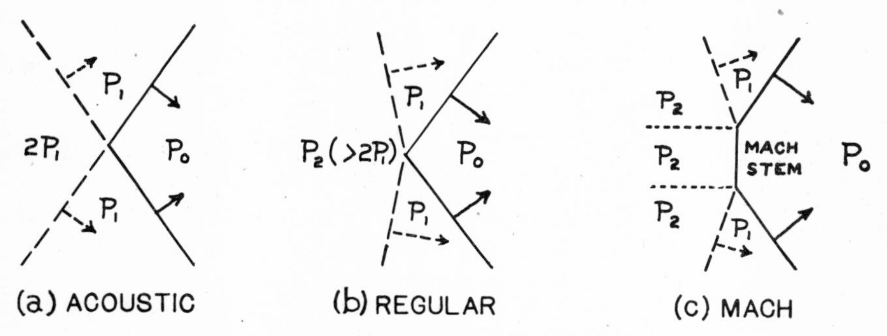

Fig. 7.16 Intersection of plane shock waves.

charge radii, the second smaller pressure discontinuity advances more rapidly than the initial shock front and the time interval between them decreases, as is to be expected for finite amplitude waves. As the shock front pressure decreases this overtaking effect is reduced, and the time interval between the peaks changes much less rapidly. For example, the double peaks observed off the end of a 300 pound cylindrical charge have a separation of 150 μsec. at 20 feet which decreases to 75 μsec. at 30 feet, but are still discernible 500 feet away.

7.7. Shock Waves from Multiple Charges

In the acoustic approximation, for which the wave equation for pressure is linear, the resultant pressure at any point in a fluid through which two pressure waves from separate sources pass is simply the sum of the pressures which either disturbance produces when unaffected by the other. Thus, the intersection of two weak pressure fronts of equal strength is represented by Fig. 7.16a in the acoustic case.

For shock waves, however, the velocity of propagation of the front increases with pressure ahead of the front, and the pressure difference across the front is large if the wave advances into fluid already under compression. The form of a shock wave therefore changes as it enters a region through which another shock wave has passed, and the pattern of Fig. 7.16a is modified into the scheme corresponding to Fig. 7.16b. This condition is described as regular reflection by von Neumann (116), as discussed in section 2.10. The case considered here, of intersecting shock waves of equal strength, is equivalent to a shock wave and its

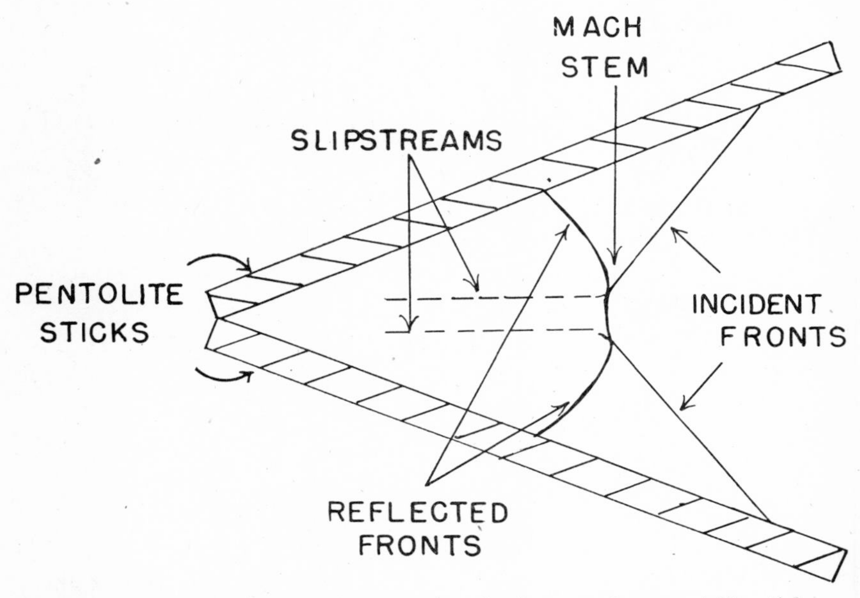

Fig. 7.17 Mach reflection of intersecting shock waves from pentolite sticks detonated at one end.

reflection in a rigid surface below the line of symmetry in Fig. 7.16, and so the analysis of this latter case is directly applicable here.

Detailed calculation, based on the Rankine-Hugoniot conditions, shows that the differences from the geometrical reflection increase with the strength of the shocks and increasing angle between the shock fronts, as measured by the half angle α between either shock and the plane of symmetry. For pressures less than about 10,000 lb./in.^{2} the differences are not large, but become so rapidly at higher pressures, particularly for values of α approaching 90°. For sufficiently large values of P_m and α, the equations at the shock fronts cannot be satisfied by the regular reflection scheme of Fig. 7.16b. Instead, the so-called Mach region reflection shown in Fig. 7.16c is found as discussed in section 2.10, with much higher pressures behind the advancing discontinuity than twice the pressure of either shock alone. The critical value α_{extr} for which this occurs decreases with the pressure, and for a pressure of 40,000 lb./in.^{2} the value $\alpha_{extr} = 40°$ is obtained, corresponding to an angle of 80° between the two fronts.

A. *Photographic examples of the Mach region.* A number of observers have obtained flash photographs of intersecting shock waves. In one

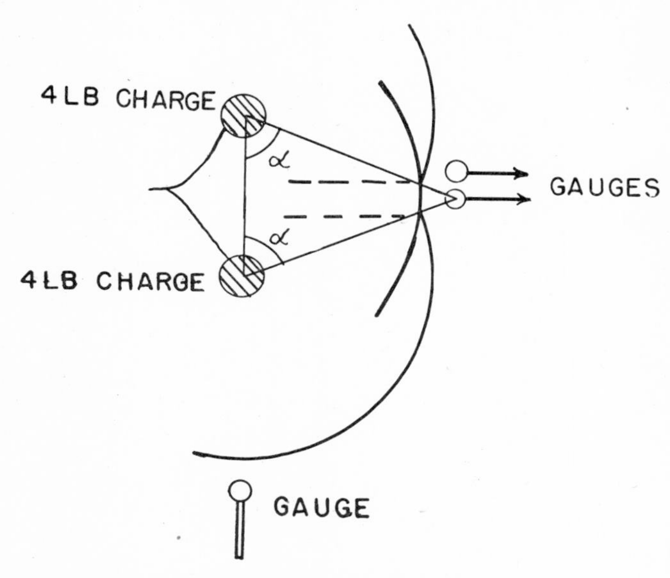

Fig. 7.18 Gauge arrangement for Mach effect measurements.

series of experiments reported by MacDougall, Messerly, and Boggs,[10] two cylindrical sticks of pentolite inclined at an angle to form a *V* were simultaneously fired at the point of the *V* to form shock waves as sketched in Fig. 7.17, illumination being obtained by explosive flash charges. The pressure at the shock front intersections for the conditions of the experiment was estimated to be 300,000 lb./in.^{2}. This corresponds to a pressure ratio of 20,000 ahead of and behind the shock front, and the maximum value of α for regular reflection at this large pressure is calculated from Eq. (2.41) to be 26°. Photographs obtained for $\alpha = 22°$ showed regular reflection, and for $\alpha = 36°$ irregular, Mach reflection as sketched in Fig. 7.17 was observed as predicted. The exact positions of the "reflected" portions of the waves travelling in the wake of the original waves are distorted because of refraction of light in passing through the waves (see section 6.3), but narrow lines observed behind the intersection as in Fig. 7.20 were believed to be real. They may thus correspond to the slipstreams predicted by von Neumann:

[10] Reported in UE 20 (114).

boundaries at which the particle velocity and density are discontinuous, but the pressures on either side are the same.

Similar photographs have been obtained at lower pressure levels, using smaller charges and explosive flash charges or high intensity stroboscopic lights for illumination. In one such experiment[11] two 50 gram charges were fired simultaneously and the intersection of the shock waves photographed when the shock waves were 12 inches from

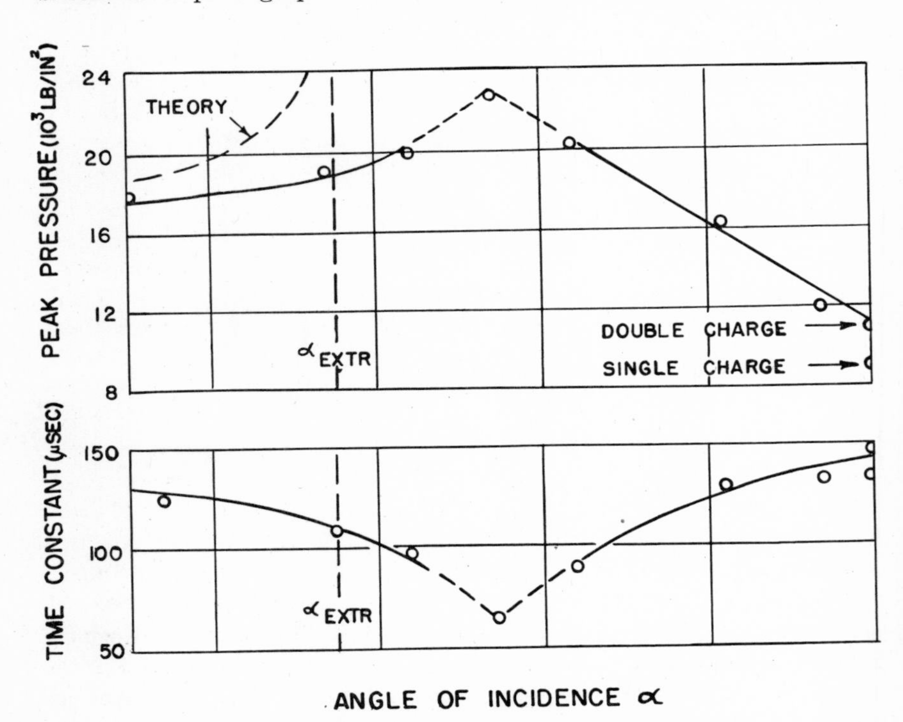

Fig. 7.19 Peak pressure and time constant of pressure at the intersection of two shock waves (the angle of incidence α is half the angle between the fronts).

the charges and the fronts made an angle of 142° at their intersection. The intersecting fronts gave rise to a Mach region very similar to that of Fig. 7.17, except for the increased obliquity and the fact that no slipstreams were visible (possibly because of the small discontinuity in velocity and density to be expected at the low pressure level). The experimental value $\alpha = 71°$ was appreciably larger than the calculated value of 55° for the experimental pressure of 11,000 lb./in.^{2}. The occurrence of the Mach region for this case is thus in agreement with the theoretical predictions.

[11] R. W. Spitzer and R. S. Price, UE 16 (114). Experiments with detonator caps exploded simultaneously have been reported by Campbell (16).

B. *Pressure measurements.* Although photographs reveal the existence of Mach regions very beautifully, no optical methods have as yet been applied to the measurement of pressure in regions of nonacoustic intersections of shock waves. Some measurements have, however, been made using piezoelectric gauges.[12] In these experiments, two 4 pound charges at varying separations were detonated simultaneously. Piezoelectric gauges were placed at points 4 feet from each charge on the plane of symmetry, and another gauge was placed on a line through the charges 4 feet from the nearer one, as shown in Fig. 7.18. The pressure-time curves on the plane of symmetry showed peak pressures of 18,000 lb./in.^{2} when the angle between the fronts was 90°, a value approximately double that due to a single charge. This pressure increased to a maximum of about 22,000 lb./in.^{2} for an angle of 135°, as shown in Fig. 7.19a. The existence of a Mach region was established for angles greater than 126° by the fact that gauges placed on, and two inches to one side of, the plane of symmetry both showed a single shock front, indicating a region of finite width with one shock only (stem of the Mach Y). The calculated maximum value of $\alpha = 60°$ for regular reflection agrees within experimental uncertainty with the value $\alpha = 63°$ observed.

The observed increase in peak pressure as α increases is in the direction predicted by theory (dashed line of Fig. 7.19a), but considerably smaller. The initial rate of decay of the curve was found to be much greater, as shown in the plot of initial time constant in Fig. 7.19b (obtained by approximating the initial part of the curve with a negative exponential). As the intersection between the two fronts was made increasingly oblique by moving the two charges together, the observed peak pressures fell off rapidly until for angles approaching 180° (coincidence of the charges) the pressure was very nearly the same as for a single charge of double the weight, as shown in Fig. 7.19a. At the same time, the time constant and duration parameters increased and approached the value for a single charge of double weight.

The fact that increases in the observed peak pressures are accompanied by decreased duration of high pressures means that the strengthening of two shock fronts by their intersection is not as pronounced as peak pressure values alone might seem to indicate. Measured values of impulse and energy flux density at the plane of intersection in fact decrease continuously, from values corresponding roughly to addition of pressures from the two charges to values comparable with those for a single charge. These estimates are complicated by the fact that, as the charges are brought close together, the later portions of the pressure wave are weakened by rarefaction waves resulting from the reflection of each shock wave off the gas bubble of the other charge! These

[12] A. M. Shanes, UE 21 (114).

changes in duration are not predicted by existing theory, which assumes shock fronts of infinite duration, but an upper limit on what can be expected must obviously be set by the fact that the total energy of the two shock waves is only double that for a single one.

The experimental results so far obtained do not indicate tremendous increases in effectiveness of shock waves by their nonlinear combination, but the various possibilities have hardly been explored in any detail, and the subject merits further investigation.

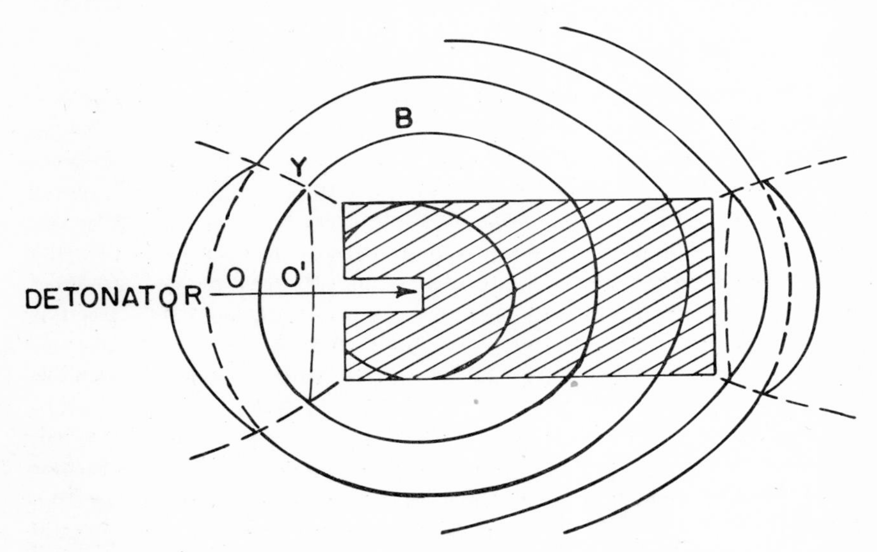

Fig. 7.20 Mach reflection scheme for a cylindrical charge.

The phenomenon of multiple shock fronts off the ends of cylindrical charges, particularly off the end near which detonation is started, must be the result of interactions of shock fronts, and a possible explanation of the effect may be similar to one suggested by G. T. Reynolds (91) for the case of air blast waves which exhibit similar properties.

The initiation by the detonator near one end of a cylindrical charge sets up a detonation wave, which on reaching the surfaces of the charge develops shock fronts in the water off the end and off the sides of the charge, as indicated schematically in Fig. 7.20. At points such as Y, however, the intersection of two shock fronts at different pressures constitutes a physically impossible situation, which must result in a new shock front. A front $O'Y$ behind OY, if this is weaker than the front YB, can provide the necessary continuity of pressure in the region behind $O'YB$.

The fronts sketched in Fig. 7.20 are similar to the Mach Y, the front OY corresponding to an incident wave, $O'Y$ to its reflection and YB to the stem, or shock front of greater strength than either of the fronts OY, $O'Y$. A similar situation is possible at the other end of the charge also, as sketched. This possible mechanism of forming multiple shock fronts is unproved for shock waves underwater, but has been quite definitely established in air blast tests. It further seems quite reasonable that intersections of shock waves of different strengths must occur in irregular shapes of charges. If this happens, the necessary continuity of pressure in the absence of a shock front requires that an extra shock front, or fronts, develop to establish this continuity.

7.8. Reflection of Shock Waves at Boundary Surfaces

The pressure-time curves and similarity curves for shock waves discussed in preceding sections represent the pressure distribution around explosive charges in free water, that is, in regions sufficiently far from boundaries that the pressure is unaffected by their presence. In all actual cases, the medium is only of finite extent, being limited in any case by the surface of the water and bottom, if not by walls or other obstructions. It is therefore necessary to consider the possible effect of the natural boundaries for any conditions, and other limits, such as targets or walls, may be equally or more important. Of the various boundaries of interest, the free surface of the water in contact with the atmosphere is the simplest and will be considered first.

A. *Surface reflections.* The reflection of a pressure wave at a free surface results from the requirement that the pressure above the surface be unchanged, and the reflected wave must therefore be one of negative pressure, or rarefaction. In the acoustic approximation, the wave is reflected as light would be from a mirror. It is convenient in considering the pressure at points below the surface to think of the reflected wave as arising from a fictitious second source of equal strength located on the other side of the surface from the actual source at the position of its optical image. The pressure at any point below the surface is then for waves of small amplitude simply the algebraic sum of the pressures in the two waves, taking into account the difference in amplitude and time of arrival.

The resultant pressure obtained by addition is the same as the original wave until the arrival of the negative reflected pressure at a later time. Although this negative peak is smaller than the positive peak, it is superimposed on a later weaker part of the positive wave, and the resulting pressure is usually less than the initial hydrostatic value, as shown in Fig. 7.21. The exact values of absolute pressure so obtained depend of course on the rate of decay of the direct wave, path differences, and the depth, but in nearly all cases a negative absolute pressure is predicted for explosive waves of short duration (the absolute pressure is of course gauge pressure plus hydrostatic pressure P_o).

Whether and to what extent predicted negative pressures, or tensions, are actually realized depends upon the experimental conditions. Various observers have succeeded in obtaining absolute tensions of as high as several hundred pounds per square inch. Values of this magnitude are obtained only for clean, air-free water, however, and under more usual explosion conditions it seems improbable that open water can sustain tensions greater than one atmosphere at most. The evidence for this belief is partly based on the estimate that a tension as low as this value will support bubbles of diameter as small as one or two microns in diameter, and partly based on the formation of cavitation bubbles in regions of water behind a rarefaction in which tensions of this order were expected. It is therefore reasonable to suppose that water

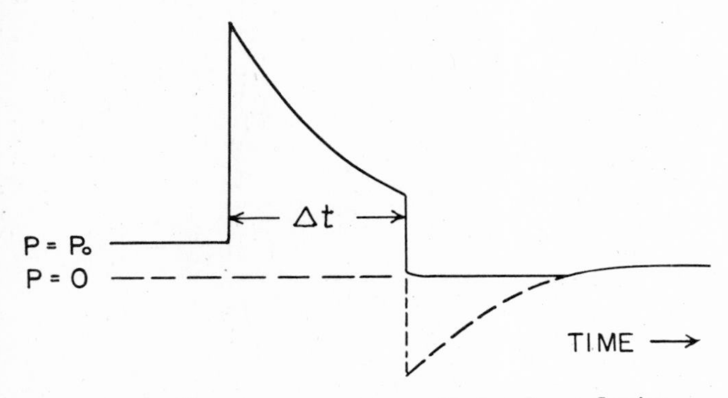

Fig. 7.21 Gauge pressure as a result of surface reflection.

not especially purified is incapable of supporting negative pressures of more than an atmosphere, and possibly much less. Instead, a region of cavitation containing many small bubbles is formed which prevents the increase in negative pressure.

Although the existence of small negative pressures in water is not ruled out, the occurrence of tensions at an interface or boundary is very much less likely, owing to the effects of oil or dirt on such boundaries. For example, gauge measurements of pressures resulting from surface reflections indicate within experimental error negative pressures equal to the hydrostatic pressure at most, as indicated in Fig. 7.21. The absolute pressure is therefore zero after arrival of the surface reflection, and the effect of the surface is to cut off the pressure wave. The impulse and energy in the resultant wave are therefore decreased below their value in free water, the effect increasing with distance from the source and at points closer to the surface. This "cutoff" effect is of importance both in measurement and in predicting the effect of charges fired near the surface. It is also sometimes useful in locating the position of an explosion (see section 10.3), the time interval Δt in Fig. 7.21 being closely equal to the value computed from the image construction if the surface of the water is smooth.[13]

B. *Bottom reflections.* If the bottom of the sea could be considered a perfectly rigid boundary, it would act as a perfect reflector of pressure. Thus the reflected wave developed at the boundary would, in the acoustic approximation, be a compression wave progressing as if it originated from an image source at the position of an optical image in the boundary of the actual source. Any real material is only imperfectly rigid as far as reflection of pressure waves is concerned, and the doubling of pressure at the boundary obtained assuming perfect rigidity is therefore an overestimate. In order to show how effectively actual bottom materials reflect the energy striking them, consider the case of a charge fired on the bottom. If no energy were absorbed by the bottom, this charge would be equivalent to a charge of double the weight fired in open water, because all the energy is confined to the water above the bottom. A charge of weight $2W$ to a first approximation gives a peak pressure at any distance increased by a factor of $2^{1/3} = 1.26$, and a duration increased in the same ratio. The impulse of the wave should therefore be increased by $2^{2/3} = 1.59$, and the energy flux density by a factor of 2 for a charge fired on a rigid bottom.

The increases in pressure and duration actually realized depend on the character of the bottom, but are seldom more than ½–⅔ of the predicted upper limits. For example, the peak pressure, impulse, and energy flux density 60 feet from a 300 pound TNT charge fired on a bottom of hard-packed sandy mud were increased by 10, 23, and 47 per cent over the values observed from a charge at mid-depth. These increases correspond to increasing the weight of a charge in free water by roughly 35–50 per cent rather than a factor of two, and an appreciable fraction of the shock wave energy was therefore transmitted to the bottom in this case.

If the bottom could be considered to be a homogeneous fluid, acoustic theory predicts that a somewhat weaker wave of compression would be reflected geometrically, the pressure being calculable from the densities and sound velocities of the water and boundary material by the formulas of section 2.8. Experimental measurements with charges fired near the bottom show, however, that the situation is not this simple. For example, an observed pressure-time curve 4 feet above the bottom and 50 feet from a 300 pound charge fired 4 feet above a sandy-mud bottom is reproduced in Plate VIII. The discontinuous shock front,

[13] It is important to realize that the geometrical reflection discussed is strictly valid only in the adiabatic approximations, and is not rigorous for waves of finite amplitude.

which arrives at the time for a direct wave from the charge, is preceded by a gradually rising pressure reaching a value of about 40 per cent of the shock discontinuity when the latter arrives.

The initial gradual rise in pressure must be the result of a disturbance which is propagated in the bottom material at higher velocity than in the water. Refraction, or possibly diffraction, of this kind implies an inhomogeneity of the bottom, as otherwise the transmitted wave in the bottom would travel in a straight line, and none of its energy would be returned to the water above it. A somewhat different mechanism, which is known to play an important part in long range transmission of sound, is the generation of seismic or earthquake waves which in turn generate waves in the adjacent water. Usually, but not always, a bottom reflection corresponding in time to the geometrical reflection path, and of rather smaller amplitude than computed for a rigid bottom, is observed for charges fired near the bottom. The time and magnitude of the ground shock depend on the nature of the bottom, and also vary considerably with the relative positions of the charge and point of observation with respect to the bottom. The character of bottom effects is thus not simply described, and their study is of more value in an understanding of the bottom structure than of the shock wave in water.

C. *Other boundaries.* The study of effects near homogeneous walls and targets should at least give more straightforward results than reflections off the sea bottom. Unless such boundaries are rigid and of large extent compared to the length of the shock wave, however, complications arise in these cases also. For example, a yielding structure may at first act as a rigid surface because of its inertia, but at later times when it has been set in motion, it acts more as a free surface. If in addition, the least dimension of such a surface is not large compared with the shock wave, the propagation of the shock wave is no longer simply described by simple reflections, and diffraction waves are established in the vicinity of the structure. The description and analysis of what happens in such cases is evidently as much a function of the structure as of the underwater pressures. Detailed discussions of problems of this kind are far beyond the scope of this book, but some general considerations are indicated in section 10.6.

7.9. The Shock Wave at Large Distances

As shock waves are propagated outward from an explosive source they become weakened by both dissipation and divergence. One might then expect that the characteristic features of finite amplitude waves would disappear rather rapidly, leaving a wave propagated essentially according to the acoustic approximation. The propagation theory of Kirkwood and Bethe indicates, however, that even at large distances from the charge the peak pressure falls off somewhat more rapidly than in the acoustic approximation and the duration continues to increase slowly. It is to be remembered in this connection that propagation theories of shock waves are based on the assumptions that dissipative processes can be neglected except at the shock front, where they are implicitly included in the Rankine-Hugoniot conditions, and that the undisturbed water has everywhere the same density and sound velocity. At short distances, where the path traversed is small and the pressure discontinuity large, both assumptions are good ones.

For longer paths and lower pressures, appreciable effects of viscosity and inhomogeneity of the fluid may well occur. The observed phenomena, although of considerable scientific and practical interest, involve to an increasing extent departures from the concept of an ideal fluid and effects of the surface and bottom. Their understanding then becomes as much a problem in oceanography and submarine geology as one in hydrodynamics, and a complete analysis is out of place here.

A. *Observed pressures.* For shock wave pressures of the order 1,000–40,000 lb./in.^{2}, the extensive investigations outlined in sections 7.3 and 7.4 show that the initial peak pressure at the shock front decreases approximately as $R^{-\alpha}$, where α has values of the order 1.15, and at the same time the initial rate of decay decreases gradually. This broadening of the profile is difficult to measure with high accuracy, but its existence is unmistakable. This is shown, for example, in Fig. 7.5. Less direct proof is contained in the more gradual decrease of impulse, which includes the effect of peak pressure and duration, as $R^{-[0.9-1.05]}$, compared with the decrease of peak pressure as $R^{-1.15}$. For these high pressures, the increase in time constant is very roughly as $R^{0.2-0.3}$.

A natural question is that of the distances and pressures to which empirical laws of this kind remain approximately valid. Less extensive investigations have been made for pressures of the order of hydrostatic. However, a number of measurements of peak pressures of the order 50–150 lb./in.^{2} for 60 to 300 pound charges at distances of 500 feet are in good quantitative agreement with the empirical formulas of section 7.4. The pressure-time curve also shows the expected broadening, as shown in Plate V for example. Although the shock front appears virtually discontinuous for the time scale of Plate V, measurements of the records give an indicated time of rise of the order of 50 μsec. at these pressure levels. This value is far in excess of that to be expected from gauge and amplifier transient response characteristics. It also seems unreasonable that such values can indicate a real loss in high frequency components of the shock wave, because the change in slope of the curve at the peak is virtually discontinuous. Measurements with smaller charges (½ to 10 pounds) at pressure levels down to pressures of the order 15 lb./in.^{2} reveal rise times of the same order but the records show a rounded peak rather than a discontinuity in slope.

Osborne (80) has found that the presence of air bubbles in coatings or on the faces of pressure gauges causes distortion of the initial portions of weak shock waves from very small charges. It is possible that the collapse of such bubbles or some other instrumental difficulty is responsible for the apparent rise time at these low pressures levels with large charges. This explanation, however, is not easily reconciled with the rapid initial response indicated at the first arrival of both the direct wave and its surface reflection, nor with the fact that calibrations of gauge sensitivity at even lower pressures by acoustic waves (see section 5.7) are in agreement with calibrations at pressures of several thousand pounds per square inch.

Osborne has measured pressures from Number 6 detonators at distances from 1 foot to 32 feet, using very small gauges found not to exhibit the peculiarity mentioned (see section 5.6 for a description). The measured peak pressures from 1,200 lb./in.^{2} down to 45 lb./in.^{2} were compared with the asymptotic pressure-distance relation of the Kirkwood-Bethe theory. Their result that the peak pressure should vary as $(1/R)$ $(\log R/a_o)^{-1/2}$, where a_o is the initial charge radius, rather than as $1/R$, and the data were found to confirm this prediction within experimental error. This result is spoiled by the fact that no measurable increase in duration with distance was observed, although the same theory predicts an increase as $(\log R/a_o)^{1/2}$, which requires an increase in duration of at least 25 per cent from 1 foot to 32 feet. Quite apart from the fact that measurements on larger charges unmistakably show increased durations, departures from the acoustic peak pressure variation without corresponding changes in duration are difficult to understand theoretically. It is important to realize, however, that accurate measurements of pressures lasting from 10 to 30 microseconds from so small a charge as a detonator cap, containing less than a gram of explosive, are extremely difficult, particularly near the charge.

The possible instrumental difficulties and scarcity of data make generalizations as to the true form of shock wave pressures below 100 lb./in.^{2} impossible at present. The predictions of the Kirkwood-Bethe theory indicate small but significant effects of finite amplitude at even lower pressures but their validity can be regarded only as probable and not proved. At these low pressures, however, long distances of propagation are involved for appreciable percentage changes in pressure, and for such long paths the mechanism of dissipation by viscosity of the fluid must be considered. The effect of viscosity on weak sinusoidal sound waves to which acoustic theory may be applied is well known to cause an absorption which increases rapidly with frequency of the wave and for plane waves leads to an attenuation which increases exponentially with the range.

A transient wave, as is generated by an underwater explosion, can to the extent that it is linear (i.e., of small amplitude) be regarded as a superposition of sinusoidal waves of suitable frequency components corresponding to the steeper portions of the pressure-time curves. These frequencies are attenuated more rapidly than the lower ones, and one should therefore expect a progressive rounding of the initial peak of the shock wave, which however becomes significant only over dis-

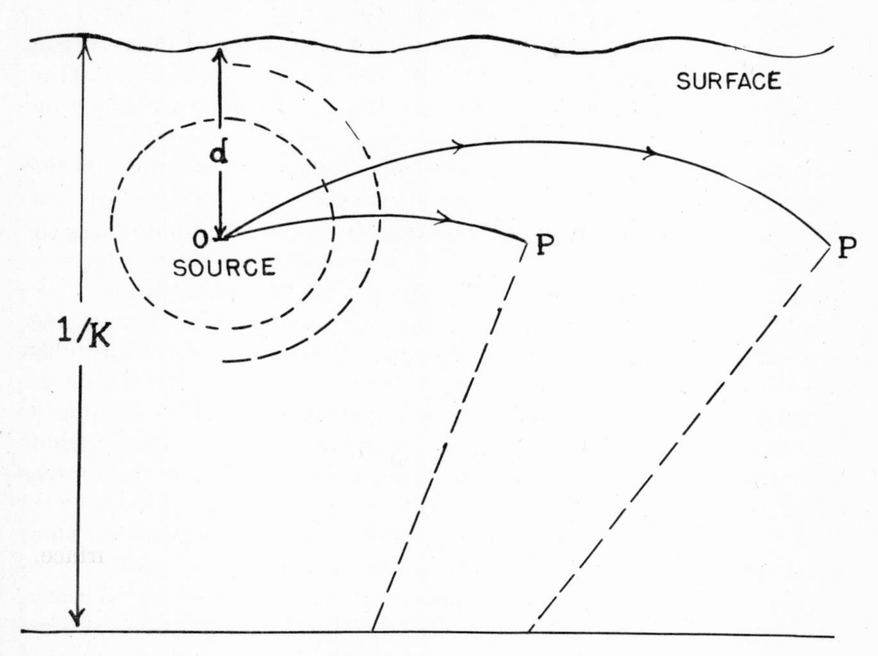

Fig. 7.22 Refraction of a sound wave as a result of sound velocity decreasing with depth.

tances of hundreds of yards for charges of reasonable size. The exact analysis of this absorption effect is complicated by a necessarily co-existing dispersion, or dependence of sound velocity of a continuous wave on frequency. Ideally, measurements of transient explosion pressures offer an excellent means of examining such effects, but instrumental and environmental difficulties in such measurements have not as yet permitted their satisfactory realization.

B. *Refraction effects.* The propagation of shock waves outward from an explosive source may be greatly modified by variations in the medium through which it travels, as well as by the presence of boundary surfaces. The effect of the medium, which results from variations in the velocity of propagation at different points, is under any practical conditions negligible for distances at which any ordinary explosion wave is sufficiently intense to inflict damage. At distances exceeding a few hundred feet, the effects may be very large for conditions encountered in the ocean, and at intermediate distances they may cause measurable effects in some investigations. A general discussion of the various phenomena belongs in a treatise on underwater sound transmission. The discussion here is intended merely to indicate the nature of the effect by a simple example which is adequate to show the differences to be expected for propagation over short ranges.

The major factor in determining the velocity of sound in the ocean at shallow depths is the temperature of the water, an increase of 10° F. increasing its value by about 40 ft./sec., the exact figure depending on the temperature, depth and saline content. In the ocean, different horizontal layers of water are not at the same temperature and the result is a vertical gradient of sound velocity. The simplest case of this kind, and the only one we shall consider, is that of a uniform negative velocity gradient, corresponding to warmer water at the surface. If, as in Fig. 7.22, a wave is initiated at point O, different parts of the front travel at different rates, the upper portions advancing faster. As a result, the wave front becomes distorted, and a sound ray drawn normal to any part of the front becomes bent downward as it progresses (except for the part of the front travelling vertically). This bending is analogous to optical refraction, and the path of the part of the front which reaches points P from O is not the straight line joining but a curved path. This path is the one of least time for a ray from O to P, and it is easily shown that Snell's law describes the inclination of successive parts of the ray, just as in geometrical optics.

If the negative gradient is a uniform one such that the velocity $c_o(d)$ at a depth d is given by $c_o(d) = c_o(1 - Kd)$ where c_o is the velocity at the surface, the path between O and any other point is a circular arc joining the points with its center in a horizontal plane at a depth $1/K$ below the surface, as indicated in Fig. 7.22. This sketch is exaggerated to show the nature of the effect; for a temperature difference of 50° F. between the surface and a depth of 1,000 feet, $K = 4 \times 10^{-5}$ ft.$^{-1}$ and the plane of centers is 25,000 feet below the surface. Even for this rather extreme condition, the actual path between points O and P a few hundred feet apart is not much longer than the straight line (the difference at 1,000 feet is 0.07 feet in this case). The fractional change in time of travel from that computed for a straight line is even smaller than the path difference, as is qualitatively evident from the fact that in time the curved path is shorter, despite its greater length in feet. The amount of difference depends on the relative position of points O and P, as well as their separation, but the percentage change is, to a first approximation, less than $(KR)^2$, where R is the separation. For $K = 4 \cdot 10^{-5}$ ft.$^{-1}$, $R = 1{,}000$ ft., the time difference is thus less than 0.04 per cent of the total time. The changes introduced by refraction increase rapidly with distance, as is evident either from this approximation or from Fig. 7.22, but for distances up to several hundred feet have a very small effect on the wave front. The only circumstances in which the small differences at short ranges may have significant effects are probably in explosion depth ranging methods depending on measurements of time differences (see section 10.3), but even in such measurements the errors are unimportant for ranges up to several hundred feet. These statements and calculations apply only to these ranges and should not be extrapolated, but do give an estimate of the order of the effects and the distances at which they must be considered more carefully.

8. *Motion of the Gas Sphere*

8.1. General Features of the Motion

After emission of the shock wave, the gaseous products of an explosion continue to expand outward at a gradually decreasing rate. As a result there are considerable radial displacements of the water, but the changes in velocity take place at a much slower rate than in the initial phases of the motion immediately following detonation. The pressures in the surrounding liquid are therefore much smaller and the whole character of the motion changes. Although the same basic hydrodynamical equations apply as in the case of the shock wave, and exact solutions would be at least as difficult to obtain, the qualitative differences in the later motion make more appropriate somewhat different approximations. In order to see the reasons for these and their limitations, it is helpful to examine briefly some experimental studies of the motion before considering the various theoretical developments.

A series of frames from a motion picture record of bubble motion (105) is shown in Plate IX; the charge used was 0.55 pound of tetryl detonated 300 feet below the surface of the water. The time intervals between these frames are all nearly equal, as indicated, and it is evident that an initial rapid expansion of the spherical bubble is gradually brought to a stop after 0.014 sec. The bubble[1] then contracts at an increasingly rapid rate until it reaches a minimum 0.028 sec later, and after an abrupt reversal again expands. Throughout the expanded phase of the bubble's motion there is very little vertical migration but near the minimum radius an appreciable upward displacement occurs. The reversal of the bubble motion at the minimum occurs so abruptly as to be virtually discontinuous on a time scale suitable for the rest of the motion; this characteristic is shown clearly in Fig. 8.1 in which the bubble radius is plotted as a function of time. A further qualitative feature of significance is the fact that the bubble retains its identity at the minimum despite the large radial accelerations and appreciable vertical motion, and remains fairly symmetrical. The motion is, therefore, a reasonably stable one in fact. A further evidence of overall stability is the continued oscillations of the bubble (if the charge is fired sufficiently deep to permit them to occur before the bubble rises to the surface): three or four complete cycles are readily observed and at least ten such cycles have been shown to exist for very deep charges.

[1] In the discussion of this and the next chapter, the gas products from an explosion are frequently described by the term "bubble" in common usage, although the term "globe" employed in Taylor Model Basin reports avoids possible confusion with other kinds of bubble.

The maximum velocity of the bubble surface as the bubble approaches minimum radius is about 200 ft./sec. in the example and over most of the motion is very much less. The maximum radius of 1.48 feet corresponds to an internal gas pressure very much less than hydrostatic (roughly one-fifth), as the equilibrium radius for this size charge is readily estimated to be 1.0 feet. This radius, for which the gas is at the hydrostatic pressure of the water, is indicated by the dashed line in Fig. 8.1 and it is evident that the pressure is less than this over some 80 per cent of the first cycle. The length and time scales will of course change with the size of charge and the depth at which it is fired, being larger for larger charges and shallower depths. In general, however, it

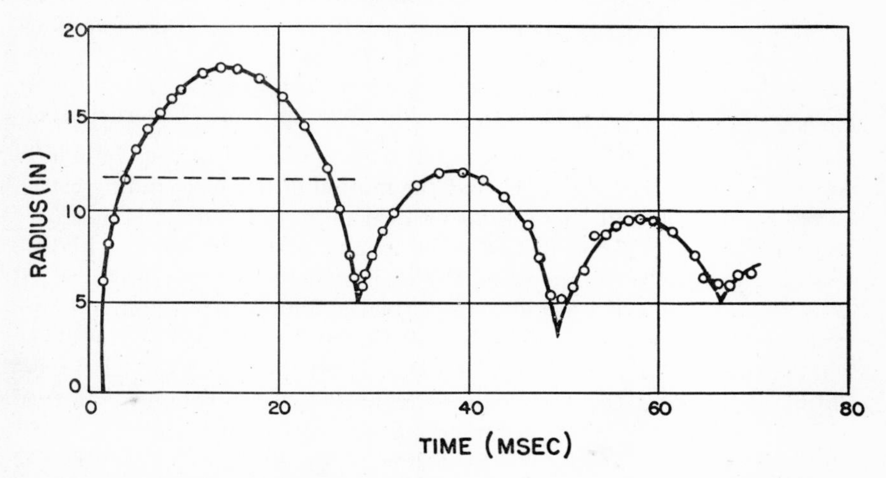

Fig. 8.1 Radius of the gas sphere as a function of time, for a 0.55 pound tetryl charge 300 feet below the surface.

is true that the radial velocities will be of the same order of magnitude, and that over most of the cycle the pressures are much smaller than hydrostatic.

These conclusions about the bubble motion are the basis of all the bubble theories which lead to general numerical predictions of bubble radius, migration and period. It is a common characteristic of such theories that changes in density of the water surrounding the bubble are neglected (the noncompressive approximation) and it is further assumed that the bubble retains a spherical form throughout the motion. From what has been said, it is evident that both these assumptions are at least plausible as far as the expanded phase of the motion is concerned. They must, however, be increasingly poor as the bubble approaches its minimum radius for which very much larger pressures and acceleration are involved. At this radius, for example, pressure waves (the secondary or bubble pulses) are emitted, which an incompressive theory can describe only as a disturbance appearing simultaneously at all points in the liquid.

From the foregoing considerations, it is to be expected that the theories based on noncompressive flow will be most successful in accounting for the general properties of the motion in its expanded phases, and in predicting properties, such as the period, primarily thus determined. The vertical displacement of the bubble at its minimum and the characteristics of the pressure wave emitted at that time are, however, more seriously affected by the approximations. In addition to the common assumptions just described all the theories involve other approximations, and the degree of refinement of the various treatments is largely a question of the extent to which these further approximations and idealizations are removed.

8.2. Noncompressive Radial Motion Neglecting Gravity

The simplest approximation to the true motion of the gas bubble is the one in which it is assumed that the motion of the surrounding water is entirely radial and there is no vertical migration. In this approximation, which has been discussed by a number of writers, the hydrostatic buoyancy resulting from differences in hydrostatic pressure at different depths is neglected. It is thus assumed that at an infinite distance from the bubble in any direction the pressure has the same value as the initial hydrostatic pressure P_o at the depth of the charge (atmospheric plus the added pressure of the water column). For a given depth of charge, the differences in pressure at the surface or near the bubble will clearly be greater the larger the charge and bubble resulting from its deformation. The neglect of differences in hydrostatic pressure should thus be more serious for large charges and small depths.

If radial flow is assumed, the equations of continuity and motion for the water are (Eqs. (2.2) and (2.4))

$$(8.1) \qquad \frac{\partial \rho}{\partial t} + u\frac{\partial \rho}{\partial r} + \rho\frac{\partial u}{\partial r} + \frac{2\rho u}{r} = 0$$

$$\rho\frac{\partial u}{\partial t} + \rho u\frac{\partial u}{\partial r} + \frac{\partial P}{\partial r} = 0$$

For pressure changes of the order 15 lb./in.^{2}, such as prevail over most of the bubble motion, the corresponding changes in density are of the order $10^{-4}\ \rho_o$, where ρ_o is the equilibrium density. Under these conditions, the derivatives of density ρ are easily seen to be negligible in the first of Eqs. (8.1), which then becomes

$$\frac{\partial u}{\partial r} = -\frac{2\rho u}{r}$$

Integrating this equation, we have

$$u(r, t) = \frac{u_1(t)}{r^2} \tag{8.2}$$

where the constant of integration $u_1(t)$ is the velocity for $r = 1$ and may depend on time. The radial velocity in noncompressive flow thus falls off as the inverse square of the distance from the origin, as is, of course, evident from elementary principles. With this result the second of Eqs. (8.1), becomes

$$\frac{1}{r^2}\rho_o\frac{du_1}{dt} + \frac{1}{2}\rho_o\frac{\partial u^2}{\partial r} + \frac{\partial P}{\partial r} = 0 \tag{8.3}$$

Integrating from the surface of the gas sphere, for which $r = a$, $u_a = da/dt = u_1/a^2$, $P = P_a$, to infinite distance where $P = P_o$ and $u = 0$, gives

$$\frac{\rho_o}{a}\cdot\frac{d}{dt}\left(a^2\frac{da}{dt}\right) - \frac{1}{2}\rho_o\left(\frac{da}{dt}\right)^2 - (P_a - P_o) = 0$$

Integrating with respect to time leads to the result

$$\frac{1}{2}\rho_o a^3\left(\frac{da}{dt}\right)^2 + \frac{1}{3}P_o a^3 - \int_o^t P_a\, a^2 da = C' \tag{8.4}$$

where C' is a constant of integration. Except for a factor 4π, the integral over a is easily seen to represent the work done by the pressure P_a in expanding the sphere to its radius $a(t)$, as the element of volume is $dV = 4\pi a^2 da$, and the integral must therefore equal the decrease in internal energy of the gas to $E(a)$ from its initial value. Absorbing this initial value into a new constant of integration Y gives after rearrangement

$$\frac{3}{2}\left(\frac{4\pi}{3}\rho_o a^3\right)\left(\frac{da}{dt}\right)^2 + \frac{4\pi}{3}P_o a^3 + E(a) = Y \tag{8.5}$$

Written in this form, it is easily seen that the first time integral of the equation of motion is merely the expression of conservation of energy, as the first term is readily shown to be the kinetic energy of radial flow outside the boundary, and the second term is the work done against hydrostatic pressure.

If the products of explosion behave as ideal gases with a constant ratio of specific heats γ and are further assumed to undergo adiabatic changes, the pressure-volume relation is $P(V/W)^{\gamma} = k$, where W is the mass of explosive products in grams and k is a constant. The internal energy $E(a)$ is then given by

$$E(a) = \int_{V(a)}^{\infty} PdV = \frac{P_a V(a)}{\gamma - 1} = \frac{kW}{\gamma - 1}\left(\frac{W}{V(a)}\right)^{\gamma-1}$$

From the last expression, it is evident that $E(a)$ decreases rapidly with increasing volume (proportional to a^3), and at sufficiently expanded stages of the motion represents a negligible fraction of the initial energy of the products. The values of radius a and corresponding pressure P_a for which $E(a)$ has a given value can be estimated from a knowledge of the adiabatic law for the products and the initial energy. For TNT, the calculations of Jones (described in section 3.5) give the adiabatic relation, valid for $P < 4{,}500$ lb./in.^{2}

$$P\left(\frac{V}{W}\right)^{1.25} = 7.8$$

where P is in kilobars, W is in grams, and V is in cm.^{3} The total energy released by 1 gram of TNT is about 1,060 cal./gm. ($= 4.44 \times 10^{10}$ ergs./gm.) of which approximately half is emitted in the shock wave (see section 4.8 for a detailed discussion).

The fraction F of the remaining energy Y which is present as internal energy at any state of expansion is

$$F = \frac{E(a)}{Y} = 0.166\, P_a{}^{1/5} = 0.42\left(\frac{W}{a^3}\right)^{1/4}$$

where P is in lb./in.^{2}, W is in lb., a in ft., and Y is taken to be 440 cal./gm. This fraction is less than 25 per cent for $P < 7.6$ lb./in.^{2}. For a 300 pound charge the corresponding radius is 13.5 feet, rather less than the maximum radius of about 20 feet for a charge detonated 50 feet below the surface. The curve of bubble expansion has the same qualitative manner of variation with time as shown in Fig. 8.1, which leads to the estimate that the internal energy of the products is less than 25 per cent of the total for more than 70 per cent of the entire cycle.

The fact that the internal energy is relatively unimportant over much of the expansion suggests, as a first approximation, neglecting it entirely. If this is done, it will be seen that the gas sphere must have a maximum radius when $da/dt = 0$. Calling this value a_m, we have from Eq. (8.5)

$$Y = \frac{4\pi}{3} P_o a_m^3 \tag{8.6}$$

This relation thus furnishes an experimental method for determining, to a rather good approximation, the total energy Y associated with the radial flow of water in terms of the maximum radius a_m of the bubble and the hydrostatic pressure P_o at the depth of the explosion. Experimental measurements based on Eq. (8.6) are described in section 8.3.

Neglecting the internal energy in Eq. (8.5) makes possible separation of the variables, and using Eq. (8.6) to eliminate Y gives the result

$$t = \left(\frac{3\rho_o}{2P_o}\right)^{1/2} \int_{a_o}^{a} \frac{da}{\left[\left(\frac{a_m}{a}\right)^3 - 1\right]^{1/2}} \tag{8.7}$$

where a_o is the initial radius of the gas sphere at time $t = 0$. The integral is not expressible in terms of elementary functions, but can be transformed to give a sum of incomplete β-functions $B_x(p, q)$, defined by

$$B_x(p, q) = \int_o^x x^{p-1}(1 - x)^{q-1} dx$$

the necessary substitution being $x = (a_m/a)^3$. Values of this function have been tabulated for discrete values of p, but unfortunately the values $p = 5/6$, $q = \frac{1}{2}$ required here are apparently not included. In general, the solution must therefore be obtained by numerical methods and some of these results are illustrated in section 8.3. In the particular case $x = 1$, corresponding to $a = a_m$, the solution is known in terms of the factorial or γ-function: $B_1\,(5/6, \frac{1}{2}) = 2.24$. These limits $a = 0$, a_m correspond to the time required for expansion from zero to maximum radius. The initial radius a_o is small compared to a_m and hence this time is approximately $\frac{1}{2}$ the period of oscillation T. Substitution in Eq. (8.7) gives the approximate result

$$T = \frac{2}{3} a_m \left(\frac{3\rho_o}{2P_o}\right)^{1/2} B_1\left(\frac{5}{6}, \frac{1}{2}\right) = 1.83\, a_m \left(\frac{\rho_o}{P_o}\right)^{1/2} \tag{8.8}$$

This result can also be expressed in terms of the total energy Y by Eq. 8.6, giving

$$T = 1.14\, \rho_o^{1/2} \frac{Y^{1/3}}{P_o^{5/6}} \tag{8.9}$$

This expression, which has been derived by a number of writers and is usually known as the Willis formula (121), shows that the "bubble period" varies as the cube root of the total energy or weight of a given explosive, and for a given weight, varies as the negative 5/6 power of the hydrostatic pressure. If the pressure at the surface of the water is one atmosphere, or 33 feet of sea water, the period therefore varies as $(d + 33)^{-5/6}$, where d is the depth in feet. The particular constant multiplying the period formula depends, of course, on the limit of integration chosen for the initial radius. The functional dependence of the period on Y, P_o, ρ_o must, however, be of the form of Eq. (8.9) from dimensional considerations, if these variables are assumed to be the only ones affecting the motion. It is therefore reasonable to expect that this form should apply more generally than the approximations of the simplified model would permit. The experimental evidence described in section 8.3 does in fact demonstrate the validity of the formula over a wide range of depths provided the multiplying factor is suitably adjusted. However, this agreement is not found for charges fired near the surface or bottom, which is to be expected as a result of the distortion of the mass flow of water by such boundaries. The modifications of the simple formula obtained by more refined analysis are considered in section 8.10.

8.3. Comparison of Radius and Period Measurements with Simple Theory

A. *Radius-time curves.* The earliest systematic measurements of the motion of gaseous explosion products are those of Ramsauer (88), who employed an ingenious electrolytic probe method to determine the position of the gas bubble boundary. In these experiments, charges of guncotton weighing one or two kilograms were fired at depths up to thirty feet in forty feet of water. A number of electrodes were supported at suitable distances from the charge by a rigid frame, and together with a common electrode formed conducting circuits with the seawater acting as electrolyte. After the charge was fired, the expanding gas bubble isolated the electrode circuits successively, and relay circuits were used to make a spark recording of the times of current interruption. In this way a displacement-time curve of the bubble motion could be determined up to its first maximum radius, and Ramsauer found that the variation of maximum radius with depth and charge weight agreed very closely with the formula

$$\frac{4\pi}{3} P_o a_m^3 = \text{constant} \cdot W \tag{8.10}$$

where W is the mass of explosive and P_o the hydrostatic pressure at the depth of explosion. By comparison with Eq. (8.6), it is seen that the right side of Eq. (8.10) should represent the energy available after emission of the shock wave, which Ramsauer computed to be about

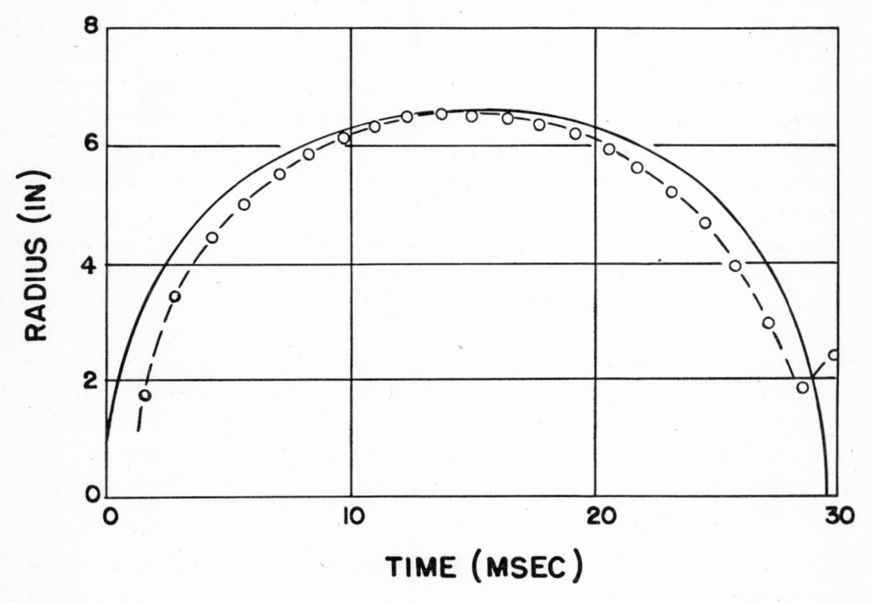

Fig. 8.2 Measured and calculated radius of the gas sphere from a detonator one foot below the surface.

forty-one per cent of the total energy. Ramsauer also found that probes placed directly below the charge indicated a maximum radius some ten per cent less than that from probes at the side, thus indicating upward migration (assuming a spherical bubble). A method basically that of Ramsauer has been used by Bancroft (114) for similar measurements on other explosives.

Ramsauer's method, ingenious though it is, suffers from the disadvantage of giving only a limited amount of information concerning isolated points on the bubble. An obviously more powerful tool is that of high speed motion pictures of the bubble motions, by methods such as those described in Chapter 6. The earliest records of this kind were taken by Edgerton of detonator caps exploded at various depths. Experimental points measured by Ewing and Crary (32) from one such record are plotted in Fig. 8.2,[2] and the solid curve is taken from Herring's calculations based on Eq. (8.7), the value of a_m having been chosen to give agreement with the experimental observations. It is seen that the simple theory predicts a curve of the same general form but somewhat broader than that found experimentally. The most likely explanations of the difference appear to be either that optical distortions in the experimental arrangement caused errors or that the proximity of the

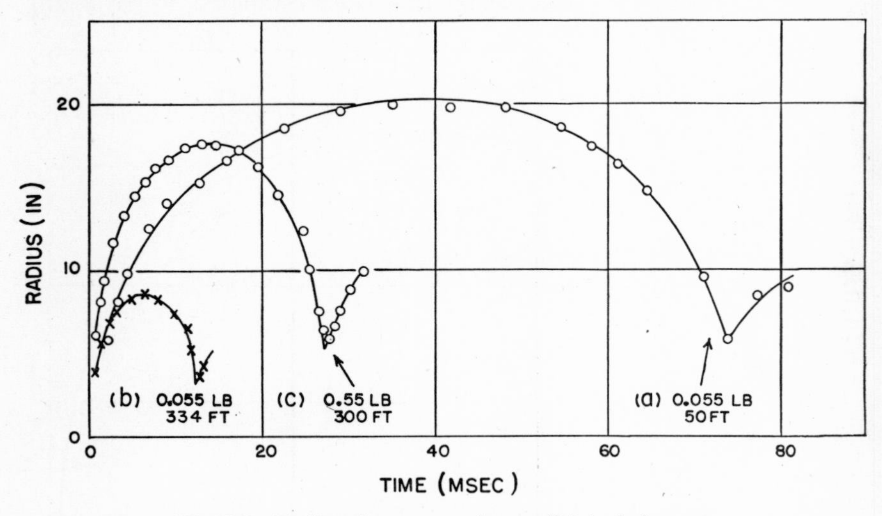

Fig. 8.3 Radius-time curves for small tetryl charges.

free surface twelve inches above the charge and a steel plate beneath it had significant effects.

These early photographs of Edgerton showed that the bubble rose slightly during its first expansion and then sank rapidly several inches below its initial position. This repulsive effect of the free water surface and a similar attraction toward a rigid surface has been further investigated by a number of workers, notably by Campbell and Wyckoff (17, 18) at the Taylor Model Basin and by Taylor and Davies (109) in England. Herring was the first to develop the hydrodynamic theory of these effects, which are discussed in more detail in section 8.8.

A considerable number of measurements of bubble radius-time curves for various small charges and depths of explosion have been made at the Woods Hole laboratory.[3] These data have been obtained

[2] The data for Fig. 8.2 are taken from Herring's report (46).

[3] These experiments, carried out in cooperation with personnel at the Taylor Model Basin (U. S. Navy), are described in a report by Arons, Borden, and Stiller (2).

for several explosives (TNT, tetryl, pentolite, etc.) at a variety of depths. The most complete data are for tetryl, and two typical examples for 0.056 pound charges at 5 feet and 330 feet are plotted in Fig. 8.3. These curves show the characteristic cusp shape of curve for each cycle, with decreasing period and amplitude of oscillation in successive oscillations. The decreased values of maximum radius and period as a result of increased hydrostatic pressure at the greater depth are also evident. A radius-time curve for a tetryl charge weighing 0.55 pound at 300 feet is shown in Fig. 8.3c, and the effect of charge weight is seen by comparing this curve with Fig. 8.3b for the 0.05 pound charge at the

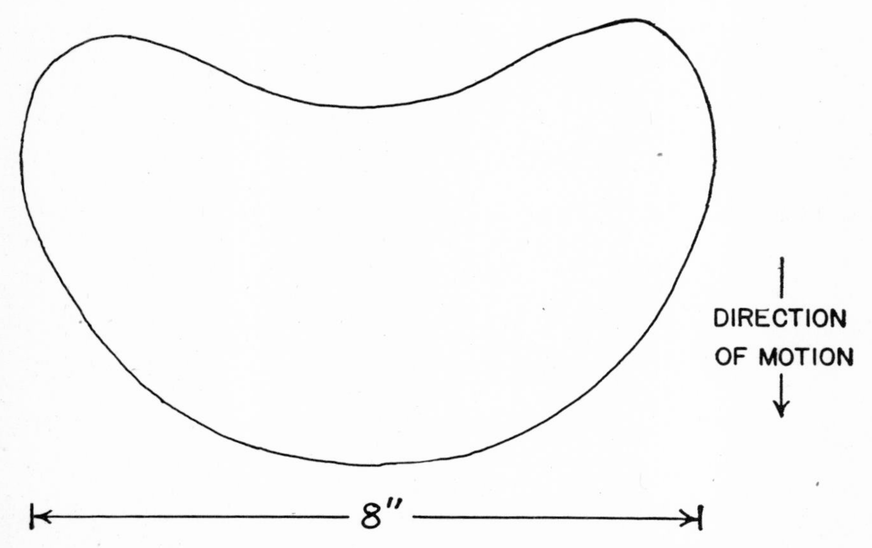

Fig. 8.4 The gas sphere near its maximum and first minimum (one ounce polar ammon gelignite charge, three feet deep).

same depth. It should be mentioned that the values of radius near the bubble minimum are somewhat uncertain, owing to "smoke" from the explosion products which obscures the bubble outline. It is also to be noted that the curves in Fig. 8.3 have been plotted without regard to vertical displacements of the bubble center and do not therefore show the upward migration under the influence of gravity. This effect is considered in section 8.5.

A difference in character of bubble shape during its motion is the distortion from spherical form near its minimum, an effect which is largest for charges fired near the surface. This difference is shown in Fig. 8.4, a sketch of the bubble outline near the minimum radius for a condition of rapid upward migration. (The sketch is based on photographic studies made by the Road Research Laboratory (95).)

At great depths, the shape near the minimum is not far from spherical, but when the bubble is moving rapidly the outline as seen from the side is kidney shaped at and near the minimum, the convex face being in the direction of motion. In the more expanded phases, the bubble is in either case closely spherical. The approximation of sphericity throughout the motion is thus seen to be least good in the contracted phase and near the free surface, where variations in hydrostatic pressure over the bubble boundary are a great part of the total.

Very few photographic records have been obtained of the bubble from large charges, and these are less satisfactory for measurement purpose than in the case of small charges, owing to technical problems in mounting the cameras and other gear. They do, however, show the same general features as the records for small charges, and there is no reason to believe that any essentially different features, apart from the change in scale, result from differences in charge weight. This conclusion is also supported by indirect measurements of period, migration, and so forth.

B. *Variation of period with depth and charge weight.* It was shown in section 8.2 on the basis of a simple approximate theory that the period T of oscillation of the gas bubble is given by

$$T = K \frac{Y^{1/3}}{P_o^{5/6}} \tag{8.11}$$

where Y is the total energy available and can ordinarily be considered proportional for a given explosive to the charge weight W, P_o is the hydrostatic pressure at the depth of explosion, and the constant k is characteristic of a given explosive. It was further argued from dimensional considerations that the functional form of this equation should be valid quite generally, despite the rather crude approximations involved in its development, if the explosion is not too close to boundary surfaces.

The predicted dependence of period with hydrostatic pressure was first verified in detail by Ewing and Crary (32) using ½ charges of TNT and small charges of SNG Oil Well explosive in depths from 40 to 675 feet. Similar period results using ½ pound tetryl charges fired at depths from 100 to 800 feet are plotted in Fig. 8.5, using double logarithmic scales, against depth +33 feet (as the water surface was at atmospheric pressure). The straight line is drawn with a slope of −5/6, and fits the experimental points within the accuracy of measurement. A similar verification of the period-depth relation for larger charges is shown in Fig. 10.3 of section 10.3, in which are plotted data for 200 pound charges fired at depths up to 900 feet in 3,000 feet of water, the data having been obtained in the course of sound ranging measurements.

These and other data are fitted by a straight line of slope $-5/6$ except for depths less than 200 feet, the points for shallower charges showing periods increasingly less than the predicted values. This difference is to be expected when the depth is not more than a few times the maximum bubble radius, and the results of more detailed investigations of the effect are discussed in section 8.9.

The variation of period in proportion to the cube-root of charge weight has not been directly verified over nearly the range as for the

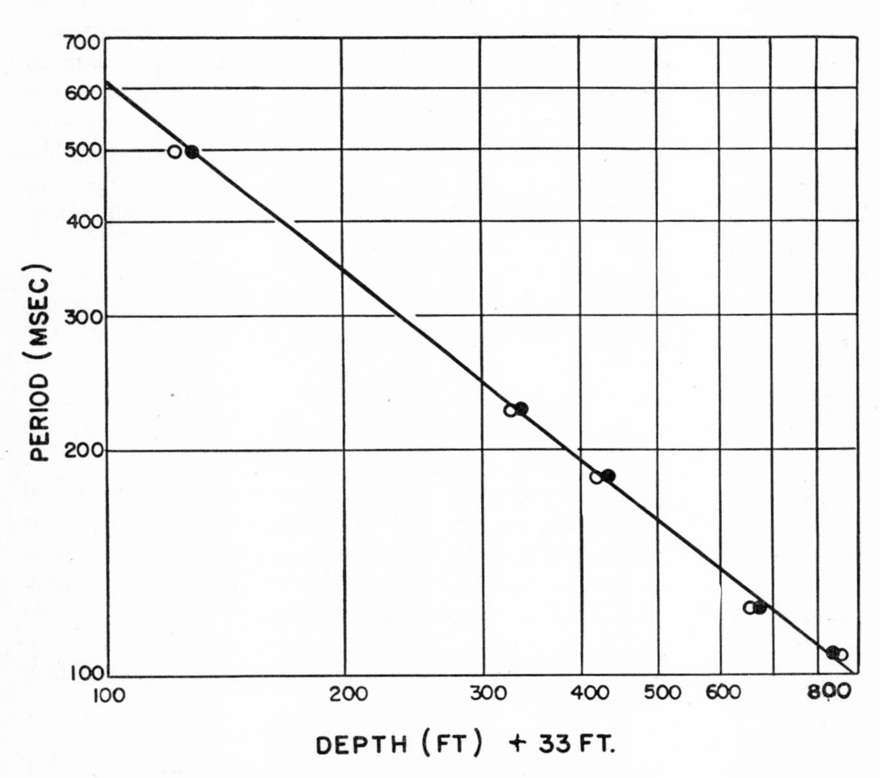

Fig. 8.5 Bubble periods against depth for 0.55 pound tetryl charges.

variation with depth, but there is no reason to doubt the correctness of the predicted law. For example, 25 and 300 gram loose tetryl charges fired at a depth of 12 feet in 24 feet of water gave values of the constant K in Eq. (8.11) differing by less than 1 per cent. The effects of the surface and bottom on period, discussed in sections 8.8 and 8.9, are not given exactly by theoretical corrections which have been developed, but except for this small uncertainty it appears that the bubble periods for ½ pound and 300 pound TNT charges require constants K in Eq.

(8.11) differing by less than 2 per cent. It is therefore possible to conclude that this bubble period formula, and bubble period measurements, can be relied upon to within two or three per cent if the charge is not too close to boundary surfaces. The values of constant K for the first bubble oscillations of several different explosives, obtained largely from small charge data, are listed in Table 8.1, the units being chosen to give the period in seconds for charge weight in pounds and depth in feet.

C. *Period of later oscillations and energy loss.* The Willis formula for period, Eq. (8.11), is based on the assumption that negligible energy losses occur in the course of the bubble pulsation except at or very near successive minima. Hence an expression of the same form should be

Explosive	Density (gm./cm.^{3})	K (sec. ft.$^{5/6}$/lb.$^{1/3}$)	Source of value (charge wt. and depth)
Loose tetryl	0.93	4.53	0.55 lb., at 39 ft.
Pressed tetryl	1.55	4.39	0.55 lb., 200 to 700 ft.
		4.35[1]	0.05 lb., 40 to 55 ft.
Cast TNT	1.55	4.36	0.62 lb., 300 to 600 ft.
	1.57	4.36[1]	290 lb., 20 to 90 ft.
Pentolite	1.6	4.35	0.55 lb., 40 to 400 ft.

[1] Values corrected for surface effect by Friedman theory (Section 8.10).

Table 8.1. Values of the constant K of the bubble period-depth relation $T = KW^{1/3}/(d + 33)^{5/6}$ for various explosives (T in seconds, W in pounds, d in feet).

applicable to second and later pulsations if a new smaller value of Y is used. The reason is simply that some of the energy in each cycle is lost at each minimum in turbulence and radiation, and the next oscillation occurs with lower energy. It is convenient to express this change by rewriting the period relation in the form

$$T_n = 0.561 \frac{(r_n QW)^{1/3}}{z_n^{5/6}} \tag{8.12}$$

where W is the charge weight in pounds, r_n is the fraction of the detonation energy Q which remains for the nth oscillation of period T_n, and z_n is the equivalent depth in feet of water for the hydrostatic pressure at the depth of the bubble at the beginning of the nth oscillation. The value of z_n will change significantly when the charge is fired relatively shallow, and the distance of migration during a cycle is an appreciable fraction of the depth. From Eq. (8.12), the ratio of the second to first bubble periods is given by

$$\frac{T_2}{T_1} = \left(\frac{r_2}{r_1}\right)^{1/3} \left(\frac{z_1}{z_2}\right)^{5/6}$$

and if the difference in initial depth can be neglected, so that $z_2 = z_1$,

$$r_2 = r_1 \left(\frac{T_2}{T_1}\right)^3 \tag{8.13}$$

Eq. (8.13) is a very useful and interesting result, for it permits an estimate of energy losses in successive contractions from the easily obtained ratio of periods. Data of this kind have been obtained from the 200 pound charge tests mentioned in part (B), which show that the ratios T_2/T_1 and T_3/T_2 are indeed nearly constant, independent of depth, for depths greater than about 320 feet. At smaller values of depth the ratios increased, this change being attributable to migration of the bubble and effect of the surface on period. Disregarding this change, the data give the values $r_2 = 0.34r_1$, $r_3 = 0.54r_2$. If a theo-

	Shock wave	First contraction	Second contraction
Fraction of original energy lost	59%	84%	92%
Fraction of available energy lost	59%	66%	46%
Fraction of original energy remaining	41%	14%	7.6%

Table 8.2. Energy losses in contractions of the gas bubble from 200 pound charges of a mixed explosive.

retical value $Q = 1.62$ kcal./gm. is used for the detonation energy of the explosive, Eq. (8.12) gives $r_1 = 0.41$. This is therefore the fraction of the original energy remaining after emission of the shock wave, and the values of r_2 and r_3, representing the fractional energies remaining after the first and second contractions, are also determined. Table 8.2 summarizes the results obtained in this way.

The period data thus indicate that more than half of the original energy of the explosion is lost before the first expansion of the bubble has developed appreciably, and that about two-thirds of the remaining energy is dissipated in the first contraction, leaving only a small residue for later oscillations. Only a relatively small fraction of the energy loss at the first contraction is accounted for by the radiated bubble pressure pulse, the rest being dissipated, as discussed in section 9.4.

Very similar results have been obtained from period ratios for small charges. In experiments on small charges of tetryl and TNT (less than ⅔ pound) fired in 23 feet of water (2), first and second bubble periods were observed for charge depths from 4 to 20 feet. In many cases, migration of the bubble had an appreciable influence on the second period T_2, but when a value T_2' corrected for the measured migration of the bubble was used, the ratio T_2'/T_1 was found to be essentially constant for the range of depths and charge weights employed, except near a bounding surface, as shown in Fig. 8.6. The values 0.7 to 0.8 indicate that from 50 to 65 per cent of the energy of the first oscillation is lost in the first contraction, and this order of magnitude is observed generally for high explosives. These data, supplemented by theoretical calculations of the detonation energy Q, also agree fairly closely in predicting that from 40 to 45 per cent of the initial energy of explosion remains in the bubble motion during its first cycle.

A further interesting application of bubble period measurements is to comparison of explosives. For two charges of equal weight fired at the same depth, Eq. (8.11) predicts that the energies Y remaining after

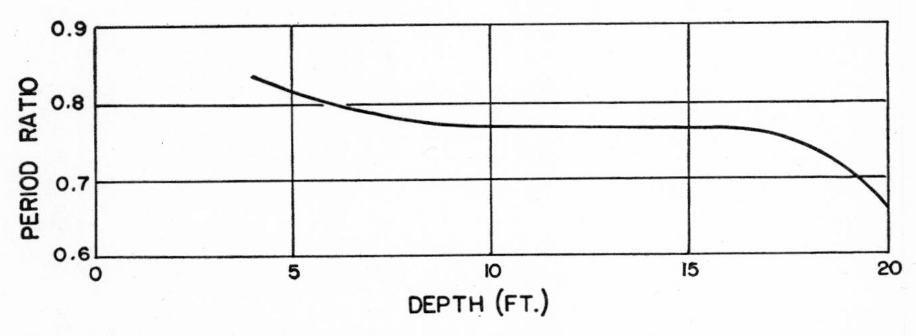

Fig. 8.6 Ratio of first and second bubble periods for 0.66 pound tetryl charges.

emission of the shock wave are in the ratio of the cubes of the observed periods. This measurement has sometimes been used as an indication of relative effectiveness of explosives. Data of this kind have been obtained for a number of explosives. A comparison with calculations of detonation energies and either calculated or measured values of shock wave energies further permits at least a partial check on the consistency of the various results. For example, if the *fractional* energy losses in the shock wave, as expressed by r_1, are the same in two explosives, the ratio of bubble energies Y should be the same as that of detonation energies. The premise to this simple comparison has been found to be approximately true in experimental data for a mixed explosive considerably more powerful than TNT. The detonation energy relative to TNT from calculations by S. R. Brinkley, Jr., agrees in this case with the ratio of periods cubed to five per cent. That the premise and comparison based on it cannot always be justified is illustrated by Pentolite. The bubble energy is (Table 8.1) virtually the same as for TNT, but the comparison of experimental shock wave energies in Table 7.4 gives a value 27 per cent greater. Adequate data to test the energy balance for Pentolite do not appear to be available at present.

Perhaps the greatest uncertainty in energy comparisons is in the ratio of shock wave energies, because the pertinent values are of course those which represent total wastage as heat after the shock wave has passed to infinity. Experimental values give neither this value nor the total of transmitted energy before any wastage has occurred, as they are obtained at a finite distance from the charge. This uncertainty considered, the comparisons of such energy ratios give rather satisfactory indications that no serious discrepancies exist in the total accounting of energies, even though this accounting is not yet as directly based on experiment as might be desired. More explicit illustration is unfortunately precluded by security restrictions applying to the more significant examples.

8.4. General Equations of Noncompressive Motion

When the effect of gravity is included, the motion of the gas sphere is symmetrical around a vertical axis through the initial position of the charge, but no longer has radial symmetry. To take account of the external force and lower symmetry, it is desirable to use the more powerful mathematical methods of potential theory rather than solve the dynamical equations by direct integration. In this section, these methods are therefore outlined as they apply to noncompressive flow of an ideal fluid incapable of supporting shearing stress (i.e., neglecting viscosity). No attempt will be made to examine their full possibilities and limitations; for such discussions, reference should be made to standard treatises.

For noncompressive motion in which the density is assumed constant, the equation of continuity expressing conservation of mass (cf. section 2.1) reduces to

$$\text{div } \mathbf{v} = 0 \tag{8.14}$$

where $\mathbf{v}$ is the vector particle velocity. For the problems of interest to us, it is possible to define a velocity potential φ at all points in the fluid, from which the components of $\mathbf{v}$ are obtained by space differentiation:

$$\mathbf{v} = -\text{grad } \varphi, \text{ i.e., } u_x = -\frac{\partial \varphi}{\partial x}, \text{ etc.} \tag{8.15}$$

With this definition, the potential φ must satisfy Laplace's equation, obtained by substituting (8.15) in (8.14),

$$\text{div}(\text{grad } \varphi) = \Delta^2 \varphi = 0 \tag{8.16}$$

where, in Cartesian coordinates, $\Delta^2 = \partial^2/\partial x^2 + \partial^2/\partial y^2 + \partial^2/\partial z^2$. In addition to satisfying this equation, the potential φ must also be so chosen that it gives correctly any prescribed values of velocity at boundary surfaces of the fluid. At rigid boundaries, the component of velocity normal to the surface must be equal to that of the surface, and in an infinite medium, the velocity must vanish at infinity at least as rapidly as $1/r^2$ if the total kinetic energy of flow is to remain finite.

In the cases to be considered, the fluid is subjected to external forces. If the force per unit mass of fluid be denoted by $\mathbf{F}$, the equation of motion becomes

$$\rho_o \frac{\partial \mathbf{v}}{\partial t} + \tfrac{1}{2}\rho_o \operatorname{grad} \mathbf{v}^2 = \rho_o \mathbf{F} - \operatorname{grad} P \tag{8.17}$$

The only forces which need be considered are conservative and hence derivable from a potential function Ω defined by $\mathbf{F} = -\operatorname{grad} \Omega$. Introducing the potentials φ and Ω in Eq. (8.17) gives

$$-\rho_o \frac{\partial}{\partial t} \operatorname{grad} \varphi + \tfrac{1}{2}\rho_o \operatorname{grad} \mathbf{v}^2 = -\rho_o \operatorname{grad} \Omega - \operatorname{grad} P$$

Space and time differentiations are interchangeable and this equation can therefore be integrated with respect to the position variables to give the generalized Bernoulli's equation

$$\frac{P}{\rho_o} + \Omega = \frac{\partial \varphi}{\partial t} - \tfrac{1}{2}\mathbf{v}^2 + F(t) \tag{8.18}$$

where $F(t)$ is a function of time only. If the motion is steady (does not change in character with time), $\partial\varphi/\partial t = 0$ and $F(t) =$ constant, and hence

$$\frac{P}{\rho_o} + \tfrac{1}{2}\mathbf{v}^2 + \Omega = \text{constant}$$

which is the familiar form of Bernoulli's theorem for steady flow.

For the motion of the gas sphere, the only external force is gravity which has the potential $-gz$, where g is the acceleration of gravity, and the depth z is measured as a positive distance below the point in the liquid at which the hydrostatic pressure would be zero. If the surface of the water is at atmospheric pressure the origin of z is therefore thirty-four feet above the surface, this distance being the length of the column of water which exerts a pressure of one atmosphere. From this equation, prescribed conditions on the pressure at boundary surfaces further define the allowable solutions for φ. For example, at the free surface of the liquid the pressure must be the same everywhere (ordinarily atmospheric), and the position of the surface must in general change in time to satisfy this condition.

A consequence of potential theory which is useful for later developments is a modified form of Green's theorem relating the variations of any potential function φ satisfying $\Delta^2\varphi = 0$ throughout a volume of fluid to its values on surfaces enclosing the fluid. As is shown in standard references,[4] this integral theorem may be written

$$\iiint \left[\left(\frac{\partial\varphi}{\partial x}\right)^2 + \left(\frac{\partial\varphi}{\partial y}\right)^2 + \left(\frac{\partial\varphi}{\partial z}\right)^2\right] dV = - \iint \varphi \frac{\partial\varphi}{\partial n}\, dS$$

where the volume integral is extended throughout the volume of fluid and the surface integral is evaluated over all boundary surfaces, $\partial/\partial n$ indicating differentiation of φ in the direction of a normal to the surface (taken as positive away from the fluid). The volume integrand, however, is double the kinetic energy of unit mass of fluid and the integral is thus $2/\rho_0$ times the total kinetic energy T of the fluid, which gives

$$T = - \frac{\rho_0}{2} \iint \varphi \frac{\partial\varphi}{\partial n}\, dS \tag{8.19}$$

The kinetic energy can therefore be evaluated from the velocity potential and its variation at the boundaries.

8.5. Motion of a Gas Sphere under Gravity

A rigorous solution from noncompressive theory for the motion of the gaseous products and surrounding water after an explosion should start from the initial form of the boundary surface and distribution of pressure and velocity, from which these quantities would be determined at later times by solution of the dynamical equations. Initially the gas boundary is spherical or nearly so, but there is no guarantee in the appropriate boundary conditions that a spherical surface is the form in equilibrium with the gas pressure of the interior. Detailed investigations show, in fact, that this is not the case if gravity is considered, and the motion may even become dynamically unstable under some conditions. Attempts to develop a solution determining the actual state of motion without artificial restrictions on the geometrical form of the gas water interface rapidly become extremely complicated and have so far only yielded qualitative information about limited phases of the motion. Some of the conclusions which have been drawn from such analysis are given in section 8.7.

From a more empirical point of view, the experimental evidence that a spherical boundary is in fact a good approximation to the actual out-

[4] See, for example, Lamb, p. 46 (65).

line over most of the motion makes reasonable the assumption that the form of the gas bubble is spherical. This procedure is not required by the basic equations, and in fact is consistent with the true boundary condition that the pressure should be the same at all points on the surface only if vertical displacements of the center of the sphere are neglected. In the general case, the spherical form must therefore be imagined as one preserved by fictitious constraint forces acting at the boundary. These forces are to a certain extent arbitrary, but if the motion is to remain dynamically reasonable, it is necessary that they shall not change the total energy of the motion and hence these forces should be restricted by the condition that they do no work.

The first reasonably complete account of the radial and vertical displacements of the bubble and of the mass flow of water, including the effects of both gravity and boundary surfaces is due to Herring (46). This theory has been amplified and extended by G. I. Taylor (107) for the case of motion under the influence of gravity alone, and by Courant and associates (102) to take into account the boundary surfaces to a better approximation. In this section, the solution for motion under gravity, as developed by Taylor, will be considered.

A. *The equations of energy and momentum.* The first task in establishing an equation of energy similar to Eq. (8.5) for purely radial motion is evaluation of the total kinetic energy of flow for the water. If, at a given instant, the center of the gas sphere of radius a has an upward velocity U (not to be confused with shock front velocity!) and the radius is increasing at the rate da/dt, the radial velocity u_r of a point P on the surface must be given by

$$(u_r)_a = \frac{da}{dt} + U \cos \theta \tag{8.20}$$

where a and θ are measured from the moving center C of the sphere, θ being the angle the radius vector makes with the vertical. A suitable velocity potential determining the flow velocities for the surrounding water must satisfy the equation

$$(u_r)_a = \left(- \frac{\partial \varphi}{\partial r} \right)_a$$

and be a solution of Laplace's equation vanishing properly at infinity. The symmetry of the problem suggests the familiar solutions of Laplace's equation in spherical harmonics, the suitable form being

$$\varphi(r, \theta) = \frac{A}{r} + \frac{B \cos \theta}{r^2}$$

Substituting in Eq. (8.20) to determine A and B, we obtain

$$\varphi = \frac{a^2}{r}\left(\frac{da}{dt}\right) + \frac{1}{2}\frac{a^3}{r^2}U\cos\theta \tag{8.21}$$

Differentiation to determine radial and tangential components of velocity gives the results

$$u_r = -\frac{\partial\varphi}{\partial r} = \frac{a^2}{r^2}\frac{da}{dt} + \frac{a^3}{r^3}U\cos\theta$$

$$u_\theta = -\frac{1}{r}\frac{\partial\varphi}{\partial\theta} = \frac{1}{2}\frac{a^3}{r^3}U\sin\theta$$

Both u_r and u_θ vanish as $1/r^3$ and are admissible solutions. The kinetic energy of the fluid is, from Eq. (8.19), given by

$$T = -\frac{\rho_o}{2}\int\int\varphi\frac{\partial\varphi}{\partial n}\,dS$$

the integral being carried out over the bounding sphere of radius a and a second outer boundary which can be considered a sphere of radius R allowed to recede to infinity. The contribution from the large sphere vanishes, while the value for the inner surface is

$$\begin{aligned} T &= \frac{\rho_o}{2}\int_o^\pi a\left(\frac{da}{dt} + \tfrac{1}{2}U\cos\theta\right)\left(\frac{da}{dt} + U\cos\theta\right)2\pi^2\sin\theta d\theta \\ &= 2\pi\rho_o a^3\left(\frac{da}{dt}\right)^2 + \frac{\pi}{3}\rho_o a^3 U^2 \end{aligned} \tag{8.22}$$

The first term, giving the energy of radial motion, is the same as the value found before (Eq. (8.5)). The second term, giving the translational energy of flow, shows that this energy is the same as if a mass $2\pi\rho_o a^3/3$, equal to one half the mass of water displaced by the gas sphere, were given the velocity U. The equation of energy is, therefore, expressed by the relation

$$2\pi\rho_o a^3\left(\frac{da}{dt}\right)^2 + \frac{\pi}{3}\rho_o a^3U^2 + \frac{4\pi}{3}\rho_o a^3 gz = Y - E(a) \tag{8.23}$$

where, as in section 8.2, $E(a)$ is the internal energy and the hydrostatic pressure $P_o = \rho_o gz$. In this equation $U = -dz/dt$, and there are therefore two unknowns, U and a. To determine these a second equation is necessary.

The second relation is simply obtained from momentum considerations. From the derivation for kinetic energy of the moving water, it was found that its effective inertia as regards translation was that of a mass $2\pi\rho_o a^3/3$ with velocity U. The vertical momentum acquired by the water should, therefore, be $(2\pi\rho_o a^3/3)U$ and is the result of the buoyant force on the gas sphere, equal to $4\pi\rho_o a^3 g/3$ by Archimedes' principle. Equating the impulse of this force to the momentum acquired, by Newton's second law, gives

$$\int_o^t F dt = \frac{4\pi}{3}\rho_o g \int_o^t a^3 dt = \frac{2\pi}{3}\rho_o a^3 U \tag{8.24}$$

and hence

$$U = -\frac{dz}{dt} = \frac{2g}{a^3}\int_o^t a^3 dt \tag{8.25}$$

a result originally given by Herring.

This rather intuitive argument may perhaps not be wholly satisfying to some. It should be pointed out that its validity is insured only if artificial external constraints on the boundary, which must strictly be introduced to keep it spherical, are such that they exert no resultant force on the boundary and hence do no work on the bubble during its motion. An equivalent condition is the requirement that the resultant effect of the fluid pressure at the boundary on its motion be zero, and if this condition is worked out, one obtains Eq. (8.24). To do this, one must first compute the pressure P_a at the boundary surface $r = a$ from Bernoulli's equation (Eq. (8.18)) and the velocity potential. This pressure acts normal to the surface and the resultant force in the direction of motion ($\theta = 0$) is obtained by integrating $P_a \cos\theta$ over the surface. Setting the result equal to zero gives Eq. (8.24).

B. *Taylor's nondimensional form of the equations.* It is readily seen that the equations of motion, (8.23) and (8.25), do not permit simple geometrical scaling, for which the equations are unchanged if all lengths and times are multiplied by the same factor. This principle of similarity, valid for shock and detonation waves, cannot therefore be applied to bubble phenomena. The reason for the failure of this scaling is of course the effect of gravity in addition to the internal gas pressure in determining the motion. The initial energy of the gas is, however, relatively small if one excludes a fairly small portion of the oscillation cycle when the bubble is near its minimum size. Over most of the cycle it is therefore reasonable, as an approximation in determining the general features of the motion, to neglect this term in the energy equation. With this approximation, the only force acting on the water is gravity and the only resistance to flow is that offered by the inertia of the water.

If the linear scale of the phenomenon is changed, the corresponding change in time scale which preserves similarity of the two motions must be such that the effect of gravity (hydrostatic pressure) is the same in either system. If this is to be true, the invariance of the acceleration of gravity g which has dimensions length/(time)^2 requires that the time scale be increased by the square root of the scaling factor for length. It is convenient to express this correspondence by using a characteristic length L in terms of which the equations can be expressed in nondimensional form. The dimensionless variable t' replacing the time t and scaling properly can therefore be written

$$t = \sqrt{\frac{L}{g}}\, t'$$

provided dimensionless variables a', z' replacing a, z are expressed by $a = La'$, $z = Lz'$. Using these definitions, Eqs. (8.23) and (8.25) become

$$2\pi a'^3 \left(\frac{da'}{dt'}\right)^2 + \frac{\pi}{3} a'^3 \left(\frac{dz'}{dt'}\right)^2 + \frac{4\pi}{3} a'^3 z' = 1 - \frac{E(a)}{Y}, \tag{8.26}$$

$$-\frac{dz'}{dt'} = \frac{2}{a'^3} \int_0^{t'} a'^3 dt'$$

if the characteristic length L is chosen to have the value $(Y/g\rho_o)^{1/4}$. With this choice of L as made by G. I. Taylor, the equations are expressed in a nondimensional form suitable for numerical integration with assigned initial values of z', a', provided the term $E(a)/Y$ is neglected. Corresponding values of measured lengths and times are then determined by the relations

$$\frac{a}{a'} = \frac{z}{z'} = \left(\frac{Y}{\rho_o g}\right)^{1/4}, \quad \frac{t}{t'} = \left(\frac{Y}{\rho_o g^5}\right)^{1/8} \tag{8.27}$$

where Y is the total energy available after emission of the shock wave.

Usually, the energy Y can be expected to be proportional to the weight of the charge and hence to the cube of its linear dimensions. The scaling laws expressed by Eqs. (8.27) thus require that, for proper scaling of bubble phenomena from one size of charge to another, the initial values of "hydrostatic" depth z and the scale of length be increased as the three-fourths power of the linear dimensions of the charge; correspondingly, the time scale must be increased as the three-eighths power of charge dimensions. Thus neither length nor time scale directly with charge dimensions, and it is not possible to preserve exact similarity for both shock wave and bubble phenomena. A further difficulty in scaling bubble phenomena lies in the fact that the depth variable z is not the depth d of the bubble center below the water surface, but is rather a measure of hydrostatic pressure at this depth. If the pressure is atmospheric at the surface and lengths are measured in feet, then $z = d + 33$ feet, and it is evident that z can be less than 33 feet only if the pressure at the surface is reduced below atmospheric. Hence model scale experiments on bubble phenomena can in many cases be made to simulate those for large charges only by reducing the pressure at the surface.

It should be emphasized that the nondimensional form (8.26) of the equations of motion and the scaling laws (8.27) are only approximations obtained by neglecting the internal energy of the gas sphere, and hence are increasingly in error as the bubble radius becomes smaller. Strictly, complete similarity of bubble motions for two different charge sizes is not possible because the scaling laws for the internal energy and hydrostatic pressure are not compatible. Any conclusions obtained on the basis of the scaling laws as expressed by Eq. (8.27) are therefore subject to some uncertainty and are particularly unreliable in considering phenomena in the region of greatest contraction. These inaccuracies may, of course, be removed at the expense of generality by restoring the term $-E(a)/Y$ to the right side of Eq. (8.26). For TNT, $E(a)/Y$ varies as $(W/a^3)^{1/4}$ from section 8.2, and expressed in Taylor's variables has the form: constant$\cdot(W^{1/16}a'^{-3/4})$. This term thus varies as $W^{1/16}$ and its inclusion requires separate solutions of the equation of motion for each charge weight.

C. *Results obtained neglecting the internal energy.* G. I. Taylor (107) has given the results of numerical integration for the initial condition that $z_o' = 2.0$, where z_o' is the initial value of the depth variable at time $t = 0$. This integration was carried out by approximate solution of Eqs. (8.26) for the initial stages of the motion, based on the fact that initially a' and the migration velocity dz'/dt' are both small. Neglecting small terms in a'^3 and dz'/dt', the first of Eqs. (8.26) gives

$$\frac{da'}{dt'} = \frac{a'^{-3/2}}{\sqrt{2\pi}}, \quad t' = \frac{2}{5}\sqrt{2\pi}a'^{5/2}$$

assuming $a' = 0$ for $t' = 0$. Using this value of t' in the second of Eqs. (8.26) and integrating gives

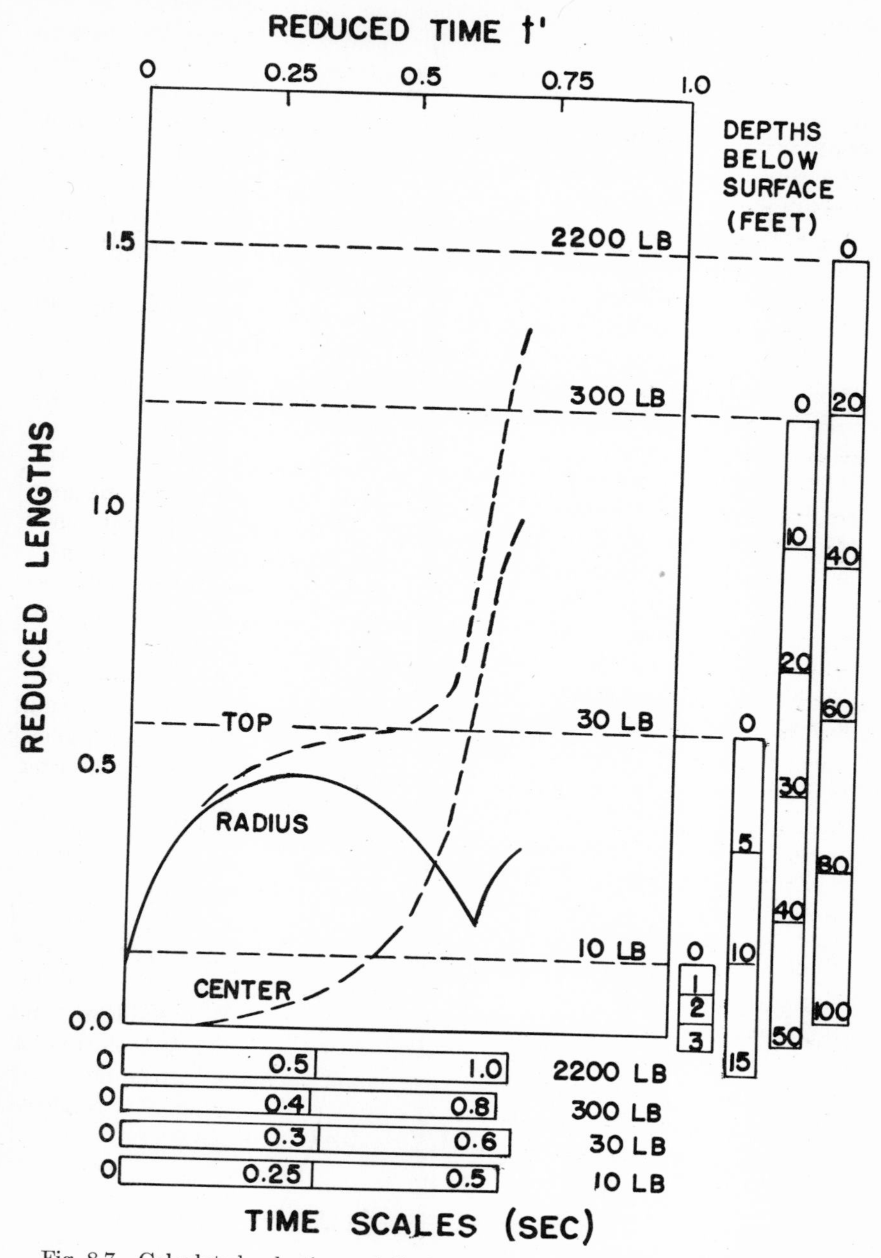

Fig. 8.7 Calculated pulsation and displacement of a gas sphere under gravity.

$$z_o' - z' = 2\int_o^{t'} \frac{1}{t''^{6/5}} \int_o^{t''} \frac{dt'}{t'^{6/5}}\, dt'' = \frac{5}{11}\, t'^2$$

an expression valid only for small radius and translational velocity. The values of z' and a' for $z_o' = 2$, $t' = 2.0$ are then used for step by step numerical integration of the complete form of Eqs. (8.26).

Taylor's results are plotted in Fig. 8.7 and show the characteristic features of the vertical migration described in section 8.1: a slow rise until the bubble passes its maximum expansion, followed by an increasingly rapid upward movement as the minimum radius is approached. The bubble has a non-zero minimum radius despite the neglect of internal gas pressure which is apparently required to stop the radial contraction. This seeming paradox Taylor resolves by noting that, as the bubble contracts, all the kinetic energy available is taken up by the vertical motion before the radius becomes zero. The computed minimum radius $\bar{a}'$ is given by $\bar{a}' = 0.21$, and the time at which this occurs is $t' = 0.64$. These figures are of course only approximate, particularly the value of $\bar{a}'$, as the effect of internal energy has been neglected.

The results plotted in Fig. 8.7 refer of course to the specific initial condition that $z_o' = 2.0$ and are in reduced units. They therefore correspond to a particular relation between charge weight and initial depth d_o, and, from Eq. (8.27), this relation is

$$d_o + 2.31P_o = z_o = 2\left(\frac{Y}{\rho_o g}\right)^{1/4}$$

where Y is the energy available, and P_o is the pressure at the surface in lb./in.^{2}. The value of Y depends of course on the weight and kind of explosive. For TNT, the total energy of explosion is approximately 1,060 cal./gm., of which approximately 40 per cent remains after emission of the shock wave. The energy Y for TNT becomes $Y = 8.4 \cdot 10^{12} W$, if the charge weight W is expressed in pounds, and the depths corresponding to given weights of explosive and reduced depth $z_o' = 2.0$ are given by

$$d_o + 2.31P_o = 20W^{1/4}$$

Similarly, the time scale for a weight W and $z_o' = 2.0$ is given by

$$t(\text{sec.}) = 0.55W^{1/8}$$

These relations can be used to find corresponding depths and charge weights, and the auxiliary scales in Fig. 8.7 illustrate the relations obtained for several charge weights. The dashed lines indicate the corresponding positions of the surface, which for atmospheric pressure is thirty-three feet below the origin of z. For the two largest charges, the first cycle always lies well below the surface, but the bubble intersects the surface before completion of one period for thirty and ten pound charges. The theory of course becomes meaningless after this occurs, as the bubble loses its identity by venting, and is unreliable in any case when the bubble approaches the surface because effects of this proximity have been neglected.

Despite the weaknesses of the theory, the examples do illustrate the necessity for reducing the pressure at the water surface if model experiments are to simulate on a reduced scale even the general features of the motion of the gas sphere from larger charges. For example, if a 10 pound charge were to reproduce the behavior for a 2,200 pound charge, the hydrostatic pressure at the surface would have to be reduced to 26 per cent of atmospheric, and the distance d_o below the surface decreased by the same factor. If the surface pressure is not scaled, it is evident that the total hydrostatic pressure near the original depth of explosion changes proportionately much less for small charges than for large, owing to the relatively larger constant pressure above the surface. Hence, the differences in pressure resulting from gravity have a smaller effect and the migrations are much smaller for explosion products of small charges. In any given case, the importance of gravity can be estimated by comparing the maximum bubble radius with the equivalent hydrostatic head at the surface. For charges of 1 pound or less of explosive with a maximum radius not exceeding 4 feet and atmospheric pressure of 33 feet of water at the surface, the variations of pressure around the gas sphere are clearly not a large fraction of the total and the migration under gravity is a small effect. For 300 pound charges, on the other hand, the radius is of the order of 30 feet and the effect of gravity is large.

8.6. Calculations of Gravity Effects and Comparison with Experiment

The first calculations of bubble motion under the influence of gravity, made by G. I. Taylor on the basis of his nondimensional formulation of the equations of motion, have been extended by other workers using more elaborate numerical methods. In addition, a number of approximate formulas and solutions representing such results have been developed, which permit calculation of the various parameters of interest with reasonable accuracy and much more simplicity.

A. *Numerical calculations.* Comrie and Hartley (24) have computed the expansion and migration of the bubble for 4 different values of initial hydrostatic pressure corresponding to values 1, 2, 3, and 4 for

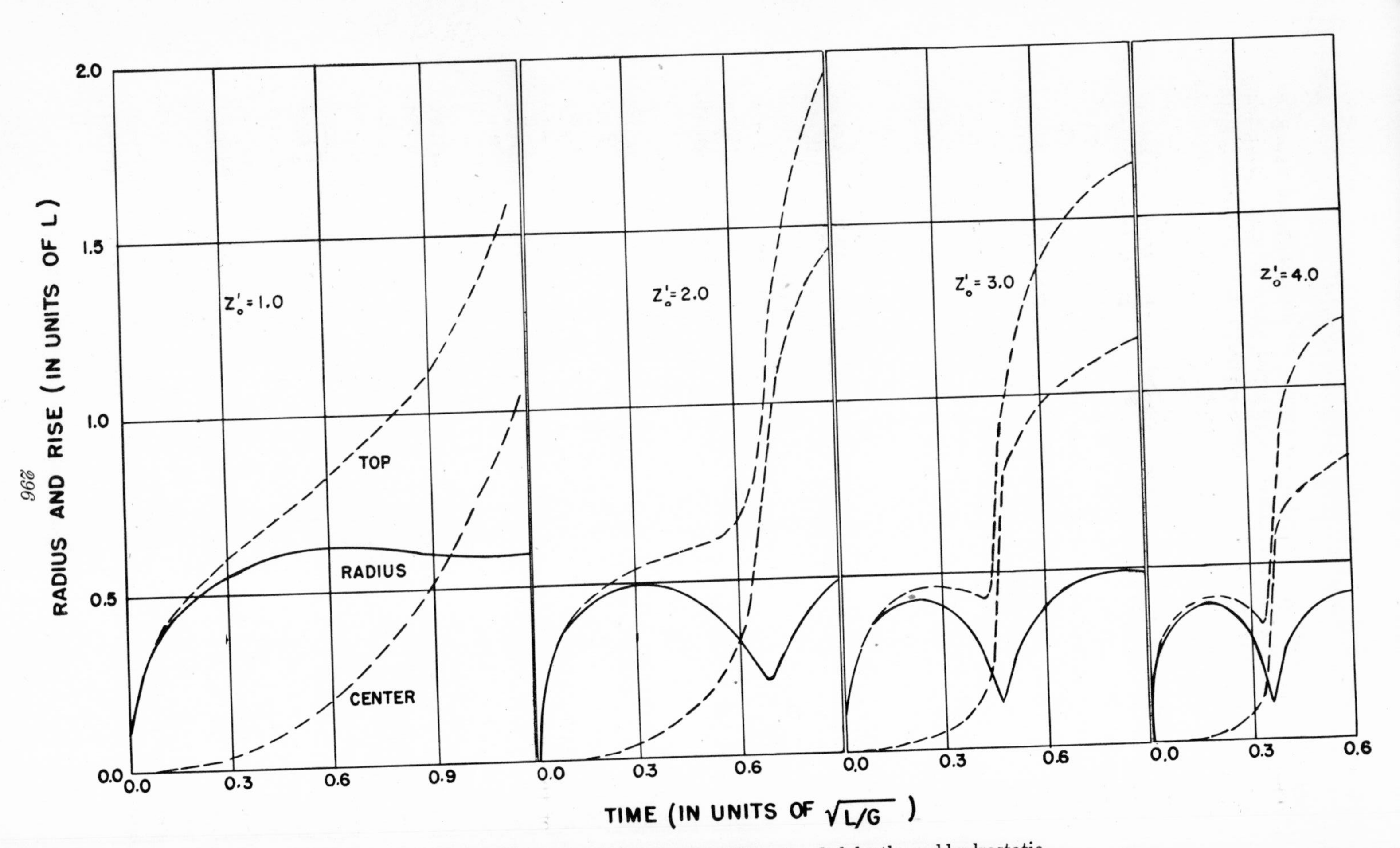

Fig. 8.8 Calculated bubble motion for different scaled depths and hydrostatic

the reduced depth z_o', the internal energy of the gas being neglected. Their results, expressed in Taylor's reduced units of length and time, are plotted in Fig. 8.8, the solid curves representing the radius, and the dashed curves the upward displacements of the center and top of the bubble. These curves apply to any desired charge weight at four depths determined by the charge weight and value of z_o'. The conversions for several weights of TNT, taken from a table given by Kennard (55), are given in Table 8.3, where the distance in feet and time in seconds corresponding to unit value of the reduced variables of Fig. 8.8 are tabulated, together with the corresponding initial hydro-

Charge weight W(lb.)	Scale factors		Pressure at center in feet of water ($= z_o' L$)[1]			
	L(ft.)	$\sqrt{L/g}$(sec.)	$z_o' = 1$	$z_o' = 2$	$z_o' = 3$	$z_o' = 4$
1/16	5	0.39	5	10	15	20
1	10	0.55	10	20	30	40
50	27	0.90	27	53	80	106
100	32	0.98	32	63	95	126
300	42	1.12	42	83	125	166
1000	56	1.30	56	112	168	225

[1] If the surface is at atmospheric pressure (33 feet of water), the explosion is at a depth 33 feet less than the equivalent pressure in feet of water.

Table 8.3. Weights of explosive and depth-time scales corresponding to different values of Taylor's reduced units.

static pressure in feet of water. The conversions are based on the formulas $L = 10W^{1/4}$ ft., $t/t' = 0.55W^{1/8}$ sec., obtained from Eq. (8.27) assuming a total energy for TNT of 880 cal./gm., half of which is left for the bubble motion ($Y = 440$ cal./gm.). It is to be noted that none of the curves can apply to the smallest weight (1/16 lb.), unless the pressure at the surface is reduced considerably below atmospheric pressure, as total pressures (surface plus water) are listed.

The qualitative effects of depth on migration and bubble contraction are clearly shown in Fig. 8.8. At the greatest depths ($z_o' = 4$), the bubble contracts to a small fraction of its maximum radius and the upward migration becomes appreciable only after the first minimum. For smaller depths, the minimum becomes less pronounced and the migration increasingly well developed at the time of the minimum. These features of the motion are readily understood if it is remembered that downward momentum of the water is acquired while the bubble is large, and changes less when the bubble radius and buoyancy are small, but that the upward velocity of the bubble for a given momentum increases as the bubble contracts. The larger vertical velocity of flow at shallow depths absorbs more of the available energy, which becomes equal to the total at a larger bubble radius, thus accounting for the very slight contraction in the case $z_o' = 1$. (It is to be observed, however, that this is the least realistic of the cases calculated, as the neglected repulsion of the free surface has a considerable effect in actual cases.)

Calculations are also given in the report of Comrie and Hartley in which the internal energy of the gas is included. These are for values $c = 0.08$, 0.09, 0.10, 0.11 in the expression $E(a')/Y = ca'^{-3/4}$, where $c = 0.075W^{1/16}$, and hence correspond to weights of TNT from 3 to 500 pounds. The curves are very similar to those of Fig. 8.8 for $c = 0$, the principal differences being that the maximum radius and migration are about 10 per cent smaller; the period and minimum radius are very nearly the same.

A number of other detailed numerical solutions have been made by H. M. Nautical Almanac Office (76), primarily for initial values corresponding to very small charges (1 gram and 1 ounce TNT); one set of results corresponds to the case $z_o' = 1.0$ first computed by Taylor (discussed in section 8.5), and shows reasonably good agreement with this more approximate calculation. In another (77), results are given for formulas developed by Temperley (111) to include an approximation to the nonsphericity of the bubble (see section 8.7 for details of this effect).

B. *Approximate formulas.* The detailed numerical calculations cited above apply only to specific relations between charge weight and depth of water, but reports from the Road Research Laboratory (England) (93) give a number of more general, approximate formulas for principal features of the motion. These results are in some cases based on approximations in the equation of motion found to be justified by the calculations, and in other cases are empirical expressions fitted to calculated points.

Examination of the calculations shows first that the period of oscillation differs only slightly from the value predicted neglecting vertical migration resulting from gravity, and in fact is influenced only slightly by including the internal energy of the gas. The effect of migration on bubble radius is also small up to the time of the first maximum. Changes in the vertical position can therefore be neglected in the equation of motion: $z' = z_o'$ and $dz'/dt' = 0$ in Eq. (8.26). If the internal energy is included in the form $E(a')/Y = ca'^{-3/4}$, the maximum radius a'_m (for which $da'/dt' = 0$) is given by

$$a'^3_m = \frac{3}{4\pi z_o'} (1 - ca'^{-3/4}_m) \tag{8.28}$$

This equation is readily solved by graphical methods and is found to agree with exact calculations to within 1 to 5 per cent.

A second fact observed about the calculations is the near-constancy of the integral $\int_o^{t'} a'^3 dt'$ during the contracted phase, beginning when the bubble has contracted to about half its maximum size and continuing beyond the next minimum. This integral, however, determines the vertical momentum acquired by the water, which thus remains nearly constant during the contraction. This constant value permits a simple expression of the velocity of the bubble center, for if we let $m = \int_o^{t'} a'^3 dt'$, the momentum equation (Eq. (8.26)) becomes

$$-\frac{dz'}{dt'} = \frac{2m}{a^3} \tag{8.29}$$

a result applicable in the contracted phase if m can be evaluated.

The value of m in Eq. (8.29) can, of course, be found exactly for the calculated cases. An approximate evaluation for other values of the parameters can be obtained if it is assumed that the shape of the radius-time curve is essentially the same, no matter what the migration, when the radius is large and the constant value of m acquired. With this assumption, $a'^3 = a'_m{}^3 F(t'/T')$, where F is a function independent of charge weight and depth, and T' is the period. We therefore have

$$m = \int_o^{\frac{3T'}{4}} a'_m{}^3 F\left(\frac{t'}{T'}\right) dt' = k' a'_m{}^3 T' = \frac{k a'_m{}^3}{z_o^{5/6}}$$

where k is a constant. The value of k required to fit numerical calculations is estimated in the RRL report to be 0.70, but a better value from calculations made by the Nautical Almanac Office would be from 0.57 to 0.67. Taking a value 0.6 gives

$$m = 0.6 \frac{a'_m{}^3}{z_o'^{5/6}} \tag{8.30}$$

With this value of m (and hence dz'/dt') known, the vertical velocity and minimum radius $\bar{a}'$ of the bubble can be computed from Eq. (8.26). Inserting the vertical velocity dz'/dt' from Eq. (8.29) and setting $da'/dt' = 0$ gives

$$\bar{a}'^3(1 - c\bar{a}'^{-3/4}) = \frac{4\pi}{3} m^2 \tag{8.31}$$

the term $(4\pi/3)\bar{a}'^3z'$ representing hydrostatic buoyancy being negligible. This equation can be solved graphically for $\bar{a}'$, and the corresponding vertical velocity $U' = -dz'/dt'$ obtained from Eq. (8.29). Results obtained in this way are in only moderately good agreement with experiment.

The displacement of the bubble at the time of the first minimum is a quantity of some importance, as it determines the position of the source of the bubble pulse. An empirical relation obtained from the numerical integration is that, to within about ten per cent, the rise $\Delta h'$ is proportional to the period of oscillation if these quantities are both expressed in Taylor's nondimensional units, the relation being

$$\Delta h' = 1.05\ T' = 1.19\ z_o'^{-5/6}$$

If the rise Δh is expressed in feet, the formula becomes

$$\Delta h = 81\,\frac{W^{11/24}}{z_o^{5/6}} \tag{8.32}$$

where the charge weight W is in pounds, the hydrostatic depth z_o in feet, and the numerical factor is obtained from Taylor's estimate of available energy for TNT.

It is to be noted that all the formulas of this section which involve numerical factors are restricted to the case of the energy for TNT. Approximate corrections for other explosives can be made either on the basis of calculated detonation energies, assuming the same fraction remains for the bubble motion, or by the observed ratio of bubble periods. The formulas and calculations are both somewhat special in that they assume a specific adiabatic law for the expansion products based on Jones' calculations for TNT. The constant c and exponent $(-3/4)$ appearing in the formulas are the result of these calculations and are not necessarily exactly correct for TNT or appropriate for other explosives. The differences resulting from better calculations would, however, probably not be large compared with other errors inherent in the theory and experimental uncertainties.

C. *Comparison with experiment.* A comparison of theoretical calculations for the effects of gravity alone on bubble motion with experiment is made difficult by the fact that these effects are appreciable only for relatively shallow explosions in which the migrations are large. A shallow explosion, however, implies the presence of a free surface of the fluid above and near the bubble, which introduces further changes in the motion. As a result, clear-cut examples of purely gravitational motion are not readily achieved. The importance of gravity compared with surfaces increases for a larger scale of bubble dimensions and surface distance, because variations in hydrostatic pressure over its surface become a larger part of the total, but the technical problems of photographic or other measurements are much more difficult for large charges. As a result, there are not many experimental data available to test the simple gravity theory, and most comparisons to experiment must be made with more elaborate theories which include surface effects. These theories and their agreement with experiment are considered in later sections.

As we have seen, the first maximum of the bubble and the period of the first oscillation are not much affected by gravity except as it deter-

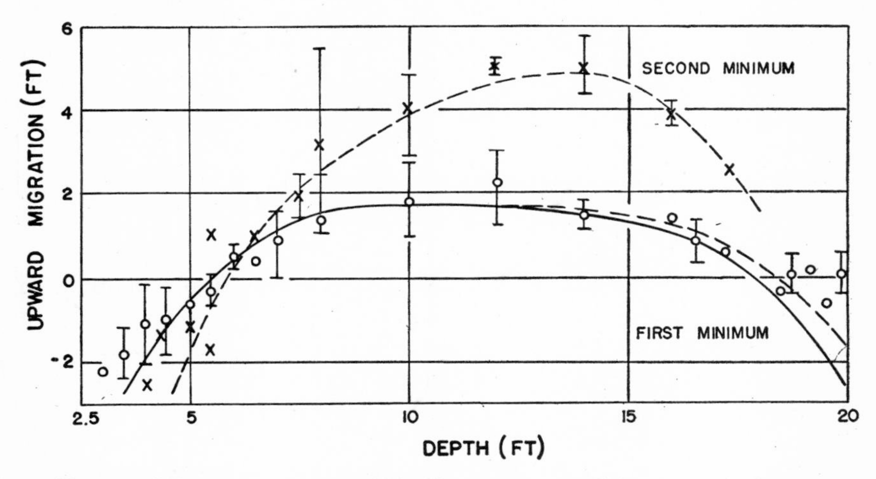

Fig. 8.9 Migrations of gas bubbles from 0.66 pound loose tetryl charges.

mines the hydrostatic pressure, and the discussion of section 8.3 neglecting gravity need not be reconsidered here. The vertical position of the bubble center as it first contracts and approaches the first minimum is affected, as are the later phases of the motion. The first systematic investigation of bubble migration for various depths and charge weights was made by H. F. Willis and Ackroyd (122). These observers measured the bubble rise at the time of the first and second minima by sound-ranging methods employing two piezoelectric gauges at known positions relative to the charge to determine arrival times of the secondary pulses at these points. Differences in these times as compared with those for the shock wave then permitted calculation of the rise. A series of charges of various weights (15 grain detonators, 1 ounce tetryl, 1 and 5 pounds blasting gelatin) were fired at wide ranges of depths in 55 feet of water.

The results of Willis and Ackroyd showed first a gradual increase in the rise as the charge depth was decreased, in fair agreement with Eq. (8.32) according to which the rise varies with depth d as $(d + 33)^{-5/6}$.

The observed rises were systematically somewhat less than the calculated ones, and close to the surface were increasingly smaller, becoming negative (i.e., a downward displacement) when the charge was a few feet below the surface. This phenomenon is shown in Fig. 8.9, which is a plot of similar data obtained at Woods Hole for 250 gm. tetryl charges (2). Migrations at the end of both the first and second cycles are shown and the solid line is the curve predicted by Eq. (8.32) for this

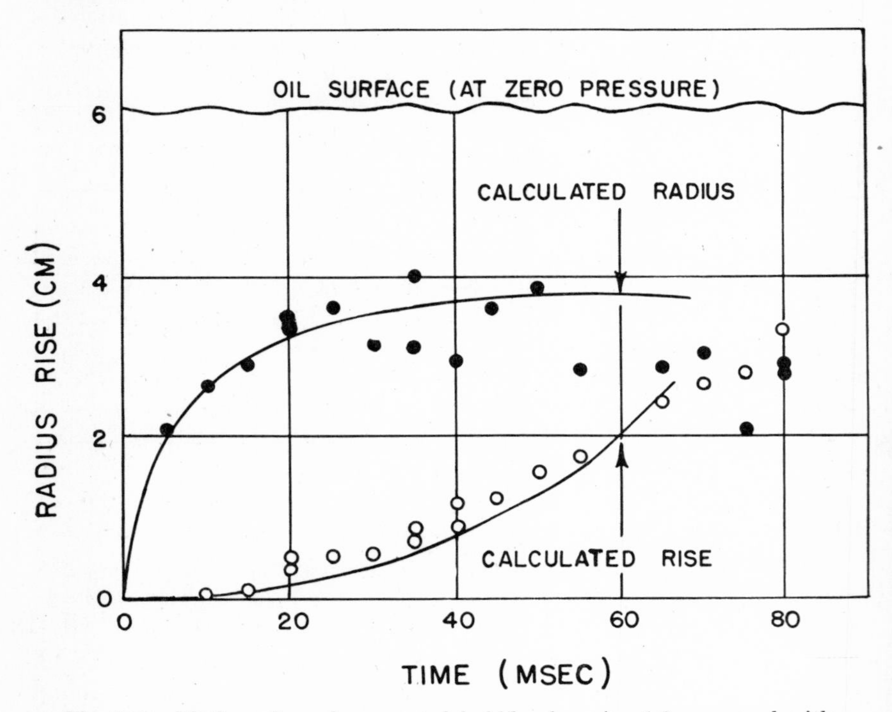

Fig. 8.10 Motion of spark generated bubbles for $z_o' = 1.0$ compared with theory. Solid circles are measured radii, open circles measured rise of the center.

charge and weight. The decreased and reversed migrations near the surface and bottom are accounted for by the surface attraction and bottom reflection and are further considered in section 8.11.

D. *Model experiments.* A more detailed comparison of Taylor's theory with experiment than has been obtained with explosive charges was made by Taylor and Davies (109) in a very beautiful series of model experiments, in which gas bubbles produced by electrical discharges in water or oil were photographed. The bubbles so obtained involved energies of the order 1/40 calory and so correspond to roughly 2.10^{-5} pounds of TNT. The bubbles had a maximum radius of a few inches, and to simulate larger scale explosions it was necessary to reduce the hydrostatic pressure above the liquid to a small fraction of atmospheric pressure.[5] In order to obtain the closest comparison with the theory so far outlined, one series of experiments was carried out under a vacuum to minimize the effect of a surface, and the spark was produced at a depth which gave $a_{\max} = 3.7$ cm., $z_o = 6.01$ cm. The ratio $a_{\max}/z_o = 0.61$ corresponds to $z_o' = 1.0$ in the calculations of Comrie and Hartley, and the experimental values of radius and rise versus time should therefore be represented by the curves for $z_o' = 1.0$ in Fig. 8.8, provided the scaling factors L and t/t' of the theory are chosen as $L = z_o = 6.1$ cm. and $t/t' = \sqrt{L/g} = 0.079$. The values of lengths l and time t are then given by

$$l = 6.05\, l', \quad t = 0.079\, t'$$

The measured radii and migrations are compared with the theory in Fig. 8.10. Although the radius values show considerable scatter at longer times, they are seen to be in rather good agreement with theory, as are the measured upward displacements. The data thus provide a very satisfactory confirmation of the gravity theory.

As already mentioned, the case just considered corresponds to shallower depths than can ordinarily be realized without large effect of the free surface. In another series of records, Taylor and Davies obtained the radial and vertical displacements for a bubble formed at the same depth, but with a surface pressure equivalent to a depth 6.5 cm. for the oil used. The resulting motion therefore corresponds to a large scale experiment at atmospheric surface pressure and the charge fired at a depth of $(6.05/6.50) \times 33 = 31$ feet. The greater pressure at the depth of the bubble leads to a smaller observed maximum radius, $a_m = 3.15$ cm., and the value $a_m/z_o = 3.15/(6.05 + 6.50) = 0.25$ corresponds to the case $z_o' = 2.0$ calculated by Comrie and Hartley. The scale length L required to convert the calculated values to experimental conditions is then $L = z_o/z_o' = 6.27$ cm., and hence $a = 6.27\, a'$, $t = \sqrt{L/g}\, t' = 0.080t'$, where a' and t' are calculated nondimensional values. Under these conditions, the bubble remained spherical during the first expansion, then flattened and became concave on its lower surface as it contracted. The subsequent expansions restored the symmetry to a considerable extent, and the bubble moved upward toward the surface with considerable velocity.

[5] In the experimental arrangement employed by Taylor and Davies, the bubble spark is produced in a cylindrical glass tank by discharge of a condenser initially charged to 4000 volts. In some of the experiments, transformer oil of low vapor pressure was used as a fluid instead of water, in order to prevent boiling at the bubble surface in its low pressure, expanded phases. Illuminating sparks at intervals controlled by a pendulum timing device gave a series of exposures on a film mounted on a rotating drum. Optical distortion by the walls of the cylindrical tank was prevented by attaching plane-walled auxiliary tanks, also filled with oil.

The observed size and displacement of the bubble are plotted in Fig. 8.11 for comparison with the predicted curves for $z_o' = 2.0$. Because of the nonspherical shape of the contracted phase, one-half the horizontal bubble diameter is plotted. It is seen that the first half cycle is in good agreement with theory, but that the later migration is much less than predicted. This difference is consistent with the greater hydrodynamic resistance of the actual flattened bubble surface as compared to the idealized sphere assumed in the calculations The ob-

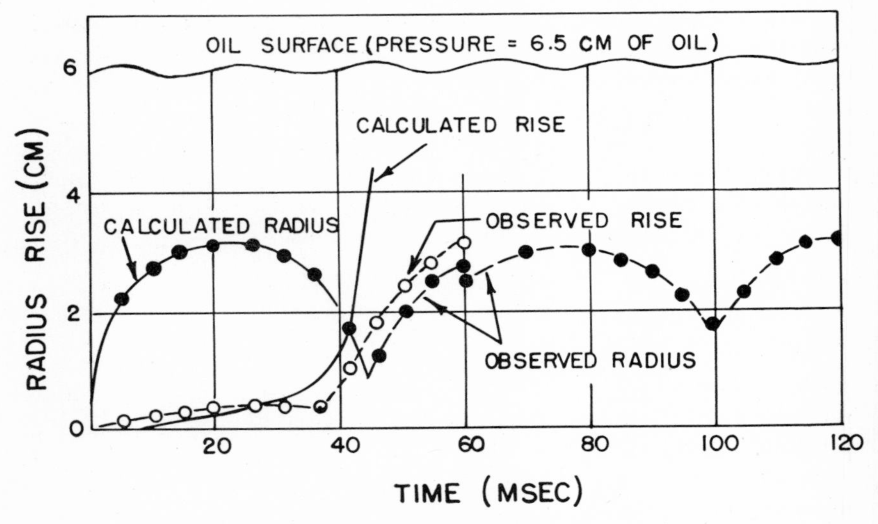

Fig. 8.11 Motion of spark generated bubbles for $z_o' = 2.0$.

served period of 45 msec. is appreciably less than the computed 54 msec., but the discrepancy is quantitatively explained by the effect of the free surface.

The model experiments under reduced pressure are seen to agree rather satisfactorily, on the whole, with Taylor's theory, the discrepancies being largest in the neighborhood of the unstable minima where the velocities are large and the bubble surface far from spherical. The differences observed are similar to those found for explosion bubbles, and this similarity makes the spark technique a useful and significant one in analyzing large scale explosions. It must be remembered that the scaling does not and cannot readily be made to include exactly to scale the viscosity and vapor pressure of the fluid and the way in which energy is released. Fortunately, these sources of error, discussed by Taylor and Davies, are not large enough to cause appreciable differences in the results. Other experiments made by Taylor and Davies include records corresponding to $z_o' = 2.8$, and a number of interesting films showing the migration of the bubble toward model rigid surfaces and targets. Similar small scale records of such motions, which are of considerable importance in assessing the role of secondary bubble pressures in explosion damage, have been taken by Campbell (17) at the Taylor Model Basin, using detonators, but without the reduction of surface pressure necessary to give a model scale of large explosions.

Campbell and Wyckoff (18) have made streak silhouette and stroboscopic pictures, using continuously running film, of half cap and Number 6 detonator cap explosions under reduced atmospheric pressures in a 24 inch water tunnel. The records obtained illustrate very beautifully the changes in maximum size and period with hydrostatic pressure, reveal interesting cavitation regions at various times between the shock wave front and the bubble surface, and also show the relation of bubble motion to displacements of water at the free surface. The energy values for these very small explosions are not too precisely known and detailed analysis of the radius and period data is not attempted here. The other observed phenomena are considered in Chapter 10.

8.7. Effects of Compressibility and Nonspherical Form on Bubble Motion

We have so far developed the approximate theories of bubble pulsation and migration on the simplifying assumptions that the bubble retained a spherical form and that the surrounding water was incompressible. Although most of the calculations of the motion suitable for comparison with experimental data have been made with these approximations, more exact formulations undertaken by Herring and others are of interest in indicating the nature of the errors to be expected from the simpler calculations. There are described here Herring's development (46) of the equations of motion for spherical symmetry, including the effect of compressibility and of gravity on the motion, and the analysis by Penney and Price (86) of the stability of the spherical form during the pulsation.

A. *The effect of compressibility.* The equation of an inviscid fluid for spherical symmetry is, from section 2.3,

$$\frac{\partial u}{\partial t} + u\frac{\partial u}{\partial r} = -\frac{1}{\rho}\frac{\partial P}{\partial r}$$

where u is the radial velocity and P the pressure at a point (r, t). Integrating this equation from the bubble surface ($r = a$) to infinity gives

$$a\frac{d^2a}{dt^2} + \frac{3}{2}\left(\frac{da}{dt}\right)^2 - a\frac{da}{dt}\lambda(a) + \int_a^\infty r\frac{\partial\lambda}{\partial t}\,dr = -\int_a^\infty \frac{dP}{\rho} \tag{8.33}$$

where the variable λ has been defined as the divergence of the velocity u, $\lambda = (1/r^2)\,\partial/\partial r\,(r^2u)$, and the following relations have been employed:

$$\int_a^\infty \frac{\partial u}{\partial t}\,dr = -\int_a^\infty \frac{\partial}{\partial t}(r^2u)d\left(\frac{1}{r}\right)$$

$$= \left| r\frac{\partial u}{\partial t}\right|_a^\infty + \int_a^\infty r\frac{\partial\lambda}{\partial t}\,dr \text{ (integrating by parts)},$$

$$\left[\frac{\partial u}{\partial t} + u\frac{\partial u}{\partial r}\right]_{r=a} = \frac{d^2a}{dt^2}$$

If we assume that the density is a known function of pressure P, neglecting irreversible processes induced by the preceding shock wave, the right side of Eq. (8.33) is a known function of $P(a)$, the pressure in the gas sphere. The departures from incompressible theory are represented by the terms in λ. The equation of continuity supplies a means for evaluating $\lambda(a)$ in terms of $P(a)$, however, for we have from section 2.3 that

$$\text{(8.34)} \qquad \lambda(a) = [\text{div } u]_{r=a} = \left[-\frac{1}{\rho}\frac{d\rho}{dt}\right]_{r=a} = \left[-\frac{1}{c^2\rho}\cdot\frac{dP}{dt}\right]_{r=a}$$

where $c^2 = (dP/d\rho)_s$ is the velocity of sound. With this relation, Eq. (8.33) becomes an ordinary differential equation for a in terms of $P(a)$, except for the integral $\int_a^\infty r\,\partial\lambda/\partial t\,dr$. This term is, moreover, the only one which can account for loss of energy by radiation of a pressure pulse, as the others are unaffected by a change in the sign of $u(a) = da/dt$. If the motion of the water is sufficiently small for acoustic theory to be applied, this integral would be given exactly by a simple analysis. For in the limit of small amplitudes, the velocity u, and hence its divergence λ, must satisfy the wave equation $\Delta^2\lambda - (1/c_o^2)\,\partial^2\lambda/\partial t^2 = 0$. Hence for spherical symmetry λ is of the form $(1/r)\,f(t - r/c_o)$. We therefore have $r\,\partial\lambda/\partial t = -c_o\,\partial/\partial r\,(r\lambda)$, and the integral in question becomes

$$\text{(8.35)} \qquad \int_a^\infty r\frac{\partial\lambda}{\partial t}\,dr = -c_o\int_a^\infty \frac{\partial}{\partial r}(r\lambda)dr = c_o a\lambda(a)$$

In the present problem, the amplitude of motion is sufficiently large that the acoustic theory is not valid, but Herring has shown that the deviations from the acoustic result of Eq. (8.35) are not large, being greatest at and near the stage of greatest contraction. This analysis, which we shall omit, shows that the most important correction term to be added to the right side of Eq. (8.35) is

$$-\frac{1}{c_o a}\frac{d}{dt}\left[a^2\left(\frac{da}{dt}\right)^2\right]$$

Substituting in Eq. (8.33), using Eq. (8.34) for $\lambda(a)$, and rearranging gives

$$a\frac{d^2a}{dt^2}+\frac{3}{2}\left(\frac{da}{dt}\right)^2-\frac{1}{ac_o}\frac{d}{dt}\left[a^2\left(\frac{da}{dt}\right)^2\right]=\frac{a}{\rho c_o}\frac{dP_a}{dt}\left(1-\frac{1}{c_o}\frac{da}{dt}\right)-\int_a^\infty\frac{dP}{\rho}$$

The left side of this equation can be written as a derivative, and if density variations are neglected so that ρ can be replaced by ρ_o, the density at the depth of explosion, we can integrate over a. Taking the lower limit to be the maximum radius a_m for which $da/dt = 0$, we obtain

(8.36)

$$a^3\left(\frac{da}{dt}\right)^2\left(1-\frac{4}{3}\frac{1}{c_o}\frac{da}{dt}\right)=\int_{a_m}^{a}\left[\frac{P_a-P_o}{\rho_o}+\frac{a}{\rho_o c_o}\cdot\frac{dP_a}{dt}\left(1-\frac{1}{c_o}\frac{da}{dt}\right)\right]2a^2da$$

In this expression, the terms in $1/c_o$ represent approximately the effect of compressibility, and if these terms are neglected the energy equation previously derived (Eq. (8.5)) is obtained. The factor $(1 - 4/3\ 1/c_o\ da/dt)$ changes sign when the bubble passes its maximum radius, as da/dt becomes negative, and hence introduces a dissymmetry in the radius-time curve. This is, however, not particularly important as da/dt, the radial velocity, is very much less than the velocity of sound c_o over nearly the entire pulsation. The term involving $dP(a)/dt$ represents the acoustic radiation of energy, and will be further discussed in section 9.4.

The equation (8.36) in principle permits a solution for the radius-time curve and the corresponding pressure at the surface of the bubble (which is determined for a given a), but no calculations of this kind have been made. It seems quite certain that the deviations from non-compressive theory will be insignificant over most of any one cycle, although the radiation term does lead to a large energy loss near the minimum.

B. *The effect of gravity.* In the theory developed by Taylor, the gas bubble is assumed to remain spherical despite its vertical migration under the influence of gravity, and in effect any constraints necessary to preserve this form are implicitly assumed present in such a way that the total energy is unaltered by them. Herring has examined the question of the form of the bubble surface when gravity is included, and he finds that to a first approximation the bubble remains spherical; but the assumption that the gravity effect is small, made in obtaining the result, breaks down in later stages.

Herring's analysis, although straightforward, will not be given in detail here. Assuming the correction for gravity to be small, the velocity potential, pressure, and radius vector to the bubble surface are expanded in powers of the acceleration of gravity g, use being made of the fact that coefficients in the expansion of φ must be solutions of Laplace's equation and hence can be expressed in spherical harmonics. The boundary condition which must be satisfied at the bubble surface is that the pressure be a function only of the volume V of gas, regardless of the shape and of g. The first order correction to the radius vector turns out to be simply an equal upward displacement of all points of the surface, the velocity U being given by

$$U = \frac{2g}{a^3}\int_o^t a^3 dt$$

The first order result for U is in agreement with the result of section 8.5, which assumes a spherical form. It shows, however, that the upward velocity increases with time, particularly when the radius becomes small in the contracting phase, owing to the factor $1/a^3$. Hence the assumption that the upward displacement due to gravity is small becomes increasingly poor as the bubble contracts, and the conclusion that the form remains spherical is no longer established. The inference that the bubble may well become unstable is supported by experimental pictures of the bubble (section 8.3), as well as by more detailed analysis of Penney and Price, and of Ward.

C. *Perturbations of the bubble form.* More elaborate approximations to the actual form of a bubble in a fluid have been developed by Penney and Price (86), in which departures from spherical symmetry are treated as small perturbations, which on account of the symmetry about a vertical axis can be simply expressed in a series of surface spherical harmonics. Ward (118) has modified this approach somewhat to obtain results expressing the departures from spherical form as a series of approximations in increasing powers of the vertical velocity U of the bubble. Ward's development is particularly interesting, as it shows rather well the stages of approximation in which the various theories already discussed are obtained, and is outlined in the following, together with some of the results of Penney and Price.

We begin by noting that for symmetry about a vertical axis the velocity potential $\varphi(r, t)$ for the fluid can be expressed as an expression in powers of $1/r$ and Legendre polynomials $P_n(\cos\theta)$:[6]

$$\varphi(r, t) = \frac{A}{r} + B_1 \frac{P_1(\cos\theta)}{r^2} + B_2 \frac{P_2(\cos\theta)}{r^3} + \cdots \tag{8.37}$$

where θ is the angle with the vertical axis, and the coefficients $A, B_1 \ldots$, are functions only of time. We further assume that the radius vector R from the center of the bubble to its surface can be written

$$R(t) = a + b_2P_2(\cos\theta) + b_3P_3(\cos\theta) + \cdots \tag{8.38}$$

where $b_2, b_3 \ldots$, are functions of time, and $b_1 = 0$ because of the choice of origin. As is shown in the development, the coefficients b_2, b_3 are of order $U^2, U^3 \ldots$.

The radial velocity u_r of the surface is given by $(-\partial\varphi/\partial r)_{r=R}$, and since the distance to any point R of the surface differs from the value a by a quantity of order U^2, the error in identifying u_r with $(dR/dt + U \cos\theta)$, to make the normal velocity in the fluid equal that of the boundary, will be of order U^4. We therefore have

$$\begin{aligned} \frac{dR}{dt} &= \left(-\frac{\partial\varphi}{\partial r}\right)_R - U\cos\theta \\ &= \frac{A}{R^2} + \frac{2B_1}{R^3}P_1 + \frac{3B_2}{R^4}P_2 + O(U^4) \end{aligned}$$

the term $U\cos\theta$ being necessary to account for the velocity U of the bubble center chosen as origin for R. The derivative dR/dt can, however, be expressed in terms of the P_n by Eq. (8.38), and if this is done coefficients of like P's must be equal, as the expression must be true for any θ. Substituting for R and dR/dt then gives the relations

$$P_o : A = a^2\frac{da}{dt}, \quad P_1 : B_1\left(1 - \frac{6}{5}\frac{b_2}{a}\right) = \tfrac{1}{2}a^3U \tag{8.39}$$

$$P_2 : \frac{db_2}{dt} + \frac{2A}{a^3}b_2 = \frac{3B_2}{a^4}, \quad P_3 : \frac{db_3}{dt} + \frac{2A}{a^3}b_3 = \frac{4B_3}{a^5} - \frac{18}{5}\frac{B_1}{a^4}b_2$$

where terms of order U^3 have been dropped.

[6] The double use of P, in connection with these polynomials and for pressure, is to be noted, but should cause no confusion.

The necessary condition on the boundary of the surface is that the pressure $P(R)$ be uniform and equal to the gas pressure P_g. The pressure is determined by Bernoulli's equation (Eq. (8.18))

$$\frac{P_o}{\rho_o} - gR\cos\theta + \left(\frac{\partial\varphi}{\partial t}\right)_R - \tfrac{1}{2}(\text{grad }\varphi)_R{}^2 = \frac{P_g}{\rho_o}$$

where P_o is the hydrostatic pressure at the bubble center. Substitution for R and φ then gives a series of terms in the $P_n(\cos\theta)$. (In carrying out this substitution, it must be remembered that the origin of R is in moving coordinates and $\partial\varphi/\partial t$ is to be evaluated in fixed coordinates.) Use is then made of the orthogonality relations for Legendre polynomials, namely that

$$\int_{-1}^{+1} P_n(\cos\theta)P_m(\cos\theta)d(\cos\theta) = \begin{cases} \dfrac{2}{2n+1} & \text{if } m = n \\ 0 & \text{if } m \neq n \end{cases}$$

Four equations therefore result from multiplying the Bernoulli equation by P_o, P_1, P_2, P_3 in turn and integrating over $\cos\theta$, these being

$$\frac{P_o}{\rho_o} + a\left(\frac{d^2a}{dt^2}\right) + \frac{3}{2}\left(\frac{da}{dt}\right)^2 - \frac{U^2}{4} + O(U^4) = \frac{P_g}{\rho_o} \tag{8.40}$$

$$\frac{1}{2}\frac{d}{dt}(a^3U) - \frac{6}{5}g\int\frac{B_2}{a^2}\,dt + O(U^3) = ga^3$$

$$\frac{1}{a^3}\frac{dB_2}{dt} - \frac{3B_2}{a^4}\frac{da}{dt} + \frac{3}{4}U^2 - \frac{1}{a^2}\frac{d^2a}{dt^2}\int\frac{3B_2}{a^2}\,dt + O(U^3) = 0$$

$$\frac{1}{a^4}\frac{dB_3}{dt} - \frac{4B_3}{a^5}\frac{da}{dt} - \frac{1}{a^2}\frac{d^2a}{dt^2}\int\frac{4B_3}{a^3}\,dt + O(U^3) = 0$$

The various stages of approximations of the theories already discussed are nicely shown from these equations. In the zeroth approximation neglecting gravity, $U = 0$ and the first of Eqs. (8.40) becomes

$$a\frac{d^2a}{dt^2} + \frac{3}{2}\left(\frac{da}{dt}\right)^2 = \frac{P_g - P_o}{\rho_o}$$

which is equivalent to Eq. (8.4), obtained for motion of a sphere neglecting gravity. The last two of Eqs. (8.40) are equivalent to the relations

$$- b_2 \frac{d^2a}{dt^2} + 3 \frac{da}{dt}\frac{db_2}{dt} + a \frac{d^2b_2}{dt^2} = 0$$

$$-2b_3 \frac{d^2a}{dt^2} + 3 \frac{da}{dt}\frac{db_3}{dt} + a \frac{d^2b_3}{dt^2} = 0$$

as can be verified by substituting the expressions for B_2 and B_3 in terms of b_2 and b_3 given by Eqs. (8.39).

Penney and Price have considered the solution of equations of this type, and they showed numerically in a special case that departures from sphericity represented by b_2 (deformation into a spheroid) were very much greater at or very near the minimum than anywhere else. This is, of course, in agreement with the observed behavior and there seems no reason to doubt that the instability of bubble contraction is perfectly consistent with noncompressive theory. These writers have also examined other cases of the first order perturbation theory, for which their paper should be consulted.

If only the first power of U is included, the second of Eqs. (8.40) gives Herring's formula for the bubble rise;

$$U = \frac{2g}{a^3} \int a^3 dt \tag{8.41}$$

as B_2 is of order U^2. Retention of U^2 terms in the first of Eqs. (8.40) gives

$$a \frac{d^2a}{dt^2} + \frac{3}{2}\left(\frac{da}{dt}\right)^2 - \frac{U^2}{4} = \frac{P_g}{\rho_o} - gz$$

which, together with Eq. (8.41) for U, is equivalent to the equations used by Taylor, as developed in section 8.5. The formula for the rise is, however, correct only for the first order in U, the term $(6/5g)\int (B_2/a^2)dt$ being neglected to obtain it.

A completely consistent theory, correct for terms of order U^2, should from the foregoing include departures from sphericity as determined by the coefficient B_2. Ward has computed the corresponding value of b_2 in the expansion of $R(t)$, using the result from Eq. (8.39) that

$$b_2 = \frac{3}{a^2} \int \frac{B_2}{a^2} dt$$

and computing B_2 by numerical integration from the third of Eqs. (8.40) with Comrie and Hartley's values for a, da/dt, etc., for $z_o' = 3.0$. The results of this approximate calculation show that b_2 remains small compared with a during the expansion of the bubble, but increases rapidly as the bubble contracts, becoming so large near the minimum (greater than a) that the whole calculation is invalid. This result, which is similar to that found by Penney and Price, thus again shows the inadequacy of assuming a spherical bubble near its minimum size, and indicates also the difficulty of obtaining better methods of approximation. The failure of theory is closely related to the existence of a large upward velocity at these times, and so should be least when the vertical motion is least: at great depths or close enough to surfaces that the upward effect of gravity is compensated.

8.8. Effects of Boundary Surfaces: The Method of Images

The mathematical developments which have so far been made were all based on the assumption that the gaseous explosion products moved in a medium of infinite extent, and the flow of water around the gas sphere was therefore restricted only by the condition that the velocity vanish sufficiently rapidly at infinity. In actual practice, however, the fluid medium is always limited in extent by the natural boundaries of the surface and bottom, and may also be limited by artificial boundaries such as the wall of a tank or the hull of a ship. The simplest cases approximating actual conditions are of an infinite rigid boundary (e.g., the sea bottom), to which the flow must be parallel, and a free surface (surface of the water), at all points on which the pressure is the same, it being supposed that any pressure differences which might otherwise develop are instantly relieved by displacement of the surface. Neither of these boundary conditions is satisfied by the results for radial and axial gravity flow around a spherical cavity in a fluid, and the motion of the cavity and surrounding water must therefore be appropriately modified to permit their being satisfied. The differences in the resulting motion will evidently be greater as the region of large flow velocities near the bubble surface approaches either the surface or the bottom. These boundaries will thus appear to exert appreciable forces on the bubble if they are near the bubble during its motion. The evaluation of such effects is most easily made by the use of the standard method of sources and images to be described.[7]

A. *Hydrodynamical sources.* The motion of the gas sphere boundary can be analyzed as the resultant of radial pulsation of the sphere superimposed on a translational velocity of its center. Considering first the radial motion, it is convenient to think of an outward radial flow of the surrounding water as being produced by a simple source of fluid at the center of the sphere, the strength M of this source being defined as the fraction $(1/4\pi)$ of the volume of fluid emitted in unit time.

[7] These methods are given in more detail in standard texts on hydrodynamics, for example the book by Milne-Thomson (74).

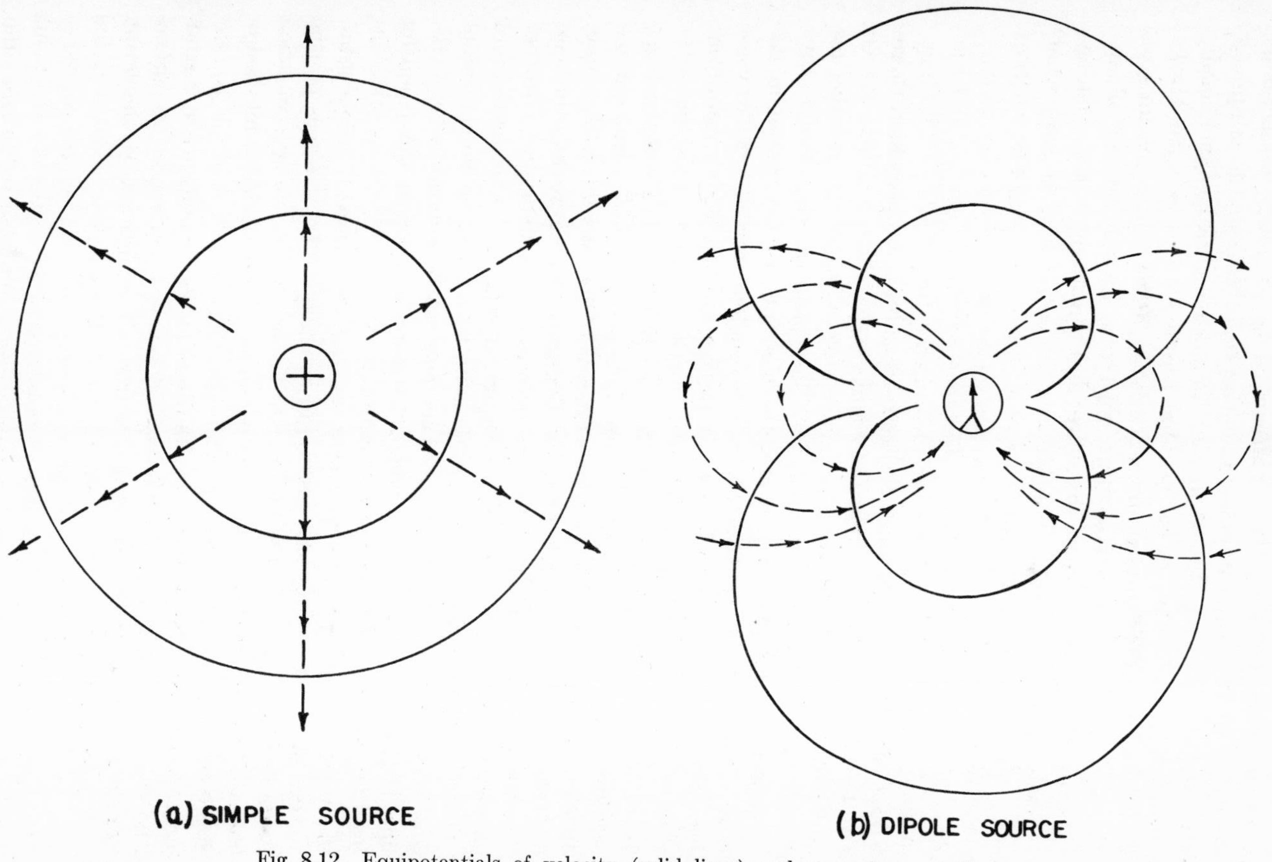

Fig. 8.12 Equipotentials of velocity (solid lines) and streamlines of flow (dashed lines) for simple and dipole sources.

An inward radial flow is similarly described by a simple sink. Such sources and sinks are of course purely conceptual devices without physical reality, and are introduced merely for mathematical convenience. If the spherical boundary of the flow has a radius a and velocity da/dt, the rate of flow across the boundary is $4\pi a^2\, da/dt$, as $4\pi a^2$ is the surface area. For incompressible flow, this must also be the rate at which fluid is emitted by a source at the center and the strength of the source is therefore $M = a^2\, da/dt$. The velocity potential $\varphi(r)$ at a point r from the center corresponding to this flow is, from Eq. (8.21), $\varphi(r) = a^2\, da/dt \cdot 1/r = M/r$. From this equation, the velocity potential is constant for a fixed value of distance r from the source, and the equipotential lines for which $\varphi =$ constant are thus circles. The flow of fluid is normal to these lines and the streamlines, across which there is no flow, extend radially from the source.

The equipotentials and streamlines shown in Fig. 8.12a are seen to be analogous to the equipotentials and lines of force about point electric charges or magnetic poles, and the concept of a hydrodynamical source may thus be used to analyze flow problems in the same way as corresponding electrical problems. A further analogy which can often be used to advantage is that of fluid flux with electric or magnetic flux. In problems with axial symmetry, this flux is expressed by the stream function ψ introduced by Stokes,[8] the value of which at a point not on the axis is defined to be $(1/2\pi)$ times the rate of flow of fluid through any surface bounded by a circle around the axis of symmetry and passing through the point. Its value is thus determined by the velocity distribution, and constant values of the function represent the streamlines (actually surfaces) across which there is no flow. Hence the condition $\psi =$ constant must be satisfied at geometrical boundaries where the flow is parallel to the surface and the stream function is conveniently used in applying such boundary conditions.

A second type of source which proves useful is formed by the combination of a source and sink of equal strength. The strength of the combination is taken to be the product of the source strength and distance between the two. If the separation is reduced to zero while keeping this product unchanged, the combination is called a dipole source analogous to an electrical dipole, its direction being defined as that of the line from the sink to the source. The velocity potential of such a dipole at any point is easily computed from its definition with the result

$$\varphi(r, \theta) = \frac{\mu \cos \theta}{r^2}$$

where θ is the angle which the radius vector r to the point from the dipole makes with its axis.

[8] See, for example, Milne-Thomson (74), Chapter XV.

The potential required to describe the flow of fluid around a sphere which has a translational velocity U is, from Eq. (8.21), given by

$$\varphi(r, \theta) = \tfrac{1}{2}a^3U \frac{\cos\theta}{r^2}$$

and hence this flow can be described by a dipole source of strength $\mu = (\frac{1}{2})a^3U$. The equipotentials are given by $\cos\theta = \text{const} \times r^2$, and the streamlines obtained from the Stokes stream function by $\sin\theta = \text{const} \times r$; these lines are plotted in Fig. 8.12b. It is seen that the dipole source acts like a small tube through which the flow is concentrated. The drawing also shows that the boundary condition for flow around a sphere is satisfied if it is remembered that the dipole must move with the velocity of the sphere.

B. *The method of images applied to a rigid boundary.* The usefulness of the concepts of simple and dipole sources lies in the fact that restrictions on the flow of fluid imposed by geometrical boundaries can be satisfied by superimposing the flows from suitable combinations of these sources placed at points inaccessible to the fluid. The velocity potentials of each such flow, and hence their sum, satisfy Laplace's equation, and from standard theorems of potential theory the resulting potential is the solution of the hydrodynamical problem. Knowing the velocity potential permits calculation of the velocity and pressure distribution, and the equations of motion can therefore be obtained.

The simplest boundary condition is that of an infinite rigid plane, representing the sea bottom or a confining wall, which requires that the flow adjacent to the plane be parallel to it, the corresponding condition on the potential being that at the plane $\partial\varphi/\partial n = 0$, where n is a normal to the plane. A pulsating sphere in an infinite medium is equivalent to a point source at its center. If a second like source is placed at a distance $2b$ from the first it is evident that the combined flow at a plane perpendicular to the line joining the sources and halfway between them will by symmetry be parallel to the plane.

This is, however, precisely the flow distribution required at a rigid boundary a distance b from the radially pulsating sphere. This restriction can thus be analyzed by replacing the boundary plane with a source located at the position of an optical mirror image of the source, i.e., at a distance b behind the plane as shown in Fig. 8.13. The flow from two simple fixed sources cannot, however, be the true one because the second image source produces a flow through the boundary of the sphere. If the sphere is small compared to its distance from the boundary, this flow is nearly uniform at all points on its surface, and the flow distribution would be nearly that around a sphere moving away from or toward the wall as it expands or contracts. It is thus seen that displacements of a pulsating sphere can maintain the required flow distributions at a rigid wall, and it is plausible that the sphere should move in this way as if the wall exerted a force on it directly. The addition of such a displacement is not, however, a true solution of the boundary condition at the sphere except in the limit of negligible size, but the required flow at this boundary can be restored by placing additional sources at suitable points inside the sphere. As this region is in fact free of fluid, fictitious sources

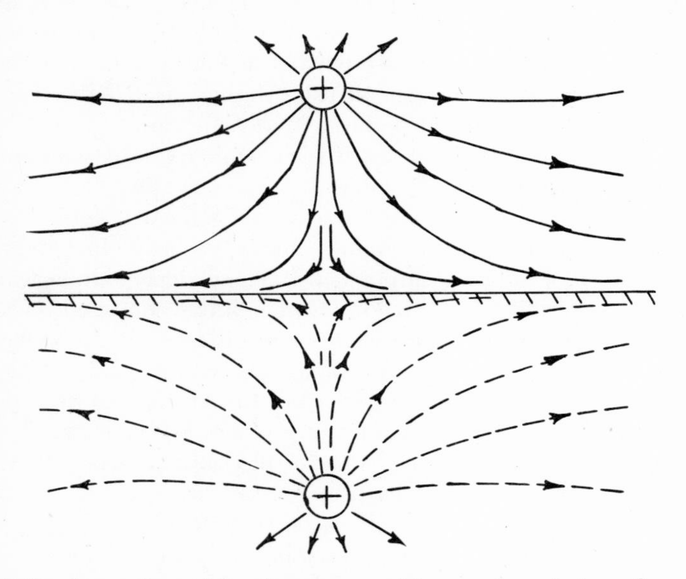

Fig. 8.13 Streamlines of flow for source and image of equal strength.

of this kind are permissible if the corresponding velocity potentials satisfy Laplace's equation at all points in the fluid.

The required sources to correct the flow at the sphere can be shown to be a point source and a continuous uniform distribution of negative sources known as a line sink, the disposition of which in the sphere is shown in Fig. 8.14. The point C at which the source is located and the line sink from the center O ends is called the inverse point to B and is defined geometrically by the relation $OB \times OC = a^2$, where a is the radius of the sphere. The required strength of the source C is $(a/OB)M$, if M is the strength of the original images at B, and the distributed line source has a strength M/a per unit length.[9] It will be observed that

[9] A proof that this distribution restores the required boundary condition at the sphere will be found in the reference of footnote 8, and only a sketch of the method is given here. The proof is most easily made in terms of the Stokes stream function Ψ, which must be constant over the sphere. The value of Ψ at a point P for a point source M at point O is easily shown to be $\Psi = M \cos \theta$, where θ is the angle between

the total strength of the combination is zero. There is therefore no net flow introduced by these sources, their effect being merely to give a redistribution of the flow pattern.

The combination of sources depicted in Fig. 8.14 thus restores the boundary conditions at the sphere, but it is evidently done at the expense of again violating the condition on the plane. It might thus appear that the complications of the method have succeeded only in restoring the problem to its original status. This, however, is not true

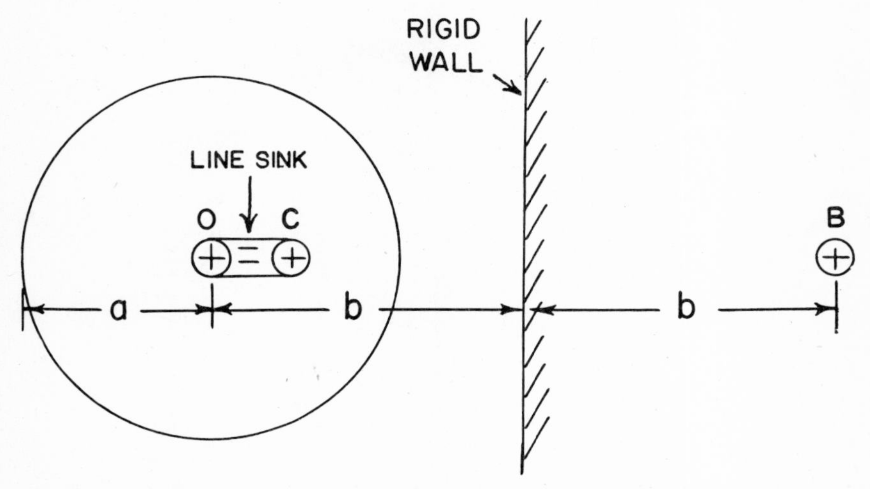

Fig. 8.14 Distribution of simple sources for a sphere of finite radius and infinite rigid wall.

because the flow pattern obtained at each stage is a better approximation that the preceding one. The method of images in this particular problem is thus one of successive approximations, as is indicated by the fact than the strengths of the sources added inside the sphere in Fig. 8.14 are decreased in strength relative to the initial ones by factors of the order $(a/2b)$. The radius a of the sphere is always less than its distance b from the boundary if the method is to apply at all, and these sources are therefore weaker as well as contributing no net flow. Successive additions of image sources to remedy boundary conditions at the plane and sphere in turn can be carried out to increase the accuracy of the approximation, the process being a convergent one, and it is evident that the number of repetitions needed will increase as the ratio $a/2b$ increases.

---

the vector OP and the axis of symmetry. Integration for a line source of strength M' per unit length gives the value $\Psi' = M' (r_2 - r_1)$ at P, r_2 and r_1 being the distances from the extremities of the line source to P. Adding the functions Ψ and Ψ' for the sources specified gives a constant value of $(\Psi + \Psi')$ on the sphere of radius a, which is therefore a streamline as required.

(a) Pulsation

In sphere			Behind plane		
Kind	Position	Strength	Kind	Position	Strength
Point	O	m	Point	$2b$	m
Point	$a(a/2b)$	$m(a/2b)$	Point	$2b - a(a/2b)$	$m(a/2b)$
Line	O to $a(a/2b)$	$-m/a$	Line	$2b$ to $2b - a(a/2b)$	$-ma$
Point	$a \dfrac{a}{2b - a(a/2b)}$	$m(a/2b) \dfrac{a}{2b - a(a/2b)}$		etc.	
Line	$a(a/2b)$ to $a \dfrac{a}{2b - a(a/2b)}$	$-(m/a)\ (2b/a)$			
	etc.				

(b) Translation

In sphere			Behind plane		
Kind	Position	Strength	Kind	Position	Strength
Dipole	O	μ	Dipole	$2b$	$-\mu$
Dipole	$a(a/2b)$	$\mu(a/2b)^3$	Dipole	$2b - a(a/2b)$	$-\mu(a/2b)^3$
Dipole	$a \dfrac{a}{2b - a(a/2b)}$	$\mu(a/2b)^3 \left(\dfrac{a}{2b - a(a/2b)}\right)^3$		etc.	
	etc.				

Table 8.4. Location and strength of images for a moving sphere and rigid plane.

The calculation of the position and strength of the successive images has been carried out by Shiffman and Friedman (102) by essentially the methods outlined above, and their results therefore permit evaluation of the velocity potential in the fluid to any desired degree of approximation. The first stage of the process, restoration at the plane boundary, evidently is accomplished by mirror images behind the plane of the sources added in the sphere, these images being matched by corresponding ones at the inverse points in the sphere, and so on. The successive stages obtained by Shiffman and Friedman in this way are indicated in Table 8.4.

With this known distribution of sources, the velocity potential at any point in the fluid can be computed. For a point source, the value of φ is given by $\varphi = M/r$ where r is the distance from the source, and the value of φ for a line source is obtained by integrating the potential $d\varphi = M'dx/r$ for an element over the length of the source. The general expression for φ becomes increasingly complicated as successive images are included. Fortunately for simplicity of calculation, this general result is not needed to determine the equation of motion of the sphere. This is because the kinetic energy of the system can be determined from the value of φ on the boundaries only (see section 8.4).

These calculations, described in section 8.8, must, however, be preceded by a determination of the system of images for a sphere moving in translation in order to take account of the migration during its radial pulsation. If the plane boundary were not present, this motion would be described by a dipole source μ at the center of the sphere. The image of this source giving tangential flow at the plane is evidently an equal but oppositely directed dipole at the image point an equal distance behind the plane. The violation of the flow condition on the sphere is rectified by adding another dipole inside the sphere at the inverse point C_1 in Fig. 8.15 and of strength $\mu(a/2b)^3$, as is easily shown from the stream function. Successive reflections thus lead to a series of oppositely directed dipoles of decreasing strength, as indicated in Table 8.4. The corresponding velocity potentials at points in the fluid are then given by $\varphi = \mu \cos\theta/r^2$, where θ is the angle the radius vector from the dipole to the point makes with the symmetry axis. It will be noted from Table 8.4 that the distances of the successive image points from the center of the spheres can be expressed by recurrence formulas as

$$b_{n+1} = 2b - \frac{a^2}{b_n}, \quad b_1 = 2b$$

$$c_{n+1} = a\frac{a}{2b - c_n}, \quad c_1 = a\left(\frac{a}{2b}\right)$$

C. *Approximate solutions for a free surface and rigid bottom.* The boundary condition to be satisfied at a free surface is that the pressure be the same at all points on the surface. From section 8.4, the pressure at any point in a fluid is given by Bernoulli's equation

$$\frac{P}{\rho_o} + gz = \frac{\partial \varphi}{\partial t} - \tfrac{1}{2}(\text{grad } \varphi)^2$$

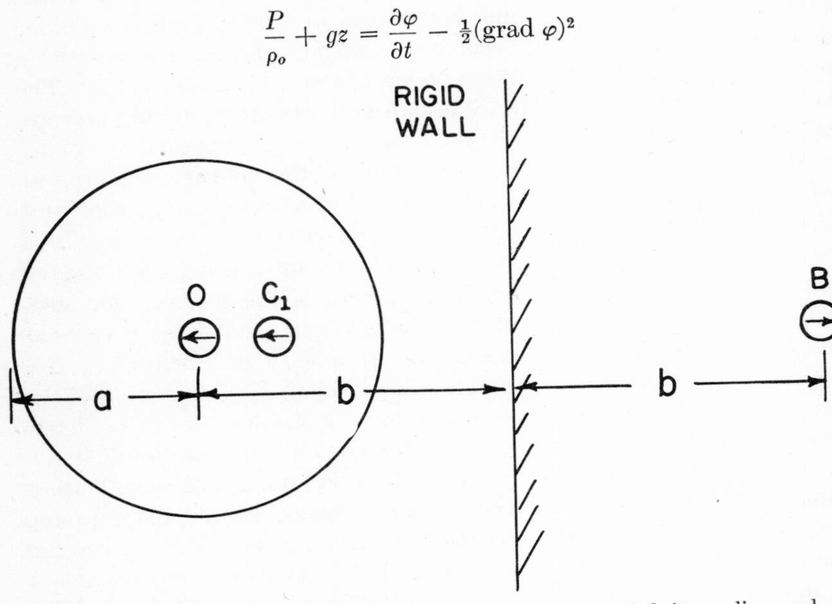

Fig. 8.15 Distribution of dipole sources for a sphere of finite radius and infinite rigid wall.

where z is the depth in feet of water which is equivalent to the hydrostatic pressure. At a free surface, $P = \rho_o gz$ everywhere and the appropriate boundary condition is therefore

$$\frac{\partial \varphi}{\partial t} - \tfrac{1}{2}(\text{grad } \varphi)^2 = 0 \tag{8.42}$$

The velocity potential in the fluid is, however, of order $1/r$, where r is the distance from the bubble center and, for sufficiently large distance d to the true surface, $(\text{grad } \varphi)^2$ is of order $1/d^4$, and hence of order $1/d^3$ compared to $\partial\varphi/\partial t$. Hence, unless the bubble is very near the surface, the boundary condition on φ at the surface can be taken to be $\varphi = 0$. This condition is equivalent to requiring that the streamlines be perpendicular to the surface, and the appropriate image of a source below the surface is a like source of equal strength at the same distance above the surface.

It is interesting to note the electrostatic analogy to the hydrodynamical situation. If sources are replaced by electrical charges, the velocity potential has its counterpart in the electrostatic potential, and streamlines of flow correspond to electric lines of force. The case of a free surface in hydrodynamics is easily seen to be equivalent to a perfectly conducting plane to which lines of force must be perpendicular, and the image of a charge in a conducting plane is a charge of opposite sign at the image point in the plane. A similar but less realistic analogy can be drawn for a rigid surface: the corresponding electrostatic surface is a boundary separating free space from a hypothetical medium of dielectric constant zero, this condition being the one for which lines of force are parallel to the surface, and the image is a charge of the same sign.

The boundary condition $\varphi = 0$ at a free surface is clearly only an approximation to the true condition expressed by Eq. (8.42) and application of the method of images to satisfy $\varphi = 0$ involves the further difficulty that, strictly, the boundary surface cannot remain fixed or even plane. This is evident if it is recalled that the streamlines ending on the surface are lines along which fluid flows. Any motion of fluid will thus displace its boundary, in general by unequal amounts at different points, and the result is formation of surface waves (this is briefly considered in section 10.1). This effect will clearly be greater when any sources required are relatively closer to the surface, and use of the condition $\varphi = 0$ at the undisturbed surface is a poorer approximation. Application of more exact boundary conditions would, however, be a matter of considerable difficulty, and as a result most developments including the effect of a free surface content themselves with the simpler condition. If these developments were to be carried out to the same degree as the treatment for a sphere and rigid boundary, a similar set of images reflected in the sphere and plane (except of course for signs) would be required, but the approximations inherent in satisfying $\varphi = 0$ at a fixed boundary do not justify such elaboration. The development for a rigid boundary does, however, permit inferring the errors resulting from neglect of the higher order images required for a sphere of finite rather than negligible radius, and this question is discussed in section 8.10.

The representation of the free surface boundary condition by a single image source is so simple that it is convenient to work out to the same approximation the more general case of a bubble pulsating at a distance b above the bottom, and distance d below the free surface, as shown in Fig. 8.16. The images of the original source O in these two surfaces must in turn have images in the other surfaces, which again must have images, and so on. As a result an infinite set of image sources is required, the first few of which are shown in Fig. 8.17.[10] This necessity for successive images indicates that surface and bottom effects are not simply additions.

The distances of the successive images from the original source are easily seen to be

$$(OP)_n = n(d + b) \quad \text{if } n \text{ is even}$$

$$= (n - 1)(d + b) + 2b \quad \text{if } n \text{ is odd}$$

$$(OQ)_n = n(d + b) \quad \text{if } n \text{ is even}$$

$$= (n - 1)(d + b) + 2d \quad \text{if } n \text{ is odd}$$

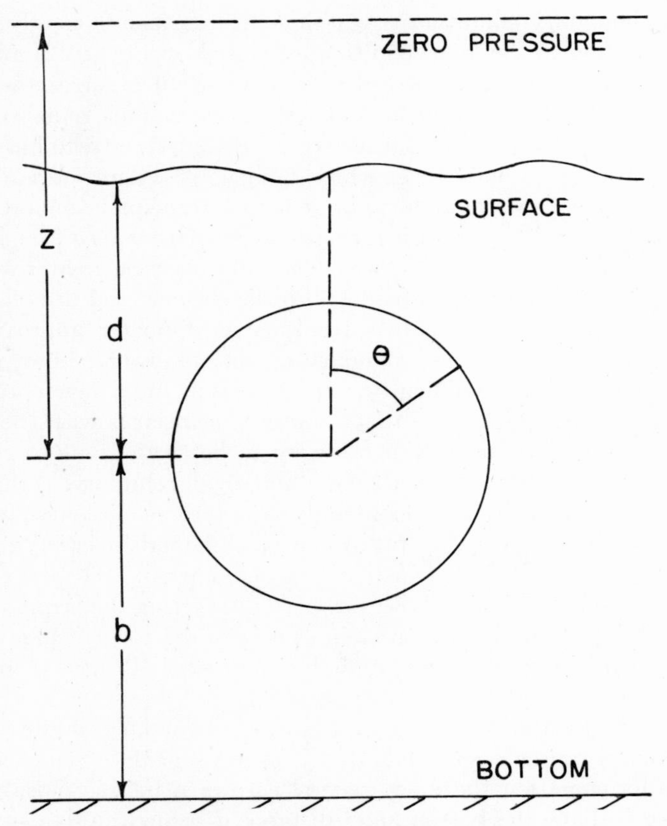

Fig. 8.16 Geometry for a sphere pulsating between the surface and bottom.

[10] The infinite set of images required for this case were first given by Kennard (55); the effect on bubble motion has been further developed by Shiffman and Friedman (102) and by Friedman (37).

The correction to the velocity potential function at any point is then obtained by adding the potentials of the image sources, and for evaluation of the kinetic energy it suffices merely to know the mean values at the original source. Since the source is, in the present approximation, a point, only the values at this point are required. Summing the values of $1/OP_n$, $1/OQ_n$ with proper regard to sign gives for the correction $\Delta\varphi_a$, remembering that the original source has a strength a^2,

$$\Delta\varphi_a = \frac{a^2}{d+b}\left[2xF(x) - \log_e 2\right] \tag{8.43}$$

where

$$x = \frac{d-b}{d+b}$$

$$F(x) = \sum_{n=0}^{\infty} \frac{1}{(2n+1)^2 - x^2}$$

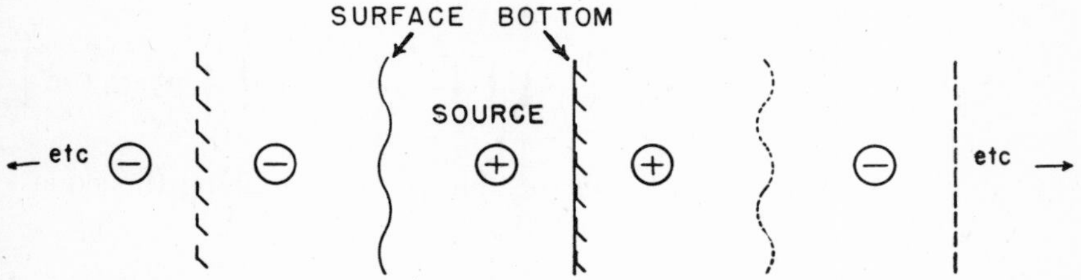

Fig. 8.17 System of images for a pulsating sphere between a rigid bottom and free surface.

8.9. The Equations of Motion for a Sphere and Bounding Surfaces

In the last section, the distributions of sources which represent the flow of a fluid between a moving sphere and infinite boundaries were obtained. In order to determine the motion of the sphere, it is next necessary to write the dynamical equations of the motion, and this is most conveniently done by use of the energy principle. The kinetic energy T of the fluid is, from section 8.4, given by

$$T = -\frac{\rho_0}{2}\iint \varphi \frac{\partial\varphi}{\partial n}\, dS \tag{8.44}$$

where the surface integral is to be extended over all boundary surfaces. The velocity potential φ is calculated from the source distributions for pulsation and translation of the sphere:

$$\varphi = \frac{da}{dt}\varphi_a - U\varphi_z$$

where φ_a and φ_z are the velocity potentials for unit velocity of translation and pulsation.

A. *Rigid surface.* The normal derivatives $\partial\varphi/\partial n$ must satisfy boundary conditions at the sphere and a rigid plane as tabulated below (see section 8.8)

	Sphere	*Plane*
Pulsation	$-\dfrac{\partial \varphi_a}{\partial n} = \dfrac{da}{dt}$	$-\dfrac{\partial \varphi_a}{\partial n} = 0$
Translation	$-\dfrac{\partial \varphi_z}{\partial n} = U \cos \theta$	$-\dfrac{\partial \varphi}{\partial n} = 0$

When these conditions are applied to Eq. (8.44), the integrals over the plane vanish, leaving

(8.45)

$$T = \frac{\rho_0}{2}\left[\left(\frac{da}{dt}\right)^2 \iint \varphi_a dS + 2U\left(\frac{da}{dt}\right)\iint \varphi_z dS + U^2 \iint \varphi_z \cos\theta dS\right]$$

the integrals to be evaluated over the sphere. In obtaining this equation, use has been made of the fact that

$$\iint \left(\varphi_a \frac{\partial \varphi_z}{\partial n} - \varphi_z \frac{\partial \varphi_a}{\partial n}\right) dS = 0$$

for any functions φ_a, φ_z satisfying Laplace's equation. The calculation of the total kinetic energy thus requires only a knowledge of the mean values of φ_a, φ_z and $\varphi_z \cos\theta$ over the sphere. Considering first the potential φ_a of radial motion, which is represented by combinations of point sources, we can easily show that the integral for a source of unit strength at any point inside the sphere is simply $4\pi a$ and for a source outside the sphere it is $4\pi a^2/b_n$, where b_n is the distance of the source from the center.

These results are simple consequences of mean value theorems (Gauss' theorem) for potential functions and can also be demonstrated by direct integration. Let the source be at a distance c_n from the center and on the axis ($\theta = 0$). Then the distance r to the ring element of length $2\pi a \sin\theta\, d\theta$ and width $ad\theta$ is given by $r^2 = a^2 + c_n^2 - 2ac_n \cos\theta$ and the desired integral for $\varphi_a = m/r$ is

$$\iint \varphi_a dS = m \int_0^{2\pi} \frac{2\pi a^2 \sin\theta d\theta}{r} = \frac{\pi m a}{c_n} \int_{r=a-c_n}^{r=a+c_n} \frac{d(r^2)}{r} = 4\pi m a$$

For a point outside the sphere at b_n, we have the same integral except for replacing c_n by b_n and changing the limits:

$$\iint \varphi_a dS = \frac{\pi m a}{b_n} \int_{r = b_n - a}^{r = b_n + a} \frac{d(r^2)}{r} = 4\pi m a \left(\frac{a}{b_n}\right)$$

The integral for a source inside the sphere is thus independent of its position, and hence line sources contribute to the same extent as a point source of equal strength. As remarked in section 8.8, each of the negative line sources introduced in the sphere is exactly cancelled by a point source and their net contribution is zero, leaving only the original source at the center. Putting $m = a^2$ gives $4\pi a^3$ for this term. This cancellation does not occur for the images behind the plane and the expressions for the line sinks and sources in this region can be obtained by direct integration and addition. Using the notation of Shiffman and Friedman (102), the result can be expressed

(8.46)

$$\iint \varphi_a dS = 4\pi a^3 + 4\pi a^3 \left[\frac{a}{2b} + \sum_{n=1}^{\infty} \left(D_{n+1} - \frac{1}{D_{n-1}} \log \frac{d_{n+1}}{d_n}\right)\right]$$

where the D_n and d_n are defined by the recurrence formulas

(8.47)

$$d_{n+1} = \frac{a}{2b - a d_n}, \quad d_1 = \frac{a}{2b}$$

$$D_n = \prod_{i=1}^{n} d_i$$

The first term comes from the original point source, the first term in brackets from its image in the plane; the D_n's are from successive point sources behind the plane, and the logarithms from the corresponding negative line sources.

The second and third integrals in Eq. (8.45) involving the dipole potential φ_z require a knowledge of the potential at the sphere directed along it. For a dipole μ located a distance s from the center of the sphere, the velocity potential on the sphere is

$$\varphi_z = \frac{\mu \cos \theta}{r^2} = \mu \frac{\cos \theta}{a^2 + s^2 - 2as \cos \theta}$$

where θ is the polar angle of the radius a with the axis of symmetry. The direct integration of this function does not lead to simple results and it is more convenient to use a series expansion in powers of s/a. For image points inside the sphere, s is one of the lengths c_n which are less than a, and an expansion in positive powers of c_n/a gives

$$\varphi_z = \frac{1}{a^2}\left[\cos\theta + \frac{2c_n}{a} P_1(\cos\theta) + \cdots + i\left(\frac{c_n}{a}\right)^i P_i(\cos\theta) + \cdots\right]$$

where the $P_i(\cos\theta)$ are the Legendre polynomials and c_n is positive for points above the origin. For image points behind the plane, s is one of the lengths b_n ($b_n > a$), and an expansion in powers of a/b_n gives

$$\varphi_z = -\frac{1}{a^2}\left[1 + 2\frac{a}{b_n} P_1(\cos\theta) + \cdots + i\left(\frac{a}{b_n}\right)^i P_i(\cos\theta) + \cdots\right]$$

Terms of degree higher than the first in powers of (c_n/a) or (a/b_n) are neglected for simplicity in applications which have been made of expansions in zonal harmonics, the justification being that these terms involve increasingly higher powers of the fraction $a/2b$, which is always less than ½. Integration of φ and $\varphi\cos\theta$ over the sphere for the series of dipoles then gives

$$(8.48)\qquad -\iint \varphi_z\, dS = 2\pi a^3\left[\sum_{n=0}^{\infty} D_n{}^3 d^2{}_{n+1}\right]$$

$$\iint \varphi_z \cos\theta dS = \frac{2\pi}{3} a^3 + \frac{2\pi}{3} a^3\left[\sum_{n=1}^{\infty} D_n{}^3\right]$$

where the D_n, d_n are defined by Eq. (8.47).

On substituting these expressions and Eq. (8.46), the expression (8.45) for the kinetic energy may be written

$$T = 2\pi\rho_o a^3(1 + f_o)\left(\frac{da}{dt}\right)^2 - 4\pi\rho_o a^3 f_1 U\left(\frac{da}{dt}\right) + \frac{\pi}{3}\rho_o a^3(1 + 3f_2)U^2$$

where the quantities f_o, f_1, f_2 are the bracketed series in Eqs. (8.46), (8.48). The equation of energy including the potential energies of hydrostatic buoyancy and internal energy $E(a)$ of the bubble is then

$$(8.49)\qquad 2\pi\rho_o a^3(1 + f_o)\left(\frac{da}{dt}\right)^2 - 4\pi\rho_o a^3 f_1 U\left(\frac{da}{dt}\right) + \frac{\pi}{3}\rho_o a^3(1 + 3f_2)U^2 + \frac{4\pi}{3}\rho_o a^3 gz = Y - E(a)$$

where z is the distance of the center below the surface of zero hydrostatic pressure and U is the upward velocity, from the bottom and toward the surface.

Comparing this result with Eq. (8.23) for motion under gravity alone, it is seen that the effect of the rigid boundary is represented by the quantities f_o, f_1, f_2, which are functions of the ratio $a/2b$ (b = distance from bubble center to the wall) and vanish for $a/2b = 0$, corresponding to negligible influence of the wall on the flow.

Energy considerations thus provide one of the two equations necessary to determine the radius and displacement of the bubble with time. In the absence of boundary effects, a second equation was obtained by momentum considerations (see Eq. (8.25) and the discussion following it). While the same approach could be used in the present more general case, it is simpler to use the more powerful method of the LaGrange formulation of the equations of motion in terms of the kinetic and potential energies of the system, employing the radius a and displacement b as generalized coordinates. The validity of this general method of mechanics for the motion of a sphere in a fluid is discussed in Lamb's Hydrodynamics (65), and is assumed here without proof.

If gravity is the only force of origin external to the sphere and V is the potential energy of hydrostatic buoyancy plus internal energy of the gas, the LaGrange function $\mathfrak{L} = T - V$ satisfies the differential equations

$$\frac{d}{dt}\left(\frac{\partial \mathfrak{L}}{\partial \dot{a}}\right) - \frac{\partial \mathfrak{L}}{\partial a} = 0$$

$$\frac{d}{dt}\left(\frac{\partial \mathfrak{L}}{\partial U}\right) - \frac{\partial \mathfrak{L}}{\partial b} = 0$$

where $\dot{a} = da/dt$, $U = db/dt$. Either of these equations may be used to give the second relation in addition to the energy integral, Eq. (8.49). The second of the two proves more convenient and gives directly the equation corresponding to the translational momentum under influence of gravity alone (Eq. (8.25)). Substitution and carrying out the differentiations gives, letting $\alpha = a/2b$,

(8.50)

$$\frac{d}{dt}\left[\frac{1}{3}a^3(1 + 3f_2)U - 2a^3f_1\left(\frac{da}{dt}\right)\right] = -\frac{a^4}{2b^2}\left[\frac{df_o}{d\alpha}\left(\frac{da}{dt}\right)^2 + \frac{1}{2}\frac{df_2}{d\alpha}U^2 - 2\frac{df_1}{d\alpha}U\frac{da}{dt}\right] + \frac{2}{3}ga^3$$

The two differential equations, (8.49) and (8.50), are more conveniently expressed in nondimensional, or reduced, variables for length and time. The variables employed by Taylor in his analysis of bubble motion under gravity could be employed, but these are not as convenient in the present problem where the position of the bottom is also a factor. Shiffman and Friedman[11] have therefore adopted somewhat different characteristic length and time units $L,^*$ C^* to effect the transformation to dimensionless variables which are defined by the relations

$$L^* = \left(\frac{3Y}{4\pi P_o}\right)^{1/3}, \quad C^* = L^*\left(\frac{3\rho_o}{2P_o}\right)^{1/2} \tag{8.51}$$

where P_o is the initial hydrostatic pressure at the depth of explosion in the differential equations. Reduced variables (denoted by asterisks) are then expressed in terms of these scaling factors by

$$a = L^*a^*, \quad b = L^*b^*, \quad t = C^*t^*$$

and initial values of hydrostatic depth z_o (corresponding to pressure P_o) and charge position b_o above the bottom by

$$z_o = L^*z_o^*, \quad b_o = L^*b_o^*$$

The scaling factors L^* and C^* so chosen are seen on comparison with the results of section 8.2 to be simply the maximum radius and two-thirds the period of oscillation obtained neglecting the internal energy and external influences of gravity and boundaries. They thus should be expected to have a close connection with these simple properties in the less approximate solutions, as proves to be the case.

Using the reduced variables, Eq. (8.49) and (8.50) become

$$a^{*3}\left[(1 + f_o)\left(\frac{da^*}{dt^*}\right)^2 + \frac{1}{6}(1 + 3f_2)\left(\frac{db^*}{dt^*}\right)^2 - 2f_1\left(\frac{da^*}{dt^*}\right)\left(\frac{db^*}{dt^*}\right)\right] = 1 - \frac{E(a^*)}{Y} - \frac{z_o^* + b_o^* - b^*}{z_o^*}a^{*3} \tag{8.52}$$

$$\frac{d}{dt^*}\left[\frac{1}{3}a^{*3}(1 + 6f_2)\frac{db^*}{dt^*} - 2a^{*3}f_1\frac{da^*}{dt^*}\right] = -\frac{a^{*4}}{2b^{*2}}\left[\frac{df_o}{d\alpha}\left(\frac{da^*}{dt^*}\right)^2 + \frac{1}{2}\frac{df_2}{d\alpha}\left(\frac{db^*}{dt^*}\right)^2 - 2\frac{df_1}{d\alpha}\left(\frac{da^*}{dt^*}\right)\left(\frac{db^*}{dt^*}\right)\right] + \frac{a^{*3}}{z_o^*} \tag{8.53}$$

[11] The notation employed here differs from that of Shiffman and Friedman in order to reduce the danger of confusion with other parts of the text. It is unfortunate that the usages and conventions of various authors are conflicting and not alway internally consistent, a difficulty which must be kept in mind when comparing different results.

The internal energy term $E(a^*)/Y$ depends on the hydrostatic pressure at the depth of explosion. If the equation of state per gram of explosive is taken to be of the form $PV^\gamma = k$ in cgs. units, the internal energy $E(a)$ in ergs for W pounds of explosive is

$$E(a) = 454W \int_v^\infty PdV = 454W \frac{PV}{\gamma - 1} = \frac{454WkV^{1-\gamma}}{\gamma - 1}$$

Expressing this in terms of the radius a in feet by the relation

$$454WV = \frac{4\pi}{3} (30.5a)^3 \qquad \text{gives}$$

$$E(a) = \frac{454k}{\gamma - 1} (3.81 \times 10^{-3})^{\gamma - 1} \frac{W^\gamma}{a^{3(\gamma-1)}}$$

The available energy Y in ergs for W pounds of explosive can be written as a fraction r of the detonation energy Q in calories per gram of explosive as

$$Y = (4.19 \times 10^7)\, rQ\, (454W)$$

and the desired fraction is therefore

$$\frac{E(a)}{Y} = \frac{(3.81 \times 10^{-3})^{\gamma-1}}{4.19 \times 10^7} \cdot \frac{k}{\gamma - 1} \cdot \frac{W^{\gamma-1} a^{-3(\gamma-1)}}{rQ}$$

To express this in terms of the nondimensional radius a^*, it is necessary to evaluate the scale factor L^* in feet, which is

$$L^*(\text{ft.}) = \left(\frac{3Y}{4\pi\rho_o g z_o}\right)^{1/3} = \left(5.17 \frac{rQW}{z_o}\right)^{1/3} \tag{8.54}$$

if z_o is in feet. The time scale factor C^* is not needed here, but is given for completeness; its value is

$$C^* = L^* \left(\frac{3}{2gz_o}\right)^{1/2} = 0.216\, L^* z_o^{-1/2} \tag{8.55}$$

where C^* is in seconds, L^* and z_o in feet. Using the scale factor gives for the internal energy term

$$\frac{E(a^*)}{Y} = k^*a^{*-3(\gamma-1)} \tag{8.56}$$

$$\text{where} \quad k^* = \frac{k(rQ)^{-\gamma}(7.36 \times 10^{-4})^{\gamma-1} z_o{}^{\gamma-1}}{(\gamma - 1)\cdot 4.19\cdot 10^7}$$

For TNT, Shiffman and Friedman have used the same constants as Taylor, namely $rQ = 440$ cal./gm., $k = 7.83 \times 10^9$, $\gamma = 1.25$. For this special case then, we have

$$\frac{E(a^*)}{Y} = k^*a^{*-3/4}, \quad k^* = 0.061\, z_o{}^{1/4} \tag{8.57}$$

where z_o is in feet. The expression $k^*a^{*-3/4}$ applies strictly only for the exponent $\gamma = 1.25$, but has been used by Shiffman and Friedman in their calculations and will be employed here. Fortunately, many of the calculations are not very sensitive to this term, and probably could be adapted to approximating other adiabatic laws than the one chosen by using suitable values of the parameter k^*.

B. *Infinite free surface and rigid bottom.* The first approximation to the combined effect of surface and bottom, neglecting the radius of the bubble in comparison with its distance from the boundary, gave a correction $\Delta\varphi_a$ expressed by Eq. (8.43) to the velocity potential φ_a. When this correction term is included, the quantity f_o appearing in the kinetic energy is easily shown to be

$$f_o = \frac{a}{d + b_o}\,[2xf(x) - \log_e 2]$$

$$\text{where} \quad x = \frac{d - b_o}{d + b_o}$$

$$f(x) = \sum_{n=0}^{\infty} \frac{1}{(2n+1)^2 - x^2}$$

For a free surface only, b_o becomes infinite, and the series $f(x)$ breaks down. For this case, however, the image is at a distance $2d$, and the value of f_o is readily seen to be merely $-a/2d$. It is interesting to note that either a free or rigid surface alone has, to a first approximation, the effect only of changing the kinetic energy of radial motion: for a free surface, there is a fractional decrease of amount $-a/2d$; and for a rigid surface there is a fractional increase of amount $+a/2b_o$. As Herring has shown, this result can be very simply obtained from the images required in the analogous electrostatic problem by calculating the work required in bringing the charge (source) and its image to their actual separation from infinite distance. This work is then the increase in kinetic energy, which is positive for like charges (rigid surface) repelling each other and negative for unlike, attracting charges (free surface). Explicit calculations using the image force $\pm e^2/4x^2$, where x is the separation, leads directly to the results given.

The hydrodynamical reason for the changes in kinetic energy is simply that for a free surface there is less inertia of surrounding fluid than in an infinite medium, while a rigid boundary interferes with the flow and increases its inertia. As Herring (46) points out, this explanation also makes obvious the observed changes in period. The effect of either surface is to change the inertia (or kinetic energy) of the oscillating system, without affecting its spring constant (or potential energy). The period is proportional to the square root of the inertia, as in the formula for a mass oscillating on a spring, and the percentage change in period is therefore approximately one-half the change in kinetic energy. Hence we might expect the modified periods T_s, T_b to be given by

$$T_s = T\left(1 - \frac{a_{\text{av}}}{4d_o}\right)$$

$$T_b = T\left(1 + \frac{a_{\text{av}}}{4b_o}\right)$$

where T is the free water period, and a_{av} is the time average of a over one complete cycle. (The justification for using a_{av} as defined is not obvious, but a detailed argument, such as Herring has given, is omitted, as more exact derivations are given in section 8.10.) The decreased inertia for a free surface thus shortens the period and a rigid surface causes an increase.

The presence of rigid and free surfaces also causes displacements of the bubble center as it pulsates, and these changes can be qualitatively understood in terms of the change in inertia of the flow. In the case of a rigid surface, the presence of the boundary interferes with radial flow of water, whether outward or inward, near a spherical surface in its vicinity. Initially, when the pressure in the gas is in excess of the hydrostatic pressure, the water on the side of the bubble surface near the wall is less readily displaced, and the bubble surface moves away from the wall. The effect is relatively small, however, because the net pressure (in excess of hydrostatic) is positive for a short part of the bubble period, and the bubble is small during this time. When the pressure falls below hydrostatic, acceleration of flow toward the bubble surface does not occur as readily on the side toward the wall, and the flow must be such as to bring the surface nearer to the wall. A considerable amount of momentum is imparted to a large mass of water in this way when the bubble is large. As the bubble contracts, the momentum acquired becomes concentrated in a smaller mass of water near the bubble, and the velocity of flow in this region increases. The bubble surface must then move toward the wall with increasing speed as if attracted to it. This effect is so much larger than the repulsion when the pressure exceeds hydrostatic that the dominant motion is an apparent attraction increasing the bubble velocity toward the wall as it contracts, even though the momentum of the flow is decreasing in the most contracted stages.

A free surface has the opposite effect on bubble migration, as in this case the water at the surface is free to move but must do so in such a way as to equalize the pressure with that of the atmosphere. As the bubble begins to expand, the water above it has less inertia and is displaced more readily than that below, and the bubble surface moves upward. When the gas pressure falls below hydrostatic, however, a downward flow of water takes place, because the water near the free surface is more easily accelerated toward the bubble. Just as for the rigid surface, the large amount of momentum acquired while the bubble is large is concentrated into larger velocities of a smaller region near the bubble surface as it contracts. This increased velocity makes it appear that the bubble is repelled downward from the surface, and the velocity increases as the size of the bubble decreases.

8.10. Calculated Motion of a Gas Sphere near Surfaces

A. *The period of oscillation.* When free or rigid infinite surfaces are present, it is necessary to integrate the equations of motion derived in the preceding section (Eqs. (8.52), (8.53)). These equations, or approximate forms of them, have been integrated numerically by several writers. On the basis of such results, a number of simplifying approximations can be made which permit an analytic expression for the period. These approximations, some of which are also discussed in other sections, are as follows:

(i). The period T and vertical momentum $\bar{S}$ at the first minimum are twice as large as the time t_m and momentum S_m at the first maximum. This assumption is exactly true in the limit of small vertical displacements, and is very nearly true from numerical integrations.

(ii). The vertical velocity is zero up to the first maximum. This is consistent with the numerical results, which show that this velocity becomes large only as the bubble radius becomes small.

Taking $db^*/dt^* = 0$, $b^* = b_o^*$, where b_o^* is the initial distance of the charge above the bottom, the energy equations (8.52) and (8.57) for $E(a^*)/Y$ give on solving for da^*/dt^*:

$$\frac{da^*}{dt^*} = \frac{\sqrt{1 - a^{*3} - k^*a^{*-3/4}}}{a^{*3/2}\sqrt{1 + f_o}}$$

This can be integrated to give the time t_m^* of the first maximum as

$$t_m^* = \int_{\bar{a}^*}^{a_m^*} \frac{a^{*3/2}\sqrt{1 + f_o}}{\sqrt{1 - a^{*3} - k^*a^{*-3/4}}}\, da^* \tag{8.58}$$

The limits of integration are the maximum and minimum radii a_m^* and $\bar{a}^*$, and are, in this approximation, the largest and smallest roots of

$$1 - a^{*3} - k^*a^{*-3/4} = 0 \tag{8.59}$$

The parameter k^*, which represents the effect of internal energy of the gas, depends on the explosive and depth. Using Taylor's assumptions for TNT as expressed by Eq. (8.57): $k^* = 0.0607\ z_o^{1/4}$, k^* is found to vary from 0.146 to 0.223 as the depth below the surface changes from 0 to 150 feet. The roots of Eq. (8.59) depend on k^* and hence vary slightly with the depth. The smallest root $\bar{a}^*$ is very nearly equal to $k^{*3/4}$, and the largest root a_m^* is somewhat less than the value unity obtained for $k^* = 0$. For $k^* = 0.20$, numerical solutions of Eq. (8.59) give $\bar{a}^* = 0.118$, $a_m^* = 0.924$. The value $a_m^* = 0.92$ shows that the assumed internal energy gives a maximum radius 8 per cent smaller than the value 1.0 neglecting the internal energy.

The evaluation of Eq. (8.58) requires insertion of the appropriate form of the function f_o, which represents the effect of surfaces. This function, obtained by the method of images in section 8.8, is an infinite series in powers of $a^*/2b_o^*$. An examination of this series shows, however, that the radical $\sqrt{1 + f_o}$ can be approximated closely by the leading terms in its expansion which gives the relation

$$\sqrt{1 + f_o} = 1 + \frac{a^*}{4b_o^*} F(x)$$

where $F(x) = 1$ for the rigid bottom only, $F(x) = -1$ for free surface only.

For both rigid bottom and free surface, the first order image theory gives

$$F(x) = (1 - x)\left[2x \sum_{n=0}^{\infty} \frac{1}{(2n + 1)^2 - x^2} - \log_e 2\right] \tag{8.60}$$

where $x = (d - b_o)/(d + b_o)$, and d is the distance below the free surface. Using the expression for $\sqrt{1 + f_o}$, Shiffman and Friedman have developed a method of evaluating the integral for $t_m{}^*$, based on approximating the integrand by suitable orthogonal polynomials. This procedure, for details of which the original paper should be consulted, gives for the period T^* the value

$$T^* = 2t_m{}^* = 1.47\left(1 + \frac{0.185\,F(x)}{b_o{}^*}\right)$$

Converting this to dimensional units gives the final result

$$T = 4.19\,\frac{W^{1/3}}{z_o{}^{5/6}}\left[1 + \frac{3.42W^{1/3}F(x)}{b_o z_o{}^{1/3}}\right]$$

the constants in the expression being based on the expression (8.56) for the internal energy of TNT, using $k^* = 0.20$.

It is convenient to write the period formula in a more general form applicable to other explosives than TNT. To do this exactly would mean recomputing the integral for T for each explosive depth, using the best available expression for $E(a^*)/Y$ in place of $k^*a^{*-3/4}$, and redetermining the limits $a_m{}^*$ and $\bar{a}^*$. Fortunately the result is insensitive to these changes, and so the evaluation need not be repeated. In a revised formula Friedman (37) has, however, used a value $k^* = 0.16$, corresponding to shallower depths, as being more representative of most available measurements, and the revised expression for the period in nondimensional units becomes

$$T^* = 1.485\left[1 + \frac{0.189\,F(x)}{b_o{}^*}\right]$$

The expression for the period T in time units does depend directly on the explosive and energy conversion. The length and time scale factors L^* and C^* were defined as

$$L^* = \left(\frac{3Y}{4\pi\rho_o g z_o}\right)^{1/3}, \quad C^* = L^*\left(\frac{3}{2gz_o}\right)^{1/2}$$

where Y is the energy remaining in the system during the cycle and d is the depth. Expressing Y as a fraction r of the total detonation energy Q in calories per gram of explosive, the scale factors L^* and C^* then become, for W pounds of explosive

$$L^* = (5.36 \times 10^3 rQ)^{1/3} \left(\frac{W}{z_o}\right)^{1/3}, \text{ if } L^* \text{ and } z_o \text{ are in feet,}$$

$$C^* = 0.216\, L^* z_o^{-1/2}, \text{ in seconds}$$

as shown in the derivation of Eqs. (8.54), (8.55).

The period T in seconds using these values of L^* and C^* is then

$$T = 0.561 \frac{(rQW)^{1/3}}{z_o^{5/6}} \left[1 + \frac{0.341(rQW)^{1/3}F(x)}{b_o z_o^{1/3}}\right] \tag{8.61}$$

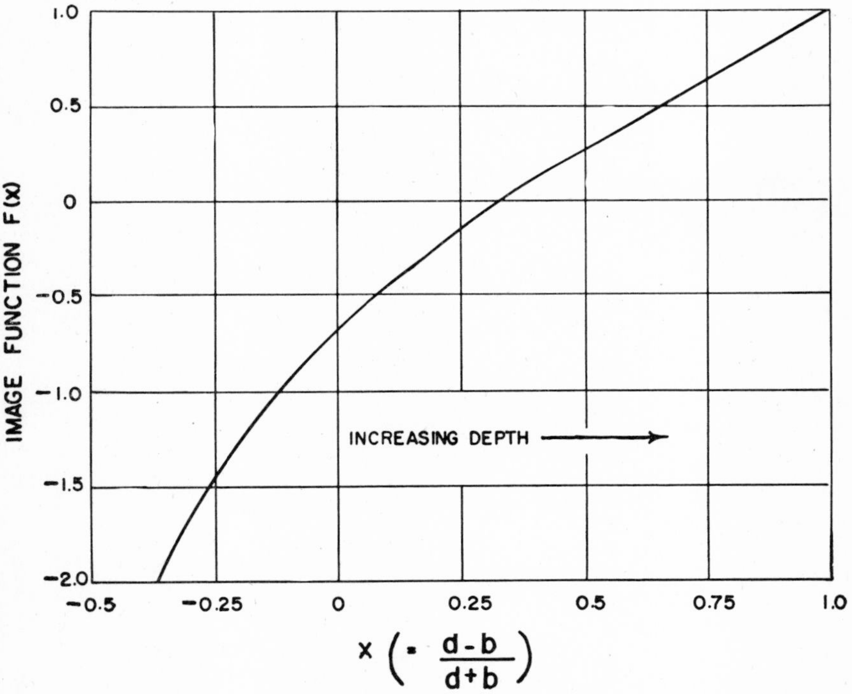

Fig. 8.1 The image function $F(x)$ for effect of surface and bottom.

The function $F(x)$, which varies between $-\infty$ and $+1$ over the range $-1 < x < +1$, is plotted in Fig. 8.18. It is interesting to note that, contrary to what one might expect, the effects of free and rigid surfaces do not cancel ($F(x) = 0$) for the value $x = \frac{1}{2}$, corresponding to a charge midway between the surface and bottom, but rather for the charge $\frac{2}{3}$ of the depth below the surface ($x = \frac{1}{3}$). Above this depth the period is decreased ($F(x)$ negative), and the period is greater closer to the bottom, in agreement with the qualitative discussion of section 8.9.

B. *The maximum radius.* The maximum radius in the first expansion is of course determined by the condition that $da^*/dt^* = 0$, which gives, if the vertical velocity at this time can be neglected,

$$1 - a_m^{*3} - \frac{E(a_m^*)}{Y} = 0 \tag{8.62}$$

The internal energy for TNT has, in almost all mathematical calculations, been taken to correspond to the adiabatic relations $PV^{1.25} = k$, where the exponent 1.25 and value of k are chosen for best fit to Jones's calculations (see section 8.5). This specialization is unnecessary, and if a more general law $PV^\gamma = k$ is used for one gram of explosive with P and V in cgs. units (8.56) gives the result that $E(a^*)/Y = k^*a^{*-3(\gamma-1)}$, where

$$k^* = \frac{k(rQ)^{-\gamma}(7.36 \times 10^{-4})^{\gamma-1}}{(\gamma - 1)\, 4.19 \times 10^7} z_o^{\gamma-1}$$

The conversion factors appearing in this equation are such that z_o is in feet, k and Q in cgs. units per gram of explosive. The largest root of Eq. (8.62) for the more general adiabatic may be obtained numerically, and is approximately given by

$$a_m^* = 1 - \frac{k^*}{3} + \frac{3\gamma - 2}{9} k^{*2}$$

Values of the radius a_m in feet are then obtained by using the scale factor L^* of Eq. (8.54).

It should be noted that the approximations by which a_m^* is obtained will often be valid only for the first maximum, as the bubble is in most cases moving with appreciable vertical velocity at later maxima. A correction could be made using measured velocities at the time of later maxima, if desired, but this procedure has not been attempted in any practical calculations. A similar difficulty applies in calculations of bubble minima with the further complication that the bubble is not in fact spherical and so such results would be of dubious value.

C. *Migration of the bubble under gravity.* The nature of the vertical motion of the bubble presents some difficulties in devising suitable approximation formulas, and attempts of this kind have been confined to determining the migration over the first period, that is, the vertical displacement when the bubble has contracted to its first minimum radius. These approximations have been based on empirical fitting of numerical calculations. One such formula has already been given in section 8.6, in which the migration was found to be roughly proportional to the translational momentum acquired by buoyancy during the first expansion, and the numerical factor of proportionality was obtained by empirically fitting values obtained by numerical calculations. Shiffman and Friedman have done essentially the same thing, using a momentum function $\bar{s}$ in reduced units. They have also added the refinement of using a relation of the form $\Delta b^* = K_1\bar{s} + K_2\bar{s}^3$ instead of $\Delta b^* = K_1\bar{s}$, where $\bar{s}$ is the value at the minimum. In order to express such a formula in usable form, it is necessary to obtain an expression for $\bar{s}$ in terms of initial conditions, and Shiffman and Friedman's derivation (102) is outlined here, as it is needed also for calculations of pressure in section 9.3.

The translational momentum $\bar{s}$ at the minimum is assumed to be twice the value s_m at the first maximum, in good agreement with numerical calculations. In the absence of surfaces, the momentum s is given by $(a^{*3}/3)\,(db^*/dt^*)$, but in their presence, the more general equation (8.53) must be used, which is

$$\frac{ds}{dt^*} = \frac{d}{dt^*}\left[\frac{1}{3}\,a^{*3}(1 + 6f_2)\,\frac{db^*}{dt^*} - 2a^{*3}f_1\,\frac{da^*}{dt^*}\right]$$

$$= -\,\frac{a^{*4}}{2b^{*2}}\left[\frac{df_o}{d\alpha}\left(\frac{da^*}{dt^*}\right)^2 - 2\,\frac{df_1}{d\alpha}\left(\frac{da^*}{dt^*}\right)\left(\frac{db^*}{dt^*}\right) + \frac{df_2}{d\alpha}\left(\frac{db^*}{dt^*}\right)^2\right]$$

$$+\,\frac{1}{z_o^*}\,a^{*3}$$

If, as in the period calculation, the vertical velocity is neglected during the first expansion, we obtain for s_m

$$s_m = -\,\frac{1}{2b_o^{*2}}\int_o^{t_m^*} a^{*4}\,\frac{df_o}{d\alpha}\left(\frac{da^*}{dt^*}\right)^2 dt^* + \frac{1}{z^*_o}\int_o^{t_m^*} a^{*3}dt^*$$

$$= -\,\frac{1}{2b_o^{*2}}\int_{\bar{a}^*}^{a_m^*}\frac{a^{*5/2}\sqrt{1 - a^{*3} - k^*a^{*-3/4}}}{\sqrt{1 + f_o}}\,\frac{df_o}{d\alpha}\,da^*$$

$$+\,\frac{1}{z_o^*}\int_{\bar{a}^*}^{a_m^*}\frac{a^{*9/2}\sqrt{1 + f_o}}{\sqrt{1 - a^{*3} - k^*a^{*-3/4}}}\,da^*$$

the second step following on eliminating t^* by Eq. (8.58). In addition to the value of $\sqrt{1 + f_o}$ already used in the period calculation, the value of $df_o/d\alpha$ is needed. This is adequately approximated by using $df_o/d\alpha/\sqrt{1 + f_o} = 1 - a^*/4b_o^*$ in the first integral. A numerical evaluation for $k^* = 0.20$ by Shiffman and Friedman, similar to the method used for the period, gives

$$(8.63) \quad \bar{s} = 2s_m = -\left(\frac{0.113}{b_o^{*2}} - \frac{0.019}{b_o^{*3}}\right) + \frac{1}{z_o^*}\left(0.704 + \frac{0.148}{b_o^*}\right)$$

As this result is written, it includes only the effect of a rigid bottom. The effect of a free surface is readily included by using the more general expressions for f_o and $df_o/d\alpha$. This will not be done here, a more approximate but simpler discussion being given in part D. It is seen from Eq. (8.63) that $\bar{s}$ can, by suitable choices of z_o^* and b_o^*, be made zero. This corresponds to a bubble which has no vertical motion at the time of the first minimum, the "stabilized" position in which the effect of gravity is compensated by the rigid bottom. The depth d^* for which $\bar{s} = 0$ is given by

$$(8.64) \quad d^* \cong 6.2b_o^{*2} + 3.3b_o^* + 0.4$$

The migration Δb^* of the bubble due to gravity and the bottom is found by Shiffman and Friedman to be given by

$$(8.65) \quad \Delta b^* = 19\bar{s}\,(1 - 62\bar{s}^2)$$

where $\bar{s}$ is determined by Eq. (8.63). These formulas for the migration and stabilized position as written here are expressed in units proportional to the maximum bubble radius. The actual values obtained for specific cases and their relation to experimental results are considered in section 8.11.

D. *Approximate formulas for migration near surfaces.* Although the analysis employed by Shiffman and Friedman for the influence of the bottom on bubble migration can be extended without difficulty to include a free surface above the bubble, the formulas are rather cumbersome. Instead of carrying out this derivation, a derivation due to Herring (46) will be outlined here which shows more explicitly the nature of the effect. The vertical momentum equation including surface terms is the starting point in this analysis, which is, from Eq. (8.53), given by

$$(8.66) \quad \frac{d}{dt^*}\left[\frac{a^{*3}}{3}(1 + 6f_2)\frac{db^*}{dt^*} - 2a^{*3}f_1\frac{da^*}{dt^*}\right] = a^{*3}\left[\frac{\partial f_o}{\partial l^*}\left(\frac{da^*}{dt^*}\right)^2 - 2\frac{\partial f_1}{\partial l^*}\left(\frac{da^*}{dt^*}\right)\left(\frac{db^*}{dt^*}\right) + \left(\frac{\partial f_2}{\partial l^*}\right)\left(\frac{db^*}{dt^*}\right)^2\right] - \frac{1}{z_o^*}a^{*3}$$

In this equation, the bracket on the left side represents the translational momentum, the bracket on the right represents the surface "forces," and the last term is the hydrostatic buoyancy. The variable l^* may represent the distance to the free surface or the bottom, and the values of f_o, f_1, f_2 as obtained by the method of images depend on which case is considered. The analysis shows that $\partial f_1/\partial l^*$, f_2, $\partial f_2/\partial l^*$ are of order $1/l^{*3}$ or higher, and will be neglected in the present approximation. The values of $\partial f_o/\partial l^*$ and f_1 for a free surface above the charge and for a rigid bottom below it are seen by inspection of the formulas of section 8.9 to be

$$\frac{\partial f_o}{\partial l^*} = -\frac{a^*}{2l^{*2}}, \quad f_1 = \pm \frac{a^{*2}}{8l^{*2}}$$

the upper sign being for a free surface and the lower for a rigid bottom. Inserting these values in Eq. (8.66) and integrating to obtain the upward velocity db^*/dt^* gives

$$\frac{db^*}{dt^*} = \pm \frac{3}{4}\frac{a^{*2}}{l^{*2}}\frac{da^*}{dt^*} - \frac{3}{2a^{*3}}\int_o^{t^*} \frac{a^{*4}}{l^{*2}}\left(\frac{da^*}{dt^*}\right)^2 dt^* + \frac{3}{a^{*3}z_o^*}\int_o^{t^*} a^{*3}dt^*$$

If the distance l^* to boundaries can be assumed fixed, the first term can be immediately integrated a second time to give a displacement of

$$\pm \frac{1}{4l^{*2}}\,[a^{*3}(t) - a^{*3}(O)]$$

This is a periodic term, which can be thought of as representing the attraction or repulsion of the bubble by its image. Its value becomes greatest at the bubble maximum, and vanishes at the minimum $(a^{*3}(T) \cong a^{*3}(O))$. Its magnitude is in any case small and will therefore be neglected.

The integrals in the last two terms increase chiefly while the bubble radius a^* is large. It is therefore reasonable to calculate them approximately for the first period T by neglecting the internal energy, in order to determine the nature of the surface correction introduced by the second term. The integrals can be evaluated by using the relation $da^*/dt^* \cong (1 - a^{*3})^{1/2}\, a^{*-3/2}$, obtained from the energy equation by neglecting the internal energy of the gas, and taking twice the integral over the first half cycle from $a^* = 0$ to $a^* = 1$ (a^* is measured in units of the maximum radius a_m obtained neglecting the internal energy). Both integrals can be expressed in terms of β-functions, giving

$$\int_o^{T^*} a^{*4}\left(\frac{da^*}{dt^*}\right)^2 dt^* = 2\int_o^1 a^{*5/2}(1 - a^{*3})^{1/2} da^* = 0.74$$

$$\int_o^{T^*} a^{*3} dt^* = 2\int_o^1 \frac{a^{*9/2}}{(1 - a^{*3})^{1/2}} da^* = 1.86$$

The ratio of the two terms is then

$$\frac{\text{Surface term}}{\text{Gravity term}} = -\,0.20\,\frac{z_o^*}{2l^{*2}} = -\,0.20\,\frac{z_o}{2l^2}\,a_m$$

the change to dimensional units introducing the scale factor for length, which is the maximum radius a_m in the present approximation. The distance $-z_o$ is just $P_o/\rho_o g$, where P_o is the hydrostatic pressure at the depth of explosion, and the migration Δb_l is roughly given by Herring's result:

$$\Delta b_l = \Delta b\left[1 - \frac{P_o}{\rho_o g l}\cdot\frac{a_m}{l}\right] \tag{8.67}$$

where Δb is the migration under gravity alone.

This formula shows that the effect of either surface falls off roughly as $1/l^2$, where l is the distance to the surface from the initial center of the bubble, and if the surface is close enough the bubble may even sink during its oscillation, despite its buoyancy. The rise due to buoyancy for charges at a given depth increases with charge weight, the effect being greater near the surface owing to the relatively greater differences in hydrostatic pressure over the larger bubble surface and the longer period of oscillation. For sufficiently large charges, the relatively greater buoyancy cannot be counteracted by the surface repulsion except for depths so shallow that the bubble breaks surface before contraction, and a rest position of zero net migration does not occur. The exact charge weight for which this first occurs has not been determined and cannot be calculated with any accuracy from the approximate formulas which have been developed, but is of the order of a few pounds.

The absence of a rest position is not indicated for a charge fired near the bottom, as the bubble is carried away from the rigid surface by gravity, and in practical cases the bubble is relatively much less buoyant than it is near the surface. It is to be noted that this case is more complex because the distance above the bottom is an independent variable for a given depth. The calculation of migration given here is of course only a crude first approximation, as only the leading terms in the surface corrections have been used, and internal gas pressure has been neglected.

Kennard (55) has made extensive calculations, based on formulas equivalent to the initial equation (8.66) of the development given here, but including the internal energy in computing the integrals. These calculations indicate that near either surface the effect varies nearly as $1/l$ when close to the surface, the variation as $1/l^2$ being approached at greater values of l. It is of course to be remembered that all such calculations become increasingly questionable too close to the surface, a criterion suggested by Kennard being that the distance l should be at least twice the maximum radius a_m for reasonable accuracy. Kennard has also considered the case of a free surface and rigid bottom both present, which is only approximately a superposition of their separate effects, and the case of an infinite rigid surface at an arbitrary orientation. For a vertical wall, a net horizontal attraction of the bubble occurs as one would expect, in addition to its rise under gravity. For further details Kennard's report (55), which gives a comprehensive discussion of the features of bubble migration under a variety of conditions, should be consulted.

It should be mentioned that a number of boundary conditions other than of infinite plane surfaces have been considered by other writers. Savic (98) has obtained approximate forms of the equations of motion under the influence of a rigid sphere and an infinite rigid cylinder. A report from the Road Research Laboratory (94) treats the case of a rigid disk, and it is found that its effect is, as would be expected, rather less than that of an infinite rigid surface. These problems all involve applications of potential theory, using the method of images or otherwise, which are similar to the cases already described. Although these calculations are of interest as idealizations of practical situations in explosion damage, their detailed developments are rather lengthy and have not been as yet of much practical application, and so will not be given here. Temperley has discussed the much more complicated case of a deformable target (112), which can appear to act as either a free or rigid surface depending on its displacement. Situations of this kind are of considerable importance in assessing the effect of bubble motion on explosion damage, a topic which is considered briefly in Chapter 10.

8.11. Measurements of Periods and Migration near Surfaces

A. *Period measurements.* A number of studies have been made of the effect of the free surface and rigid bottom on the period of oscillation for small charges (½ pound of explosive). If the boundaries are far from the bubble, theory predicts a variation of period with hydrostatic pressure according to the formula $T = KW^{1/3}(d + 33)^{-5/6}$, where W is the charge weight in pounds, d is the distance below the surface in feet, and the pressure at the surface is 33 feet of water. The predicted decrease in period near the sea surface and increase near the bottom are shown by the data of Fig. 8.19, in which the periods for 0.55 pound charges of tetryl fired at various depths in approximately 24 feet of water (2) are plotted against $d + 33$. Also plotted for comparison is the period-depth relation $T(\text{sec.}) = 4.53W^{1/3} (d + 33)^{-5/6}$ obtained for charges fired in deep water.

The most accurate calculation of periods when both surface and bottom must be considered is the result of Friedman (37) given by Eq. (8.61), which for the present purpose may be written

$$(8.68) \quad T(d + 33)^{5/6} = 0.561(rQW)^{1/3}\left[1 + \frac{0.341(rQW)^{1/3}F(x)}{b_o(d + 33)^{1/3}}\right]$$

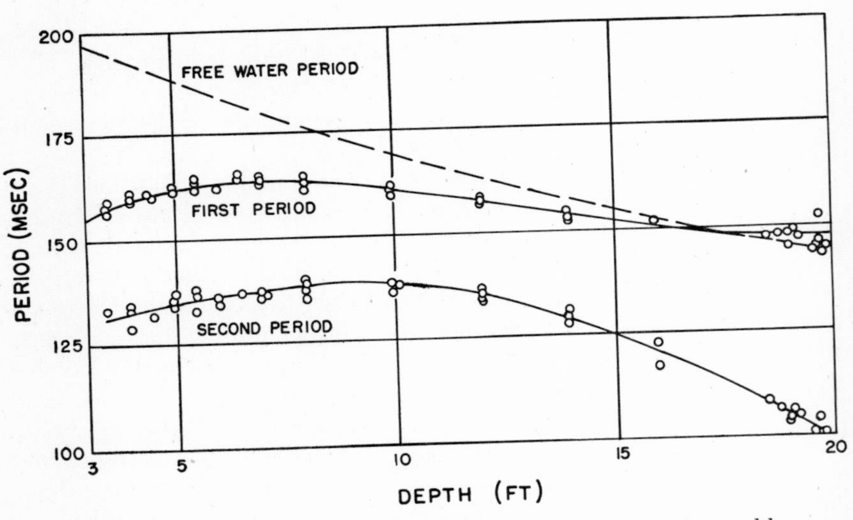

Fig. 8.19 First and second periods of bubble pulsations for 0.66 pound loose tetryl charges fired in 24 feet of water.

where W is the charge weight in pounds, and $F(x)$ is the function of distances d and b_o from the surface and bottom defined by Eq. (8.60). The argument x is defined by the ratio $(d - b_o)/(d + b_o)$, and $F(x)$ takes on the limiting values $-\infty$ near the surface, $+1$ near the bottom,[12] and zero for a charge fired at a distance equal to ⅓ of the depth of water

[12] The divergence of the function $F(x)$ for $d \ll b_o$ makes its use less convenient than that of the function $F'(x) = -(1 + x)\ F(x)$ in terms of which the period is given by

$$T = \frac{0.561\ (r_1QW)^{1/3}}{(d + 33)^{5/6}}\left[1 - \frac{0.341\ (r_1QW)^{1/3}\ F'(x)}{d(d + 33)^{5/6}}\right]$$

Values of $F'(x)$ as a function of x are as follows

x	−1.0	−0.8	−0.6	−0.5	−0.4	−0.3	−0.2	−0.1
$F'(x)$	1.00	0.998	0.985	0.970	0.946	0.911	0.861	0.791

above the bottom ($x = \frac{2}{3}$). At this latter depth the theory thus predicts no correction to the period, the opposite effects of the two surfaces cancelling.

The validity of Eq. (8.68) can be tested in various ways, one of the simplest being to calculate the value of $(r_1QW)^{1/3}$ from it, using the observed distances and period. If the theory is correct, it should then give a constant value of $(r_1QW)^{1/3}$ regardless of depth and surface distances, and this value should agree with the one obtained from measurements in deep water. Values of $K = (rQ)^{1/3}$ obtained in this way are plotted against depth of explosion in Fig. 8.20, and lie within 0.5 per cent of the straight line representing the free period value except when the charge is less than 6 feet from the bottom. The formula thus represents the observed deviations remarkably well near the surface, but overcorrects for the effect of the bottom.

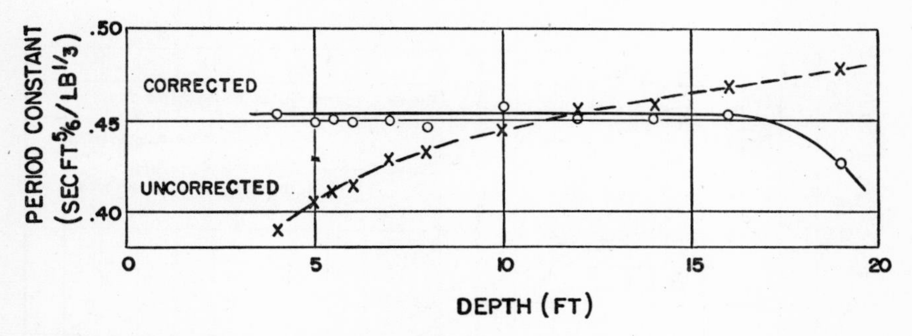

Fig. 8.20 Values of period constant for 0.66 pound loose tetryl charges after correction for surface effects.

The same general type of deviation is found for other small charges fired in the same total depth of water, thus leading to the conclusion that the effect of the bottom is overestimated by representing it as an infinite, plane, perfectly rigid surface. This conclusion, insofar as period is concerned, is also reached by Willis and Willis (123) from measurements on one and five pound charges of blasting gelatin, their data showing little or no real deviations from five/sixth power law near the bottom.

The available evidence from large charges for further tests of the theory is much more meager, but leads to the same general conclusions. The most comprehensive data for the purpose were obtained in an investigation (103) of bubble phenomena from 300 pound TNT depth charges fired in about 100 feet of water. The observed periods are plotted in Fig. 8.21 against $(d + 33)$ feet. It is evident that the marked deviations from the straight line for period varying as $(d + 33)^{-5/6}$ in an infinite body of water are in the proper direction. No precise data suitable for directly determining the constant in the simple period expression were available, and its value was therefore obtained by assuming that 43 per cent of the computed detonation energy of 1,060 cal./gm. for TNT remained in the first pulsation.

Friedman (37) has outlined a procedure whereby the agreement of the observed periods with the theory including surface effects can be checked, and which also permits estimates of the energy in the bubble motion if the theory applies. The theoretical expression, Eq. (8.68), can be written

$$T(d + 33)^{5/6} = \alpha - \beta \frac{F'(x)}{d(d + 33)^{1/3}}$$

where $\alpha = 0.561(rQW)^{1/3}$, $\beta = 0.192(rQW)^{1/3}$, and

$$F'(x) = -(1 + x)F(x) \text{ (see footnote 12)}$$

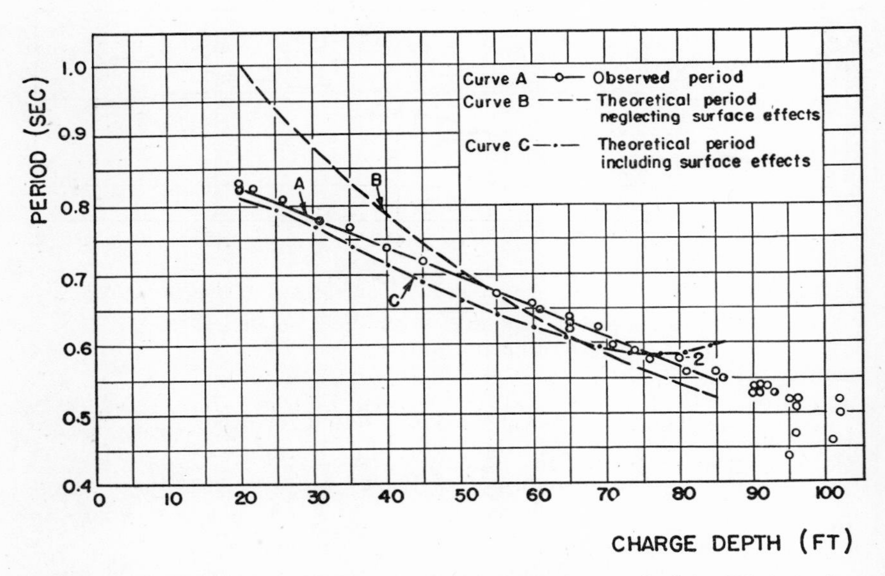

Fig. 8.21 Bubble periods for 300 pound TNT charges fired in 100 feet of water.

A plot of $T_1\ (d + 33)^{5/6}$ against $F'(x)/d(d + 33)^{1/3}$ should therefore be a straight line of slope β and intercept α, from which $(rQW)^{1/3}$ can be evaluated. The result is shown in Fig. 8.22. For depths greater than 60 feet, the points depart considerably from a straight line, indicating failure of the rigid bottom correction. Fitting the points at shallower depths by the straight line indicated gives $rQ = 424$ cal./gm. from the value of β, and $rQ = 490$ cal./gm. from the value of α. The mean value of 456 cal./gm. corresponds to the values $r = 0.43$, $Q = 1{,}060$ cal./gm. used to plot the straight line depth-period relation of Fig. 8.21. The theoretical relation including surface effects, also plotted in this graph, lies about 3 per cent below the experimental points for d less than 60 feet, but is otherwise an excellent representation of the free surface effect.

The overestimate of the effect of the sea bottom for both small and large charges, obtained by assuming the bottom to be a perfectly rigid plane surface, is surprisingly great. The actual bottoms in these investigations were of course not perfectly rigid, being fairly hard packed

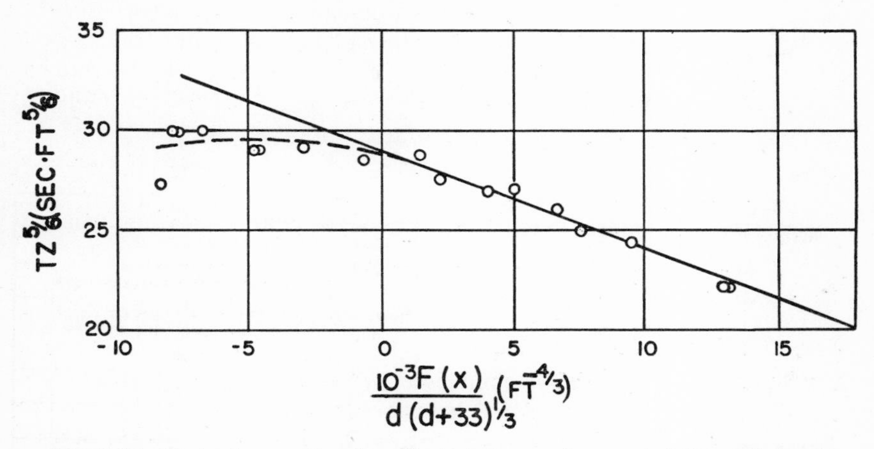

Fig. 8.22 Test of correction for effect of free surface on bubble periods of 300 pound TNT charges.

sandy mud. During most of the bubble pulsation, however, the pressures are low and the assumption of a rigid bottom, which should refer primarily to resistance against mass flow of water rather than its acoustic properties, ought on this basis to be at least a fair approximation. A possible explanation of the discrepancy for charges fired near the bottom may be that the bottom below the charge is hollowed out, or cratered, sufficiently by the action of the shock wave and initial large velocities of outward flow that the subsequent motion of the bubble is not affected as much as it would be by an undisturbed bottom. It is interesting in this connection to note Ward's comparison of craters (119) formed by charges fired under the ground with the maximum bubble size from underwater explosions, which he found to be not very different in radius. Whatever the explanation,[13] there appears to be little

[13] The effect of a nonrigid bottom has been taken into account in unpublished calculations by B. Friedman by assuming that its effect is that of a fluid of arbitrary density ρ, i.e., in the noncompressive approximation a medium is characterized by its inertia to flow. Small charge period data are predicted quite accurately by assuming a bottom density of 3.0, but this picture is hardly very realistic if cratering occurs.

doubt of the discrepancy, as predictions regarding migration of the bubble and its effect on bubble pressure pulses are likewise not well confirmed (see part (c) and section 9.6).

B. *Photographic studies of bubble migration.* The most complete set of data on motions of explosion bubbles as affected by free and rigid surfaces is undoubtedly contained in the records obtained by Lt. D. C. Campbell at the Taylor Model Basin (17). In these experiments, most of the bubbles were formed by the explosion of Du Pont Number 8 detonator caps, although a few results are given for larger mine detonators and for ½ ounce tetryl charges. The bubbles were photographed using the high-speed stroboscopic motion picture camera technique developed by Edgerton, most of the exposures being made at 1,500 frames/sec. In Campbell's report a large number of sequence records are shown taken under the following conditions:

(1) Charge at various depths, surfaces remote.
(2) Charge fired at several distances from a rigid vertical concrete wall.
(3) Charge fired at several depths beneath a steel sheathed wooden bottom boat.
(4) Charge fired at several distances above a concrete block.
(5) Charge fired midway between two steel plates with several separations.

The charges fired under condition (1) served essentially as control shots for the later series, in which surfaces were present, and show the characteristic radius-time curves already discussed. In some cases as many as 17 oscillations of steadily decreasing amplitude were observed, but after two or three cycles the changes in amplitude become so small that detection of a cycle is very difficult. The charges placed near the vertical wall show very strikingly the attractive effect appearing at the minimum radius. Displacement-time curves obtained from the films are reproduced in Fig. 8.23 and the rapid motion at the time of the minimum is clearly evident, the maximum velocity of the bubble center toward the wall being of the order of 200 ft./sec. for a charge fired 9 inches from the wall. At close distances, this velocity decreases, and the bubble becomes flattened on the side toward the wall, being very nearly a hemisphere when initiated by a contact explosion. Much the same phenomena were observed for charges fired near the concrete bottom.

Charges fired close to the free water surface showed the expected repulsion. A downward motion of part of the gas volume occurred even for a charge fired 3 inches under the surface, although the bubble lost part of itself through the surface during the first expansion. This free water repulsion was also observed for a charge fired 12 inches below the wooden boat, but a charge fired 6 inches beneath it acquired an upward velocity of 180 ft./sec. during the contraction and shortly later came in contact with the boat.

The most spectacular result of all was obtained when the charges were fired midway between steel plates. If these were twenty-four inches apart, the bubble was not affected appreciably. For a twelve inch separation, the bubble expanded spherically, but as the contraction phase set in, it began to stretch out toward each vertical plate, and two

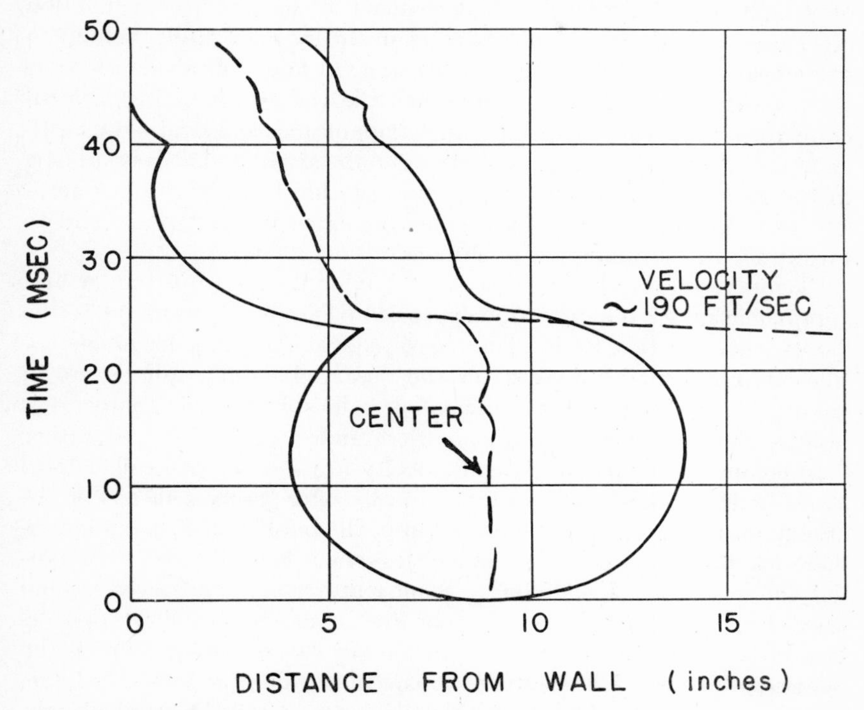

Fig. 8.23 Migration of the bubble from a detonator placed nine inches from a vertical wall.

or three milliseconds before its maximum contraction divided rapidly in two globes, one for each rigid surface, so to speak. The two globes contracted separately and moved rapidly toward the two plates where they expanded in contact and flattened out again. This bubble division was found to be critical to charge position, deviations of one-quarter of an inch from exact centering causing the bubble to move intact toward the nearer plate. This result is not surprising in view of the strongly distant-dependent nature of the phenomenon.

Although Campbell's results give a beautiful confirmation of the predicted effects of surfaces on bubble displacements, they are not well suited to comparison with theory owing to the very small size and uncertain energy content of the charge. Measurements made at the Road Research Laboratory (95) using one ounce polar ammon gelignite charges three feet below the surface and a single flash photographic technique have been compared with calculations based on Taylor's theory, corrected for surface effect according to Herring's analysis. The observed maximum radius of 1.65 feet was sufficiently large compared with the depth of 3 feet for a large, downward migration of the bubble to occur. Measured displacements of the center as a function of time agree with values calculated from theory moderately well up to times near the minimum. The discrepancy at later times is most strikingly shown if velocities are compared, the theory predicting a maximum value of 900 ft./sec. while the maximum measured value was estimated at 180 ft./sec. This difference is to be attributed at least in part to the greater hydrodynamic resistance of the flattened surface actually realized, as compared to a sphere assumed in the theory, and to turbulence in the wake of this surface.

Temperley (111) has modified the Taylor theory to include a first approximation for the distortion expressed in surface harmonics (see the discussion of section 8.7). The measured values of the harmonic coefficient, obtained by analysis of the observed bubble sphere, are in qualitative agreement with calculated values based on Temperley's results, the agreement being somewhat improved if observed rather than computed values of vertical velocity are used in this calculation.

Although photographic methods ideally offer the best means of obtaining information regarding the form of the bubble near its minimum, there are numerous practical difficulties which have limited the quantity and usefulness of such data. Some features of the observed motion which have not been brought out in the discussion so far are the formation of smoke-like streamers and narrow spiked projections on the bubble periphery. These are not conspicuous when the bubble is large, but become more evident as the bubble contracts and increasingly obscure its outline, making determination of the shape and displacement rather uncertain. An example of this is shown in the motion picture frames of Plate IX for times near bubble minima. The general fog and obscuration is probably due to the explosion products and stray foreign matter such as charge case, if any, detonator fragments, and so on.[14] The projections have been attributed by Penney to the effect of increasing instability, as the bubble contracts, of high order harmonic terms expressing local deviations from spherical form.

The experimental difficulties of obtaining good underwater photo-

[14] The difficulties with solid products can be minimized by using an explosive such as Polar Ammon Gelignite, as has been done in British work, the better oxygen balance resulting in much less solid carbon. A further advantage is decreased contamination of water in experimental tanks.

graphs, discussed in Chapter 6, make necessary carefully controlled light conditions which are not readily achieved for small scale explosions and are very difficult to obtain for large charges. As a result, not very many experiments have been done except on a very small scale. Some recent experiments (105), done in extremely clear water off Cuba, should provide useful information about migrations of bubbles from charges up to ½ pound in weight, but at the time of writing these data had been only partially analyzed. Photographic records were also obtained of the bubble from 300 pound TNT charges, but uncertainty as to the camera position and orientation make analysis of migrations a very dubious possibility.

C. *Measurements of migration by sound ranging.* The difficulties in obtaining photographic records of bubble motion have led to the use of less satisfactory methods based on determining bubble position by sound ranging methods. In such work, arrival times of the secondary bubble pulses emitted at successive bubble minima are recorded; the only information so obtained is the position of the pressure source at these times. The method is limited in accuracy by uncertainties as to gauge positions and suffers from more fundamental errors. Most sound ranging methods depend on measurement of time differences in reception of the secondary pulses at two or more different gauges (see section 10.3). As contrasted with the shock wave, however, bubble pulses ideally have no abrupt changes or discontinuities in outline and frequently have durations comparable with the time differences in feasible gauge arrangements, which inevitably limits the accuracy obtainable.

Frequently, recorded bubble pulses show small spikes or other minor discontinuities superimposed on the general profile, and these local disturbances have commonly been used as reference points. This procedure is not free from objection, however, as gauges differently oriented often show rather different forms of bubble pressure-time curves, even though corresponding reference points are quite obvious. These differences imply a lack of spherical symmetry in the pressure wave, and make dubious the assumption that the spikes have their effective origin at the bubble center. Measurements on different pairs of spikes in two gauge records confirm the likelihood of such difficulty by sometimes giving quite different indications of the bubble positions. Examination of such data makes reasonable the proposition that these spikes may have their origin in the observed irregularities of the bubble surface when near its minimum size. The computed source of sound may thus differ in position from the actual center by distances of the order of the minimum bubble radius. Both these inherent difficulties and errors in establishing the geometry of the experiment limit the accuracy and reproducibility of sound ranging data, and these limitations should be kept in mind when considering the results. A further limitation, common to all migration measurements, is the instability of the bubble motion near the minimum. This is particularly evident near surfaces to which the motion is sensitive, and small uncertainties in position, such as waves on the sea surface or bottom irregularities introduce, can well have a large effect.

Willis and Willis (123) have examined by sound ranging methods the migration of bubbles from 1 and 5 pound charges of blasting gelatin fired near the surface and bottom in 66 and 150 feet of water. The observed effect of the free surface is found to agree fairly well with theory both as to the rest position of no displacement and as to amount of migration at greater depths. Near the bottom, however, the upward migrations, although somewhat less than predicted for gravity rise only, show no systematic attraction toward the bottom except for charges fired actually on the bottom. This absence of predicted attraction for a perfectly rigid surface is in agreement with the results of period observations. Indicated downward migration of the bubble for charges fired on the bottom, which Willis and Willis believe not entirely attributable to errors, supports the hypothesis of a crater into which the bubble is attracted.

The Willis' measurements for one pound charges show much the same phenomena near the upper rest position, but the observed migrations near the bottom are in better agreement with the predicted surface effect. Data obtained at Woods Hole (2) for TNT, tetryl, and other charges of various weights up to one-half pound fired in twenty-three feet of water agree in showing the existence of both upper (free surface) and lower (bottom) rest positions. The migrations during the first pulsation were found to be comparable with, although smaller than, the predicted rise under gravity. The results for 0.55 pound tetryl charges are plotted in Fig. 8.9. The curve is calculated from Kennard's analysis based on Taylor's theory as corrected for the free surface, and the curve near the bottom is the bottom correction obtained from Shiffman and Friedman's analysis. The latter is seen to fit the points near the sea bed quite well. Experimental points for the migration at the end of the second oscillation are also plotted in Fig. 8.9. These are roughly twice as great as at the end of the first oscillation, but very much less than would be predicted by extending the spherical bubble theory to the second cycle. It is interesting to note that there is little or no migration for the second cycle if there is none at the end of the first; a bubble which is initially balanced thus remains so for a considerable time.

The conclusions which can be drawn from the period and migration data so far discussed are that upward migrations are somewhat smaller than calculated values for a spherical bubble rising under gravity, that the surface repulsion observed is in rather good agreement with theory, and that for charges of 1 pound or less the predicted attraction of the bottom is fairly well confirmed. The data of the Willis' for migrations and periods of 5 pound charges and period data for 300 pound TNT charges fail to indicate more than a fraction of the predicted bottom effect, and migration results for 300 pound charges fired in 100 feet of water are consistent with this conclusion. These results (105) are less accurate than the values for small charges, owing to errors of ranging and ambiguity of ranging reference marks on the pulses, but the ob-

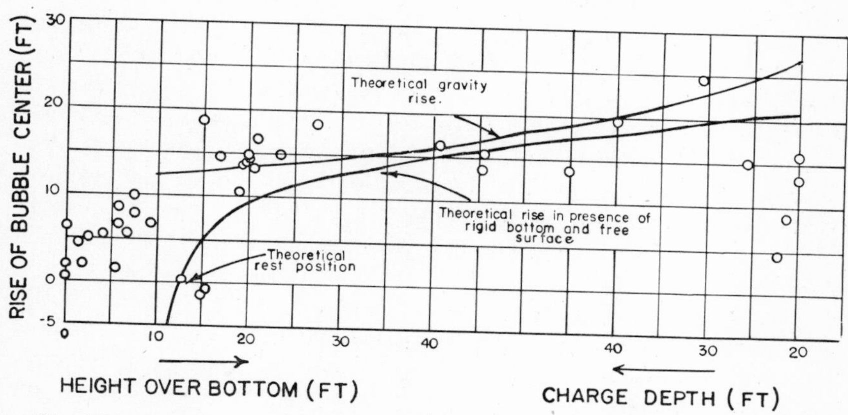

Fig. 8.24 Migration of the bubble from 300 pound TNT charges fired in 100 feet of water.

served values plotted in Fig. 8.24 show no reproducible stabilized position of zero migration at the predicted point 15 feet above the bottom.

The gear used in setting the 300 pound charges was suspended from the surface and precise measurements of the charge-bottom distance were not possible. As a result, the depths corresponding to the experimental values may be subject to random errors, but the failure to obtain consistent zero or downward migrations argues against the existence of a stabilized position. Three points do indicate zero migration at about the predicted position, but failure to reproduce these values makes their accuracy somewhat dubious. At shallower depths between 35 and 75 feet, gravity is the main factor in determining the upward rise, and the experimental points are seen to be in good agreement with the solid line predicted by theory. The smaller migrations observed near the surface are appreciably less than predicted by the small theoretical correction (dashed line), but it is believed that the differences are probably attributable to systematic ranging errors of as much as 5 feet resulting from the indicated sound source lying on the undersurface of the bubble.

This same sort of error may make the bubble appear too high when near the bottom. Even so, the absence of a rest position over the sea bottom of these tests seems indicated, and this result is consistent with bubble period and pressure measurements (the latter are discussed in section 9.6).

9. Secondary Pressure Waves

Theoretical calculations of the pressures in water following emission of the primary shock wave have been without exception based on the incompressive approximation, in which it is assumed that changes in density of the water with pressure can be neglected. While this assumption leads to a fairly accurate description of the main features of gas sphere motion, particularly of the maximum radius and the period of pulsation, it becomes an increasingly poor approximation as the bubble contracts to its minimum size with increasing velocities of flow and higher pressures. This part of the motion is, however, the one which is of interest as far as pressures are concerned, for the secondary pressure waves, often referred to as "bubble pulses," are emitted while the bubble is at or near its states of maximum contraction.

The weakness of noncompressive theory in describing secondary pressure waves will become evident in the discussion of this chapter. However, it may be remarked here that it is exemplified by the requirements of such theory that changes in pressure at any point be propagated instantaneously to all parts of the fluid rather than with the finite sound velocity observed. Differences in pressure in this approximation are associated with changes in flow velocity, and can be a reasonable approximation to the true pressures only for steady or slowly changing flow.

A better approach to the analysis of these pressures is that of supplementing the noncompressive theory with a wave theory, and considerations of this kind are necessary to estimate the energy radiated near the bubble minimum in the wave of compression. A detailed examination involves great mathematical difficulty, if done at all rigorously. If, on the other hand, the simpler procedure were adopted of attempting to keep the two types of approximation reasonably distinct, serious physical problems would arise in suitably patching together basically incompatible solutions. Moreover, the investigations of Herring and of Penney and Price, described in section 8.7, bring out the more fundamental difficulty that the motion in the critical region of minimum size is dynamically unstable. It is therefore reasonable to expect that turbulent flow with dissipation of energy in the formation of eddies will set in, and this conclusion is supported by comparison of the total energy loss at the minimum with that radiated as a compression wave (see section 9.4). The much more complicated problem then presented has, understandably, not been attempted except in very rough approximation.

When these difficulties are considered it should be no surprise that the theoretical results for bubble pressures are qualitatively useful rather than quantitatively reliable. A heavier burden is therefore placed on experimental investigations in evaluating the phenomena and it is unfortunate that the theoretical difficulties have their counterpart in serious experimental problems of measurement and analysis. These problems, which arise largely from the relatively small magnitude and long duration of the secondary pulse, are too numerous and technical to be adequately summarized here (see section 9.5). In spite of both types of difficulty, theoretical and experimental, a fairly satisfactory picture has been developed which makes it possible to predict at least roughly what will happen in most circumstances of interest.

9.1. The Generalized Form of Bernoulli's Equation

The pressure P at any point in an incompressible fluid of density ρ_o without viscosity is most conveniently found from the generalized form of Bernoulli's equation derived in section 8.4, which is

$$\frac{P}{\rho_o} = -\tfrac{1}{2}u^2 + \frac{\partial\varphi}{\partial t} - \Omega + F(t) \tag{9.1}$$

In this equation, u is the particle velocity at the point, φ the velocity potential, Ω the potential of external force on unit volume of the fluid, and $F(t)$ is an arbitrary function of time. The physical significance of $F(t)$ lies in the fact that the distribution of flow in an incompressible fluid is unaffected by changes in time of absolute pressure at all points in its interior. These would be produced, for example, by variations of atmospheric pressure at the free surface, which are propagated instantaneously to all points in this approximation. In the applications we shall make, such changes need not be considered, and the function $F(t)$ is therefore a constant, the value of which is specified if the pressure and velocity are known at some point at any time. It is, for example, determined by the hydrostatic pressure existing initially before the motion of the gas sphere and surrounding flow is initiated. The term $\partial\varphi/\partial t$, which would be zero for steady flow, gives the effect of velocity changes.

The Bernoulli equation (9.1) bears a close relation to the principle of conservation of energy for any element of the fluid, and in its more elementary forms is usually derived by energy considerations from the work done by pressure differences and external forces in increasing kinetic energy of flow. In the problems to which we shall apply the equation it is necessary to include the conditions at boundaries of the fluid. This is most easily accomplished by use of the velocity potential which is calculated by standard methods of potential theory, as has been done in Chapter 8 for the problems of interest here. In applying these results to the evaluation of Eq. (9.1), one condition must be respected: that the differentiation $\partial\varphi/\partial t$ be performed for a point fixed in space. The velocity potentials obtained in Chapter 8 are, however, expressed for convenience in coordinates moving with the velocity U of the center of the sphere, and a point fixed in space therefore has a velocity $-U$ in this system. The desired derivative $\partial\varphi/\partial t$ is then

$$\frac{\partial\varphi}{\partial t} = \left(\frac{\partial\varphi}{\partial t}\right)_m - U\left(\frac{\partial\varphi}{\partial x}\right)_t$$

where $(\partial\varphi/\partial t)_m$ is for fixed positions in the moving coordinates and where x is in the direction of U (positive upward in the cases to be considered). By definition of the velocity potential, $u^2 = (\text{grad } \varphi)^2$ and Eq. (9.1) becomes

$$\text{(9.2)} \quad \frac{P}{\rho_o} = -\tfrac{1}{2}(\text{grad } \varphi)^2 + \left(\frac{\partial\varphi}{\partial t}\right)_m - U\left(\frac{\partial\varphi}{\partial x}\right)_t - \Omega + \text{constant}$$

9.2. The Pressure Distribution for Gravity Alone

If boundary surfaces are assumed sufficiently far away from the gas sphere to have a negligible effect, the velocity potential φ for a hollow sphere of variable radius a is given by (Eq. (8.21))

$$\varphi = \frac{a^2}{r}\frac{da}{dt} + \tfrac{1}{2}U\frac{a^3}{r^2}\cos\theta$$

where U is the upward velocity of the center and the coordinates r, θ are measured from this center and the axis x drawn upward as positive. The derivative $(\partial\varphi/\partial x)_t$ in Eq. (9.2) is given by

$$\left(\frac{\partial\varphi}{\partial x}\right)_t = \cos\theta\left(\frac{\partial\varphi}{\partial r}\right) - \frac{\sin\theta}{r}\left(\frac{\partial\varphi}{\partial\theta}\right)$$

The potential Ω for gravity is $\Omega = -gz$, where, as in section 8.5, z is the distance below the level of zero pressure. Substitution in Eq. (9.2) then yields Taylor's result (107):

$$\text{(9.3)} \quad \frac{P}{\rho_o} = gz + \frac{1}{r}\frac{d}{dt}\left(a^2\frac{da}{dt}\right) + \frac{1}{2}\frac{a^2}{r^2}\left(a\frac{dU}{dt} + 5U\frac{da}{dt}\right)\cos\theta$$

$$+ \frac{a^3}{r^3} U^2(\cos^2\theta - \tfrac{1}{2}\sin^2\theta)$$

$$- \left[\frac{1}{2}\frac{a^4}{r^4}\left(\frac{da}{dt}\right)^2 + \frac{a^5}{r^5} U \frac{da}{dt}\cos\theta + \frac{1}{2}\frac{a^6}{r^6} U^2(\cos^2\theta + \tfrac{1}{4}\sin^2\theta)\right]$$

The pressure distribution in the surrounding water is determined by Eq. (9.3) if the kinematic history of the bubble is known, i.e., its radius and displacement as a function of time, and the solutions obtained in Chapter 8 can therefore be used directly to find the pressure. It is to be noted that r and θ are moving coordinates and calculations of the pressure at a point fixed in space and different times must strictly be found using different values of these variables in order to allow for the motion of the center of the sphere. The complicated dependence of pressure on the angle from the vertical is, of course, the result of the vertical displacement which leads to radial asymmetry of flow and corresponding pressure variations near and on the surface of the sphere. The indicated variation of pressure over the surface ($r = a$) is of course not real, and results from the artificial constraint of the bubble to spherical form (see section 8.5). These effects are most pronounced near the surface of the sphere and become larger for more rapid migration. If this upward motion is neglected by setting $U = 0$ in Eq. (9.3), one obtains

$$\frac{P}{\rho_o} = gz + \frac{1}{r}\frac{d}{dt}\left(a^2\frac{da}{dt}\right) - \frac{1}{2}\frac{a^4}{r^4}\left(\frac{da}{dt}\right)^2 \tag{9.4}$$

The first term on the right of Eq. (9.4) represents the hydrostatic pressure P_o at the depth z, and the last term is important only at points near the bubble surface by virtue of the factor $(a/r)^4$. Hence except for points close to the charge, the variations of pressure are determined primarily by the term $(1/r)\, d/dt\, (a^2\, da/dt)$. This term is also the dominant one at sufficiently great distances when the velocity U cannot be neglected, as the terms involving U in Eq. (9.3) all vary as the second or higher powers of $1/r$.

A. *The pressure in early stages of expansion.* In the initial stages of expansion, the vertical momentum of the gas sphere is small, and its radial velocity da/dt is large. The approximate relation Eq. (9.4) obtained for $U = 0$ may therefore be used, and the last term is unimportant except near the bubble surface. If the additional approximation is made of neglecting the internal energy, the analysis of section 8.5 gives the result

$$\frac{da'}{dt'} = \frac{1}{\sqrt{2\pi a'^{3/2}}}, \quad t' = \frac{2}{5}\sqrt{2\pi}a'^{5/2} \tag{9.5}$$

for the initial rate of expansion. In this equation, a' and t' are the dimensionless variables introduced by Taylor and related to the radius a and time t by $a = La'$ and $t = \sqrt{L/g}\, t'$, where $L = (Y/\rho_o g)^{1/4}$ and Y is the total energy. Substituting, Eq. (9.4) becomes

$$P - P_o = L^2(g/2\pi)^{1/2} \frac{1}{r}\frac{d}{dt} a^{1/2} \tag{9.6}$$

$$= \frac{1}{(1250)^{1/5}} \left(\frac{g}{2\pi}\right)^{3/5} L^{12/5} \frac{t^{-4/5}}{r}$$

As in the acoustic approximation for compression waves, the pressure at a given time varies inversely as the distance r in this approximation and decreases slowly with time. The time dependence is the same as that predicted by the complete pressure formula of the Kirkwood-Bethe theory for the shock wave and initial motion of the gas sphere (see section 3.8).

It is of interest to compute the pressure obtained in this noncompressive theory and compare it with the observed pressure-time variation after the initial shock front. If Eq. (9.6) is expressed in English units it becomes

$$(P - P_o)\ (\text{lb./in.}^2) = 1.74 \cdot 10^{-3} L^{12/5} \frac{t^{-4/5}}{r} \tag{9.7}$$

L and r being in feet. The scaling factor $L = 0.47Y^{1/4}$ if Y is in calories and it is seen that the initial pressures are independent of depth. Using the example given by Taylor (107) of the pressure 14 feet from a 4.66 pound TNT charge gives

$$(P - P_o)\ (\text{lb./in.}^2) = 0.39t^{-4/5} \tag{9.8}$$

where t is time after detonation.

A basic difficulty in comparing this result with the measured pressure-time curve lies in the fact that noncompressive theory assumes an instantaneous propagation of pressure to all points in the fluid. Actually a time of 2.8 msec. is required for the pressure wave to reach a radius of 14 feet, and the noncompressive approximation is an inadequate description of pressures changing significantly in this interval. A crude accounting for the wave propagation is to assume that the pressure is still given by Eq. (9.8) but occurs a time R/c_o later, where R is the distance from the source. If this is done, the upper dashed curve of Fig. 9.1 is obtained, measured shock wave pressures being indicated by the circled points and solid line. If the calculated pressure is plotted without correction for the propagation time, the lower dashed curve is obtained.

The pressures which would exist if the initially exponential decay continued are indicated by the dotted line on the semilogarithmic plot in Fig. 9.1. It is evident that the noncompressive result is qualitatively of the form necessary to account for the shock wave tail, and the ambiguity of the calculation is suggested by the difference of the two

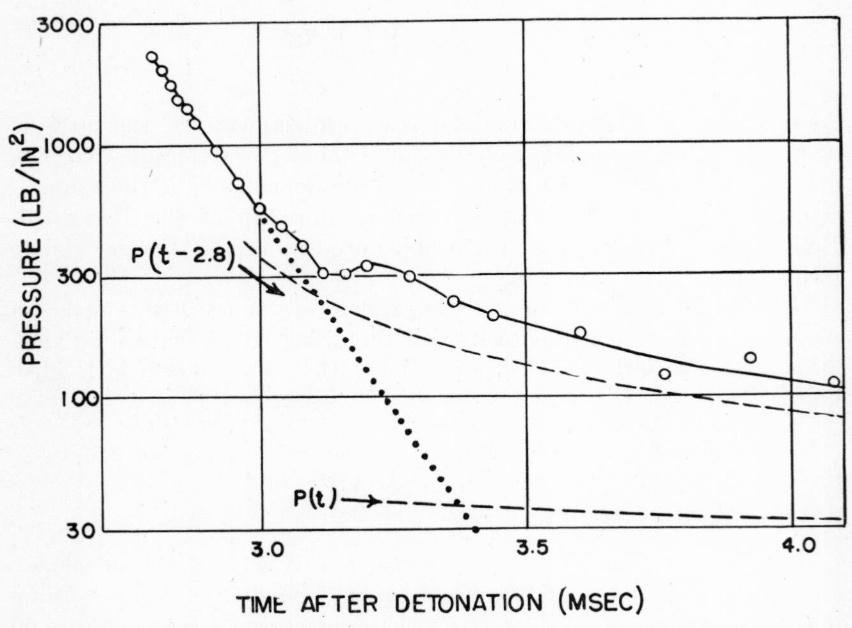

Fig. 9.1 Pressures in later portions of a shock wave compared with values calculated from the noncompressive approximation.

plotted curves. The result is further inaccurate because the internal energy of the explosion products was neglected in the derivation of Eqs. (9.7), (9.8). The effect of gas pressure has been taken into account by Taylor in calculations based on numerical integration of the energy equation (Eq. 8.23) backward to zero time. This leads to somewhat higher pressures in the water and to the paradoxical result that the minimum radius is attained at a negative time. Difficulties of this kind are, as Taylor remarks, inevitable in such a calculation, for the reason that the initial shock wave is taken into account only by the energy Y remaining after its emission. Because of this inadequacy and the related difficulty of the propagation time, attempts to patch together shock wave and noncompressive solutions, as suggested by the curves of Fig. 9.1, are clearly makeshifts to avoid the formidable difficulties of a more inclusive single solution. The analysis does, however, illustrate the relative parts played by shock pressure and "kinetic" pressure of the outward flow.

B. *Afterflow energy.* If the vertical motion of the bubble is neglected, Eq. (9.4) for the excess pressure $P - P_o$ becomes

$$\frac{P - P_o}{\rho_o} = \frac{1}{r}\frac{d}{dt}\left(a^2\frac{da}{dt}\right) - \frac{1}{2}\left(\frac{a}{r}\right)^4\left(\frac{da}{dt}\right)^2 \tag{9.9}$$

The second term is simply $-\frac{1}{2}u^2$ from the equation of continuity for noncompressive flow

$$u = \frac{dr}{dt} = \frac{a^2}{r^2}\left(\frac{da}{dt}\right)$$

This term, which decreases rapidly with distance, represents the Bernoulli pressure $-\frac{1}{2}\rho_o u^2$ for flow velocity u.

The outward flow has associated with it kinetic energy, and is often referred to as the kinetic wave to distinguish it from the shock wave of compressional energy. The energy of the noncompressive flow is often described as afterflow energy, and as the "schubenergie" in German research. The relation of this energy to the pressure as given by Eq. (9.9) is made evident by calculating the work done by pressure P on a spherical shell of radius r. This is given by

$$W = 4\pi \int r^2 P u dt$$

Substitution for P in terms of particle velocity u and distance r gives

$$W = 4\pi\rho_o \int \left[\frac{1}{r}\frac{d}{dt}r^2 u - \frac{1}{2}u^2\right] r^2 u dt + 4\pi P_o \int r^2 u dt$$

By rearrangement and use of the relation $u\,dt = dr$, we obtain

$$\begin{aligned} W &= 4\pi\rho_o \int \left(\frac{3}{2}u^2 r^2 dr + u r^3 du\right) + 4\pi P_o \int r^2 dr \\ &= 2\pi\rho_o \int d(u^2 r^3) + 4\pi P_o \int d\left(\frac{r^3}{3}\right) \\ &= \Delta(2\pi\rho_o r^3 u^2) + P_o \Delta\left(\frac{4\pi}{3}\rho_o a^3\right) \end{aligned}$$

The first term is easily shown to be simply the increase in kinetic energy of the fluid exterior to the surface of radius r (see, for example, section 8.2), and the second term is the increase in potential energy from work done against hydrostatic pressure. The total afterflow energy of the noncompressive flow thus goes, as of course it must, into mechanical energy of the water.

The kinetic energy of unit volume of water is given by $\frac{1}{2}\rho_o u^2 = \frac{1}{2}\rho_o(a/r)^4 \cdot (da/dt)^2$, and thus falls off inversely as the fourth power of distance from the origin. The maximum value at a given point occurs when $d/dt\ (a^2\ da/dt) = 0$, and from Eq. (9.4)

$$(\tfrac{1}{2}\rho_o u^2)_{\max} = -\ (P - P_o)$$

The pressure at this time is therefore less than hydrostatic, and the maximum value of the kinetic energy of outward flow occurs when the bubble is expanded beyond its equilibrium radius. The negative gauge pressure would of course be measured only by a device offering no opposition to the outward (or inward) flow.

The total kinetic energy in the water when at a maximum represents a great part of the energy remaining after emission of the shock wave, as the gas pressure in the products is less than hydrostatic and the work done against gravity in expansion to the radius at this time is of the order of thirty per cent of the total. This energy is concentrated in a region near the gas sphere, and the large values of energy density in this region have led many writers to assign great importance to it as a factor in damage from near contact explosions. This viewpoint, however, is certainly erroneous, insofar as it implies the existence of a "kinetic wave" as separate and distinct from the pressure wave. As Kennard (57) has emphasized, the effect of underwater explosions on targets is determined by the pressure field; the pressure is physically inseparable from motion of the water and must include any effects of this motion. This does not mean, of course, that the pressure on a surface is the same as that in free water, because the water and target together constitute a single dynamical system, each part of which affects the other.

The significance to be attached to afterflow velocity and afterflow energy in a spherical pressure wave can perhaps be made more clear by the simple example suggested by Kennard of an infinite, plane, rigid plate in the field of an explosion. A spherical wave striking this plate must be modified in such a way that the resultant normal velocity of the water in contact with the plate, including the afterflow velocity, is zero. This condition is, however, exactly satisfied for acoustic waves by a reflected spherical wave of the same intensity, proceeding as if it originated at the same distance behind the plate as the source of the original wave. The component of velocity normal to the plate in the wave, including the afterflow velocity, is exactly equal and opposite to that in the incident wave, and as for plane waves the excess pressure at the plate is doubled by reflection. Similar considerations must hold for more complicated structures and finite amplitude waves, the effect of a spherical or any other wave being determined by the resulting pressure.

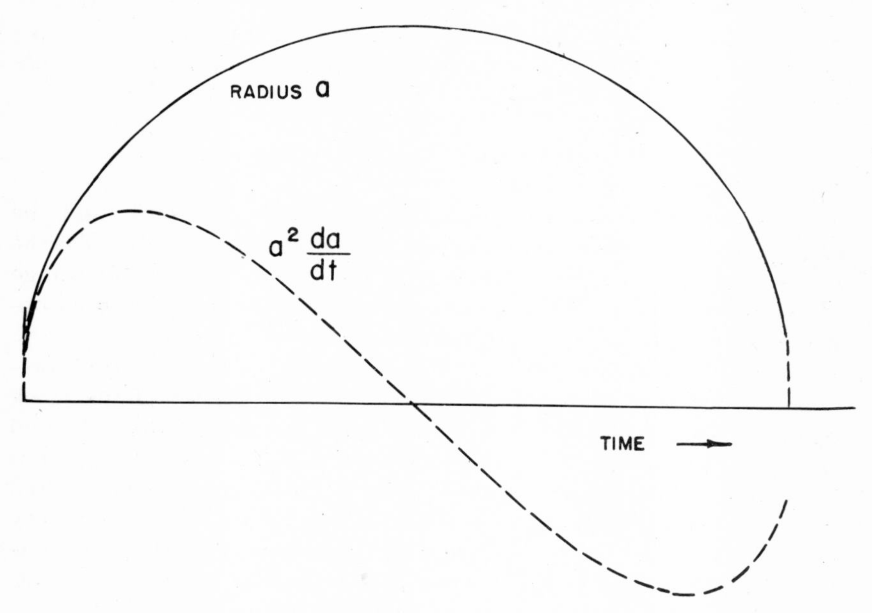

Fig. 9.2 Variation of the radius a and pressure function $a^2 \frac{da}{dt}$ during the bubble pulsation.

This does not imply that the later portions of a pressure wave can be disregarded; their effect is determined by the magnitude of the pressure and the dynamical properties of the structure.

C. *The pressure in later phases of the motion.* As the gas sphere expands, the pressure falls off more and more gradually, reaching a minimum value at times near the instant of maximum expansion. At this time, the pressure at points in the vicinity of the bubble is less than hydrostatic, corresponding physically to the limit of radial oscillation in which the gas pressure has fallen below the equilibrium hydrostatic value. Mathematically, the minimum pressure, except in the immediate vicinity of the bubble surface, is determined by the minimum value of $d/dt\ (a^2da/dt)$. The nature of variation of this function is shown qualitatively by Fig. 9.2. The radial velocity da/dt decreases continuously from its initial positive value to a large negative value at the instant of greatest contraction and the function a^2da/dt has a rounded sawtooth outline as indicated. Over most of the cycle its slope, and hence the pressure difference (gauge pressure) is negative, and only near the times of maximum contraction is the slope positive and the pressure greater than hydrostatic. The pressure builds up very rapidly, however, and the pressure variation is thus in the nature of relatively large, but short-lived, excess pressures near the time of maximum contraction between much longer intervals of negative gauge pressures.

The pulses of positive pressure are naturally of primary interest, and it is evident that their character must be determined by the internal energy and pressure inside the gas sphere. The pressure in the gas is, however, considerably modified by the translational velocity of the sphere and energy associated with it. As a result, the secondary pressure pulses vary greatly in magnitude and duration for different velocities of migration at the times of their emission. A further effect of migration is the displacement of the bubble during the cycle between the short intervals of appreciable pressure, which displaces the effective origin of these pulses vertically from the original location of the charge. Thus, if gravity alone is considered, the secondary pulses always appear to be emitted from a point closer to the surface than the origin of the shock wave, the magnitude of the displacement being greater for larger charges. If there are boundaries sufficiently near to modify the flow appreciably, both the total displacement before the compression sets in and the extent to which it is developed are materially affected. The secondary pulses from explosions and their effectiveness at any point in producing damage must therefore be evaluated not only in terms of the charge and distance from it, but also in terms of the depth of water, proximity of boundary surfaces, and the orientation relative to the charge of the point at which the pulse is of interest.

D. *The optimum peak pressure.*[1] From the analysis of the function $d/dt\,(a^2\,da/dt)$, it is reasonable to expect that the peak value of pressure due to this term occurs when the bubble has contracted to its minimum, for in this case there is no kinetic energy of radial flow. It is also reasonable to expect that the peak pressure will be maximized when the bubble has no vertical motion, as in this case all the energy is concentrated in compression of the gas products. Anticipating these results, proved in section 9.3, the optimum peak pressure P_{max} at a point r is, from Eq. (9.4),

$$P_{max} - P_o = \rho_o \frac{1}{r}\frac{d}{dt}\left(a^2 \frac{da}{dt}\right)_{a=\bar{a}}$$

[1] The analysis here and in section 9.3 follows closely the work of Shiffman and Friedman (102).

where $\bar{a}$ is the minimum radius of the gas sphere at the time for which $da/dt = 0$. At this time, the pressure on the surface of the sphere must equal the gas pressure $P(\bar{a})$, and hence

$$P_{\max} - P_o = \frac{\bar{a}}{r} P(\bar{a}) \tag{9.10}$$

if P_o is neglected in comparison with $P(\bar{a})$. At the time of the minimum, the excess pressure in the water thus falls off inversely as the distance r from the value $P(\bar{a})$ on the gas sphere.

In order to compute the value of $P_{\max}$ in a given case, it is necessary to evaluate $P(\bar{a})$ and $\bar{a}$ for the explosive products. This is conveniently done by noting that when $a = \bar{a}$, all of the available energy Y is in the form of internal energy $E(a)$ of compression of the gas. For product gases obeying the law $P_a V^\gamma = k$, where V is the volume per gram of explosive and P_a is in dynes/cm.², we therefore have

$$E(\bar{a}) = Y = \frac{P_{\bar{a}}\bar{V}}{\gamma - 1} = \frac{P_{\bar{a}}}{\gamma - 1}\left[(30.5)^3 \frac{4\pi}{3} \bar{a}^3\right]$$

if $\bar{a}$ is the minimum radius in feet. Solving for $P_{\bar{a}}$ and substituting in Eq. (9.10) gives

$$P_{\max} - P_o = \frac{3}{4\pi(30.5)^3} (\gamma - 1)\, Y\bar{a}^{-2} \tag{9.11}$$

The value of $\bar{a}^{-2}$ can also be determined from the condition $E(\bar{a})/Y = 1$. The expression for $E(a)/Y$ in terms of a has been worked out in section 8.9, and substitution of the value of $\bar{a}$ obtained by setting $E(\bar{a})/Y = 1$ gives

$$P_{\max} - P_o = \left(\frac{3 \cdot 454}{4\pi(30.5)^3}\right)^{1/3} k \left(4.19 \cdot 10^7 rQ \frac{(\gamma - 1)}{k}\right)^{1 + \frac{2}{3(\gamma-1)}} \frac{W^{1/3}}{r}$$

As written, this equation is expressed in the units of $P(a)$ (dynes/cm.²), r is the fraction of the detonation energy Q expressed in cal./gm. of explosive, W is the charge weight in pounds, and r is in feet. Converting the equation to express pressure in the more generally employed units of lb./in.² gives

$$P_{\max} - P_o = 2.26 \cdot 10^{-6} k \left(4.19 \cdot 10^7 rQ \frac{(\gamma - 1)}{k}\right)^{1 + \frac{2}{3(\gamma-1)}} \frac{W^{1/3}}{r} \tag{9.12}$$

Substituting the parameters usually employed for TNT ($\gamma = 1.25$, $k = 7.83 \times 10^9$, $rQ = 440$) gives

$$(9.13) \qquad P_{\max} - P_o = 2590 \frac{W^{1/3}}{r}$$

For comparison with this result, the peak shock wave pressure P_m is experimentally found to be given by (see section 7.4)

$$P_m = 2.16 \times 10^4 \left(\frac{W^{1/3}}{R}\right)^{1.13}$$

At the same distance r, the maximum pressure in the first secondary pulse is thus less than 20 per cent of the shock wave peak pressure if the latter exceeds 1,000 lb./in.^{2}, and later pulses are much weaker, owing to energy losses in successive contractions. One might therefore be inclined to dismiss the effects of secondary pulses as of little importance, particularly when it is considered that the most favorable case for generation of high pressures has been computed. This conclusion is, however, not justified, because the duration of the bubble pressure pulse is much greater and the impulse, or time integral of pressure, is comparable to the impulse of the shock wave. Both the magnitude and duration of impulsive pressures need to be considered in most situations, and the large impulse of the first secondary wave, plus the possibility of its originating nearer a target than the primary shock wave, may make it comparable in effect to the shock wave. The later pulses are so much weaker that it is doubtful if they are ever of great importance in causing damage, although their magnitude and time of occurrence are of considerable interest in analysis of energy losses (see, for example, section 9.4).

9.3. Pressures During the Contraction Phase

The preceding discussion has shown that the interesting times, in which appreciable pressures are developed, occur when the bubble radius a is small. With this condition, the general energy equation (8.23) for incompressible motion can be simplified by neglecting the potential energy of buoyancy of the cavity in comparison with the total energy. We therefore have, using the dimensionless form of the equation of motion introduced by Shiffman and Friedman (102),

$$(9.14) \qquad a^{*3}\left[\left(\frac{da^*}{dt^*}\right)^2 + \frac{1}{6}\left(\frac{db^*}{dt^*}\right)^2\right] + k^*a^{*-3/4} = 1$$

where $k^* = 0.0607\, z_o^{1/4}$ for TNT, z_o being the hydrostatic depth. The buoyancy of the bubble should, for consistency, be neglected also in Eq. (8.25) for the translational momentum, which therefore becomes

$$\frac{d}{dt^*}\left[\frac{1}{3} a^{*3} \left(\frac{db^*}{dt^*}\right)\right] = 0 \tag{9.15}$$

The bracketed term represents the vertical momentum acquired during the expanded phase of the motion, which thus remains essentially constant near the minimum radius, when the buoyant force of the gas is small. (See section 8.5 for a more detailed discussion.)

In order to determine the pressure, it is necessary to evaluate the quantity $d/dt^*\,(a^{*2}\, da^*/dt^*)$, which may be written

$$\frac{d}{dt^*}\left(a^{*2}\frac{da^*}{dt^*}\right) = a^{*2}\frac{d^2a^*}{dt^{*2}} + 2a^*\left(\frac{da^*}{dt^*}\right)^2 \tag{9.16}$$

It is convenient, in solving the energy equation (9.14) for da^*/dt^*, to eliminate the velocity db^*/dt^* by introducing the constant of vertical momentum $\bar{s} = \frac{1}{3}\, a^{*3}\, db^*/dt^*$. Solving Eq. (9.14) for da^*/dt^* then gives

$$\left(\frac{da^*}{dt^*}\right)^2 = \left(\frac{1}{a^{*3}}\right) - \frac{3}{2}\cdot\frac{\bar{s}^2}{a^{*6}} - k^*a^{*-15/4} \tag{9.17}$$

Substitution in Eq. (9.16) for da^*/dt^* and d^2a^*/dt^{*2} (obtained by differentiation) gives for the dominant term in pressure at a point r

(9.18)

$$P - P_o = \frac{2P_oL^*}{3}\frac{1}{r}\frac{d}{dt^*}\left(a^{*2}\frac{da^*}{dt^*}\right) = \frac{L^{*3}}{C^{*2}}\frac{a^*}{r}\left[\frac{1}{2a^{*3}} + \frac{3\bar{s}^2}{2a^{*6}} - \frac{k^*}{8a^{*15/4}}\right]$$

where L^* and C^* are the scaling factors for length and time defined by Eq. (8.51).

A. *The peak pressure.* The condition on the radius a for the peak value of pressure is easily shown from this equation to be that a^* have its minimum value. The derivative dP/da^* is given by

$$\frac{3r}{2L^*P_o}\cdot\frac{dP}{da^*} = -\left[\frac{1}{a^{*3}} + \frac{15}{2}\frac{\bar{s}^2}{a^{*6}} - \frac{11}{32}\cdot\frac{k^*}{a^{*15/4}}\right]$$

from which, using Eq. (9.17),

$$\frac{3r}{2L^*P_o}\cdot\frac{dP}{da^*} = -\left[\left(\frac{da^*}{dt^*}\right)^2 + \frac{9\bar{s}^2}{a^{*6}} + \frac{21}{32}\cdot\frac{k^*}{a^{*15/4}}\right]$$

The negative value of the derivative requires that P decrease with increasing a^*, and the peak pressure therefore occurs for the smallest value of bubble radius a^* which is attained.

The peak pressure realized under any conditions can be determined from Eq. (9.18) from the minimum value of radius a and the translational momentum $\bar{s}$ at this time. The factors influencing its value become more evident if its pressure is expressed in terms of the radial velocity da/dt and vertical velocity U at the time of minimum radius. Substitution in Eq. (9.18) from Eq. (9.17) gives

$$P - P_o = \frac{P_o L^*}{3} \frac{a^*}{r} \left[\frac{1}{2} \left(\frac{da^*}{dt^*} \right)^2 + \frac{1}{4} \left(\frac{db^*}{dt^*} \right)^2 + \frac{3k^*}{8a^{*15/4}} \right]$$

and restoring dimensional variables (see section 8.9) gives

$$P - P_o = \frac{a}{r} \left[\frac{1}{2} \rho_o \left(\frac{da}{dt} \right)^2 + \frac{1}{4} \rho_o U^2 + kV^{-\gamma} \right] \tag{9.19}$$

The first two terms in this equation represent the kinetic energies associated with radial and translational motion of the bubble, and thus contribute what might be called hydrodynamic pressures resulting from these two types of flow. The last term is simply the internal pressure P_a in the gas sphere of specific volume V.

The dynamic pressure terms in Eq. (9.19) decrease for smaller values of the minimum radius, da/dt being zero at the minimum, and the term resulting from compression of the gaseous products increases. The optimum value of pressure therefore depends on the relative importance of gas pressure and dynamic pressure from translation. The principle of stabilization, as formulated by Friedman and Shiffman, is simply the statement that the optimum value of peak pressure in the surrounding water is obtained for very nearly the conditions which maximize the gas pressure. These occur for initial conditions such that the gas sphere has no vertical velocity at the time of greatest contraction and the radius at this time has its smallest value.

At the time of peak pressure, the radial velocity da^*/dt^* is zero and it is convenient in establishing the principle of stabilization to express the peak pressure P_m in terms of the momentum factor $\bar{s}$ which is proportional to the vertical velocity U^* of the bubble center. Setting $da^*/dt^* = 0$ in Eq. (9.17) gives

$$\frac{1}{\bar{a}^{*3}} - \frac{k^*}{\bar{a}^{*15/4}} = \frac{3}{2} \frac{\bar{s}^2}{\bar{a}^{*6}} \tag{9.20}$$

It is convenient to express this relation in terms of the fraction $E(\bar{a})$ of the total energy Y which is in the form of internal energy of the gas at the minimum. Denoting this fraction by $\bar{e}$, we have $\bar{e} = E(\bar{a})/Y = k\bar{a}^{*-3/4}$, and Eq. (9.20) becomes

$$\frac{1 - \bar{e}}{\bar{e}^4} = \frac{3}{2}\left(\frac{\bar{s}}{k^{*2}}\right)^2 \tag{9.21}$$

and $\bar{e}$ is therefore a function only of $\bar{s}$ for a given explosive material. Expressing Eq. (9.18) in terms of $\bar{e}$ gives

$$P_m - P_o = \frac{1}{r}\left(\frac{P_o L^*}{4k^{*8/3}}\right)\bar{e}^{8/3}(4 - 3\bar{e}) \tag{9.22}$$

and the pressure is therefore also determined in terms of $\bar{s}$ by Eq. (9.21). Differentiation of Eq. (9.22) shows that P_m is maximized for $\bar{e} = 32/33$, corresponding to 97 per cent of the total energy returned to the gas. The corresponding value of $\bar{s}$ is then, by Eq. (9.21), $\bar{s} = 0.151\ k^{*2}$. The quantity k^* is, however, so small that the maximum is obtained for all practical purposes at $\bar{s} = 0$, corresponding to the stabilized position of zero velocity of the gas sphere at the time of minimum size.

It is sometimes convenient to express the peak pressure P_m in terms of the minimum radius $\bar{a}^*$, with the result that

$$P_m - P_o = \frac{1}{r} P_o L^* \frac{1}{\bar{a}^{*2}}\left[1 - \frac{3k^*}{4}\bar{a}^{*-3/4}\right] \tag{9.23}$$

The optimum peak pressure P_{max} derived more directly in section 9.2 is readily obtained from this equation, for when da^*/dt^* and db^*/dt^* are zero, $\bar{a}^* = k^{*4/3}$ very nearly, and hence

$$P_{max} - P_o = \frac{P_o L^*}{4} \cdot \frac{\bar{a}^{*-2}}{r} = \frac{3}{4\pi} \cdot \frac{Y}{4} \cdot \frac{\bar{a}^{*-2}}{r}$$

which is Eq. (9.11) for the case $\gamma = 5/4$ if a is expressed in cm.

The underlying reason for the result that optimum peak pressure occurs when the bubble is stationary lies in the fact that only for this condition is all of the originally available energy of the gas products returned to them. If the bubble is in motion, however, part of the total energy remains in the surrounding water as kinetic energy of flow and is not returned. The pressure of the gas proves to be the more important factor in determining the peak pressure in the water and the latter is therefore largest when the flow of water is very nearly stopped.

The variation of the peak pressure factor $\bar{e}^{8/3}/(4-3\bar{e})$ in Eq. (9.22) with vertical velocity is such that the pressure has a rather flat maximum near the stabilized position, but is considerably smaller than the optimum value for sufficiently large values of $\bar{s}$. The pressure is, from Eq. (9.21), an even function of $\bar{s}$, and the curve for negative values of $\bar{s}$ is therefore the reflection of the one for positive $\bar{s}$. The value of $\bar{s}$ for given initial conditions of charge weight and depth must be obtained by integration of the equation of motion for vertical momentum. Integrations of this kind have been carried out in special cases by Shiffman and Friedman for the more general problem where the effects of the surface and bottom are included, and their result, given in section 8.10, is compared with experiment in section 9.5. Approximate calculations for various charge weights and depths, neglecting surface effects, are given in graphical form in Road Research Laboratory reports (93).

The effect of vertical velocity in decreasing the maximum peak pressure has been illustrated by Taylor (107) in a calculation for 4.65 pounds of TNT fired 20 feet below the surface. If the vertical motion is neglected, the calculated internal pressure in the gas sphere has a maximum value of 8,300 lb./in.^{2} Because of the vertical momentum and energy acquired, however, the minimum radius actually calculated is 3.3 times that obtained assuming no migration and the internal pressure of the gas is reduced, by a factor $(3.3)^{15/4}$, to about 1 per cent of its optimum value. The peak pressure will not be reduced by as large a factor because of the vertical flow velocity, on which the pressure also depends. Other factors must also be kept in mind. The first is that, although the peak pressure is considerably decreased, the duration is longer and the time integral of pressure, or impulse, is not affected to nearly the same extent (see part B). The second factor is the repulsion of the bubble by the free surface which, although insufficient in this case to prevent net migration, would reduce it somewhat and hence partially neutralize the effect of gravity. The calculation should not, therefore, be taken literally as demonstrating that pressures in the water are reduced by factors of a hundred in representative cases,[2] but rather as an illustration that the peak pressure may be appreciably modified by vertical migration. The actual motion of the gas sphere in cases encountered in practice is nearly always appreciably modified by the proximity of either the surface or bottom, if not other boundaries. Further discussion of pressure variations with the migration is therefore deferred to section 9.6, where these factors are considered.

B. *The impulse.* It has been remarked in various places that the peak value of a transient pressure is not the only criterion for comparison of different conditions, nor is it even necessarily a significant one. If the duration of the wave under consideration is very much longer than other

[2] Statements of this kind must be properly interpreted, as they refer to values at equal distances from the source at the time the pressure is developed. This position will, if there is migration, not be the initial position of the charge.

times of interest (for example, time of deformation and damage of a structure), then the wave is effectively a hydrostatic pressure and its maximum value is of primary interest. If the opposite situation of a relatively short duration applies, then the peak value loses this predominant importance and the duration becomes equally significant. It is therefore necessary to consider the form of the secondary pulse in greater detail.

The discussion in part (A) of this section shows that the excess pressure, primarily determined at points not too near the bubble by $d/dt\ (a^2\, da/dt)$, is negative and small over most of the cycle of pulsation. The duration of positive pressure, which occurs during the contraction, is conveniently estimated by determining the radius a_o for which $d/dt\ (a^2\, da/dt) = 0$ and the pressure is hydrostatic (except for the Bernoulli term). The condition is equivalent to requiring that $(a^2\, da/dt)^2$ be maximized, and this quantity is conveniently obtained from the equation of motion as expressed in terms of vertical momentum $\bar{s}$ by Eq. (9.17). Setting the derivative $d/da^*\ (a^{*2}\, da^*/dt^*)^2$ equal to zero gives

$$1 - \frac{k^*}{a^{*3/4}} + \frac{3\bar{s}^2}{a^{*3}} - 4a^{*3} = 0$$

If the migration and internal energy are neglected by setting $\bar{s}$ and k^* equal to zero, the solution is $a^* = (\frac{1}{4})^{1/3} = 0.63$, corresponding to 63 per cent of the maximum radius. If values $k^* = 0.2$, $\bar{s} = 0.06$, corresponding to TNT and a large vertical momentum, are used, the value of a^* is 0.61. This shows that the point in the oscillation at which the pressure exceeds hydrostatic is insensitive to the vertical motion, as would be expected from the fact that the bubble is comparatively large at these times in the first cycle, and has not acquired appreciable velocity of translation.

The fact that the value of a^* for zero excess pressure occurs at times in the pulsation which are insensitive to any later migration implies that these times are approximately constant fractions of the period of oscillation. This fraction would be nearly constant under any conditions if the effect of migration is primarily to change the scales of length and time proportionately. Examination of numerical solutions show that this assumption leads to reasonably accurate results, and from such solutions the interval between minima and the nearest time of hydrostatic pressure is found to be approximately eleven per cent of the period.

A rough measure of the second pulse duration can be taken to be the time during which the pressure exceeds hydrostatic, which is thus twenty-two per cent of the period if the second cycle of pulsation is assumed to be of the same amplitude and duration as the first. This assumption, which ignores any acoustic radiation or energy loss in turbulence near the minimum, is evidently not a very good one. The duration as defined, however, is useful as a qualitative indication of the extent of times of interest rather than as an accurately measurable quantity of quantitative value, and a more precise estimate is hardly warranted. The calculations so far considered thus show that the peak pressure is fairly sensitive to the migration, but that the duration of positive gauge pressure depends primarily only on the period and is proportional to it. The pressures to be expected at other times are not indicated and must therefore be considered in more detail.

Explicit calculation of the pressure-time curve can of course be made by numerical evaluation of the quantity $d/dt\ (a^2\, da/dt)$ from computed values of a and t. This tedious and rather inaccurate process would have to be repeated for each case of interest, and does not therefore readily give a general view of what to expect under different conditions. A simpler approach, at the expense of less information, can be made by considering the positive impulse I, or time integral of excess pressure, for the interval of positive pressure. From its definition, this is given by

$$I = \int_{t_1}^{t_2} (P - P_o)dt = \frac{2P_oL^*C^*}{3} \cdot \frac{1}{r} \left(a^{*2} \frac{da^*}{dt^*} \right) \Bigg|_{t_1}^{t_2}$$

where the limits can be taken to correspond to the values of the integrated function when $P = P_o$. If the motion is assumed symmetrical about the minimum, the impulse is twice the value for the limits $P = P_{\max}$, $P = P_o$. For the first limit the bubble radius is a minimum and $da^*/dt^* = 0$. Hence we have

$$I = \frac{4P_oL^*C^*}{3} \frac{1}{r} \left[a^{*2} \frac{da^*}{dt^*} \right]_{P=P_o}$$

and expressing the bracketed quantity in terms of a^* and vertical momentum $\bar{s}$ gives

$$I = \frac{4P_oL^*C^*}{3} \frac{1}{r} \left[a^{*2} \left(\frac{1}{a^{*3}} - 1 - \frac{k^*}{a^{*3\gamma}} - \frac{3}{2} \cdot \frac{\bar{s}^2}{a^{*6}} \right) \right]_{P=P_o}$$

The radius a^* corresponding to $P = P_o$ was found to be approximately 0.61, and for this large a radius, the term in $\bar{s}^2$ is negligible and can be set equal to zero. The term $k^*/a^{*3\gamma}$ depends on the equation of state for the products, but the impulse will not be seriously in error for most conditions if the values $k^* = 0.2$, $\gamma = 5/4$ used for TNT are inserted. With these substitutions, we obtain $I = 0.73P_oL^*C^*/r$, and substituting values of L^* and C^* (from section 8.9) gives

$$I = 0.47(rQ)^{2/3} \frac{P_o}{z_o^{7/6}} \cdot \frac{W^{2/3}}{r} \tag{9.24}$$

where the energy rQ is in cal./gm., W is charge weight in pounds, and r and z_o are in feet. The impulse is then expressed in the same units as P_o times seconds. If P_o is chosen to be in lb./in.², $P_o = 0.446\, z_o$ for z_o in feet and the impulse in lb. sec./in.² is

$$I = 0.21(rQ)^{2/3} z_o^{-1/6} \cdot \frac{W^{2/3}}{r} \tag{9.25}$$

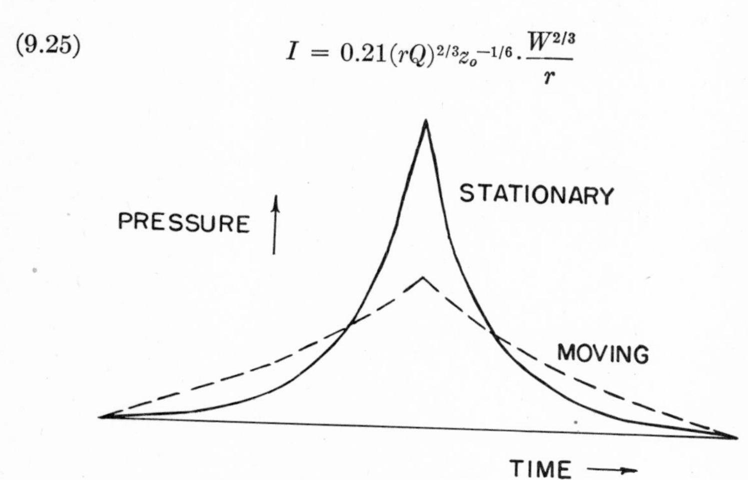

Fig. 9.3 Effect of bubble motion on form of the bubble pulse.

This equation shows that the impulse varies with charge weight and distance in the manner predicted for acoustic waves, as is to be expected, and also decreases slowly for increasing depth from the factor $z_o^{-1/6}$. As already noted, the total positive impulse obtained here depends only very slightly on migration, because the radius for which the pressure is hydrostatic is large and but little affected by migration. This fact, plus the fact that the peak pressure decreases considerably if the bubble has appreciable vertical velocity in its minimum, shows that the shape of the pressure-time curve in the region of the maximum must also depend considerably on the vertical velocity, in such a way as to keep constant the total area between the curve and the hydrostatic pressure P_o. Thus, if the bubble acquires appreciable vertical momentum the curve must be broader and lower near the maximum than if the bubble remains stationary, as sketched in Fig. 9.3.

A simple method has been used in a Road Research Laboratory

report (93) for illustrating the change in the bubble pressure pulse shape. This consists in computing a shape factor, defined as the ratio of the positive impulse to the area under a triangle of height equal to the peak pressure and base equal to the duration of positive pressure:

$$\text{Shape factor} = \frac{2I}{(P_m - P_o)\cdot \text{duration}} = \frac{2I}{0.22T(P_m - P_o)}$$

where T is the period of pulsation. A number of calculated values of these quantities for different weights and depths of TNT charges are given in Table 9.1, some of these values being taken from the Road Research Laboratory report already mentioned. These values have been computed neglecting the effect of the free surface and assuming infinite depth of water, in order to illustrate the effect of migration which decreases with increasing depth approximately as $z_o^{-5/6}$. They are not therefore directly applicable to most situations actually encountered. The increase in peak pressure and decrease in shape factor with greater depth, corresponding to a sharper pulse, and the slowly changing impulse are evident from the entries, which also show the orders of magnitude of these quantities for various conditions.

It is of interest to compare the computed impulse from Eq. (9.25) with values of impulse for shock waves. Assuming $rQ = 440$ cal./gm. for TNT at a depth d_o of 40 feet gives

$$I(\text{lb. sec./in.}^2) = 6.0\,\frac{W^{2/3}}{r}$$

Shock wave similarity curves for TNT give for the impulse, integrated to five times the time constant,

$$I(5\theta)\ (\text{lb. sec./in.}^2) = 1.3\,\frac{W^{0.63}}{R^{0.89}}$$

The total positive impulse from the secondary pulse at a depth of 40 feet is thus of the order of four times the value $I(5\theta)$ for the shock wave. This by no means implies that the secondary bubble pulse is four times as effective as the shock wave, even if impulse is the proper criterion for effectiveness in the circumstances of interest. In the first place, it is to be remembered that the secondary impulse calculation includes all pressures greater than hydrostatic. Much of this impulse comes from long intervals of low, slowly changing pressure, and hence ordinarily is both unimportant and difficult to observe.

On the other hand, the shock wave impulse quoted is an underestimate because the pressure-time curve is integrated to a relatively

Charge weight (lb.)	Depth (ft.)	Peak pressure x distance (lb./in.^{2} x ft.)	Duration (sec.)	Impulse x distance (lb. sec./in.^{2} x ft.)	Shape factor
1.0	10	830	0.041	8.0	0.47
	20	1,160	0.034	7.8	0.39
	40	1,750	0.026	7.4	0.32
16	20	1,300	0.087	50	0.89
	40	2,250	0.066	47	0.63
	80	4,130	0.046	43	0.45
300	40	2,100	0.175	314	1.71
	80	5,300	0.092	294	1.21
	160	11,500	0.079	268	0.59

Table 9.1. Calculated parameters of the first bubble pulse for several charge weights and depths.

very short time after the initial peak. At this time, the pressure is both fairly large and slowly decreasing, and if the integration were extended to the time of hydrostatic pressure a considerable increase in impulse would result. This increase must in fact be an appreciable fraction of one-half the secondary impulse computed above, for the reason that the slowly decreasing pressures in the tail of the shock wave are due largely to noncompressive flow and hence are to be computed in the same way as the secondary pressures. The proper conclusion from the impulse calculations is therefore that the secondary pulse has an impulse comparable to the shock wave because of its long duration, but is inferior to the shock wave in both peak pressure and energy flux. The question of energy radiated acoustically in the secondary pulse is of interest in this connection and is considered in the next section.

The variation of pressure with time during the phase of the motion when the bubble is large and the pressure less than hydrostatic presents no unusual features. It is evident from the slow variation of the function $a^2\, da/dt$ at these times that the pressure differences are small and smoothly varying. It is interesting, however, that the negative impulse is large, owing to the long time interval for which $P < P_o$. In fact, this negative impulse must equal the positive impulse when the pressure exceeds hydrostatic because the integral $\int (P - P_o)\, dt$ depends only on the difference in the values of $(a^2\, da/dt)$ at the two limits of integration. If these are corresponding points in successive cycles, the difference is zero, requiring that positive and negative areas during the cycle be equal. This negative impulse or suction phase is comparatively feeble, as the pressure difference never is greater than hydrostatic. The fact that the total impulse is so large illustrates further the danger of taking large values of impulse at face value without examining the pressures and durations in detail.

9.4. Energy Losses in the Pulsations

The discussion of fluid motion and pressures during motion of the gas sphere have so far been based on noncompressive theory. The explosive energy left after emission of the shock wave is thus assumed to exist entirely as potential and kinetic energy of flow of the water plus internal energy of the gaseous products. In this approximation, no mechanism has been provided for energy loss during the pulsations, the only change possible being reversible redistribution of the total energy between the gas and the surrounding water. The energy losses actually occurring in the course of the motion cannot therefore be estimated without taking into account, to some degree of approximation, mechanisms by which energy can be dissipated: the compressibility of the water, by which energy is radiated as a wave and ultimately dissipated in heat as the wave passes to infinite distance, and turbulent flow around the bubble.

A. *Energy loss by radiation.* Herring (46) has shown that, if vertical motion and internal energy of the gas sphere are neglected, an integral of the equation of motion can be obtained which takes account of compressibility of the water to a first approximation. This expression, the derivation of which is outlined in section 8.7, is

$$(9.26) \quad 2\pi\rho_o a^3 \left(\frac{da}{dt}\right)^2 - 4\pi \int_{a_m}^{a} [P_a - P_o] a^2 dt = \frac{4\pi}{c_o} \int_{a_m}^{a} \frac{dP_a}{dt} a^3 da$$

where a_m is the maximum radius of the bubble, P_a is the pressure on the gas sphere of radius a, and c_o is the velocity of sound. In deriving this result, a number of approximations have been made. In particular terms of order $(1/c_o)(da/dt)$ have been dropped, which requires strictly that the flow velocities everywhere be much less than the velocity of sound.

The terms on the left side of Eq. (9.26) represent the kinetic energy of the gas as the radius contracts from a_m to a. Except for a constant, this side is thus the conservative energy of noncompressive motion, and changes in its value determined by the right hand side, which is always negative, must represent energy radiated by compression of the water. An exact solution of Eq. (9.26) would require a second equation determining $P(a)$ in terms of radius a. If, however, it is assumed that the difference from noncompressive motion is not great, the solution already obtained for $a(t)$ in this approximation may be used to evaluate the right hand side and thus obtain a rough estimate of the energy loss. It is evident from the preceding section that this term can only be significant while the bubble is near its minimum size, and the major contribution to the integral can come only at these times. This conclusion is readily verified explicitly by substituting expressions for $dP_a/dt = dP_a/da \cdot da/dt$, and $a(t)$, but this analysis is omitted here and only the approximation to $a(t)$ suitable for small a will be considered.

The gas pressure $P(a)$ for an adiabatic equation of the form $P_a \cdot a^{3\gamma}$ = constant is conveniently written as $P_a a^{3\gamma} = P_{\bar{a}} \cdot \bar{a}^{3\gamma}$, where $P_{\bar{a}}$ is the maximum pressure corresponding to the minimum radius $\bar{a}$, and hence $dP/dt = (-3\gamma/\bar{a})\, P_{\bar{a}}\, (a/\bar{a})^{-3\gamma-1}\, (da/dt)$. Near the minimum, the buoyancy of the gas sphere is negligible and the energy equation is $2\pi\rho_o a^3\, (da/dt)^2 + E(a) = Y$, and hence

$$\frac{da}{dt} = \sqrt{\frac{Y}{2\pi\rho_o}}\, a^{-3/2} \sqrt{1 - \frac{E(a)}{Y}}$$

We have, however, that

$$E(a) = \frac{4\pi}{3}\frac{Pa^3}{\gamma - 1} \quad \text{and} \quad \frac{E(a)}{Y} = \left(\frac{a}{\bar{a}}\right)^{-3(\gamma-1)}$$

since $E(\bar{a}) = Y$, there being no kinetic energy when $a = \bar{a}$. Substituting in the right side of Eq. (9.26) gives

$$\Delta Y = \text{Energy radiated}$$
$$= -\frac{4\pi}{c_o}\left(\frac{Y}{2\pi\rho_o}\right)^{1/2}\frac{3\gamma}{\bar{a}}P_{\bar{a}}\int a^{-3/2}\left[1 - \left(\frac{a}{\bar{a}}\right)^{-3(\gamma-1)}\right]\left(\frac{a}{\bar{a}}\right)^{-3\gamma-1} a^3 da$$

The limits on the integral should strictly be from the first maximum radius a_m to the next maximum to get the total energy radiated. In the noncompressive theory a is symmetrical about the minimum radius $\bar{a}$ and we can therefore take twice the integral from a_m to $\bar{a}$. The integrand is large only near $a = \bar{a}$ and the limit a_m can be replaced by $a = \infty$ without appreciable error. Letting $a/\bar{a} = x$ then gives

$$\Delta Y = \frac{24\pi\gamma}{c_o}P_{\bar{a}}\left(\frac{Y}{2\pi\rho_o}\right)^{1/2} a^{-3/2}\int_1^\infty [1 - x^{-3(\gamma-1)}]^{1/2}x^{1/2-3\gamma}dx$$

which can be evaluated in terms of gamma functions to give the fractional energy loss

$$(9.27) \qquad \frac{\Delta Y}{Y} = \left[2\sqrt{6}\,\frac{\gamma}{\gamma-1}\cdot\frac{\Gamma\left(\frac{3}{2}\right)\Gamma\left(\frac{2\gamma-1}{2\gamma-2}\right)}{\Gamma\left(\frac{5\gamma-4}{2\gamma-2}\right)}\right]\frac{P_{\bar{a}}{}^{1/2}}{c_o\rho_o{}^{1/2}}$$

where the relation $Y = (4\pi/3)P_{\bar{a}}\bar{a}^3/\gamma - 1$ has been used.

This expression should also represent the energy flux through any sphere drawn around the bubble, and Herring's result, Eq. (9.27), has independently been derived by Willis (121) in this way. Willis calculates the energy flux using the acoustic expression relating excess pressure and particle velocity $u = (P - P_o)/\rho_o c_o$, which gives for the energy flux across a sphere of radius r

$$\Delta Y = \int 4\pi r^2 u(P - P_o)dr = \frac{4\pi}{\rho_o c_o}\int (P - P_o)^2 dt$$

Willis' approximation in evaluating the integral is that the pressure P is given by the noncompressive theory; substitution of $P - P_o = (\rho_o/r)$

$d/dt\,(a^2 da/dt)$ and elimination of t by using $da = (da/dt)\,dt$ then leads to a result equivalent to Eq. (9.27).[3]

According to Eq. (9.27), the fractional energy radiated in the secondary pulse varies as the square root of the maximum gas pressure, and is therefore largest when the bubble attains its smallest minimum volume without vertical motion. If the value $\gamma = 1.25$ is used, the expression becomes

$$\frac{\Delta Y}{Y} = 1.87 \frac{P_{\bar{a}}^{1/2}}{c_o \rho_o^{1/2}}$$

Taylor's calculation that at the first minimum $P_{\bar{a}} = 8300$ lb./in.² for TNT if there is no migration gives $\Delta Y = 0.31Y$, which is a significant fraction of the total energy. As the pressure $P_{\bar{a}}$ varies as $\bar{a}^{-3\gamma}$, a minimum radius 3 times its smallest value, as computed in the example of section 8.5, gives $\Delta Y/Y = 0.03$. Hence only 3 per cent of the bubble energy is calculated to be lost in the acoustic pressure pulse in this example.

Measurements of periods of successive oscillations of the gas bubble give a direct measure of the total fractional energy loss in successive contractions, as shown in section 8.3. The most extensive data of this kind show that for 200 pound charges of a mixed explosive fired at depths from 60 to 800 feet about ⅔ of the energy Y of the first pulsation is lost in the first contraction, this figure showing little systematic variation with depth when the effect of the free surface on the period is taken into account. Comparable figures are obtained for other explosives and charge weights, and the figure of 30 per cent for acoustic radiation accounts for less than half of the total energy loss during the first contraction. In addition, it should be remembered that this calculated figure is the result of rather crude approximations, and experimental values obtained by integration of observed pressure-time curves give smaller values (see section 9.6).

B. *Energy loss by turbulence.* A detailed accounting for the balance of the energy loss in terms of dissipation in turbulent flow would be a difficult task which has not so far been attempted. It is clear that these losses will be largest near the minimum, particularly while the bubble is in its extreme stage of instability while contracting, and the form is distorted from its originally spherical shape. The energy loss by thermal conduction for the short times and small temperature gradients is readily shown to be negligible, even in the gas sphere, if there is no turbulence. The remaining mechanism of viscosity is a perfectly reasonable one, and there appears no reason to doubt that it, together with acoustic radi-

[3] The mathematical equivalence of the two approaches is readily demonstrated by transformation of Eq. (9.26).

ation, is capable of accounting for the dissipation indicated by period measurements.

One result which is roughly accounted for by theory is the fact that the total energy loss near the first minimum changes very little regardless of migration, although the acoustic energy loss is greatly decreased if the bubble has appreciable upward velocity. Under these conditions, the bubble is considerably larger and photographs of such bubbles show a flattening of the after surface (which may often be concave). A turbulent wake is also formed as a result of water breaking away from the bubble surface, so to speak, and forming eddies in which energy is dissipated by work done against viscosity.

The hydrodynamic drag due to viscosity must, if the bubble moves upward with a constant velocity $\dot{U}$, be equal to the buoyant force of the hollow, and can be expressed in terms of a drag coefficient C_d defined by

$$\text{Buoyant force} = \rho_0 g V = C_d[\tfrac{1}{2}\rho_0 U^2 A]$$

where A is the projected area of the bubble normal to the direction of motion and V its volume. Taylor and Davies (110) have measured velocities of air bubbles released under liquids and found values of C_d of the order 1.0 sec.$^{-1}$, which had rather large scatter, for volumes ranging from 1.5 to 30 cm.^{3} The time rate of energy dissipation in such motion must be $U \times$ buoyant force, and hence is given by

$$\text{Rate of dissipation} = \tfrac{1}{2}\rho_0 A C_d U^3$$

Taylor and Davies were able to show that the loss so obtained was of the same order as values calculated from a theory of distortion of a bubble in an assumed turbulent field of viscous flow.

The magnitude of energy losses from turbulence in upward bubble motion has been estimated by the above equation for the Road Research Laboratory measurements (95) of the bubble from a one ounce charge detonated three feet below the surface. The rate of dissipation increases as the cube of the velocity and hence is appreciable only near the minimum radius, more than eighty per cent of the loss occurring in the last millisecond of the eighty msec. period. The total loss up to the first minimum for the estimated velocity at this time amounted to only four per cent of the total bubble energy. The velocity during the short time near the minimum is difficult to determine accurately, too low a value being probable, and the drag coefficient might well be larger for the irregular explosion bubble than for the air bubbles measured by Taylor and Davies. It is therefore not unreasonable to suppose that the energy losses in vertical motion can be explained, at least semi-empirically, in this way. (It should be noted that, in the case considered, the kinetic energy of vertical flow at the minimum was estimated to be about six per cent of the total.)

We have already, in section 8.7, discussed the instability of the spherical bubble form assumed in the approximate theories which have led to fairly complete estimates of the motion. The theories of such departures apply strictly only to small perturbations, but do at least make reasonable the observed flattening of the bubble as it contracts. This flattening implies an increase of hydrodynamic inertia to translation, and is thus consistent with the small observed upward velocity of the bubble as compared with calculated results for a spherical form. The energy considerations, however, seem definitely to rule out the possibility of accounting for the decreased migration by this mechanism alone. The observed migrations during the second contraction are comparable to those during the first, which suggests that the vertical motion is very nearly stopped at the end of the first contraction, the bubble starting its second expansion practically from rest. The only accounting for the energy losses therefore seems to be in turbulence near the bubble of the general type described. If this is true, attempts to refine the noncompressive theory without including effects of this kind are pointless. It should further be no surprise to find the observed bubble pressures in poor agreement with calculations based on a spherical bubble in an inviscid fluid if the predicted velocities of translation near the minimum are large.

9.5. Problems in Measurement of Secondary Pulses

Measurements of pressure in the secondary, or bubble, pulses from underwater explosions present a number of difficulties, both in measurement and in proper evaluation of the experimental data. Unless these problems are understood, conclusions as to the significance of experimental results are apt to be inaccurate, if not misleading. It is therefore important to consider the various factors involved in the measurements before examining the specific results in detail.

A. *Experimental errors.* The nature of the experimental problems can be brought out by comparing the form of the pressure-time curve for the secondary pulse with that for the shock wave. The bubble pressures are very much smaller than the peak shock wave pressure, have a relatively slow initial rate of rise rather than a discontinuous shock front, and have an appreciable fraction of their peak value for much longer times than the shock wave. Time and pressure scales in which the complete bubble pulse is reproduced to a reasonable scale thus represent the shock wave as a very high and narrow spike, as sketched in Fig. 9.4.

The principal difficulties in piezoelectric gauge recording of bubble pressures are spurious gauge noise, cable signal, microphonic response of amplifiers, and need for high sensitivity and good low frequency response. It has been found that placing gauges too close to the expanding gas sphere gave rise to spurious random noise on the records, presumably as a result of disturbance of the gauge and associated gear by the shock wave and later outward flow of water. Satisfactory working distances depend on the size of charge; a rough estimate from experience is that the gauge should be at least 2 maximum bubble radii from the charge, 60 feet from a 300 pound charge having been found satisfactory.

A gauge system which is to measure bubble pressures must also withstand the force of the shock wave, as it can hardly be inserted in place during the time between the two events. This obvious fact presents a number of difficulties and limitations. Tourmaline piezo-

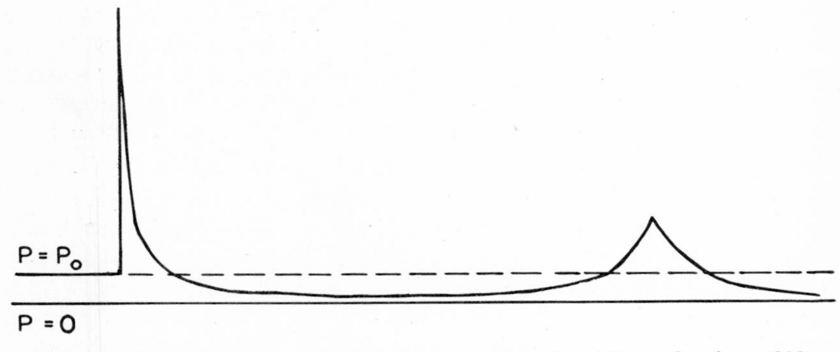

Fig. 9.4 Pressure-time curve for shock wave and bubble pulse from 300 pounds of TNT fired at a depth of 50 feet.

electric gauges, which are the simplest and most rugged, are comparatively insensitive, and their use introduces the further complication of cable signal. This phenomenon, discussed in section 5.8, is particularly disturbing in bubble pulse measurements. This is because the large stresses, induced by the shock wave as well as by the pulse of interest, develop this spurious charge which cannot leak away rapidly, owing to the necessity for good low frequency response of the electrical system if the long duration pulses are to be faithfully recorded.

The preceding shock wave also occasions difficulties with electrical recording equipment in large charge work, as it often must strike the recording site before the desired gauge signal occurs. Unless the amplifiers used are especially protected against microphonics, the recording system introduces serious distortion and may even become inoperative. This is a particularly difficult problem because of the need for amplifiers of high sensitivity and, by ordinary standards, superlative low-frequency response (see sections 5.8, 5.9).

Perhaps less sophisticated, but equally serious, problems arise in assuring known, satisfactory, and reproducible experimental conditions.

These difficulties are particularly evident in large charge work, which must usually be conducted under conditions somewhat less than perfect, and lead to errors in determining relative positions of gauge, charge, and water boundaries. The total depth of water, character of the bottom, and migration of the bubble are some of the factors not ideally determined, and the consequence is scatter of the data.

B. *Problems and errors of analysis.* The analysis of pressure-time records is complicated by a number of factors, the most important of which are choice of a reference axis for pressure and correction for the reflected pressures from the surface and bottom. The experimental results for pressure and derived quantities (energy flux and impulse) must be measured with respect to some chosen pressure level. For comparison with theory, this value is best taken as hydrostatic pressure (zero gauge pressure). An accurate determination of this level in records such as the one reproduced in Plate X is simple for the shock wave, but difficult for the bubble pulse because of its smooth, gradual rise from values less than hydrostatic while the bubble is large. A true baseline is thus not accurately defined except for times preceding the shock wave, and judgment must be used in drawing a reference line. The resulting errors are not large for calculation of peak pressure, but may seriously affect values of impulse and energy flux density (proportional to $\int(P - P_o)dt$ and $\int(P - P_o)^2dt$ respectively), owing to the long intervals of small excess pressure which contribute large areas. Errors of this kind in the investigations summarized in section 9.6 introduced scatter of the order of five per cent in peak pressure and ten per cent in impulse and energy.

The second problem in analysis of the recorded pressure-time curves is that the pressure existing at a given point almost always is the superposition of the "direct" wave and reflected waves from the surface and bottom. In acoustic theory, the pressure wave striking a free surface is geometrically reflected as a wave of equal magnitude but negative pressure. The resultant absolute pressure at other points is then the algebraic sum of the pressure in this diverging reflected wave and the pressure existing as a result of the wave travelling in a straight line from the charge. It is readily calculated by simple superposition taking into account the later time of arrival and increased attenuation in the reflected wave. The simple picture may be inaccurate if the absolute pressure at any point becomes negative and cavitation sets in. The bottom reflection occurs as a wave of positive pressure with somewhat less initial amplitude than the direct wave, owing to the bottom not being perfectly rigid, acoustically speaking. In addition, there may be a ground wave, corresponding to signals transmitted through the sea bed and re-emitted into the water (see section 7.8). The proper amount of correction is thus not simply determined.

The observed pressures in actual situations are thus those of a composite wave formed by the direct wave and its reflections, as sketched in Fig. 9.5 for a gauge and charge position nearer the bottom than the surface. If comparison with theory is to be made simply, the direct wave profile is to be preferred, although the composite wave corresponds to the actual state of affairs for the particular positions of the point of observation and boundary surfaces. The durations of bubble pressures are so long that the difference between the two profiles is appreciable in many interesting or necessary situations, and some means of correction for reflections is necessary in order to compare with theory.

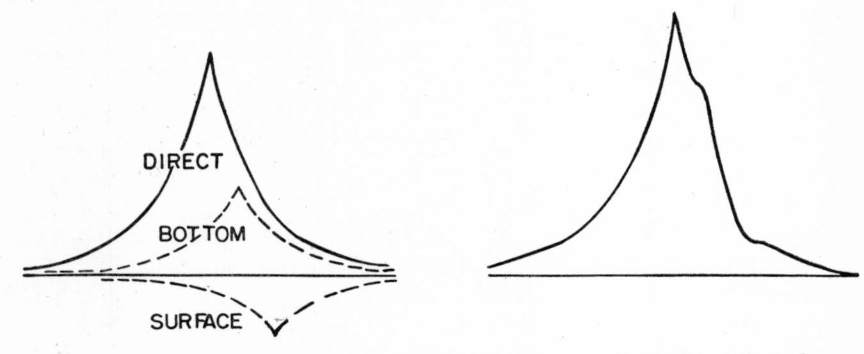

Fig. 9.5 Effect of surface and bottom reflections on observed bubble pulse pressures.

Rather rough corrections can be made using point-by-point calculations, starting at the first rise in pressure and adding positive or negative contributions at later times determined by the path differences. The errors in this procedure are evidently emulative, but are usually not serious up to the maximum. At later times, the errors rapidly become worse if any boundary is close, and if cavitation occurs near the surface no correction is possible. An expedient sometimes adopted is to measure to the peak and multiply impulse and energy flux density values by two, which is roughly equivalent to theoretical assumptions. The correction process and estimation of the peak time introduce random errors of the order of ten to fifteen per cent, being smaller for peak pressure.

The errors in analysis and interpretation are in addition to experimental errors arising from cable signal and incorrect distances. The magnitude of errors from all known causes is of the order five to ten per cent for peak pressure, ten to thirty per cent for impulse and energy, being greater for large charge measurements and the "direct" wave, corrected parameters. These errors would occur to some extent in measurements even if the phenomenon itself were perfectly reproducible. This, however, is not the case, as the discussion of part C makes clear.

C. *Lack of reproducibility of pressures.* Although measurements of bubble migrations and secondary pulse pressures are subject to some scatter because of the various errors enumerated, there is considerable evidence from these and other measurements that several aspects of the phenomena are not very reproducible under presumably identical conditions. For example, bubble pulse pressure-time curves on successive shots often show distinct qualitative differences in their general outline and occasionally exhibit one or more sharp narrow "spikes" of fairly large amplitude superimposed on the profile. This latter feature is plausibly explained by the hypothesis of collapse of local pockets of gas or other distortions formed on the contracting surface. This explanation is consistent with the observation that the spikes and other irregularities are more frequently observed for the hydrodynamically less stable condition of rapid vertical motion near the minimum. It is also in accord with the result that indicated positions of the sources of different spikes on a given record obtained by sound ranging measurements often disagree, sometimes by amounts comparable with the minimum radius of the bubble.

Comparison of pressure and migration results show that the scatter is generally much greater in measurements near the surface or bottom. As mentioned in section 8.11, these variations are not surprising in view of the sensitiveness of motion near the minimum to the exact conditions, such as surface waves and irregularities of the bottom. Willis and Ackroyd (122) have commented on evidence of lack of reproducibility of secondary pressures, particularly when the charge is fired on or near the bottom, weak and irregular pressures being observed. The conclusion that the motion during the short interval near the time of minimum radius is most critically affected by slight differences in experimental conditions is easily understood from the hydrodynamics of the motion. It is also supported by the fact that period and maximum radius measurements, which are quite insensitive to changes in this time, show very much less scatter. Period measurements, for example, are rather easily reproduced to plus or minus one per cent, and the chief uncertainty is selection of the time of minimum radius on the pressure-time record.

It is reasonable to conclude that the extreme precautions which are necessary to achieve much greater precision in experimental methods would be for many purposes unwarranted, owing to variations likely to be inherent in the phenomenon under many conditions and the rather approximate nature of theories so far developed.

9.6. Experimental Measurements of Secondary Pulse Pressures

A. *Small charge results.* It is to be expected that experimental pressure-time curves for bubble pulses will be in better agreement with theoretical predictions in the case of small charges. This is because the migrational velocities due to gravity are smaller, assuming that the surface pressure above the charge is atmospheric and not scaled to correspond to large charge conditions. Small charge data therefore provide the most favorable test of simple theory, and are experimentally much easier to obtain with some precision. Investigations of this kind are therefore considered first, although generally speaking these conditions are of less practical interest, and investigations with larger charges are taken up later.

(1). Shallow 1 ounce charges. The pressure-time curves for 1 ounce charges of Polar Ammon Gelignite at depths from 1.5 to 6 feet have been measured by the Road Research Laboratory (96). In the first series of measurements reported, gauges were mounted above, and at the same depth as the charge at distances from 1.5 to 2.5 feet. At a depth of 6 feet, the migration during the first pulsation was found to be about 10 inches upward, as compared with the theoretical estimate from Taylor's theory of 11 inches. The observed pressure pulse for this condition of relatively small migration was found to be in good agreement with one calculated from the approximate formula $P - P_o = (\rho_o/r)\, d/dt\, (a^2\, da/dt)$ for the same weight of TNT, as shown in Fig. 9.6, which is reproduced from the RRL report. (These two explosives are nearly equivalent, weight for weight, in calorific value.) This pressure was measured at the side of the charge; measurements above it indicated higher values, both because of the closer proximity of the bubble as a result of migration and because at very close distances flow pressures must be considered. Interestingly, the peak pressure at a fixed point above the charge was found to be higher in the second pulse than in the first, which further indicates the effect of bubble migration and the need for considering its influence in explosion damage.

For a charge depth of three feet, no migration effects were observed, indicating the repulsion of the free surface, and the peak pressures in the first pulse were some thirty per cent higher than at six feet, as is to be expected for a motionless sphere at the time of maximum contraction. In a second report, measurements were made at several depths which indicated a maximum peak pressure at about three and one-half feet, being smaller on either side of this value, while the impulse increased somewhat with increasing depth. These results were for a gauge two feet from the charge and at the same depth. Gauges mounted above and below the charge showed somewhat different variations explainable in terms of migration and surface reflections.

(2). Charges up to ½ pound in 24 feet of water. An extensive program of measurements, carried out at Woods Hole in cooperation with the Taylor Model Basin (2), has provided data on secondary bubble pulse pressures from charges of tetryl (25, 50, 120, 300 gm.), cast TNT (250 gm. plus 44 gm. tetryl booster), and other explosives which were fired at various depths in 24 feet of water. Only a few of the results, which were not completely analyzed at the time of writing, are given here.

The peak pressure and positive impulse for 50 gm. tetryl charges are plotted in Fig. 9.7. The peak pressures show a pronounced maximum near the upper rest position (measured experimentally as described in section 8.11 and indicated by the vertical line). The calculated theoretical value for the first pulse of 430 lb./in.^{2} at this depth and gauge distance of 3 feet (from Eq. 9.12) is comparable with the experimental

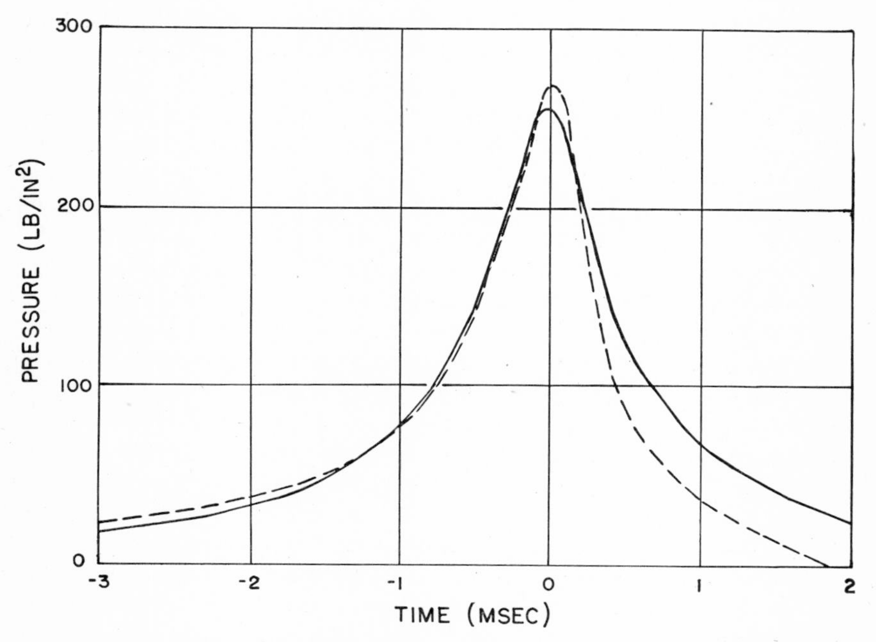

Fig. 9.6 Observed and calculated bubble pulses for a one ounce Polar Ammon Gelignite charge fired six feet below the surface.

value of 600 lb./in.^{2}. There should also be a lower rest position of maximum peak pressure at about 20 feet, but this region was not measured in sufficient detail to reveal it for this charge weight. The impulse curves show less pronounced maxima and curves computed from Eq. (9.25), included for comparison, give somewhat higher values than are observed.

Similar measurements for larger TNT charges (300 gm. equivalent weight) are plotted in Fig. 9.8. Both upper and lower rest positions of maximum peak pressure, corresponding to the vertical lines indicating zero measured migration, are evident, and some similarity is evident in the impulse data. The measured peak pressures at the upper rest position of about 800 lb./in.^{2} is in reasonably good agreement with the value 600 lb./in.^{2} from Eq. 9.12. The measured value of 400 lb./in.^{2} at the lower rest position is somewhat less than the computed value of 600 lb./in.^{2}, indicating that assumption of a rigid boundary overestimates the effect of the bottom even on this small scale. The comparison of impulse with theory shows an unexpected rise at greater depths, but it should be pointed out that no correction of experimental data for the added pressures from bottom reflections was made (surface corrections were applied when possible).

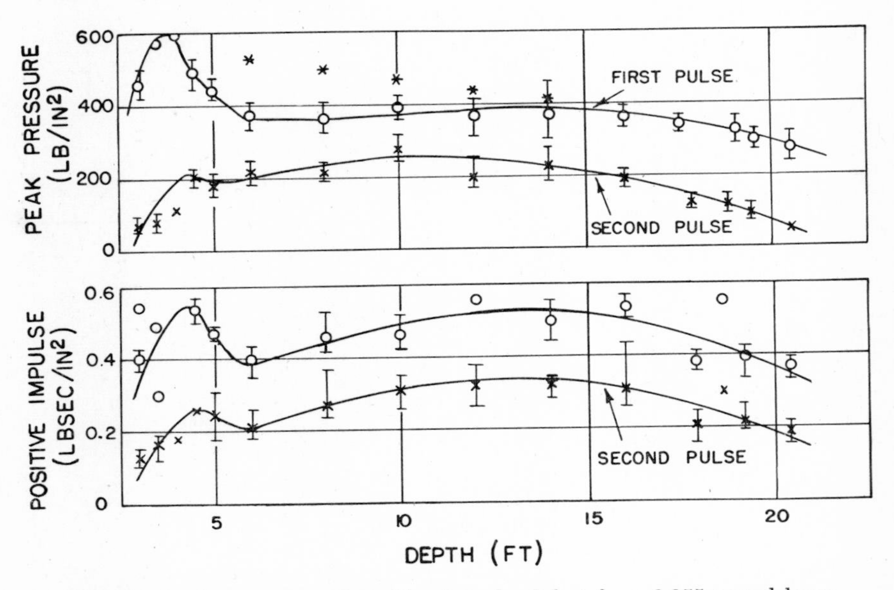

Fig. 9.7 Peak pressure and positive impulse 3 feet from 0.055 pound loose tetryl charges fired in 24 feet of water.

The most surprising feature of the peak pressure and, to a lesser extent, impulse curves is the sharp maximum for a depth slightly less than that of the upper rest position. The observed pulse in this narrow region is of very large amplitude and relatively short duration and the optimum peak pressure is more than double the level at greater depths. Reproductions of experimental pressure-time curves for 280 gm. cast charges in Plate X show the difference between this pulse at 4.3 foot depth with a normal pulse at 8 foot depth. The reality of this pulse has been confirmed for several sizes and weights of small charges and for 300 pound TNT charges, but its occurrence is extremely sensitive to charge depth and is realized experimentally with some difficulty. The existence of this "anomalous pulse" is not predicted by any of the theories which have been considered and the reason for its occurrence is somewhat of a puzzle.

It is probably significant that the anomalous pulse occurs for depths such that the calculated maximum radius of the bubble is very nearly equal to the depth of the charge below the surface. Kirkwood has suggested that at the stage of maximum expansion the bubble sucks in air from the atmosphere above; further reaction of the gaseous products then evolves energy which is available for the later motion. Photo-

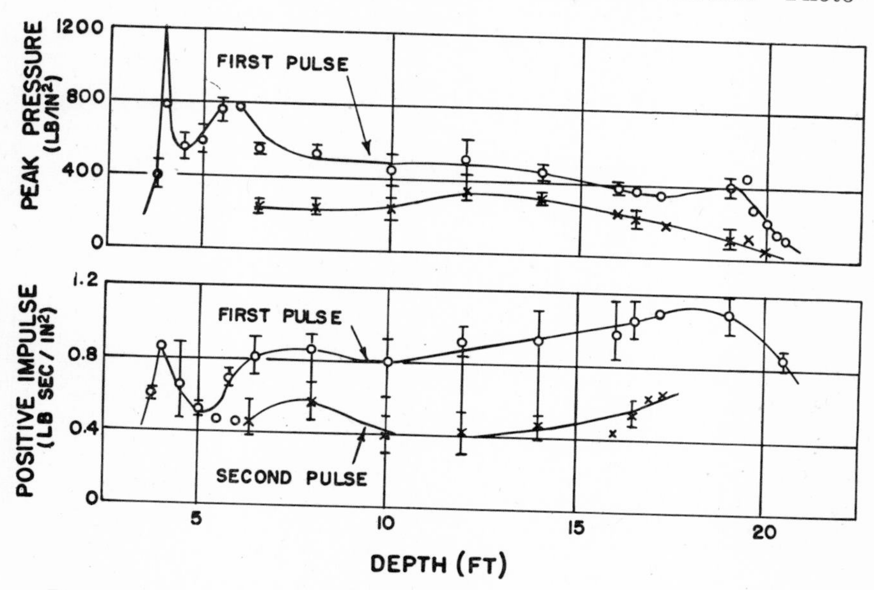

Fig. 9.8 Peak pressure and positive impulse 4 feet from 0.55 pound TNT charges fired in 24 feet of water.

graphs of the bubble under these conditions show a funnel shaped region between the top of the bubble and the water surface, which may indicate air venting into the interior of the bubble, but conclusive proof of the true reason for the phenomenon has not been found. The critical nature of the effect makes its practical significance somewhat dubious.

B. *Large charge results.* The experimental difficulties in making bubble pulse measurements under controlled conditions present a serious problem in large charge measurement, as discussed in section 9.5. As a result, only meager data have been collected in comparison with small charge results, although a number of scattered observations have accumulated as a by-product of other tests on service weapons.[4] The

[4] A number of references to such work will be found in the report by Slichter, Schneider, and Cole (103).

only reasonably comprehensive set of results at the time of writing is from a series of measurements performed at Woods Hole. In these tests, 300 pound TNT charges were fired at various depths in approximately 100 feet of water (the period and migration data from these tests are given in section 8.11). The results (103) give, within rather large limits of error, a good indication of the effects to be expected for large charges and so will be described in some detail.

In these tests, three tourmaline gauges were positioned at depths of twenty, fifty, and eighty feet below the surface. The experimentally

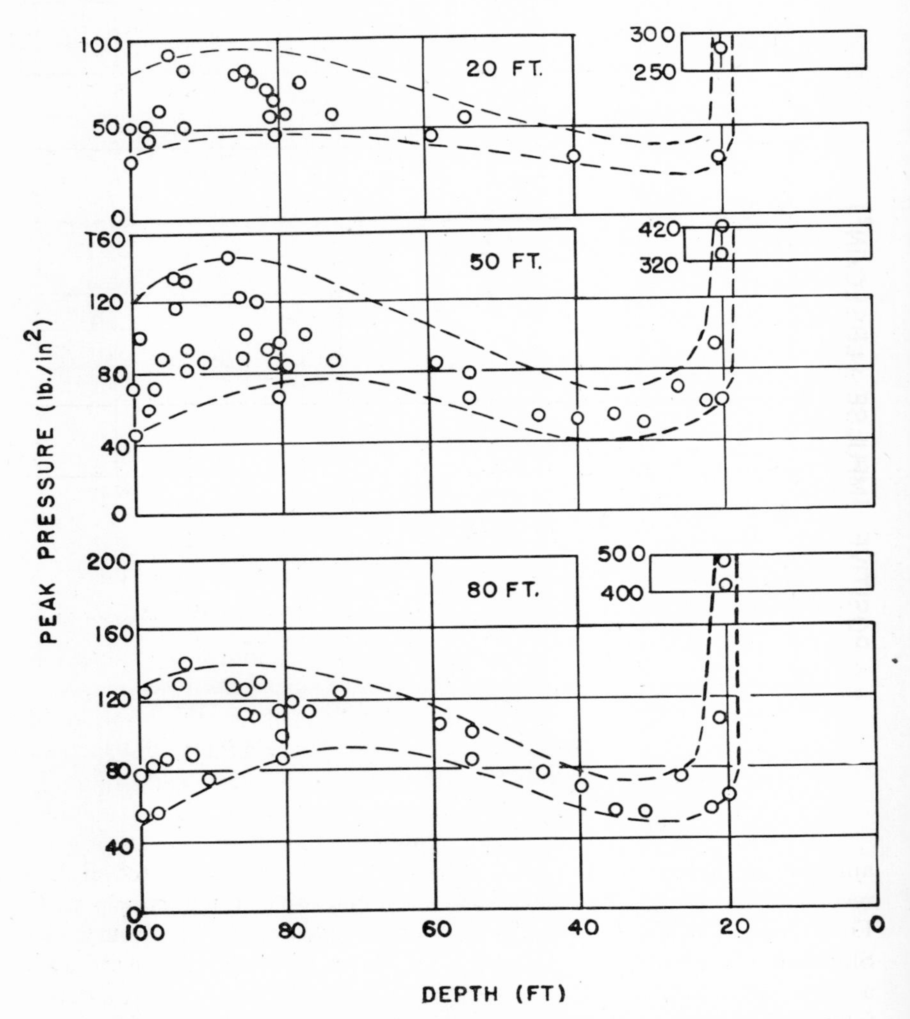

Fig. 9.9 Observed peak pressure 60 feet from 300 pound TNT charges fired in 100 feet of water.

measured peak pressures, corrected to a gauge distance of sixty feet assuming a distance variation as $1/r$, are plotted in Fig. 9.9. The values plotted include the effects of surface and bottom reflections, which account in large part for the large pressures near the bottom. When corrections are made for these reflections, this effect largely dis-

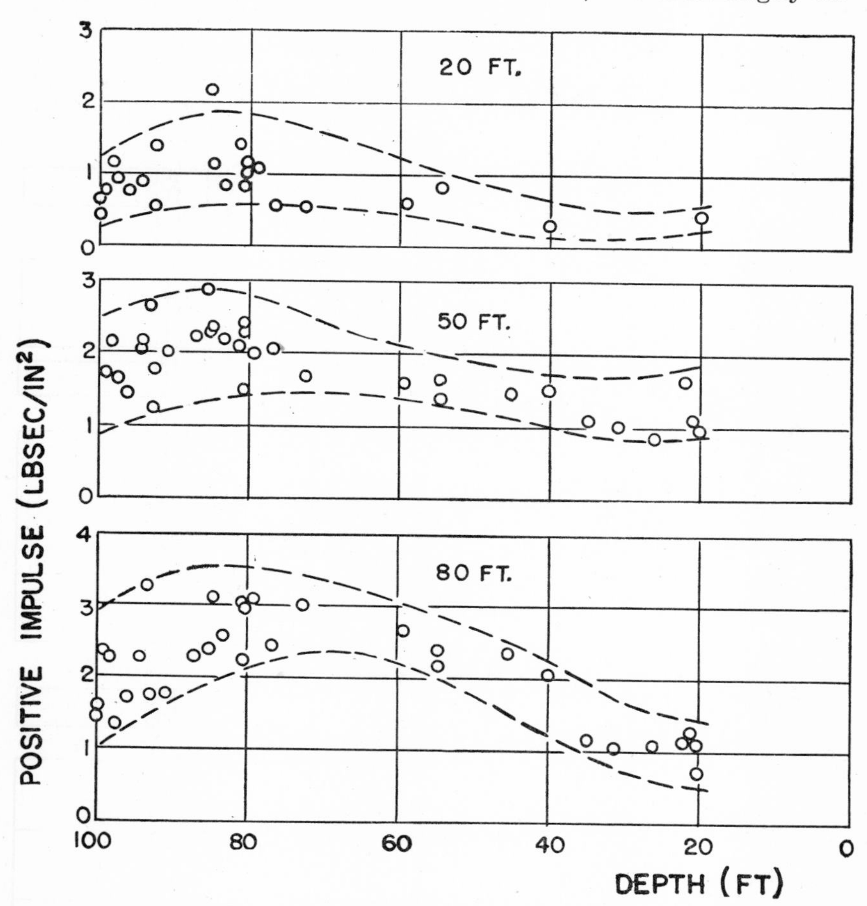

Fig. 9.10 Observed positive impulse 60 feet from 300 pound TNT charges fired in 100 feet of water.

appears, as shown in Fig. 9.11. This curve was constructed from records of the gauge fifty feet deep, for which the corrections were least. For comparison, the peak pressure curve predicted by the theory of Shiffman and Friedman is plotted. The sharp maximum predicted for a lower rest position is not realized, the data showing instead a gradual falling off of peak pressure as the charge is placed closer to the bottom.

As for small charges, an anomalous pulse is observed, in this case for a charge depth close to twenty feet, and its peak value was of the order three to five times that for the level at greater depths. The exact magnitude of the optimum pulse is uncertain for two reasons. First, its occurrence is so critical to depth that it is not very reproducible, and second, the observed pulse is diminished considerably by surface re-

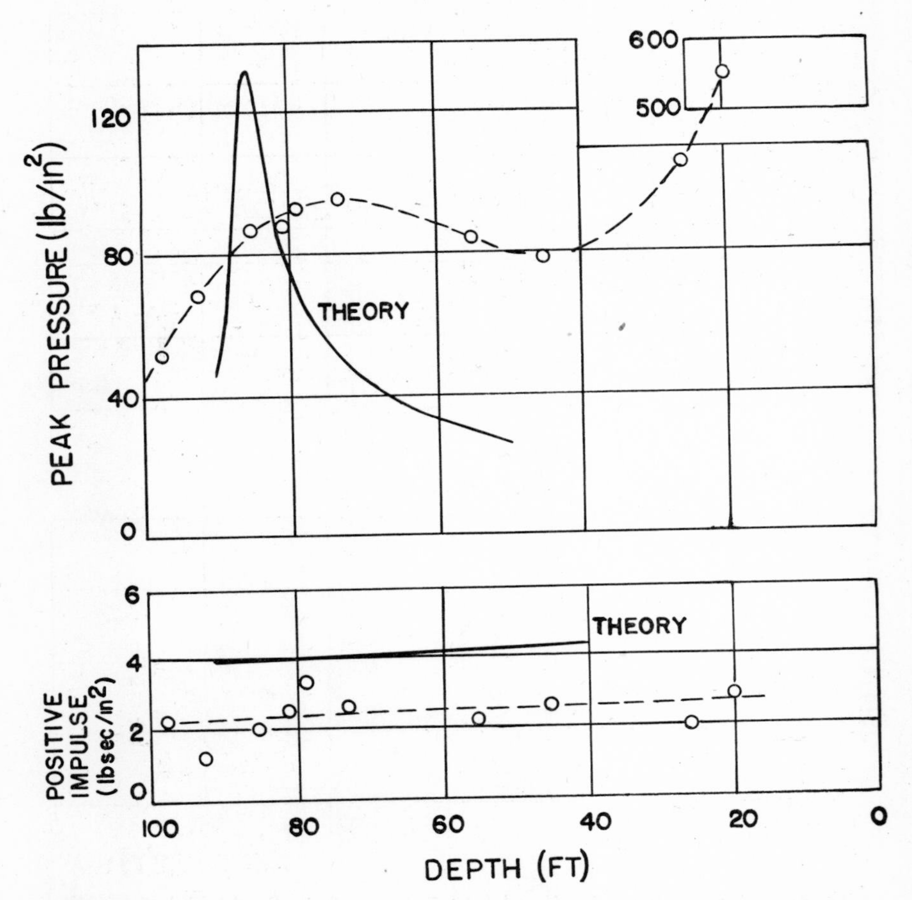

Fig. 9.11 Peak pressure and positive impulse, corrected for surface and bottom reflections, from 300 pound TNT charges fired in 100 feet of water.

flection, for which a large, rather inaccurate correction was necessary in these data. It has, however, been estimated that the corrected peak pressure is at least one quarter of the shock wave pressure. The duration of the pulse, however, is relatively short, as impulse data (see the following discussion) show no significant increase over values at greater depths. The total energy flux is, in virtue of the high peak pressure, at

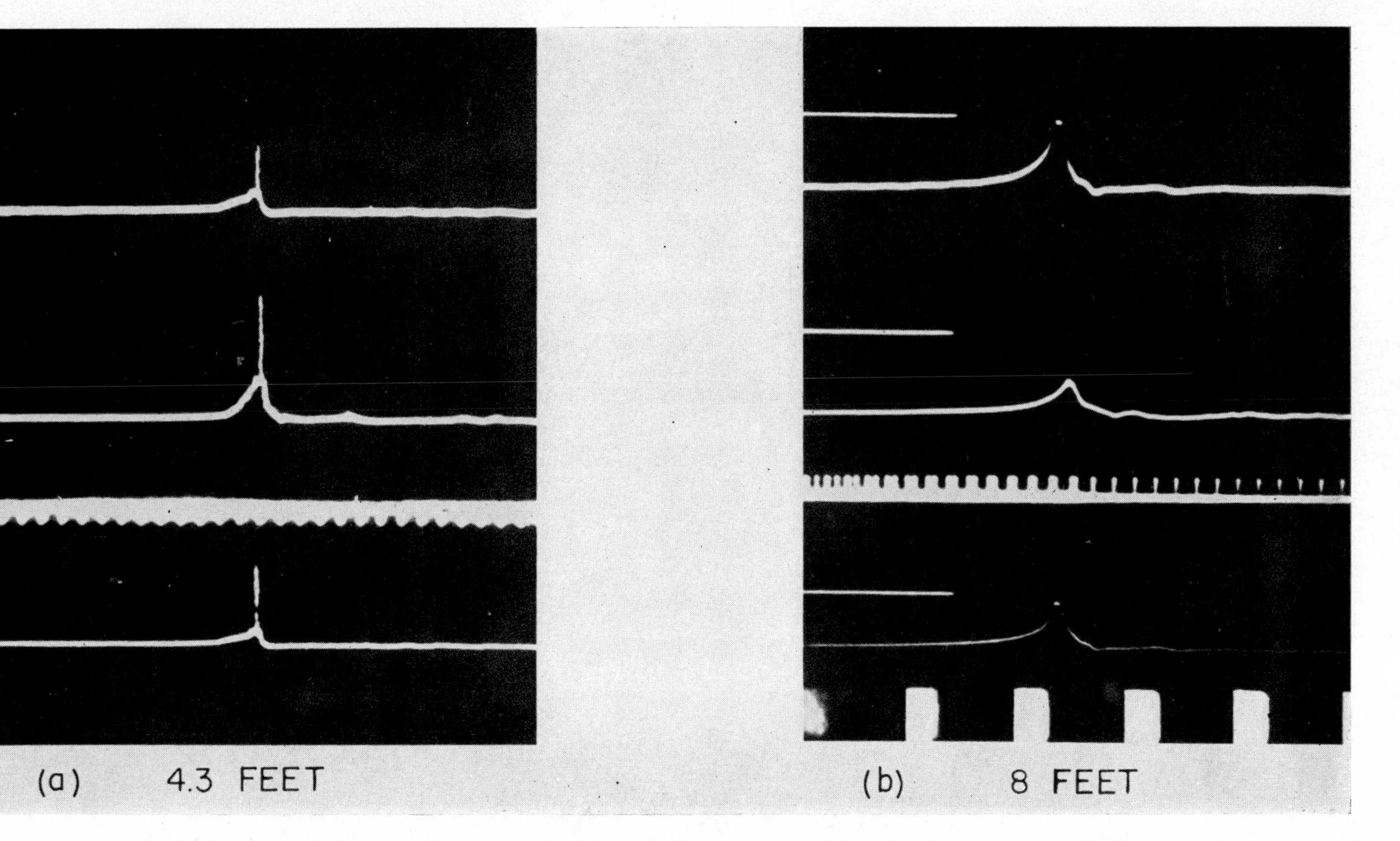

Plate X. Bubble pulse pressure-time curves recorded by 3 gauges for 0.64-lb. cast charges at two depths. The timing marks (overlapping) are 1.0 msec. apart.

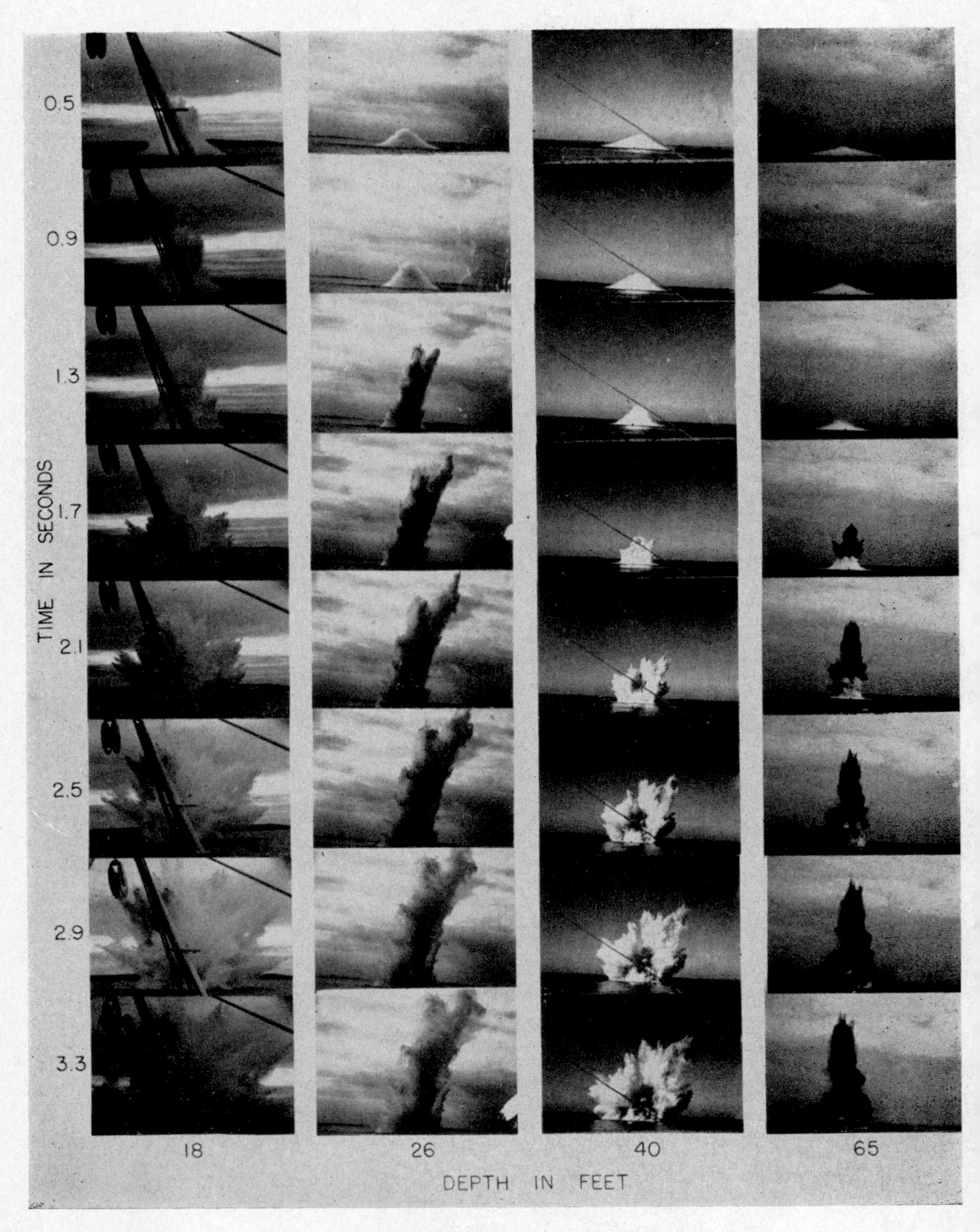

Plate XI. Dome and plume formation by 300-lb. TNT charges fired at four depths. The times are in seconds after detonation.

least forty per cent of the calculated total energy of the first bubble pulsation.

The measured impulse is plotted against charge depth for the three gauge positions in Fig. 9.10, and shows only the increases with depth to be expected from the inclusion of surface and bottom reflections. When corrections are made for these, the points of Fig. 9.11 are obtained, which indicates values of the order of one-half the theoretical curve (from Eq. (9.25)).

The experimental results do not provide any support for the predicted increase in effectiveness of the bubble pulse as a result of stabilization of the bubble against migration. The failure to realize this state

	Depth (ft.)	Peak pressure (lb./in.^{2})	Positive impulse[1] (lb. sec./in.^{2})	Energy density[1] (in. lb./in.^{2})
Shock wave	40	1770	1.15	170
	Bottom	1940	1.41	250
Composite Bubble pulse	20	428	1.1	47
	26	71	0.84	5.9
	45	56	1.5	9.5
	65	84	4.0	18
	96	81	1.2	11
Direct Bubble pulse	20	555	2.9	130
	26	106	2.1	16
	45	79	2.6	19
	65	93	3.4	28
	96	68	1.2	7

[1] Impulse and energy values for the shock wave obtained by integration to 2.0 msec. after shock front, for the bubble pulse by integration over times of pressure in excess of hydrostatic.

Table 9.2. Comparison of shock wave and bubble pulse pressures 60 feet from 300 pound TNT charges fired in approximately 100 feet of water.

is to be attributed to the failure of the bottom to act as the rigid, undisplaced surface assumed in theoretical analysis, as discussed in section 8.11. At other depths than that for the anomalous pulse, the peak pressure and energy of even the first bubble pulse are both very much inferior to the corresponding experimental values for the shock wave. The positive impulse is comparable with usually quoted values for the shock wave, but the danger, emphasized in section 9.3, of drawing hasty conclusions from such a comparison must be remembered. The relative magnitudes of shock wave and bubble pulse parameters for the 300 pound charges used in this investigation are summarized in Table 9.2, which includes shock wave data for 2 charges fired on the same type of bottom as in the bubble measurements. It may be added that the measured energy flux in the first and second pulses, except for the anomalous pulse already discussed, are of the order of 5 and 1 per cent of the shock wave energy flux.

10. *Surface and Other Effects*

10.1. Phenomena above the Water Surface

The most obvious and spectacular features of underwater explosions are the disturbances of the water surface above the charge. Several distinct phenomena are observed which change in magnitude and appearance with depth, but are most prominent for charges fired near the surface.[1] The first effect of the explosion to arrive at the surface is the underwater shock wave (assuming that the initial products of explosion are completely under the surface). The shock wave arrival at different points is visible as a rapidly expanding ring of apparently darkened water, often called the "slick." This ring owes its contrast against the background to the fact that ripples or small waves are calmed by the reflected shock wave pressure and a layer of cavitated, bubbly water is formed.

The slick can be seen on a reasonably calm surface out to large distances compared with the size of the charge, even if the charge is quite deep. Its speed depends on its position and the depth of the charge, but is always greater than the speed of the shock wave, because at any point the shock wave velocity is the projection of the slick velocity in the direction of propagation of the shock wave (see section 10.3). The slick is thus lost to view in a few hundredths of a second, if visible at all, and can be resolved only by high speed photographs.

Following the slick is the growth of the spray dome, a whitish mound of broken water thrown up by the reflection of the shock wave at the surface. The initial velocity is proportional to the pressure in the incident wave and hence is largest directly over the explosion. The extent to which a visible upward motion can be seen around this point is determined by the pressure and the effect of cavitation, and for sufficiently deep charges the dome is not visible at all. The water thrown up near the center of the explosion rises more rapidly, and continues to rise for a longer time than the water further away, and as a result the dome grows steeper sided as it rises. The time during which the rise continues and the maximum height, reached when the opposing forces of gravity and air resistance have reduced the velocity to zero, depend on the initial pressure and velocity, which in turn depend on the weight of charge and its depth below the surface. The analysis in more detail

[1] A number of writers have discussed surface phenomena from various points of view. A complete list of references bearing on the subject is not attempted here, but two reports by Kennard (54, 56) may be mentioned here and others are cited in section 10.2. The qualitative introductory discussion of this section is based largely on photographic records obtained at Woods Hole, especially the results reported by Slichter, Schneider, and Cole (103).

is taken up in section 10.2. An interesting observation is the occurrence of later domes caused by secondary or bubble pulses, if the charge is fired at sufficient depth for the gas bubble to contract one or more times before venting. These secondary domes are much smaller and less regular than the shock wave dome, because of the smaller pressure and their superposition on earlier disturbances, which obscure them.

For shallow charges, the products of explosion retain their identity in the gas bubble until this bubble reaches the surface and venting occurs. The size and state of motion of the bubble at this time both change greatly with the initial depth, and as a consequence the effects of venting also vary markedly with depth. The result of venting of the bubble gives rise to plumes of spray mixed with explosion products, two distinct types of plume being formed under different conditions:

(1). A relatively narrow vertical plume of variable height and velocity, depending on the charge depth.

(2). Radial plumes projected outward in all directions through the spray dome, the number and development of which depend on the depth of the charge. These plumes, if present, appear at the same time as or later than the vertical plume.

The differences in plume formation with depth are clearly shown in the sequence photographs in Plate XI for 300-pound TNT charges fired at depths of 18, 26, 40, and 65 feet. These photographs are reproduced for the same intervals of time after firing, but the distance scale is not the same for the different sequences. The first phenomenon of the spray dome appears at all depths, its height and duration decreasing with depth as is to be expected. At a charge depth of 18 feet, the gas bubble for this size of charge reaches the surface before its contraction begins and has a small upward velocity of migration. The venting therefore is nearly radial and the radial plumes of Plate XI result. At a depth of about 26 feet, however, the bubble reaches the surface when smallest and moving upward most rapidly, and almost all the water above it is thrown up vertically to form the narrow high plume of Plate XI.

At greater depths the vertical plume becomes increasingly less developed and the radial plumes reappear. This change in plume formation is simply understood as a result of the fact that venting of the bubble occurs at progressively later stages of its second oscillation; the vertical motion becoming smaller and the bubble motion more nearly radial as its size increases. This development reaches its maximum at a depth of about 40 feet, where the plumes are nearly all radial, as shown in Plate XI. At greater depths the vertical plume reappears and reaches maximum development at a depth of 65 feet, corresponding to venting at the second contraction of the bubble with maximum upward velocity. This plume (Plate XI) is similar to the one at 26 foot depth, but smaller because of the energy lost in the first contraction.

The relation between the time of appearance and initial upward velocity of the vertical plume is shown in Fig. 10.1 In this figure, estimated time at which the plume leaves the position of the surface before the explosion, and its estimated relative velocity at this time, are plotted against depth. Also plotted are the times, from bubble pulse period measurements, at which the bubble reaches or would reach minimum size at each depth. The intersection of these curves with the plume-time curve is seen to correspond closely with the maximum up-

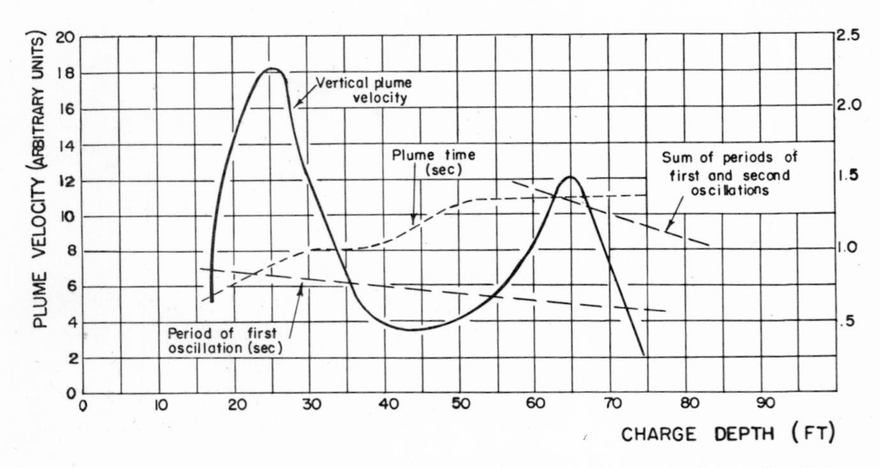

Fig. 10.1 Initial velocity and time of formation of vertical plumes by 300 pound TNT charges fired at various depths.

ward velocity, and confirm the explanation, originally proposed by Butterworth, of plumes as a result of venting of the explosion products.

At increasingly great depths, the same cycle might be expected to repeat itself but on a much smaller scale because of greater energy losses. After more than two cycles, however, the bubble has an insignificant fraction of its original energy and loses its identity into a mass of turbulent water, which appears at the surface a considerable time after the explosion.

At very shallow depths, the bubble vents while expanding rapidly and a vertical plume appears almost immediately, which is very narrow and rises to great heights. If the charge is just below the surface, the exact depth makes little difference, and for 1,000 pounds of TNT for example the maximum height is of the order of 1,150 feet. The height and nature of the plumes at other depths depend in a straightforward but rather complicated way on the depth and charge weight, which can be estimated from the bubble migration or, conversely, used to infer the migration.

The results so far discussed are evidently rather incomplete from several points of view. A more thorough analysis should include consideration of the lateral extent of dome formation, the internal structure of the dome, and the effect of charge depth. The phenomenon in its entirety is a fairly elaborate one, as yet imperfectly understood. Some of the factors involved are considered in section 10.2, but much of the existing evidence has had to be omitted from the present discussion for reasons of space and unavailability of the reports.

The venting of the gas bubble and other disturbances also lead to the development of surface waves, as distinct from waves of compression. These have been the subject of some study and are found to be insignificant for all but extremely large charges. For 1,800 pounds of TNT fired at optimum depth, the crest to trough height is of the order of 5 inches 500 feet from the charge and falls off rapidly with increasing distance.

10.2. Dome Formation

A. *Initial formation of the spray dome.* The particle velocity at a free surface can be thought of as the resultant velocity due to the incident shock wave and the reflected tension wave. The particle velocities in these waves have equal and opposite horizontal components, which therefore cancel, and equal upward components of magnitude $P \cos \delta / \rho_o c_o$, where δ is the angle with the vertical of a line from the charge to the point on the surface. The resultant velocity u_z of a particle at the surface is therefore

$$u_z = \frac{2P \cos \delta}{\rho_o c_o} \tag{10.1}$$

a result which of course is also obtained from the acoustic formula of section 2.2.

If the incident wave is exponential, the pressure P falls rapidly from its peak value P_m and it would appear from Eq. (10.1) that u_z would decrease correspondingly and give rise to very small displacements. However, the head of the rarefaction wave reflected into the water falls progressively behind the direct wave at increasing depths. The resultant pressure just behind the rarefaction front thus decreases below the hydrostatic pressure P_o and soon becomes a negative pressure, or tension. Water can, however, withstand only a limited tension, and at some value (P_b) of resultant pressure, cavitation bubbles form which prevent the pressure from decreasing further. The depth at which cavitation first forms corresponds to the time t' at which the incident pressure has fallen from its peak value P_m by an amount $(P_o + P_b)$ and hence $P = P_m - (P_o + P_b) = P_m e^{-t'/\theta}$. If P_m is large, t'/θ is small and expanding the exponential gives $t' = (P_o + P_b)\theta/P_m$. If P_b is only a few atmospheres and P_m is large, t' is a small fraction of the time constant, and the depth Δ is only a small fraction of the length of the incident wave.

If the pressure is assumed to vary with distance as $R^{-\alpha}$, where R is the distance from the real source at a depth d, an approximate calculation setting the resultant pressure excess at a depth d equal to $-(P_o + P_b)$ gives the result obtained by Pekeris (82) that

$$\Delta = \frac{(P_o + P_b)Rc_o\theta}{2P_m d\left(1 - \alpha c_o \dfrac{\theta}{R}\right)}$$

This is obtained by expanding the difference and ratio of the paths of the two waves to point Δ in series and using only the first terms. The distance Δ comes out to be very small if small values of P_b, of not more than a few atmospheres, are assumed. For example taking $P_b = 3P_o$, $P_m = 2250$ lb./in.², $d = 50$ feet (corresponding to 300 pounds of TNT 50 feet deep) gives $\Delta = .02$ feet. On this picture then, a thin layer of water will become detached and rise with virtually the peak velocity of the surface.

The initial velocity predicted by Eq. (10.1) should, if the foregoing analysis is correct, agree with experimental values if the measured peak pressure P_m is used. This comparison is simply made for the velocity and pressure at a point directly above the charge. A series of measurements of this kind made by Pekeris from motion picture records for large charges fired at 40 foot depth agreed with calculated values to within 7 per cent in the average. In another experiment, streak pictures of the rise of the dome above 5.1 and 5.5 pound cast TNT charges fired at a depth of 4 feet gave initial velocities of 216 and 230 ft./sec., the values computed from gauge measurements of P_m being 220 and 228 ft./sec. It should be mentioned, however, that anomalously high apparent initial velocities have been observed on motion picture records of large charges, the cause for which is not known.

As has been noted by Pekeris, several approximations are involved in the use of Eq. (10.1) to relate peak pressure and initial particle velocity at the upper surface of the dome. The first use is the fact that, as written, Eq. (10.1) refers to acoustic waves of infinitesimal amplitude travelling with the acoustic velocity c_o. For finite amplitude shock waves, the Rankine-Hugoniot conditions at the shock front should be used, which from Eq. (2.28) gives Eq. (10.1) if c_o is replaced by the shock front velocity U which is a known function of pressure.

This correction has been made in obtaining the small charge velocities quoted in the preceding paragraph, the difference being eight per cent in this case. A second small error is in the fact that the use of an image source of equal strength to represent the surface reflection is not exactly correct, because the resulting particle velocity u_z has associated with it a Bernoulli pressure $-(\frac{1}{2})\rho_o u_z^2$. The necessary correction to compensate this tension is evidently of order $(1/\rho_o c_o u)$ $(\frac{1}{2}\rho_o u_z^2) = u_z/2c_o$, which is ordinarily negligible.

The influence of cavitation so far considered is only a part of the entire story. Although it has been sometimes assumed that tensions amounting to as much as 600 lb./in.[2] are necessary before cavitations can form, this estimate appears from a variety of evidence to be much too high (see section 10.4), at least for water in the open sea. If cavitation occurs as close as one inch from a free surface in a typical case, it is not unreasonable to think that the upper surface of the dome breaks up into a spray rather than as a solid mass of water, particularly if this surface is not absolutely smooth.

What happens underneath the surface spray is a more complicated question, which has been discussed by Kennard (56) in detail. Kennard has shown that when cavitation is once established the isobaric surface of pressure (P_b) at which it first forms must either spread with a velocity exceeding that of sound, or not advance at all if the pressure gradient and particle motion do not permit this rapid an advance. Initially, the cavitation below the free surface must therefore spread rapidly downward in the water, and at a depth where the negative pressure P_b can no longer, so to speak, keep up with itself, the cavitated region ends. A mass of cavitated water at negative pressure (not necessarily as large as P_b in magnitude, its value depending on whether the bubbles formed contain air or saturated water vapor only) and upward velocity is therefore realized. The vertical extent of this region is uncertain, because of the lack of precise knowledge as to the proper value of P_b, but it seems clear that it must be only a fraction of the charge depth.[2] This water will therefore rise upward behind the surface of the spray dome and then fall back. Estimates of the details of the motion can only be made by rather tedious calculations from the position and magnitude of the direct and reflected pressure waves for specific values of P_b, and in addition the problem remains as to what extent the cavitated region is broken up and the pressure in its interior increased to atmospheric. These and related questions will be found taken up in detail in Kennard's report.

B. *Development of the dome.* The initial velocity of the upper sur-

[2] It should be pointed out, however, that measurements of pressure at a depth of 600 feet from a charge at the same depth have shown a surface cutoff in the tail of the shock wave; the pressure at this time could not be determined with any accuracy.

face of the dome as a function of distance r along the surface from the point above the charge is easily obtained from Eq. (10.1) if the peak pressure P_m is known as a function of distance R from the charge d feet deep. Assuming a power law $P_m = P_m(d)\,(d/R)^\alpha$ where $\alpha \sim 1.15$, gives

$$u(r) = \frac{2P_m(d)}{\rho_o c_o} \cdot \left(1 + \frac{r^2}{d^2}\right)^{-\frac{\alpha+1}{2}} \tag{10.2}$$

$$\frac{u(r)}{u(o)} = \left(1 + \frac{r^2}{d^2}\right)^{-\frac{\alpha+1}{2}}$$

The outline of the dome is thus a smooth curve, the center of which rises most rapidly, the contour at other points depending on the ratio r/d. Measurements of the dome shape therefore permit a calculation of the depth of explosion d if a scale for r is known. This and related deductions have formed the basis of a number of methods for determining depths of underwater explosions. The first of these, proposed by Shaw (101), consisted in determining the distance r' for which the rise of the dome is one-half its central value. Shaw assumed the value $\alpha = 1$ for peak pressure variation with distance, which from Eq. (10.2) gives $r' = d$, the depth of explosion. This result is modified slightly for $\alpha > 1$, but the method can still be applied. Shaw's method has the disadvantage that a distance scale is needed, and there are practical difficulties in accurate measurement of small initial heights, to which the method strictly should be applied. Pekeris has for this reason outlined a procedure (82) based on extrapolating more accurate values at later times, and has also analyzed a method in which the initial velocity of the dome center is obtained by extrapolation of the measured velocity-time curve. A critical study of these and other methods has been made by Halverson and co-workers, which is discussed in section 10.3.

The initial upward velocity of the dome of course decreases as it rises, because of gravity, air resistance and differences in pressure between the upper surface and interior of the dome. If gravity alone acted, the deceleration would be 32 ft./sec.², but Pekeris has found an average deceleration for domes above depth charges of about 85 ft./sec., nearly 3 times this value. This increased deceleration could be attributed either to drag resistance of drops of spray, or to the effect of pressure excess on an unbroken surface. As Pekeris points out, the observed deceleration is a strong argument against the existence of an unbroken surface of the dome, as the deceleration of a sheet of thickness z feet with pressure difference of one atmosphere (14.7 lb./in.²) would be $14.7 \times 32/0.433\, z = (1100/z)$ ft./sec. The layer would on this basis have to be at least 10 feet thick to give the observed deceleration. It must therefore be concluded, both from calculation and from the observed appearance, that the upper surface at least is spray, and the deceleration must be attributed to the drag of the drops of spray.

10.3. Determination of Depths of Explosion

A. *Dome analysis.* The methods of Shaw and Pekeris mentioned in section 10.2 have been extensively applied to measurement of depths of explosions. Critical comparisons have shown that measurements based on the first few frames of motion picture records are very unreliable, but that extrapolation of velocities to zero time gave results with scatter of the order of five per cent. Further tests with charges at a variety of known depths gave calculated peak pressure-distance curves agreeing with piezoelectric gauge measurements to within six per cent. An anomaly was found in that the apparent measured velocities from the first few frames were significantly higher than extrapolation of the smooth curves for later times. The reason for this discrepancy, not observed in small charge trials (see section 6.3), is not known.

Other methods of an empirical nature have been developed by Halverson (43). The first of these is simply measurement of the dome height at an arbitrary fixed time and use of calibrated curves made with charges of the same weight at known depths. A second method eliminates the need for either time or a distance scale by employing the ratio of the dome height to its width at the time the first plume from the gas bubble breaks through. As in the first method, calibration curves are obtained by measurements with charges of the same weight at known depths, and it is found that for a given charge this ratio varies with depth to the $-3/2$ power. The reason for using "plume time" as the time of measurement is to provide a simple criterion as to time of measurement which gives a calibration sensitive to depth. Simple as it is, the method gives remarkably reproducible values, the only complication being the necessity for empirical calibration.

B. *Shock wave spread.* Although the modified dome analysis methods are simple to use, their application becomes difficult or impossible at great charge depths, because of the smallness of the disturbance. Another method based on surface measurements is determination of the rate of spread of the shock wave across the surface, which is often visible at considerable depths. The distance r the front travels across the surface from the point above the charge is given by

$$r^2 = (d + ct)^2 - d^2 = 2dct + c^2t^2 \tag{10.3}$$

The velocity dr/dt along the surface is thus

$$\frac{dr}{dt} = \frac{d}{r}c + \frac{ct}{r}c$$

and is therefore always greater than the shock wave velocity c. Various methods of applying this result, such as plotting r^2/t or $r\,dr/dt$ against t, can be used from which a distance scale can be inferred and d determined, or if a distance scale is known, d can be found by fitting Eq. (10.3) to the observed (r, t) curve. None of these procedures is

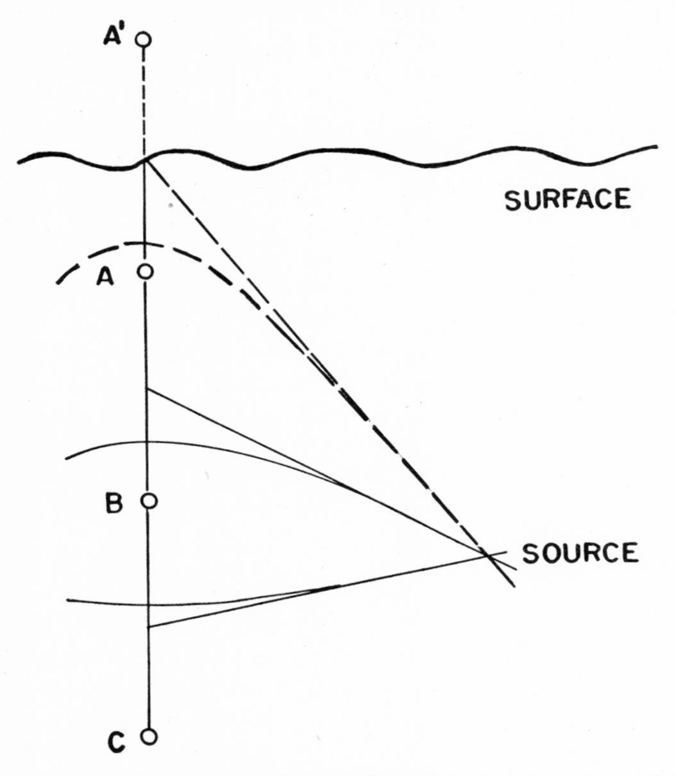

Fig. 10.2 Vertical gauge array for sound ranging of underwater explosions.

sufficiently accurate to recommend it if other methods are available, and they require high speed motion pictures, preferably taken from overhead. Perhaps the most valuable result of investigations of such methods is their identification of the slick as a product of the shock wave.

C. *Sound ranging.* A powerful method of determining explosion depths is by sound ranging, which depends on measurements of differences in arrival times of the shock wave at three or more gauges in known positions. The method has many possible variations; one of the simplest which illustrates the basic principle is illustrated in Fig. 10.2. Gauges at known separations on a vertical line as at A and B pick up the shock wave at intervals depending on the paths from the charge.

The observed time, and hence path, difference for any pair requires that the source lie on a hyperboloid of revolution around the gauge line, the intersection with the plane of Fig. 10.2 being a hyperbola as shown. Another pair gives a second hyperboloid intersecting the first in a circle around the line, and the depth and perpendicular distance from the gauge line are therefore determined. Not all of the gauges need be real ones if surface or bottom reflections of the shock wave recorded by any gauge are used, for these correspond to virtual image gauges at points such as A^1 above the surface, from which hyperbolas of constant path difference are also obtained, as shown by the dashed curve for the gauge pair AA^1.

It is evident that a vertical array of gauges is only one of many possible arrangements. It is usually the simplest and most accurate for depth measurement, because known vertical distances are more readily established, but tides and current may even so cause displacements from the vertical. Surface image gauges are free from this error, but become inaccurate if there are appreciable waves (bottom image gauges are usually unreliable). The vertical array is obviously indifferent to the horizontal direction of the explosion, and must be supplemented by gauges off the line if this quantity is to be determined. Despite these limitations, the vertical sound ranging system is a proven one, and has been successfully used to measure explosions at depths from 25 to 800 feet.

D. *Bubble period measurement.* The discussion of Chapter 8 shows that, for a given charge, the period of oscillation of the gas bubble is a simple function of depth. The measurement of this period by the difference in arrival times of shock wave and bubble pulses at a gauge therefore provides a method of depth determination, which has been found to give results of high accuracy when properly used. If the charge is sufficiently far from boundary surfaces, the period T is given by

$$T = KW^{1/3}/(d + 33)^{5/6} \tag{10.4}$$

where K is a constant characteristic of the explosive, W is the charge weight, and d the depth in feet. Although the constant K is known quite accurately for a number of explosives, it is desirable for good precision to determine it empirically for the charge of interest under the conditions of measurement. This is particularly true if effects of boundary surfaces modify Eq. (10.4) appreciably (see section 8.10).

As an example of such calibrations, the data of Fig. 10.3 for period versus depth were obtained by firing four 200-pound charges suspended at depths up to 800 feet in 3,000 feet of water. The plot of period versus depth is fitted within experimental error by a function of the form of Eq. (10.4), as shown by the straight line. Points are also plotted in

Fig. 10.3 which were obtained by a vertical sound ranging system comprising five gauges at 200-foot intervals to a depth of 815 feet, and illustrate the order of agreement found.

As compared to a sound ranging method, bubble period measurements offer the advantage that very much simpler equipment and gear

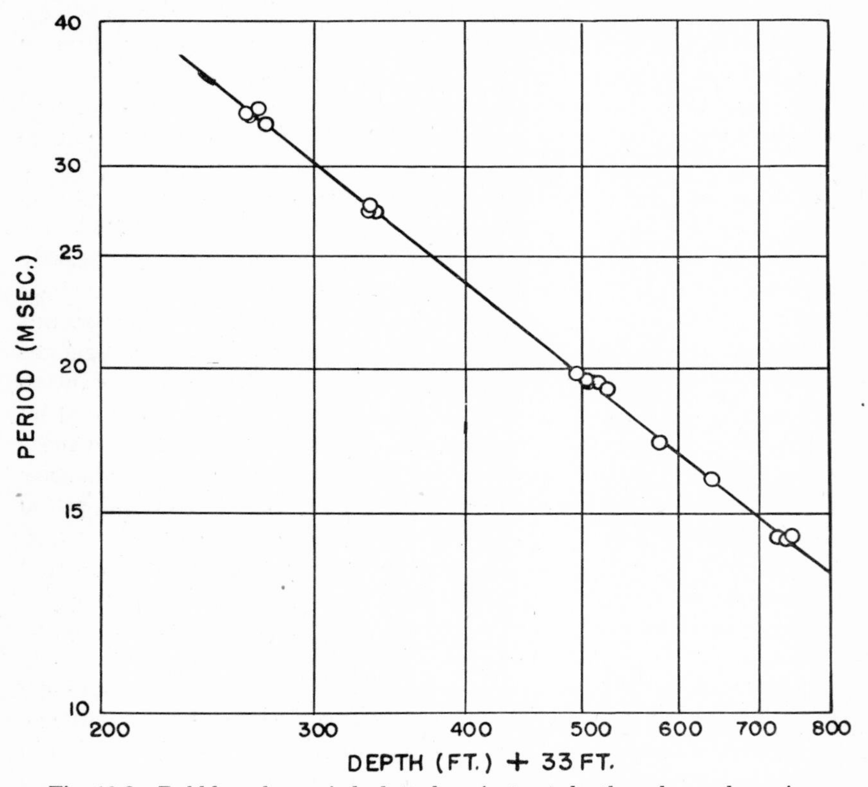

Fig. 10.3 Bubble pulse period plotted against set depth and sound ranging depth.

is needed. Only a single gauge need be used and it can be placed at any point which is close enough for the pulses to be picked up and not so close to boundary surfaces that reflections distort the pulse. Sound ranging gauges, however, must be relatively close to the source and accurately placed relative to each other and to reference points. If these conditions can practically be satisfied (and this is not easy in open water), the sound ranging method is more powerful because calibration is not as essential and distances other than depth can be determined.

10.4. Pressures near Yielding Surfaces

The analysis of damage caused by underwater explosions is, as has been suggested at several points, a very complicated one in all its ramifications. Clearly, the pressures developed by an explosion both affect and are affected by structures or targets in its neighborhood, and the motions of the water and structure must be treated together as a single dynamical problem. Even approximate treatments of simplified practical cases involve a large number of complications and variations, and any attempt to present a reasonably complete picture of what is known, and needs to be known, about such cases would require a volume at least the size of the present one. Rather than ignore the subject entirely, however, this and the next section are devoted to consideration of greatly simplified and idealized situations. It is not to be presumed that these correspond to actual problems, the intent being rather to illustrate more simply some of the factors involved in more complicated cases.

A. *Analysis of an infinite free plate.* A simple example, which we consider first, is that of an infinite free plate acted upon by the shock wave from an explosion.[3] By free is meant that there is no constraint to motion of the plate except that offered by its own inertia, and the actual occurrence of transmissions and internal reflection of elastic waves in the plate will be neglected, the plate being assumed to move bodily or not at all. (This effectively amounts to assuming an infinite velocity of the elastic or plastic waves in the plate.) If this plate is initially stationary, a plane pressure wave striking it will give the plate an initial velocity because of its finite mass for unit area of its surface. A reflected wave of pressure will at the same time be transmitted back in the water, which must be of such magnitude that the resultant particle velocity of the water in contact with the plate is equal to the velocity of the plate.

Let the plate have a mass m per unit area and velocity u, and assume the incident pressure P_1 to be a plane acoustic wave. If P_2 is the pressure in the reflected wave, Newton's second law for the motion of the plate requires that

$$m \frac{du}{dt} = P_1 + P_2 \quad (x = 0) \tag{10.5}$$

if distance x is measured from the plate into the water. The boundary conditions at the plate are

$$\begin{aligned} u &= u_1 + u_2 && (x = 0) \\ &= 0 && (x = 0, t = 0) \end{aligned}$$

[3] The motion of a free plate under various conditions has been considered by several writers, notably G. I. Taylor (108), E. H. Kennard (57), and Emily Wilson (see Reference (31)).

if u_1 and u_2 are the particle velocities in the incident and reflected waves. For plane waves, $u_1 = P_1/\rho_o c_o$, $u_2 = -P_2/\rho_o c_o$, and hence we have

$$u = \frac{P_1 - P_2}{\rho_o c_o} \quad (x = 0) \tag{10.6}$$

$$= 0 \qquad (x = 0,\ t = 0)$$

The incident and reflected plane waves must be of the form $P_1(t + x/c_o)$ and $P_2(t - x/c_o)$, and a specified function P_1 determines the values of P_2 and u from Eqs. (10.5) and (10.6). Assuming an exponential wave $P_1 = P_m e^{-(t+x/c_o)/\theta}$ and solving for $P_2(x = 0)$ and u on the boundary surface (assumed not to move appreciably) by standard methods gives

$$u = \frac{2P_m\theta}{m(1-\beta)} [e^{-\beta t/\theta} - e^{-t/\theta}] \ (t > 0) \tag{10.7}$$

$$P_2(x = 0) = \frac{P_m}{1-\beta} [(1+\beta)e^{-t/\theta} - 2\beta e^{-\beta t/\theta}] \ (t > 0)$$

where $\beta = \rho_o c_o \theta/m$.

The displacement s of the plate in the direction of the incident wave obtained by integrating Eq. (10.7) is

$$s = \frac{2P_m\theta^2}{m(1-\beta)} \left[\frac{1}{\beta}(1 - e^{-\beta t/\theta}) - (1 - e^{-t/\theta}) \right]$$

and the final displacement for $t = \infty$ is therefore

$$s = \frac{2P_m\theta^2}{m\beta} = \frac{2P_m\theta}{\rho_o c_o} \tag{10.8}$$

The final displacement is thus proportional to the impulse $P_m\theta$ of the incident wave, and is in fact just twice the particle displacement of the water by this wave. This result is true regardless of the mass of the plate, because the motion in this case is limited only by the inertia of the plate. If the plate is subject to other constraints, as it must be in nearly any practical case, their effect must restrict the motion, and only for a structure of relatively slow response compared to the duration of the incident wave will the impulse be the controlling factor. A further restriction is the fact that the analysis leading to Eq. (10.8) requires in most cases that the water in front of and in contact with the plate withstand large negative pressures (tensions).

The reflected pressure P_2 at points in the water ahead of the plate must be a function of $(t - x/c_o)$, and from Eq. (10.7) the resultant pressure P in the water is therefore

$$(10.9) \qquad P = P_1 + P_2 = P_m\left[e^{-\left(t+\frac{x}{c_0}\right)/\theta} + \frac{1+\beta}{1-\beta}e^{-\left(t-\frac{x}{c_0}\right)/\theta} - \frac{2\beta}{1-\beta}e^{-\beta\left(t-\frac{x}{c_0}\right)/\theta}\right],\ t > \frac{x}{c_0}$$

The quantity β, which expresses the effect of the plate on the pressure in the water, is zero for an infinitely thick plate ($m \to \infty$) and infinite for a plate of zero mass (i.e., a free surface), and Eq. (10.9) for pressure

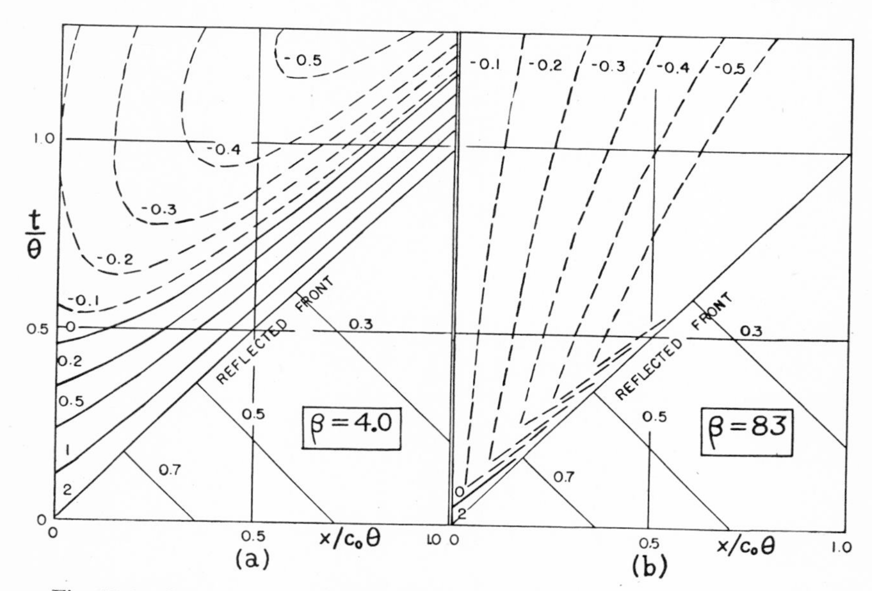

Fig. 10.4 Contours of equal pressure in front of accelerated free plates for two different relative inertias.

P is easily seen to reduce to the results of section 2.8 in these cases. In intermediate cases, the solution must be carried out explicitly.

It is evident, however, that for sufficiently large β, P will become negative at points in the fluid and may acquire large negative values for a sufficiently light plate or long duration of the incident wave. This corresponds to a rarefaction wave reflected from the accelerated surface of the plate, and the water in front of the plate may therefore be called upon to withstand large tensions. If, instead, the water cavitates, the subsequent motion of the plate will be modified.

The pressure field in any particular case can conveniently be illustrated by plotting isobars (contour lines of equal pressure) as a function of position and time. This is done in Fig. 10.4a[4] for the case $\beta = 4.0$,

[4] The data for these plots are taken from Reference (31).

corresponding to a relatively large inertia and slow response of the plate in comparison with the duration of the incident wave. Reduced units of length and time in terms of the time constants of the incident wave are used, and contours are drawn for various values of P/P_m, given by Eq. (10.9). The diagonal lines $t/\theta = -x/c_o\theta$ and $t/\theta = x/c_o\theta$ then represent the fronts of the incident and reflected waves. It is seen that the excess pressure at the plate decreases rapidly, becoming zero for $t = 0.46\theta$ and reaching a maximum negative value of about $-0.2P_m$. In this case ($\beta = 4$), negative pressures first occur at points in the water. The contours for larger β, corresponding to a lighter plate, are compressed downward toward the line for the reflected front, as shown in Fig. 10.4b for the case $\beta = 83$. Comparison with Fig. 10.4a shows that the least pressure at the plate becomes increasingly larger. For example, the value $P = -0.05P_m$ is just reached at the plate for $\beta = 83$ and the value $P = -0.10P_m$ is first realized at $x = 0.04c_o\theta$.

The results in this simple problem are of interest because they illustrate some of the factors to be considered in motion of the water near a target. For one thing, the region in which cavitation occurs in front of such a plate can be used to estimate the necessary tension to produce it. Experiments of this kind (31) have been performed using a thin (.020 inch) sheet of cellulose acetate 6 inches in diameter backed by air and photographing the water in front of the plate after reflection of a shock wave. The photograph reproduced in Plate XII shows cavitation bubbles extending from a point within ⅛ inch of the plate nearly to the front of the reflected wave (approximately at B in the photograph).

The mass of the plate used in the experiment corresponds to $\beta = 83$ in the preceding analysis, and the pressure ⅛ inch from the plate never becomes less than $-0.1P_m$ from Fig. 10.4b ($x/c_o\theta = 0.04$ in this case). The peak pressure at the experimental distance of 24 inches from a 10 gram loose tetryl charge is approximately 900 lb./in.^{2} The hydrostatic pressure was 18 lb./in.^{2} and a tension of 70 lb./in.^{2} is therefore indicated as an upper limit on the tension sustained by the open sea water of this experiment before cavitation occurred.

A similar experiment with a weaker incident wave gave an even smaller upper bound of 40 lb./in.^{2} for the tension. These results are of course only approximate because of the simplifying assumptions of the analysis from which they were obtained, namely that the plate was of infinite extent and rigidity and that the incident wave was plane. These experiments and others of a similar nature do, however, show quite conclusively that the necessary tension for cavitation, in open sea water at least, can hardly exceed a few atmospheres and is probably much less. This being the case, it is usually a good approximation to assume that cavitation begins practically at the surface of a free plate

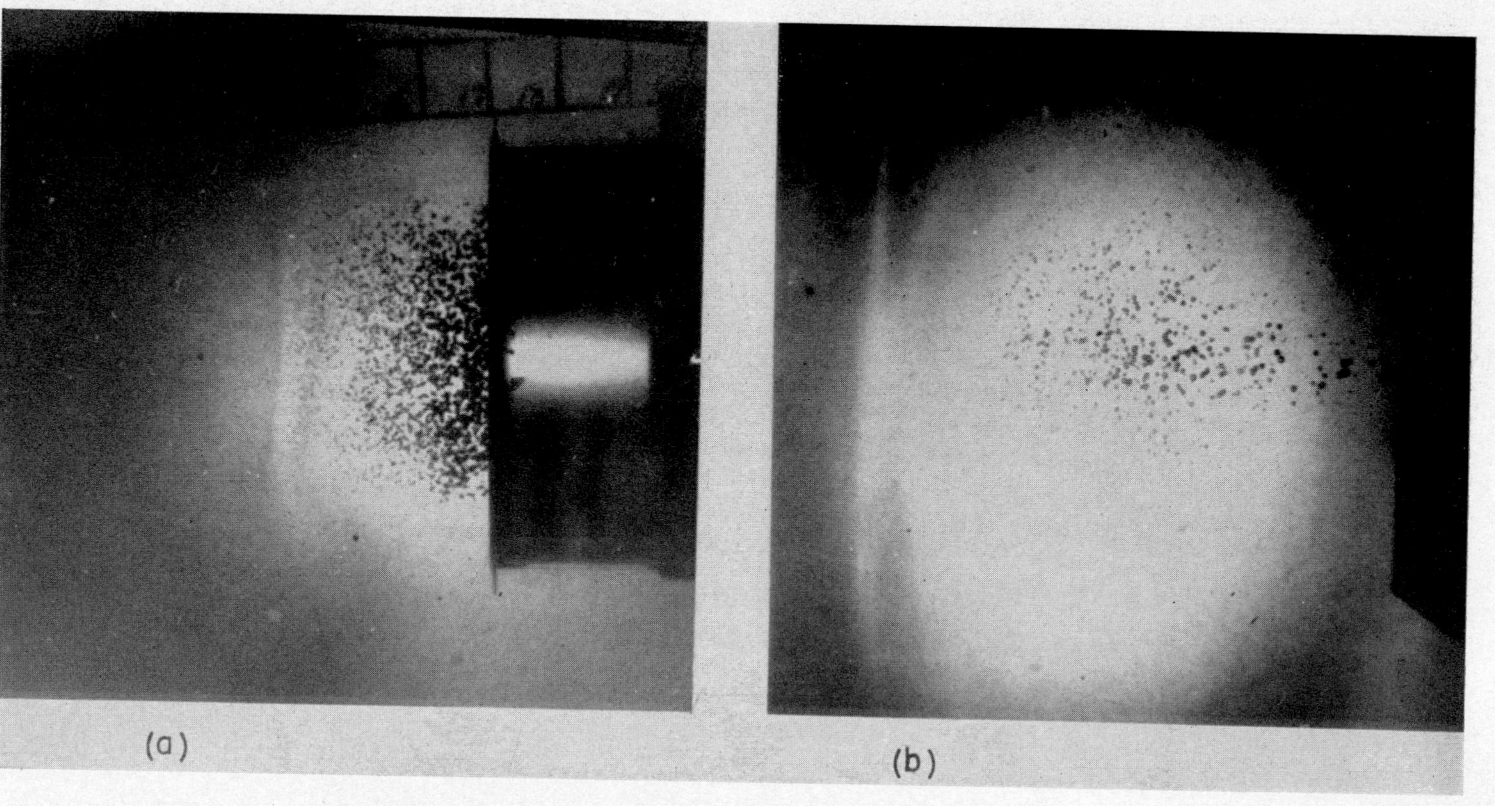

Plate XII. Cavitation in front of circular plates after reflection of a shock wave.

when the excess pressure first reaches zero. If this is true, the time θ_c at which cavitation begins can be obtained by setting $P = 0$ in Eq. (10.7) with the result

$$\theta_c = \theta \left(\frac{\log_e \beta}{\beta - 1} \right) \tag{10.10}$$

B. *Effect of cavitation on displacement.* The occurrence of cavitation in front of a target makes the analysis of part (A) cease to apply after a time θ_c, for after this time the water in front of the plate is at a constant pressure very nearly zero. Under these circumstances the plate will be subjected to a very small loading by the pressure behind the plate, and its velocity at this time, which can be calculated from Eq. (10.7), will only very slowly be reduced to zero.

No structure can in actual fact be considered as infinite without constraint or support. To be more realistic, the simple model of a free plate must therefore be modified to include the resistance of the structure to deformation and also include diffraction of the incident pressure wave resulting from the finite extent of the targets. These more elaborate considerations are briefly discussed in the next section. The results so far considered, although of only very limited application in practical problems, do illustrate simply the importance of cavitation as a determining factor in underwater explosion damage. The time at which cavitation may be expected to set in is conveniently estimated in many cases by Eq. (10.10), which serves as a convenient although rough criterion in more complex problems than the simple case for which it was derived.

10.5. Plastic Deformation and Diffraction Effects

The analysis of the preceding section concerns the simplest possible type of structure or target, the only complication introduced being the inertia of a plate assumed to undergo no deformation by action of applied pressure. Although the analysis of this simple case does illustrate the significant role which cavitation may play in determining explosion damage, it does not bring out other factors which are equally significant in more realistic cases. The nature and effects of some of the more important of these factors are illustrated in the relatively simple case of a thin circular diaphragm supported at its circumference. The large amount of experimental and theoretical work which has been done on this problem gives a rather clear picture of the important phenomena which may be expected to take place in other problems also, and the circular diaphragm is thus of interest in a general understanding of explosion damage.

The problems encountered in analyzing the motion of a diaphragm or plate may be broadly grouped under determination of the elastic or plastic resistance to deformation, and analysis of the disturbances in the pressure field resulting from the finite area of the structure. Each of these affects and is affected by the other and the two must therefore be considered together in obtaining solutions for the motion, but we shall begin by first considering the resistance of the structure.

A. *Plastic deformation.* The elastic properties of a material such as steel are characterized by the relation of applied forces to the charges in shape which they produce. If a plate clamped at its edges undergoes an increase in area by load applied on one face, standard methods of the theory of elasticity can be applied to formulate the differential equations of its motion in terms of its elastic moduli and inertia, and the applied load. The cases of most interest, however, are those in which the material is stressed beyond its elastic limit and a permanent deformation results, which may become so large that the plate thins to the point of rupturing. The plastic state, considered in its full generality, is complicated and incompletely understood, particularly under the conditions of dynamic loading of interest here. Fortunately, a simple first approximation to the true situation can be made which is illuminating and reasonably accurate. In this approximation, it is assumed that a definite elastic limit exists, below which Hooke's law applies and above which the plate acts like a membrane under a constant tension determined by the yield stress under tension at the elastic limit.

The nature of this plastic region depends on the material, but steel and copper can withstand very large strains before finally rupturing. The assumption that a definite value of the yield stress σ_o exists and is further independent of the rates of strain is an idealization. It is well established that, at the large strain rates of interest in explosion damage, σ_o is much greater than for statically applied loads, but this value does not increase rapidly for large strain rates. It is reasonable to assume a figure for σ_o which is independent of this rate and somewhat greater than the static value, and such an assumption is common to virtually all the theoretical treatments.

With the assumption that the plastic deformation can be described by a tangential stress equal to the yield stress, the behavior of a plate becomes the same as that of a film under tension of magnitude $\sigma_o h$ for unit length of a cut in the plate, h being its thickness. The deformation of the plate into a curved profile therefore develops a normal stress component σ_n for unit area of the plate as shown in Fig. 10.5a. The magnitude of σ_n can be calculated in the same way as for a film with surface tension, with the result

$$\sigma_n = -\frac{2\sigma_o h}{r}$$

if the radius of curvature r is assumed independent of the direction of a tangent line to the surface. The normal stress thus increases with the curvature. If a circular diaphragm of radius a is assumed to be deformed into a spherical profile, as shown in Fig. 10.5b, the radius of curvature is for small central deflection z_c given by $a^2/2z_c$, and the normal stress becomes

$$\sigma_n = -\frac{4\sigma_o h}{a^2} z_c \tag{10.11}$$

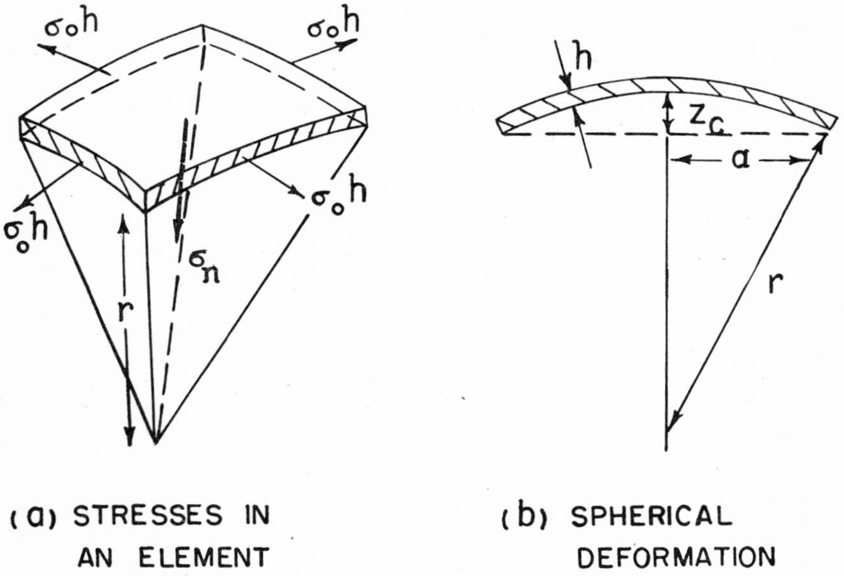

(a) STRESSES IN AN ELEMENT

(b) SPHERICAL DEFORMATION

Fig. 10.5 Tangential and normal stress for a plate in "plastic tension."

The plate is therefore in equilibrium with an outward normal pressure equal to σ_n. If this pressure is released the plate will not, however, return to its original undeformed profile, but will acquire a permanent set less than the plastic deformation by the amount of elastic recovery from the stress σ_n. The initial displacement of the plate will also of course be in the domain of elastic strain, but since the cases of interest involve large plastic displacements, the elastic effects can be neglected in an approximate treatment.

The permanent plastic deformation is the result of plastic work W_p done by the applied forces, and if the increase in area as a result of tension $\sigma_o h$ is ΔA this work is given by

$$W_p = \sigma_o h \Delta A$$

For spherical deformation as in Fig. 10.5b the increase ΔA for a small deflection z_c is given by $\Delta A \cong \pi z_c^2$ and the plastic work is

$$W_p = \pi \sigma_o h z_c^2$$

The energy absorbed by the plate thus varies as the square of its central displacement.

An analysis based on the idealization of a uniform plastic tension is clearly very approximate, and neglects such effects as anisotropic stresses, resistance of the plate to local bending, and tangential propagation of plastic effects under dynamic loading.[5] The stress and energy relations do, however, provide a rough description of the plastic resistance to deformation, and together with a formulation of the pressure in the adjacent water permit at least a crude analysis of the resulting motion.

B. *Diffraction effects at a target.* The pressure in the water at a deformable surface must differ from that in the incident wave in such a way that the normal component of particle velocity is the same as the corresponding velocity of the plate. If the incident pressure is a plane wave and the target is an infinite one, all parts of which undergo the same motion, this requirement is satisfied in the linear (acoustic) approximation by superimposing reflected plane waves. If, however, the target is of finite extent, the pressure and velocity must vary over its face in such a way as to be compatible with the velocity and stress in the plate. The latter must be different, for example, at the center and at the supported edge of a plate, and some system of waves other than plane waves of infinite extent must be necessary. The mathematical analysis of such cases, originally due to Kirchhoff, is described by the physical concept of diffracted spherical waves originating at points on the boundary, which combined with the incident wave satisfy the boundary conditions and hydrodynamical equations. A linear superposition of this kind is only possible if the hydrodynamical equations are linear, and the analysis to be described therefore applies exactly only to waves of acoustic intensity. The presentation given here is adapted from Kirkwood's discussion.[6]

In the linear approximation, wave propagation in the water is characterized by the velocity potential φ, from which the excess pressure and particle velocity are obtained by

$$u = -\operatorname{grad} \varphi, \quad P - P_o = \rho_o \frac{\partial \varphi}{\partial t}$$

[5] A discussion of such effects in circular diaphragms is given by Hudson (48).

[6] Reference (61), see also the more general discussion by Lamb (65), pp. 498ff.

If we consider an infinite boundary surface A in the water the velocity at any point of which is given by $u(\mathbf{r}', t)$, where $\mathbf{r}'$ is the vector displacement from a suitable origin, it can be shown that the velocity potential at any point $\mathbf{r}$ on the surface is given by the diffraction integral

$$\varphi(\mathbf{r}, t) = -\frac{1}{2\pi}\iint_A \frac{u_n\left(\mathbf{r}', t - \dfrac{|\mathbf{r} - \mathbf{r}'|}{c_o}\right)}{|\mathbf{r} - \mathbf{r}'|}\, dA' \tag{10.12}$$

where $u_n(\mathbf{r}', t)$ is the component of velocity normal to the surface at $\mathbf{r}'$ and directed away from the water. This equation can be interpreted as giving the effect, at a point $\mathbf{r}$, of spherical waves emitted from points $\mathbf{r}'$ over the surface. Each such diffracted wave, which travels with a velocity c_o in the water, requires a time $|\mathbf{r}-\mathbf{r}'|/c_o$ to reach the point $\mathbf{r}$ from $\mathbf{r}'$, since the absolute value $|\mathbf{r}-\mathbf{r}'|$ is the distance between the points. The effect at $\mathbf{r}$ and time t is thus associated with the velocity u_n at $\mathbf{r}'$ for an earlier, retarded time $\tau = t - |\mathbf{r}-\mathbf{r}'|/c_o$. The diffraction integral thus bears a close relation to Huyghens' principle for light waves, by which perturbing effects of boundaries are represented by diffracted spherical waves, and is derived by similar mathematical methods.

The value for the present purpose of the velocity potential formulation is the fact that it permits calculation of the pressure at any point on the surface A. In order to do this, it is convenient first to express the perturbed potential φ in terms of the unperturbed value φ_o which would exist if the boundary A had no effect and the normal component of velocity were the value u_{on} in the unperturbed incident wave. The value of φ_o is then given by

$$\varphi_o(\mathbf{r}, t) = -\frac{1}{2\pi}\iint_A \frac{u_{on}(\mathbf{r}', \tau)}{|\mathbf{r} - \mathbf{r}'|}\, dA' \tag{10.13}$$

In the linear approximation, velocity potentials are additive, and taking the difference of Eqs. (10.12) and (10.13) gives

$$\varphi(\mathbf{r}, t) = \varphi_o(\mathbf{r}, t) - \frac{1}{2\pi}\iint_A \frac{u_n(\mathbf{r}', \tau) - u_{on}(\mathbf{r}', \tau)}{|\mathbf{r} - \mathbf{r}'|}\, dA_o$$

The pressure at any point on A is given by $P = \rho_o\, \partial\varphi/\partial t$, and if P_1 is the pressure in the incident wave, we have

(10.14)

$$P(\mathbf{r}, t) = P_1(\mathbf{r}, t) - \frac{\rho_o}{2\pi}\iint_A \frac{1}{|\mathbf{r} - \mathbf{r}'|}\left[\frac{\partial}{\partial t}u_n(\mathbf{r}', t) - \frac{\partial}{\partial t}u_{on}(\mathbf{r}', t)\right]_\tau dA'$$

the subscript τ indicating evaluation at time $\tau = t - |\mathbf{r} - \mathbf{r}'|/c_o$.

If we consider the pressure on a diaphragm or plate of finite extent, it is evident that the pressure at any point is modified by the pressure at other points on the plate and external to it after the times required for hydrodynamic signals to arrive from these points. The method of support and boundary external to the plate must therefore be considered. If the plate is assumed set in a rigid structure, or baffle, of infinite extent, the velocity u_n must be zero at all points external to the plate. The integral over u_{on} then represents the reflected pressure on an infinite surface and Eq. (10.14) becomes

$$(10.15)\quad P(\mathbf{r}, t) = 2P_1(\mathbf{r}, t) - \frac{\rho_o}{2\pi} \iint_{A_o} \frac{1}{|\mathbf{r} - \mathbf{r}'|} \left(\frac{\partial u_n}{\partial t}\right)_\tau dA'$$

where the integral is extended over the area A_o of the plate. This equation represents the doubling of pressure which would occur if the surface were perfectly rigid, less the effect of motion of the surface in the direction of the incident wave.

If no baffle at all is assumed and boundaries behind the plate are ignored, we have $u_n = u_{on}$ external to the area A_o and $\rho_o\ \partial u_{on}/\partial t = -\text{grad } P_1$, giving

$$(10.16)\quad P(\mathbf{r}, t) = P_1(\mathbf{r}, t) + \frac{\rho_o}{2\pi} \iint_{A_o} \frac{1}{|\mathbf{r} - \mathbf{r}'|} \left(\frac{\partial u_n}{\partial t}\right)_\tau dA' + \frac{1}{2\pi} \iint_{A_o} \frac{1}{|\mathbf{r} - \mathbf{r}'|} (\text{grad}_{r'}\ P_1)_\tau\, dA'$$

If the incident wave varies slowly enough with distance from the surface, the second integral can be neglected. In this limit, the difference between presence and absence of a baffle is that in the latter case the pressure P_1, rather than the doubled pressure $2P_1$, is modified by the rarefaction or damping term from motion of the surface. The effect of diffraction from the unbaffled periphery of the surface is thus to reduce and ultimately destroy the doubling of pressure resulting from a rigid surface. The effect of an imperfectly rigid baffle or boundary can be taken into account by suitable assumptions as to the velocity distribution on such boundaries, as has been done by Kirkwood (61), and Kennard (57).

C. *The pressure at the center of a circular plate.* Only the simple case of a circular plate of radius a in an infinite rigid baffle will be considered further here, and a further simplifying assumption will be made that the plate deflection is always paraboloidal. This assumption implicitly requires a suitable constraint by which this profile is artificially maintained, as a more exact solution obviously determines what the profile must be. This question has been examined by Kirkwood, who finds that the initial phases of deformation are more nearly conical in form, and a change in shape occurs with time of deformation and time variation of incident pressure, as is also found experimentally. The paraboloid form is, however, fairly typical, and we assume for simplicity that the deflection $z(r, t)$ is given by

$$z(r, t) = z_c(t)\,(1 - r^2/a^2) \qquad t > 0$$

$$= 0 \qquad t < 0$$

where the central deflection $z_c(t)$ depends only on time and the edge is fixed ($z(a, t) = 0$). Substituting this expression for z in Eq. (10.15) gives for the pressure at the center of the plate

$$P(t) = 2P_1(t) - (\rho_o/2\pi) \int_o^a \left(\frac{d^2 z_c}{dt^2}\right)_\tau \left(1 - \frac{r^2}{a^2}\right) dr$$

since $u_n(r, t) = \partial z/\partial t$. Changing the variable of integration to $\tau = t - r/c_o$ and integration by parts gives finally Kirkwood's result that

(10.17)

$$P(t) = 2P_1(t) - \rho_o c_o \left[\frac{dz_c}{dt} + \frac{2}{\theta_d} z_c(t - \theta_d) - \frac{2}{\theta d^2} \int_{t-\theta_d}^{t} z_c(\tau)d\tau\right]$$

where the diffraction time $\theta_d = a/c_o$ is the time for a wave to be propagated from the edge of the plate to its center. The pressure for other points on the plate can of course be similarly evaluated, but involves greater mathematical complications which is unnecessary for the present purpose.

The significance of Eq. (10.17) for the pressure at the center of a plate is more clearly brought out by considering the limiting cases to which it reduces under special conditions. In the initial phases of the motion for which $t < \theta_d$, we have $z_c(t - \theta_d) = 0$ and the integral is small because $z_c(\tau) = 0$ for $\tau < 0$. The pressure is then very nearly

$$P(t) = 2P_1(t) - \rho_o c_o \frac{dz_c}{dt} \tag{10.18}$$

This, however, is simply the pressure in front of an infinite plate moving with velocity dz_c/dt, the term $2P_1(t)$ representing doubling of pressure at a fixed surface and the term $\rho_o c_o \, dz_c/dt$ the rarefaction resulting from its motion. As time increases this pressure is modified by the arrival of diffracted waves from adjoining points on the plate and baffle, these being represented by the last two terms of Eq. (10.17), which act to reduce the pressure gradient.

A second limiting case is that in which the acceleration of the plate changes slowly enough that the propagation time can be neglected. This corresponds to evaluating d^2z_c/dt^2 in the integral preceding Eq. (10.17) at the same time τ for all points, as if the motion and pressure at all points were the same as in noncompressive flow. In this limit, Eq. (10.17) becomes

$$P(t) = 2P_1(t) - \frac{2\rho_o a}{3} \frac{d^2z_c}{dt^2} \tag{10.19}$$

The second term on the right has the form of an inertial reaction of the accelerated flow near the plate, which in the equation of motion for plate and water becomes equivalent to an added mass $(2\rho_o a/3)$ per unit area (at the center of the plate). The load on the plate thus approaches a doubling of pressure less the inertia of noncompressive flow. This transition is a simple example of what Kennard has called the reduction principle, by which "the action of a wave tends continually to change into or reduce to the type of action that is characteristic of incompressible liquid." Considerations of this kind are often useful in simplified analysis of the pressure distribution, but it is important to observe that this limiting behavior is brought about by compressive equalization of pressure differences and the limiting case results only if the time for this to occur can be neglected. The noncompressive approximation has been treated by Butterworth (15), and by Kennard (57) in a comprehensive discussion of the effects of pressure waves on diaphragms.

10.6. Motion of a Circular Plate

A. *The equation of motion for a plate under plastic tension.* If a deforming plate is assumed to have geometrically similar profiles at all times, only the equation of motion for an element of the plate need be solved, as this solution determines the rest of the profile. As an illustration, consider the assumed parabolic deformation of a circular plate, subject to a pressure $P(t)$ at its center, negligible restraint by the medium behind it, and a stress $\sigma_n(t)$ normal to the surface as a result of its deformation. If m is the mass of unit area of the plate, the equation of motion for unit area at the center of the plate is

$$m \frac{d^2z_c}{dt^2} = P(t) - \sigma_n(t)$$

where for an infinite baffle $P(t)$ is given by Eq. (10.17). If the plastic deformation is approximated by membrane tension, the value of σ_n is approximately $(4\sigma_o h/a^2)z_c$ from Eq. (10.11) (this result is strictly for small spherical deformations, but the difference in profile for small values of z_c/a is negligible). Substitution in the equation of motion gives

$$(10.20)\qquad m\frac{d^2z_c}{dt^2} + \frac{4\sigma_o h}{a^2} z_c = 2P_1(t) - \rho_o c_o\left[\frac{dz_c}{dt} + \frac{2}{\theta_d} z_c(t-\theta_d) - \frac{2}{\theta^2{}_d}\int_{t-\theta_d}^{t} z_c(t)dt\right]$$

The deformation of a plate actually involves thinning, as it is approximately true that the density ρ remains unchanged in plastic flow, and hence neither m nor h is strictly constant. If this thinning is neglected, Eq. (10.20) is a linear integro-differential difference equation which is most readily solved for known $P_1(t)$ by use of Laplace transform theory. Kirkwood (61) has obtained the solution for the case of an exponential incident wave of the form $P_1(t) = P_m e^{-t/\theta}$, and has also outlined a procedure by which empirical pressure-time curves can be used for $P_1(t)$, if the incompressive approximation of Eq. (10.19) for $P_1(t)$ is used. The solution for $z_c(t)$ so obtained can apply only up to the time t_m of maximum deflection for which $dz_c/dt = 0$. The final solution using the more exact expression for the diffracted wave given by Kirkwood is rather complicated. The physical nature of the solutions can, however, be seen by considering the solutions obtained in the limiting cases of no diffraction effect at the center, and of noncompressive action.

For times less than θ_d the last two terms on the right of Eq. (10.20) can to a first approximation be neglected. The inertial term for sufficiently thin plates is relatively small, and if it is dropped, the equation of motion for an exponential wave $P(t) = P_m e^{-t/\theta}$ is

$$\frac{dz_c}{dt} + \frac{4\sigma_o h}{\rho_o c_o a^2} z_c = \frac{2P_m}{\rho_o c_o} e^{-t/\theta}$$

The solution of this equation is

$$z_c = \frac{2P_m\theta}{\rho_o c_o}\cdot\frac{\theta_p}{\theta - \theta_p}\left[e^{-t/\theta} - e^{-t/\theta_p}\right], \quad \text{where } \theta_p = \frac{\rho_o c_o a^2}{4\sigma_o h}$$

The characteristic time θ_p is the time for the plate to deflect to a fraction $1 - 1/e = 0.63$ of its final set in response to a step pressure, and is called the plastic time by Kirkwood. The final deflection $z_c(t_m)$ for which $dz_c/dt = 0$ is given by

$$z_c(t_m) = \frac{2P_m\theta}{\rho_o c_o} e^{-t_m/\theta_p}, \quad \text{where} \quad t_m = \frac{\theta\theta_p}{\theta - \theta_p}\log\left(\frac{\theta}{\theta_p}\right)$$

The maximum deflection $z_c(t_m)$ thus is proportional to the impulse $P_m\theta$ of the incident wave, but unless the plastic time is large only a fraction of the impulse is utilized. If θ_p is much smaller than the time constant θ, the deflection approaches a limiting value $z_c(t_m) = 2P_m\theta_p/\rho_o c_o$ as t_m becomes large compared to θ_p, and the deformation is proportional to the peak pressure P_m. The time θ_p for a diaphragm 10 inches in diameter and 0.1 inch thick with a yield stress of 60,000 lb./in.² is 600 μsec. For small charges with θ less than 100 μsec. (less than 5 pounds of explosive), the deformation would thus be determined by impulse. For charges of the order of 300 pounds or more, with θ of the same order of magnitude as θ_p, only the earlier portions of the pressure-time curve would be effective. These considerations give an indication of the changes in effect of the incident wave. They are, however, often inapplicable, as they neglect diffraction effects, which are important for large charges, and the possible development of cavitation, which makes the solution invalid in many practical cases.

The effect of diffracted waves is simply illustrated in the noncompressive approximation, for which the equation of motion is

$$\left(m + \frac{2\rho_o a}{3}\right)\frac{d^2 z_c}{dt^2} + \frac{4\sigma_o h}{a^2}\, z_c = 2P_m e^{-t/\theta}$$

The solution of this equation is

$$z_c(t) = \frac{2P_m\theta}{\rho_o c_o}\cdot\frac{\omega_o\theta_p}{1 + (\omega_o\theta)^2}\,[\omega_o\theta e^{-t/\theta} + \sin\omega_o t - \omega_o\theta\cos\omega_o t]$$

where $\omega_o^2 = 4\sigma_o h/ma^2(1 + 2\rho_o a/3m)$ and θ_p is the plastic time previously defined. The quantity ω_o has the characteristics of an angular frequency of motion of the plastic diaphragm loaded by the moving water near its surface. The time t_m for maximum deformation (for which $dz_c/dt = 0$) is given by the transcendental equation

$$e^{-t_m/\theta} = \cos\omega_o t_m + \omega_o\theta\sin\omega_o t_m$$

and the deflection $z_c(t_m)$ is then

$$z_c(t_m) = \frac{2P_m\theta}{\rho_o c_o}\,\omega_o\theta_p\sin\omega_o t_m$$

This maximum deformation is for the case of an infinite rigid boundary external to the plate. If there is no baffle, the pressure doubling from reflection is reduced by diffraction, and in the noncompressive approximation the deformation is very nearly half as great.

If the natural period $2\pi/\omega_o$ of the structure is large compared with θ, t_m is nearly ¼ of this period or $\pi/2\omega_o$, and the deformation is proportional to the impulse $P_m\theta$ of the incident wave. If ω_o is large, one has $\sin \omega_o t_m \cong 1/\omega_o\theta$ giving the result that $z_c(t_m) \simeq 2P_m\theta_p/\rho_o c_o$. This is just the expression found by neglect of diffraction in the corresponding limit of short plastic times and, as before, the deformation is determined by peak pressure, as of course it must be. It is interesting to note that the equation of motion and its solution are very similar to those for the ball-crusher gauge (see section 5.1).

The results of the noncompressive approximation have been extended by Kirkwood to include baffles of finite rather than infinite extent. These results have been found to describe the deformations of small steel diaphragms in the diaphragm gauge described in section 5.3 with considerable accuracy, both as to magnitude of central deformation, and its variation with charge weight and distance. The observed values average some fifteen per cent lower than the calculated ones, but this difference is partly removed by the use of experimental pressure-time curves rather than the actually employed values from the Kirkwood-Bethe shock wave theory.[7] When this is considered, the agreement obtained is perhaps fortuitously good, considering the many approximations necessary in the theory. It is to be emphasized that the theory cannot always be applied this successfully, particularly if cavitation occurs.

Kirkwood and Richardson have made a number of investigations aimed at removing some of the arbitrary and approximate features of the simple theory which has been presented here, such as the parabolic constraint, the neglect of tangential plastic waves, and other idealizations of the plastic state. For an account of these results, reference should be made to the original reports (61).

B. *The effect of cavitation on deformation.* The analysis of diaphragm motion which has been given assumes that the resultant pressure on the surface is at all times given by superposition of the incident and diffracted waves. In many cases, however, this resultant pressure will become negative because of the deformation, which would require that the water withstand tension if the analysis is to continue to apply. Except under laboratory conditions water cannot withstand appreciable tensions, as discussed in section 10.4, and when cavitation takes place the results already discussed and others similarly obtained must fail. Some criterion as to whether cavitation actually occurs is thus obviously desirable. An essential factor in considering cavitation is the fact that the circumstances which lead to cavitation in front of an infinite plate are modified by the equalizing effect of diffraction from the edge of a finite plate. For example, the decrease in pressure by acceleration at the center of the plate will be counteracted by the arrival of diffracted waves which act to equalize the pressure with the

[7] The calculations are summarized in a report by Fye and Eldridge (39).

values at points external to the plate. If the plate is circular with radius a, the equalizing effect from the edge of the plate can occur only after a time $\theta_d = a/c_o$ after arrival of the incident wave at the edge.

Kirkwood (62) has proposed as a criterion for the formation of cavitation that the diffraction time must exceed the cavitation time at the center of the plate. This cavitation time depends of course on the motion of this part of the plate, but an approximation to the initial motion is obtained by assuming the center to act as an element of an infinite free plate of mass m per unit area (see section 10.5, part C). If cavitation occurs when the resultant pressure on this element first becomes zero, the time after arrival of an exponential pressure wave is the Taylor cavitation time θ_c obtained in section 10.4, which is

$$\theta_c = \frac{\theta}{\beta - 1} \log_e \beta, \quad \text{where} \quad \beta = \frac{\rho_o c_o \theta}{m} \tag{10.21}$$

Kirkwood's criterion may therefore be expressed for a circular plate by the proposition that if θ_d exceeds θ_c cavitation will occur. Corresponding estimates for other types of target and incident pressure wave can be formulated in a similar way by suitable estimate of diffraction time from the dimensions of the structure and its response to pressure.

Although the simple criterion of comparing the diffraction and free plate cavitation times is admittedly rough, its usefulness and approximate validity has been well established experimentally for small circular plates. A number of experiments have been performed, for example, in which photographs were taken of the water in front of diaphragms after arrival of a shock wave from a 50 gm. tetryl charge. Varying cavitation and diffraction times were obtained by changing the radius and thickness of the diaphragms. In all except borderline cases, cavitation was found only when the time for arrival of a diffraction wave at the center exceeded the cavitation time computed from Eq. (10.21).

In the case illustrated in Plate XIIb, a steel diaphragm gauge plate of 3.3 inch diameter, 0.084 inches thick, was struck by an exponential wave of time constant 55 μsec. In this case the diffraction time $\theta_d = a/c_o = 29\mu$sec. was somewhat greater than the cavitation time of 21 μsec., and the predicted cavitation near the center was observed, as shown in Plate XIIb.

It has also been found that positions of the cavitation boundary calculated by similar methods are in good agreement with observed values, and the simple cavitation criteria thus furnish a quite reliable guide for analysis. The cavitation time is smaller for thin plates and short durations of the incident wave, and the diffraction time increases with the lateral dimensions. Cavitation therefore occurs for relatively thin plates, or short durations of shock wave pressure. Many structures of interest have a sufficiently small ratio of thickness to distance between regions of support that cavitation will occur for shock waves, and consideration of cavitation effects is therefore important.

If cavitation forms in front of a plate, the water ceases to exert a significant load. If the cavitation persists, the plate is effectively detached from the water after the time θ_c, and is brought to rest only by its resistance to plastic deformation. The plastic work done on the plate should then be very nearly the whole of the kinetic energy acquired before cavitation occurred, as none of this energy is returned to the water. If the cavitation time is short, a first approximation to the kinetic energy acquired can be obtained by assuming that the plate acts as a free plate without significant resistance. The velocity of a free plate at time θ_c for an exponential wave $P_m e^{-t/\theta}$ is readily shown from Eq. (10.7) of section 10.4 to be

$$\left(\frac{dz}{dt}\right)_{\theta_c} = \frac{2P_m}{\rho_o c_o} \beta^{\frac{1}{1-\beta}}, \quad \text{where} \quad \beta = \frac{\rho_o c_o \theta}{m}$$

The total kinetic energy for a plate of radius a is then

$$KE = \tfrac{1}{2}\pi a^2 m \left(\frac{dz}{dt}\right)^2_{\theta_c} = \frac{2P_m{}^2\theta}{\rho_o c_o} \pi a^2 \beta^{\frac{1+\beta}{1-\beta}} \tag{10.22}$$

and is thus proportional to the shock wave energy density $P_m{}^2\theta/2\rho_o c_o$, a fraction of which is absorbed. If the plate is assumed to act as a membrane under tension $\sigma_o h$, the plastic work done for a central deflection z_c is given by $\pi\sigma_o h z_c{}^2$ from section 10.5. The final deflection $z_c(t_m)$ is obtained when the plastic work is equal to the kinetic energy acquired, and is proportional to the square root of shock wave energy. The value of $z_c(t_m)$ from Eq. (10.22) is

$$z_c(t_m) = \frac{P_m a}{\rho_o c_o}\sqrt{\frac{2m}{\sigma_o h}}\, \beta^{\frac{1}{1-\beta}} = 2\sqrt{\frac{\beta}{\sigma_o h}}\, \beta^{\frac{\beta}{1-\beta}}\, aE^{1/2} \tag{10.23}$$

where E is the shock wave energy density for the assumed acoustic plane wave.

This result is evidently rather crude, as it neglects diffraction effects or the possibility that, in later stages of the motion, the plate may be reloaded by disappearance of cavitation as the plate decelerates. It is therefore not surprising that deformations calculated from Eq. (10.23) come out to be considerably smaller than observed values. As an example, experiments made by Goranson (41) may be cited, in which steel diaphragms 21 inches in diameter and 0.11 inches thick were deformed by the pressure wave from 1 pound TNT charges. The central depressions computed from Eq. (10.23) were of the order of half the observed values. These varied from 1.2 inches at 6 feet to 4.2 inches at 2.5 feet, indicating that the plates must actually have absorbed 4 times as much energy from the shock wave as is accounted for by the kinetic energy at the time of cavitation. Goranson showed in fact that the energy absorbed was very nearly equal to the total available shock wave energy, by a calculation of plastic work similar to the one outlined here. This approximate equality suggests the existence of some mechanism by which the kinetic energy of cavitated water in front of the plate can be delivered to the plate and increase the deformation by doing further plastic work on it.

A mechanism for providing this reloading has been suggested and examined by Kirkwood (62). The physical basis of the theory lies in the fact that, as the diaphragm decelerates, the pressure in front of it increases in a manner calculable from its equation of motion. When this pressure becomes larger than the cavitation pressure, the cavitation is destroyed and a reflected wave of compression moves back into the cavitated regions. The water between the plate and the front of recompression is moving with the plate in the noncompressive approximation, and the kinetic energy of both this layer and the plate is ultimately dissipated as plastic work. If the reloading wave is idealized to be a plane wave front, moving into the cavitated water with forward velocity at essentially zero pressure, its velocity and hence the thickness of the reloading layer of water can be computed from the pressure behind the front and the Rankine-Hugoniot condition at the front.

Kirkwood's application of these considerations gives the result that the deformation predicted by Eq. (10.23) should be increased by a factor $(1 + \beta/4)^{1/2}$, his equation being

$$z_c(t_m) = \frac{P_m a}{\rho_o c_o} \sqrt{\frac{2m}{\sigma_o h}}\, \beta^{\frac{1}{1-\beta}} \left(1 + \frac{\beta}{4}\right)^{1/2}$$

This result, when applied to the 21 inch diaphragms for which $\beta = 6.7$, gives central deflections in much better agreement with experiment for an assumed yield stress $\sigma_o = 45{,}000$ lb./in.^{2}, but still somewhat smaller. Kirkwood has suggested that the secondary bubble pulse may account for the increased damage. If this is to be true, the peak pressure P_m' in the secondary pulse must be greater than the normal stress corresponding to the final shock wave deflection $z_c(t_m)$, as otherwise the later displacement will be elastic without permanent set. The necessary pressure is therefore given by $P_m' = (4\sigma_o h/a^2)\, z_c(t_m)$, which for this case gives the result that P_m' must be greater than 6 per cent of the shock wave peak pressure. Secondary pressures of at least this magnitude would be expected in this case, and the explanation is thus a tenable one. An actual calculation of the supplementary damage would of course necessitate developing a theory appropriate to the bubble pulse pressure-time curve.

The calculations outlined above for circumstances when cavitation occurs are clearly idealized, as they take no account of differences in character of the loading at different points on the diaphragm and further involve approximations in analysis of the spreading of the reloading wave. More refined analytical treatments would evidently be difficult to carry out. The significant result in the idealized case is the relation of damage to absorption of energy from the incident wave if cavitation occurs, and it is not unreasonable to expect that at least qualitatively similar results would be obtained from such treatment.

The analytical treatments for damage to circular diaphragms which have been given here are based largely on theories developed by Kirkwood and by Kirkwood and Richardson.[8] A comprehensive discussion of damage to circular diaphragms and plates, which treats similar problems in more detail than has been possible here, has been given by Kennard (57). The reports of these workers should be consulted for more complete analytical developments. A number of experimental investigations under varying conditions have been carried out at several laboratories. Among these may be mentioned experiments on 10 inch plates by Hudson (48) at the Taylor Model Basin, on 21 inch plates (Goranson (41)), and Modugno gauges (see section 5.3), and on 3.3 inch diameter diaphragm gauges by Fye and Eldridge at Woods Hole (39). Another important series of investigations has been carried out by the British Admiralty Department of Naval Construction and by the Road Research Laboratory, employing box models with rectangular air-backed steel plates having areas of several square feet.[9] The discussion of these and other larger scale damage trials is beyond the scope of the present development, which has for its purpose only an indication of the factors which must be involved in an understanding and evaluation of such experimental work.

10.7. General Considerations in Underwater Explosion Damage

The discussion of the last three sections shows that damage from underwater explosions involves a number of factors, which depend both on the pressure waves and the nature of the structure. As a result, no single answer could be given to the question as to what characteristic of underwater explosions is decisive in causing damage, even if our knowledge of the subject were complete. Enough of the phenomena are sufficiently well understood, however, that some useful generalizations can be stated. Most of these conclusions can be understood in terms of,

[8] Reports of these calculations are given in References (61) and (62).

[9] The work along these lines is presented in a series of reports too long to be listed here; many of them have been listed as Undex Reports (115).

and to a large extent are drawn from, analyses of idealized cases such as the ones treated in the preceding sections. It is to be remembered that any simple conclusions must lose in precision with their extension to other than the simple cases for which they strictly apply. Even so, these qualitative conclusions can at least serve to suggest the relative importance of the various phenomena in the water for specific cases of explosion damage.

A. *Relative importance of shock wave and later pressures.* A first question is of course whether the shock wave, bubble pulse, or pressures at other times are of major importance. The more important high explosives are remarkably similar in that roughly half of the available chemical energy is radiated in the primary shock wave. Although a third or more of this energy may be lost as heat within distances of significance for damage, the fact remains that the shock wave constitutes the largest single source of available energy throughout the surrounding water. Under many conditions, the shock wave is of decisive importance and in no case should its effect be dismissed as negligible without good reason.

As compared with the shock wave, secondary pressure pulses from gas sphere contractions are of much greater duration, but the total energy radiated is much smaller, being certainly of the order of one-third or less of the radiated shock wave energy. At the same distances from their effective source, the maximum pressures in the secondary pulses are a small fraction of that in the shock wave. The bubble pulse may well be responsible for supplementary damage, however, particularly if migration of the bubble places it close to the target when the pulse is emitted. Examples of greatly increased damage as a result of this migration are well established by both direct and indirect evidence. Because of the long duration of these pulses, if they are not seriously attenuated by possible loss of energy to the free surface of the water, their effect in many cases is more nearly that of static pressure and can be gauged by the peak pressure. The magnitude of this pressure and the possible importance of the bubble pulse are strongly affected by the position and state of motion of the gas sphere, and both the explosive and its position relative to the target must be considered. It is reasonable to conclude, however, that the bubble pulse is of major or decisive importance only if the geometry of the charge, target, and nearby surfaces is particularly favorable.

A frequently recurring question in underwater explosion damage is that of the importance of the essentially noncompressive outward flow of water following emission of the shock wave. Most of the remaining energy appears as kinetic energy of outward flowing water at a later stage of expansion of the explosion products. This fact has presumably been the reason leading a number of writers to ascribe major significance to this kinetic or afterflow energy of the "kinetic wave" as a factor in damage, particularly close to the charge. This conclusion is surely both misleading and erroneous, if there is implied the existence of effects not determined by the associated pressures (see section 9.2). There is, however, the perfectly valid possibility that, in the region of high pressures close to the charge, pressures following the initial peak value of the shock wave are large enough to be effective for much longer times than at greater distances. This fact, and the fact that a greater fraction of the radiated shock wave energy is available for mechanical work in near-contact explosions, must be considered in analysis of any data or arguments presumed to establish the importance of afterflow energy, "kinetic waves," or "schubenergie."

B. *Dependence of damage on nature of applied pressure.* In idealized examples of a circular plate undergoing plastic deformation, it is found that the damage resulting from pressure waves depends very much on the relative dimensions and characteristic times of the plate and the pressure wave. Similar differences must be expected for other structures, and the property or properties of the pressure wave which are of importance more generally can be qualitatively inferred from these results. Characteristic times which must be significant are: the duration of the incident pressure (e.g., time constant of an exponentially decaying wave), the natural response times or frequencies of the structure (e.g., the plastic time of a circular plate), and the diffraction times, determined by dimensions of the structure, required for equalizing diffracted pressure waves to be propagated over the structure.

If the duration of the pressure wave is much shorter than other times, the effect of pressure is essentially that of a local impulsive blow on all parts of the structure, which then moves in a manner determined by the impulse of the wave. This impulse varies for a given explosive roughly as $W^{2/3}/R$, where W is the charge weight and R the distance. This state of affairs can be realized only for relatively large structures with slow response, and will not be realized unless the structure is sufficiently rigid to prevent cavitation in the time interval before rarefaction pressures can be equalized by diffraction waves. Some criterion as to occurrence of cavitation must therefore be applied to determine whether analysis based on superposition of pressures, which asymptotically involves the impulse of the wave, can be applied.

In the extreme that the duration of the wave is much longer than the response time or diffraction time, the wave will act much more like an applied static pressure. In the case of a shock wave, the initial peak pressure will come to be the controlling factor, as will the peak value of bubble pulse pressure (although its rate of rise may also be significant). The peak shock wave pressure varies with charge weight and distance roughly as $W^{0.38}/R^{1.14}$ for ranges of interest, whereas slightly smaller exponents are probably appropriate for bubble pulse pressures. The dependence of damage on peak pressure or impulse in limiting cases, with so to speak a fractional utilization of impulse in intermediate cases, is well illustrated by Modugno or diaphragm gauge plates, which for small charges undergo deformations determined by impulse, and for charges of several hundred pounds respond much more nearly to the initial shock wave peak pressure. The dynamical properties of the gauge and its surroundings must of course also be considered. For example, increased rigidity of surrounding or supporting structures has the effect of increasing the equalizing diffraction pressures, and hence leads to increased deformations.

If cavitation occurs in the water adjacent to the structure, a somewhat different state of affairs occurs in which a large amount of energy can be trapped as kinetic energy of the structure and cavitated water, and later expended in plastic work of deformation. The resultant deformation may then be roughly in proportion to the square root of incident shock wave energy, and hence vary with charge weight and distance roughly as $W^{1/2}/R$. Experimental evidence in some cases indicates approximate equality of this energy with plastic work of deformation. It is important, however, to observe that the total energy absorbed by a plate of finite area may considerably exceed the incident shock wave energy over the same area, the excess coming from other parts of the wave by the agency of diffraction.

The various empirical laws as to variation of pressure wave parameters with weight and distance are frequently helpful in analysis of damage, but these laws should not be taken too literally. Observed variations of damage may be determined in part by one parameter more than another, but a considerably different weight-distance law for damage may, and often does, result from the dynamical properties of the structure. Furthermore, the differences in the characteristic laws are not always clear cut. A weight exponent corresponding to deformation varying as $W^{0.42}$ may perfectly well be the result of peak pressure proportional to $W^{0.38}$, or square root of energy varying as $W^{0.5}$. Another consideration is the fact that shock waves near a charge are propagated with considerable energy losses by dissipation, and conclusions valid at greater distances should not be extrapolated to near-contact explosions without careful analysis or supporting data.

Appendix.

APPROXIMATIONS IN THE PROPAGATION THEORY OF KIRKWOOD AND BETHE

An exact estimate of the errors involved in any approximate propagation theory for shock waves requires either an exact theory for comparison or appropriate experimental data. The former does not exist and the latter are treated in some detail in Chapters 4 and 7. It is of interest, however, to form some theoretical estimates of the nature of the errors in Kirkwood and Bethe's theory, as outlined in Chapter 2 and developed more fully in Chapter 4. This analysis is conveniently made in a somewhat indirect manner in terms of the Riemann function $Q = \frac{1}{2}(\sigma - u)$. For acoustic pressures the hydrodynamical equations for the velocity potential function Φ reduce to the wave equation

$$\frac{\partial^2 \Phi}{\partial r^2} - \frac{1}{c_o^2}\frac{\partial^2 \Phi}{dt^2} = 0 \tag{A.1}$$

The enthalpy and particle velocities are obtained by the relations

$$u = \frac{\Phi}{r^2} - \frac{1}{r}\frac{\partial \Phi}{\partial r} \tag{A.2}$$

$$\Omega = \frac{1}{r}\frac{\partial \Phi}{\partial t}$$

which are also valid in the general case if Φ/r satisfies Eq. (2.23) of the text.

The solution of the wave equation (A.1) for an outgoing wave is evidently

$$\Phi = \Phi\left(t - \frac{r}{c_o}\right)$$

and the first of Eqs. (A.2) may be written

$$u = \frac{\Phi}{r^2} + \frac{\Omega}{c_o}$$

The kinetic enthalpy Ω and enthalpy ω differ only by the term $u^2/2$ which in the acoustic approximation is negligible. The Riemann σ and the enthalpy ω are defined as

$$\sigma = \int_{\rho_o}^{\rho} c\frac{d\rho}{\rho} \qquad \omega = \int_o^P \frac{dP}{\rho} \tag{A.3}$$

We can therefore write, c being independent of ρ in the acoustic case,

$$\sigma = \frac{1}{c_o}\int_o^P \frac{dP}{\rho} = \frac{\Omega}{c_o}$$

The function Q therefore becomes

$$Q = \tfrac{1}{2}(\sigma - u) = -\tfrac{1}{2}\cdot\frac{\Phi}{r^2}$$

an approximation also valid for incompressive flow ($c \to \infty$).

The basic approximation of the Kirkwood-Bethe theory can be considered to be the assumption that the function $r(\omega + \sigma^2/2)$ is propagated with velocity $c + \sigma$ behind the shock front. This is readily shown to be equivalent to the assumption that the difference $Q = \frac{1}{2}(\sigma - u)$ is given by

$$Q(r, t) = -\frac{\varphi}{2r^2} - \frac{1}{r}\int_{t_o}^{t} \frac{\partial}{\partial r}\cdot\frac{r(\sigma^2 - u^2)}{4}\,dt \tag{A.4}$$

where the integration over t is performed at constant r and t_o is the time of arrival of the shock wave at r.

The propagation of $r(\omega + \sigma^2/2)$ with velocity $c + \sigma$ if Q is given by Eq. (A.4) can be shown as follows. Substitution from Eq. (A.2) in $Q = \frac{1}{2}(\sigma - u)$ gives

$$Q = \frac{\sigma}{2} - \tfrac{1}{2}\frac{\Phi}{r^2} + \tfrac{1}{2}\frac{1}{r}\frac{\partial \Phi}{\partial r}$$

and equating this relation to Eq. (A.4) gives

$$r\sigma + \frac{\partial \Phi}{\partial r} + \tfrac{1}{2}\int_{t_o}^{t}\frac{\partial}{\partial r} r(\sigma^2 - u^2)\,dt = 0$$

Differentiating with respect to t, we have

$$r\frac{\partial \sigma}{\partial t} + \frac{\partial}{\partial t}\frac{\partial \Phi}{\partial r} + \tfrac{1}{2}\frac{\partial}{\partial r} r(\sigma^2 - u^2) = 0 \tag{A.5}$$

In order to express this equation in terms of $r(\omega + \sigma^2/2)$ we note that $\partial\omega/\partial t = c\,\partial\sigma/\partial t$, and hence

$$\frac{\partial}{\partial t} r\left(\omega + \frac{\sigma^2}{2}\right) = r(c + \sigma)\frac{\partial \sigma}{\partial t}$$

$$\frac{\partial}{\partial r} r\left(\omega + \frac{\sigma^2}{2}\right) = \frac{\partial}{\partial r}\frac{\partial \Phi}{\partial t} + \frac{\partial}{\partial r} r\,\frac{\sigma^2 - u^2}{2}$$

the last step following from Eq. (A.2). Substituting in Eq. (A.5) gives

$$\left[\frac{\partial}{\partial r} + \frac{1}{c + \sigma}\frac{\partial}{\partial t}\right] r\left(\omega + \frac{\sigma^2}{2}\right) = 0$$

which is the equation for propagation of $r(\omega + \sigma^2/2)$ with velocity $c + \sigma$.

The question of importance with regard to this result is, of course, what error is incurred by the approximation that the difference $\frac{1}{2}(\sigma - u)$ is given by Eq. (A.4). In the acoustic approximation, for which we have shown that $(\sigma - u)/2$ is given by $-\Phi/2r^2$, the error in Q is measured by the integral in Eq. (A.4), which is readily shown to be at least of order $1/r^3$. The error in Q is therefore small for increasingly large r. Kirkwood has estimated the value of Q from Eq. (A.4) on the basis of the Kirkwood-Bethe theory and obtains values of the same order as those computed by Penney by direct integration of the Riemann equations (2.17). This is hardly conclusive evidence in favor of the approximation, as both calculations are approximate, but does furnish some support for it. More exact theoretical estimates of the error in the enthalpy have not been made and would, as mentioned at the outset, imply an exact solution of the propagation equations.

Bibliography

This bibliography should not be assumed to give a complete or even fully representative index to work on underwater explosives and explosions. The primary cause of its shortcomings lies in the requirements imposed by military security. Many classified reports of work during World War II are excluded entirely and others described only by a code number. Many of the unclassified reports were at the time of this compilation rather inaccessible, thus diminishing the value of their listing.

It is to be hoped that detailed accounts of publishable material will soon appear in the open literature, and that the abridged accounts given in this book will in the meantime give a useful summary and discussion of such work.

1. H. L. Abbot, Experiments to develop a system of submarine mines, Number 23 of the *Professional Papers*, U. S. Army Engineers Corps.
2. A. B. Arons, A. Borden, and B. Stiller, OSRD 6578 (1946).
3. A. B. Arons and R. H. Cole, OSRD 6239 (1946).
4. A. B. Arons and R. R. Halverson, Hugoniot calculations for sea water at the shock front, OSRD 6577 (1946).
5. A. B. Arons and P. F. Smith, report to be issued as a NavOrd report by U. S. Navy Bureau of Ordnance.
6. A. H. Bebb, Undex 131 (1945).
7. R. Becker, *Zeits f. Physik*, *8*, 321 (1922).
8. H. A. Bethe, The theory of shock waves for an arbitrary equation of state, OSRD Report 545 (1942).
9. A. Borden, TMB Report R–269 (1944).
10. S. R. Brinkley, Jr., and J. G. Kirkwood, Tables and graphs of the theoretical peak pressures, energies, and impulses from explosive sources in sea water, OSRD 5649 (1945).
11. S. R. Brinkley, Jr., and E. B. Wilson, Jr., Calculation of detonation pressures for several explosives, OSRD 1231 (1943).
12. R. H. Brown, Consistency of the NOL ball crusher gauge, U. S. Navy Bureau of Ordnance, Explosives Research Report #1 (1944).
13. F. P. Burch, *Phil. Mag.* (7), *13*, 760 (1932).
14. Bureau of Ships, U. S. Navy, Underwater Explosion Report 1942–3 (1942).
15. S. Butterworth, Damage to ship's plates by underwater explosions, Admiralty Research Laboratory Report ARL/S/10 (1924).
16. D. C. Campbell, TMB Report R–203 (1943).
17. D. C. Campbell, TMB Report 512 (1943).
18. D. C. Campbell and C. W. Wyckoff, TMB Report 520 (1943).
19. D. L. Chapman, *Phil. Mag.* (5), *47*, 90 (1899).
20. R. H. Cole, The use of electrical cables with piezoelectric gauges, OSRD 4561 (1944).
21. R. H. Cole, D. Stacey, and R. M. Brown, Electrical instruments for study of underwater explosions and other transient phenomena, OSRD 6238 (1945).
22. J. S. Coles, OSRD 6240 (1946).
23. J. S. Coles, OSRD 6241 (1946).
24. L. J. Comrie and H. O. Hartley, British Report SW26 (1942).

25. T. L. Davis, *The Chemistry of Powder and Explosives* (Wiley, 1941).
26. J. C. Decius and P. M. Fye, OSRD 6247 (1946).
27. E. Dietze, Calibration of electro-acoustic transducers for hydrophonic systems, OSRD Report C4–sr212–101 (1942).
28. E. Dietze, NDRC Division 6 Report 6.1–1130–1971 (1944).
29. H. B. Dixon, *Trans. Roy. Soc. London*, A*126*, 97 (1893).
30. Durand, *Aerodynamic Theory*, Vol. III, the mechanics of compressible fluids (Guggenheim Fund, 1933).
31. J. E. Eldridge, P. M. Fye, and R. W. Spitzer, Photography of underwater explosions I, OSRD 6246 (1946). See also Reference 105.
32. M. Ewing and A. Crary, Multiple impulses from underwater explosions, Woods Hole Oceanographic Institution (1941).
33. M. Ewing, A. C. Vine, and J. L. Worzel, *J. Opt. Soc. Am.*, *36*, 307 (1946).
34. Explosives Research Laboratory, Bruceton, Pa. The flash photography of detonating explosives to May 1, 1943, OSRD 1488 (1943).
35. Explosives Research Laboratory, Bruceton, Pa., Interim reports of detonation, fragmentation, and air blast (DFA and DF Reports). These reports, issued monthly over the period August 1943 to September 1945 are preliminary accounts of research in progress at the time. Much of the material in these reports has since been incorporated in formal OSRD reports, listed in this bibliography by authors.
36. J. C. Fletcher, W. T. Read, R. C. Stoner, and D. K. Weimer, Final report on shock tube, piezoelectric gauges, and recording apparatus, OSRD 6231 (1946).
37. B. Friedman, Bubble periods, NDRC Applied Mathematics Panel, AMP–NYU Note 88 (1945).
38. C. Frondel, Construction of tourmaline gauges for piezoelectric measurement of explosion pressures, OSRD 6256 (1946).
39. P. M. Fye and J. E. Eldridge, OSRD 6248 (1946).
40. M. F. Gardner and J. L. Barnes, Transients in linear systems (Wiley, 1942).
41. R. W. Goranson, U. S. Navy Bureau of Ships Underwater Explosion Report 1942–4 (1943).
42. M. Greenfield and M. Shapiro, TMB Report 523.
43. R. R. Halverson, W. G. Schneider, and P. C. Cross, OSRD 6258 (1946).
44. G. K. Hartmann, U. S. Navy Bureau of Ordnance Memorandum for File A16 (CND) (1942).
45. G. K. Hartmann, TMB Report 531 (1946).
46. C. Herring, Theory of the pulsations of the gas bubble produced by an underwater explosion, NDRC Division 6 Report C4-sr20 (1941).
47. H. W. Hilliar, (British) Department of Scientific Research and Experiment Report RE 142/19 (1919).
48. G. E. Hudson, TMB Report 509 (1943).
49. H. Hugoniot, *J. de l'Ecole Polytech*, *57*, 3 (1887). *58*, 1 (1888).
50. H. Jones, British Report RC–166 (1941).
51. H. Jones, British Report RC–212 (1941).
52. H. Jones, British Report RC–383 (1943).
53. E. Jouget, *Comptes Rendus*, *132*, 673 (1901); *J. de Math* (liouville), *6*, II, 5 (1905–6).
54. E. H. Kennard, TMB Report 480 (1941).
55. E. H. Kennard, TMB Report R–182 (1943).
56. E. H. Kennard, TMB Report 511 (1943).
57. E. H. Kennard, TMB Report 527 (1944).

58. D. A. Keys, *Phil. Mag.*, *42*, 473 (1921).
59. J. G. Kirkwood et al., The pressure wave produced by an underwater explosion.
    I. J. G. Kirkwood and H. A. Bethe, OSRD 588 (1942), Basic propagation theory.
    II. J. G. Kirkwood and E. Montroll, OSRD 676 (1942), Properties of pure water at a shock front.
    III. J. G. Kirkwood and J. M. Richardson, OSRD 813 (1942), Properties of salt water at a shock front.
    IV. J. G. Kirkwood, S. R. Brinkley, Jr., and J. M. Richardson, OSRD 1030 (1942), Calculations of initial conditions, superseded by V and OSRD 3949.
    V. J. G. Kirkwood, S. R. Brinkley, Jr., and J. M. Richardson, OSRD 2022 (1943), Calculations for thirty explosives.
    VI. O. K. Rice and R. Ginell, OSRD 2023 (1943), Calculations for cylindrical symmetry.
    VII. O. K. Rice and R. Ginell, OSRD 3950 (1944), Final report on propagation theory for cylindrical symmetry.
60. J. G. Kirkwood and S. R. Brinkley, Jr., Theory of the propagation of shock waves from explosive sources in air and water, OSRD 4814 (1945).
61. J. G. Kirkwood and J. M. Richardson, The plastic deformation of circular diaphragms under dynamic loading by an underwater explosion wave, OSRD 4200 (1944). This report is a summary of most of the work presented in two earlier OSRD reports by Kirkwood under the title "Plastic deformation of marine structures by an underwater explosion" (OSRD Reports 793 and 1115).
62. J. G. Kirkwood, UE Report No. 10 (114).
63. G. B. Kistiakowsky, OSRD 5401 (1945).
64. G. B. Kistiakowsky and E. B. Wilson, Jr., The hydrodynamic theory of detonation and shock waves, OSRD Report 114 (1941).
65. H. Lamb, *Hydrodynamics*, 6th edition (Cambridge University Press, 1932).
66. C. W. Lampson, Cable compensation for piezoelectric gauges, OSRD 1179 (1943).
67. B. Lewis and Friauf, *J. Am. Chem. Soc.*, *52*, 3905 (1930).
68. B. Lewis and G. von Elbe, *Combustion, Flames and Explosions of Gases.*
69. P. Libessart, British Report RC–417 (1944).
70. L. Mach, *Sitzungberichte der Wiener Akademie*, Ab. II, Band 101 (1892).
71. A. Marshall, *Explosives* (J. and A. Churchill, 1917).
72. G. H. Messerly, OSRD 1219 (1943).
73. M. Meyer, *The Science of Explosives* (Thomas Y. Crowell, 1943).
74. L. M. Milne-Thomson, *Theoretical Hydrodynamics* (Macmillan, 1938).
75. P. M. Morse, *Vibration and Sound* (McGraw-Hill, 1936).
76. Nautical Almanac Office, Undex 25 (1943), also Reports NAO/10.
77. Nautical Almanac Office, British Report NAO/3.
78. Naval Ordnance Laboratory, Report NORL–751.
79. P. Newmark and E. L. Patterson, OSRD 6259 (1946).
80. M. F. M. Osborne, Relative pressure measurements in shock waves from underwater explosions, Naval Research Laboratory Report S–2305.
81. E. L. Patterson, OSRD 6260 (1946).
82. C. L. Pekeris, NDRC Division 6 Report 6.1–sr–1131–1433 (1944).
83. W. G. Penney, British Report RC–142 (1941).
84. W. G. Penney, Undex 72 (1943).
85. W. G. Penney and H. K. Dasgupta, British Report RC–333 (1942).

86. W. G. Penney and A. T. Price, British Report SW–27 (1942).
87. H. Polachek and R. J. Seeger, Regular reflection of shocks in ideal gases in water-like substances, Bureau of Ordnance Explosive Research Reports #13, #14.
88. C. Ramsauer, *Ann. d. Phys.* (4), *72*, 265 (1923).
89. W. J. M. Rankine, *Phil. Trans.*, *160*, 277 (1870).
90. Lord Rayleigh, *Proc. Roy. Soc.* (A), *84*, 247 (1910).
91. G. T. Reynolds, Charge orientation tests, OSRD 1532 (1943).
92. Road Research Laboratory, Undex 65 (1943); Undex 66 (1943); Undex 90 (1944).
93. Road Research Laboratory, Undex 10 (1942); Undex 115 (1944); Undex 134 (1945).
94. Road Research Laboratory, Note ADM/218/ARB (1945).
95. Road Research Laboratory, Undex 78 (1944); Undex 108 (1944).
96. Road Research Laboratory, Undex 16 (1943); Undex 105 (1944).
97. P. Savic, Undex 110 (1944).
98. P. Savic, Undex 50 (1943); Undex 95 (1944).
99. R. L. Scorah, *J. Chem. Phys.*, *3*, 425 (1935).
100. F. Seitz, A. W. Lawson, and P. H. Miller, NDRC Division 2 Report A–63 (1942); OSRD 619.
101. R. A. Shaw, British Report SW–22 (1941); British Report SW–23 (1942).
102. M. Shiffman and B. Friedman, On the best location of a mine near the sea bed, NDRC Applied Mathematics Panel Report 37.1 R (1944).
103. C. P. Slichter, W. G. Schneider, and R. H. Cole, OSRD 6242 (1946).
104. Stanolind Oil and Gas Co., Development of explosion pressure gauges and recording equipment, OSRD 1739 (1943).
105. E. Swift et al., Photography of underwater explosions II, to be issued as a NavOrd report by U. S. Navy Bureau of Ordnance.
106. G. I. Taylor, British Report RC–178 (1941).
107. G. I. Taylor, British Report SW19. This has been reprinted by the Taylor Model Basin as TMB Report 510.
108. G. I. Taylor, British Report RC–235 (1941).
109. G. I. Taylor and R. M. Davies, Undex 13 (1943).
110. G. I. Taylor and R. M. Davies, Undex 88 (1944).
111. H. N. V. Temperley, Undex 64 (1943).
112. H. N. V. Temperley, Undex 92 (1944).
113. L. H. Thomas, *J. Chem. Phys.*, *12*, 449 (1944).
114. Underwater Explosives Research Laboratory (Woods Hole, Mass.), Interim Reports of underwater explosives and explosions (UE Reports). These reports, issued monthly over the period August 1942 to September 1945, are preliminary accounts of research in progress at the time. Much of the material in these reports has since been incorporated in a series of formal OSRD reports, listed in this bibliography by authors.
115. Undex Reports. Underwater explosives research in England during the war was coordinated by the Undex Panel, the detailed functions being largely referred to the Undex Sub-Panel. Reports by various British agencies on underwater explosion research have since 1942 been catalogued by this group and assigned Undex report numbers. More than 200 reports have been so classified, only a few of which could be included in this bibliography.
116. J. von Neumann, Oblique reflection of shocks, Bureau of Ordnance Explosives Research Report #12 (1943).

117. J. von Neumann, On the theory of stationary detonation waves, OSRD 1140 (1942).
118. A. B. Ward, Undex 20 (1943).
119. A. B. Ward, Undex 40 (1943).
120. F. J. Weyl, Analytical methods in optical examination of supersonic flow, U. S. Bureau of Ordnance NavOrd Report 211–45 (1945).
121. H. F. Willis, Underwater explosions, time interval between successive explosions, British Report WA–47–21 (1941).
122. H. F. Willis and R. T. Ackroyd, Undex 36 (1943).
123. H. F. Willis and M. I. Willis, Undex 52 (1943).
124. A. B. Wood, (British) Admiralty Research Laboratory Report ARL/S/12, pp. 442–616.

Note added in Proof: Since this bibliography was compiled in the summer of 1946, a number of papers have appeared which deal with the subject of this book. Citation of these in the text and bibliography has not been feasible, but a supplementary, partial list of papers up to January 1948 is given in the following. References in the text or bibliography which these papers supplement or make more readily available are indicated in parentheses.

(a) *Explosives and Detonation*

1a. P. W. Bridgman, Effect of high mechanical stress on certain solid explosives, *J. Chem. Phys.*, *15*, 311 (1947).
2a. S. R. Brinkley, Jr., Note on the conditions of equilibrium for systems of many constituents, *J. Chem. Phys.*, *14*, 563 (1946) (Reference 11).
3a. S. R. Brinkley, Jr., Calculation of the equilibrium composition of systems of many constituents, *J. Chem. Phys.*, *15*, 107 (1947). (Reference 11).
4a. M. A. Cook, An equation of state for gases at extremely high pressures from the hydrodynamic theory of detonation, *J. Chem. Phys.*, *15*, 518 (1947).

(b) *Properties of Water*

1b. H. B. Briggs, J. B. Johnson, and W. P. Mason, Properties of liquids at high sound pressures, *J. Acous. Soc. Am.*, *19*, 664 (1947).
2b. E. N. Harvey, W. D. McElroy, and W. H. Whitely, On cavity formation in water, *J. App. Phys.*, *18*, 162 (1947).
3b. J. M. Richardson, A. B. Arons, R. R. Halverson, Hydrodynamic properties of sea water at the front of a shock wave, *J. Chem. Phys.*, *15*, 785 (1947) (Reference 4).
4b. H. N. V. Temperley, The behavior of water under hydrostatic tension III, *Proc. Phys. Soc. Lond.*, *59*, 199 (1947).

(c) *Shock Wave Theory*

1c. A. B. Arons and D. R. Yennie, Partition of energy in underwater explosions, to be submitted to the *Reviews of Modern Physics*. (Sections 4.8 and 9.4.)
2c. S. R. Brinkley, Jr., and J. G. Kirkwood, Theory of the propagation of shock waves, *Phys. Rev.*, *71*, 606 (1947). (Reference 60.)
3c. S. R. Brinkley, Jr., and J. G. Kirkwood, Theory of the propagation of shock waves from infinite cylinders of explosive, *Phys. Rev.*, *72*, 1109 (1947).
4c. R. Finkelstein, Normal reflection of shock waves, *Phys. Rev.*, *71*, 42 (1947). (Footnote 10, Chapter 2.)
5c. M. F. M. Osborne and A. H. Taylor, Non-linear propagation of underwater shock waves, *Phys. Rev.*, *70*, 322 (1946). (Reference 80.)
6c. A. H. Taub, Refraction of plane shock waves (in air), *Phys. Rev.*, *72*, 51 (1947).

7c. G. I. Taylor, The air wave surrounding an expanding sphere, *Proc. Roy. Soc. A, 186*, 273 (1946).

(d) *Shock Wave Measurements*

1d. A. B. Arons and R. H. Cole, Design and use of piezoelectric gauges for measurement of transient pressures, to be submitted to the *Journal of Applied Physics*. (Reference 3).

2d. R. H. Cole and J. S. Coles, Propagation of spherical shock waves in water, *Phys. Rev., 71*, 128 (1947).

3d. F. E. Fox, K. F. Herzfeld, and G. D. Rock, The effect of ultrasonic waves on conductivity of salt solutions, *Phys. Rev., 70*, 329 (1946).

4d. J. H. McMillen and E. N. Harvey, A spark shadowgraphic study of body waves in water, *J. App. Phys., 17*, 541 (1946).

5d. M. F. M. Osborne and J. L. Carter, Transient analysis of linear systems, using underwater explosion waves, *J. App. Phys., 17*, 871 (1946).

6d. M. F. M. Osborne and S. D. Hart, Transmission, reflection, and guiding of an exponential pulse by a steel plate in water. II. Experiment, *J. Acous. Soc. Am., 18*, 170 (1946).

7d. W. Payman and W. C. F. Shepard, Explosion and Shock Waves, VI, The disturbance produced by bursting diaphragms with compressed air, *Proc. Roy. Soc. A, 186*, 293 (1946).

(e) *Gas Globe Motion and Pressure Pulses*

1e. A. B. Arons, Secondary pressure pulses due to gas globe oscillations in underwater explosions. I, Experimental data (with J. P. Slifko and A. Carter). II, Selection of adiabatic parameters in the theory of oscillation. Submitted to the *Journal of the Acoustical Society of America*.

2e. B. Friedman, Theory of underwater explosion bubbles, *Report IMM-NYU 166* of the Institute for Mathematics and Mechanics, New York University. (A summary of material in References 37 and 102.)

See also References 1c, 1d.

Index